Mechanics, Solid State and Engineering Materials

Edited by
Fei Hu
Beibei Wang

Mechanics, Solid State and Engineering Materials

Selected, peer reviewed papers from the
2011 International Conference on
Mechanics, Solid State and Engineering Materials
(ICMSSEM 2011),
September 1-2, 2011, Hangzhou, China

Edited by

Fei Hu and Beibei Wang

TTP

Trans Tech Publications Ltd
Kreuzstrasse 10
CH-8635 Durnten-Zurich
Switzerland
http://www.ttp.net

Volume 279 of
Advanced Materials Research
ISSN 1022-6680

Full text available online at *http://www.scientific.net*

Distributed *worldwide by*

Trans Tech Publications Ltd
Kreuzstrasse 10
CH-8635 Durnten-Zurich
Switzerland

Fax: +41 (44) 922 10 33
e-mail: sales@ttp.net

and in the Americas by

Trans Tech Publications Inc.
PO Box 699, May Street
Enfield, NH 03748
USA

Phone: +1 (603) 632-7377
Fax: +1 (603) 632-5611
e-mail: sales-usa@ttp.net

printed in Germany

Preface

It is a great honor and pleasure for us to invite our distinguished guests and participants to attend 2011 International Conference on Mechanics, Solid State and Engineering Materials (ICMSSEM 2011) held on September 1-2, 2011 in Hangzhou, China.

As we all know, in the last few years, mechanics, solid state and engineering materials have played an important role in the progress of human society and research of scholars in the related field has accumulated a treasure of accomplishments and promoted the development of academic studies, meanwhile, researchers are faced with a lot of problems. Under such background, ICMSSEM 2011 is held, with the purpose of being a great platform for researchers and engineers in the areas of mechanics, solid state and engineering materials to report the latest achievements of research efforts, to discuss the wide range of problems encountered at present face to face, and to be inspired and enlightened by the works of others.

This collection of 84 peer-reviewed papers is divided into sections of solid materials, solid mechanics and related enlightening topics. We believe the volume will be essential reading for those in the related areas and will provide an inspiration for future studies and advancement.

We would like to thank Prof. Vadim Hudramovich, Head of Department of Dynamics and Strength of Structures, National Academy of Sciences of Ukraine and State Space Agency of Ukraine for his being our keynote speaker. Thanks also go to Trans Tech Publications for the perfect support and service in publishing this periodical. We are grateful to all those who have contributed to the success of ICMSSEM 2011.

Organizing Committee

September, 2011

2011 International Conference on Mechanics, Solid State and Engineering Materials (ICMSSEM 2011)

Sponsored by

Institute of Mechanical, Material, Computer and Electrical Engineers

Organizing and Technical Committee

Fei Hu, Institute of Mechanical, Material, Computer and Electrical Engineers, Hong Kong

Beibei Wang, Institute of Mechanical, Material, Computer and Electrical Engineers, Hong Kong

Peng Shi, Yongkang Weiye Die-casting Research Center, China

Qian Wang, Institute of Mechanical, Material, Computer and Electrical Engineers, Hong Kong

Basim A. Khidhir, Universiti Industri Selangor, Malaysia

Nuraini Abd. Aziz, Universiti Putra Malaysia, Malaysia

Muhamad Ridzuan Abdul Latif, Universiti Teknologi PETRONAS, Malaysia

Abuzar Abid Siddiqui, Bahauddin Zakariya University, Pakistan

Table of Contents

II. Solid Mechanics

III. Related Enlightening Topics

I. Solid Materials

Advanced Materials Research Vol. 279 (2011) pp 3-9

doi:10.4028/www.scientific.net/AMR.279.3

Dielectric Properties of Graphite Nanosheets doped Poly(vinylidene fluoride)/Silicon Carbide Hybrid

Yuchao Li[1, a], Xiangcai Ge[1, b], Sie Chin Tjong[2, c]

[1]Department of Materials Science and Engineering, Liaocheng University, Liaocheng, Shandong, PRC

[2]Department of Physics and Materials Science, City University of Hong Kong, Kowloon Town, Kowloon, Hong Kong

[a]cityu.lyc@gmail.com, [b]gexiangcai@lcu.edu.cn, [c]aptjong@cityu.edu.hk

Keywords: Dielectric constant, PVDF, Silicon carbide, Hybrid

Abstract. Binary poly(vinylidene fluoride)/silicon carbide (PVDF/SiC) composites and its graphite nanosheets (GN) doped ternary hybrids were fabricated using simple solution mixing and hot compression molding process. The dielectric behavior of such hybrids was studied over a wide frequency range. Additions of graphite nanosheets with their concentrations close to the percolation threshold were found to be very effective to enhance the dielectric permittivity of PVDF/SiC system. The dielectric constant for PVDF/SiC/GN 80/20/2 system was ~ 200, being twenty times higher than that of neat PVDF. Furthermore, the dielectric behavior of such hybrid displayed strong frequency dependence.

Introduction

Development of electronic devices for high pressure, large temperature fluctuation, high frequency and high radiation environment applications is of practical interest. One major technical challenge for developing such electronic devices is the attainment of high dielectric permittivity. In general, piezoelectric perovskites ceramics such as $BaTiO_3$ and $PbTiO_3$ can meet this requirement readily, but they are quite brittle. In contrast, polymers exhibit superior low dielectric loss and flexibility but poor dielectric permittivity. Therefore, effective enhancement in permittivity of polymers is of great importance in electronic industry. Commonly, there are two approaches in obtaining high permittivity of polymers. One is by incorporating high permittivity ceramics into polymer matrix [1, 2]. However, large loading levels of ceramic microparticles are needed to achieve the desired dielectric properties which lead to the loss of flexibility and poor processability of polymers. On the contrary, scientists found that the permittivity of polymers can be enhanced by adding low volume fractions of conductive fillers [3-5]. A significant rise in dielectric constant of such three-component 0-3 composites occurs near the percolation threshold. This is due to the formation of a large quantity of minicapacitors associated with the separation of conducting filler surfaces by a polymer layer. A defect of this method is that the large permittivity is always accompanied with a large dielectric loss.

Silicon carbide (SiC), a very hard and strong ceramic material having excellent thermal and electrical properties is commonly used to improve the mechanical and thermal properties of metallic or polymeric materials [6, 7]. Moreover, SiC-based electronic devices are presently being developed for use in severe conditions [8, 9]. In addition, poly(vinylidene fluoride) (PVDF) is an ever interesting polymer due to its unique high piezoelectric properties. Thus, PVDF and its copolymers find widespread industrial applications in transducers, transistors and capacitors [10, 11]. Herein, graphite nanosheets (GN) doped PVDF/SiC hybrids were fabricated via solution mixing method. The purpose of this study is to investigate the electrical performances of polymer composites containing both SiC ceramic and conductive GN.

Experimental

PVDF resin (Kynar 740) was supplied by Atofina Chemicals Inc.. Microsize SiC (~ 5μm) particles were received from Goodfellow Company. Graphite nanosheets (GN) with a large aspect ratio of ~ 250 were kindly supplied by the Institute of Polymer & Nanomaterials, Huaqiao University (Quanzhou, Fujian, China) [12].

PVDF was dissolved initially in a flask containing N, N-Dimethylformamide (DMF) solvent. SiC with contents varying from 0 to 50 wt% was added to the precursor solutions for fabricating PVDF/SiC composites. For fabrication of 3-component hybrid composites, the PVDF/SiC weight ratio was kept at 80/20. GN with concentration from 1 to 4 phr was also added respectively to the precursor solution. The composite solution was mechanically stirred for 5h and then followed by ultrasonic oscillation for another 5h. The mixture was placed in an oven at 70 °C for solvent removal. Finally, the resulting products were compression-molded at 230 °C for characterizations.

The microstructure of composites was observed in a scanning electron microscope (SEM, JEOL JSM-820). X-ray diffraction (XRD) measurements were performed in a Philip X'pert diffractometer in the 2θ range of 10-70° and a scanning speed of 3 °/min. Electrical properties were measured with an impedance analyzer (Agilent model 4294). Therefore, a complex permittivity of the composites can be obtained: $\varepsilon^* = \varepsilon' - j\varepsilon''$, where ε' is the real part of permittivity and ε'' is the dielectric loss.

Results and Discussions

Fig. 1(a)-(b) show representative SEM images of PVDF/SiC 80/20 and PVDF/SiC/GN 80/20/2 hybrid. Solution enables homogeneous dispersion of silicon carbide and graphite nanosheets. In Fig. 1(b), the presence of SiC microparticles somewhat prevents the formation of GN conducting network.

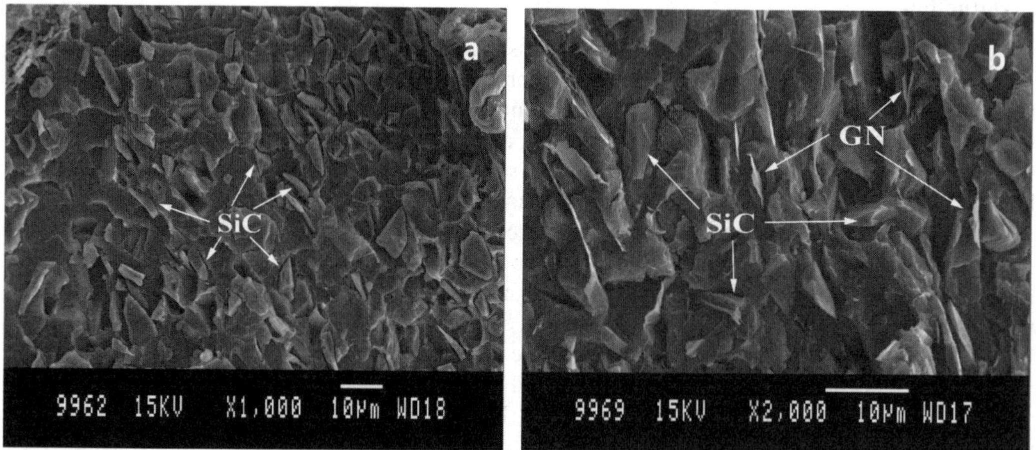

Fig. 1: Representative SEM images of (a) PVDF/SiC 80/20 and (b) PVDF/SiC/GN 80/20/2 hybrid

Fig. 2 shows the XRD patterns of PVDF, SiC, PVDF/SiC 80/20 and PVDF/SiC/GN (8/20/2) composites. The peaks located at 2θ = 18.4°, 19.9° and 26.9° are characteristic reflections of the α-PVDF. No β-PVDF reflection peak (2θ = 20.5°) was observed after high temperature compression molding. A distinct peak at 2θ = 26.6° is assigned to the (002) graphite characteristic reflection for the PVDF/SiC/GN (80/20/2) composites. Other peaks, such as 34.2°, 35.7°, 38.1°, 41.4° and 60°, are characteristic peaks of silicon carbide.

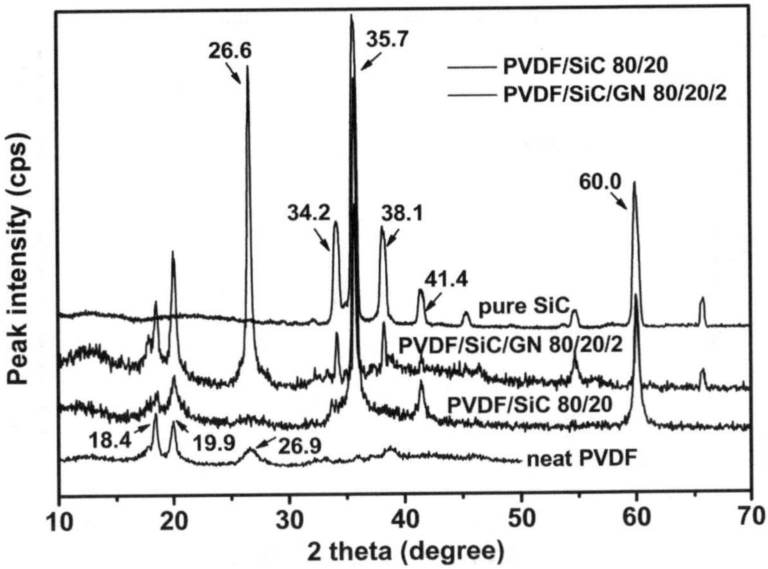

Fig. 2: XRD patterns of PVDF and its composites

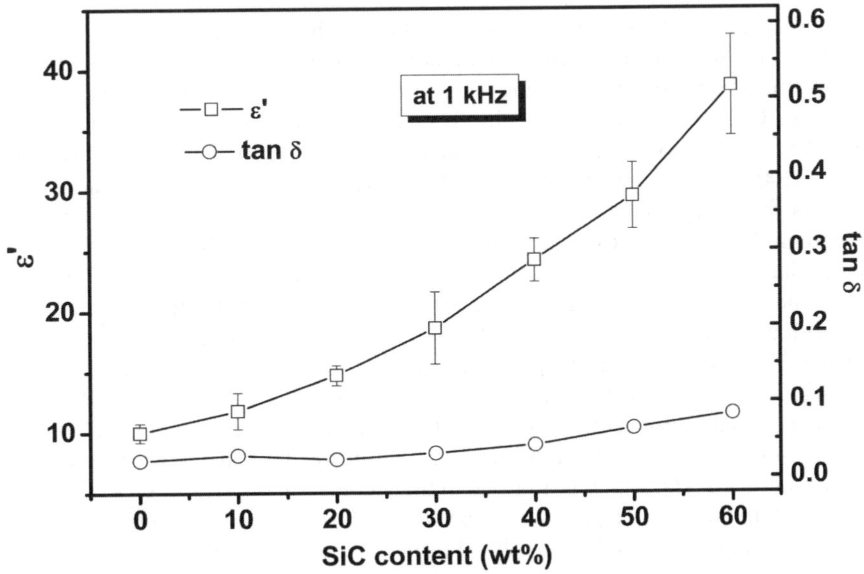

Fig. 3: Dielectric constant as a function of SiC content in PVDF/SiC composites

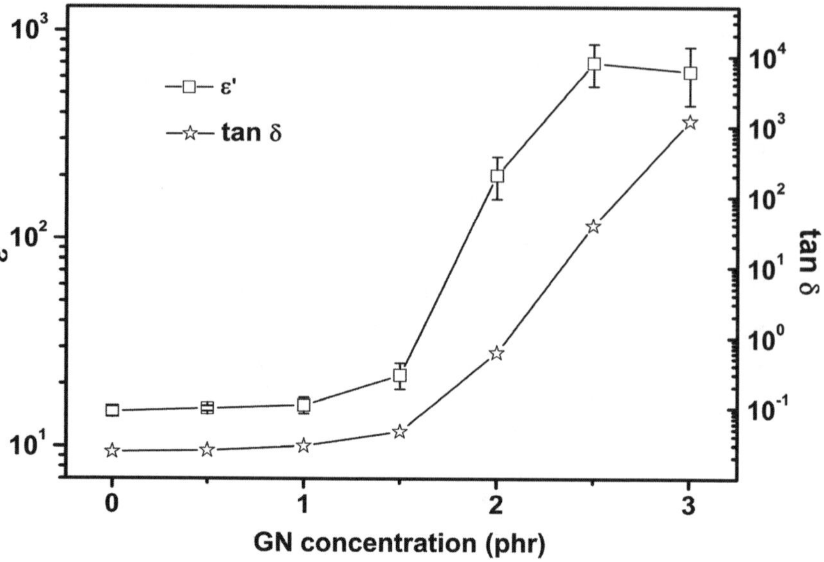

Fig. 4: Dielectric properties of graphite nanosheets doped PVDF/SiC (80/20) hybrid

Fig. 3 shows the dielectric properties of PVDF/SiC system. As expected, the dielectric constant increases gradually with increasing SiC content in the studied binary PVDF/SiC system. The dielectric constant for PVDF/SiC 20 wt% is 14.7, a little higher than that of neat PVDF (~10). It is not until 50 wt% of SiC addition, the permittivity reaches ~30, being three times higher than that of neat PVDF. However, in GN doped PVDF/SiC system (Fig. 4), the dielectric constant increases slowly with increasing GN content up to 1.5 phr. Thereafter, they increase sharply as the GN content approaches the percolation threshold. The PVDF/SiC/GN (80/20/2) hybrid exhibits a large dielectric constant of ~ 200 and low dielectric loss of 0.6 at 1 kHz. The dielectric constant is thirteen times larger than that of binary PVDF/SiC 80/20 system ($\varepsilon' = 14.7$). This composite also shows superior electrical and mechanical properties than that of the above studied PVDF/SiC 50/50 ($\varepsilon' = 38.6$) system. An extremely large value of $\varepsilon' = \sim 700$ can be achieved by adding 2.5 phr GN loading. However, this hybrid is unsuitable for energy storage applications due to its extremely large dielectric loss ($\tan\delta = 39$) associated with the formation of GN network. The formation of conducting network enables trapped carriers to pass through readily, thereby dissipating the stored energy considerably.

Fig. 5(a)-(b) illustrate the respective plots of ε' and $\tan\delta$ vs frequency for the PVDF/SiC system. The dielectric constant of each composite (Fig. 5(a)) increases with increasing SiC concentration, but slightly decreases with increasing frequency. A significant drop in permittivity occurs when the frequency reaches ~ 10^6 Hz, which is ascribed to the fact that dipole relaxation of the composites can not catch up with frequency changes of external fields. In this regime, characteristic relaxation peaks of PVDF matrix can be observed (Fig. 5(b)). Further, these relaxation peaks slightly shift to higher frequencies with increasing SiC concentrations indicating the promotion of dipole relaxations of PVDF after adding SiC microparticles.

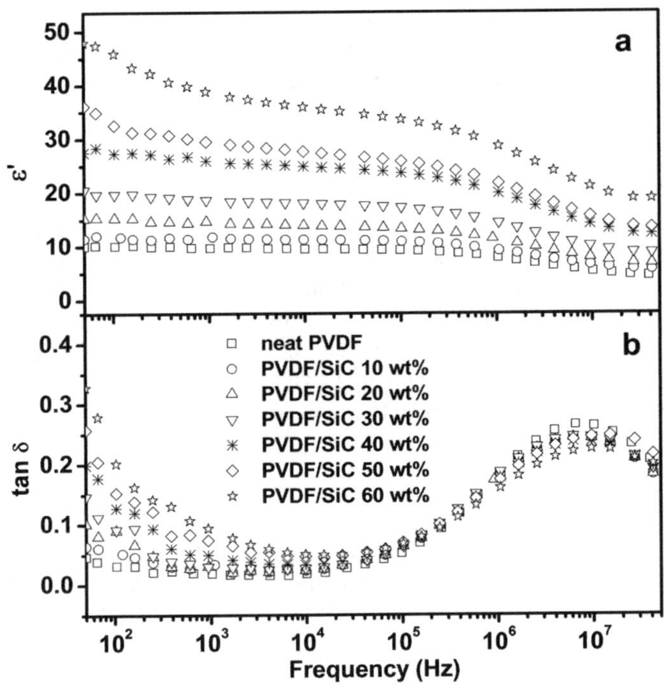

Fig. 5: (a) ε' and (b) tan δ as a function of frequency for PVDF/SiC composites

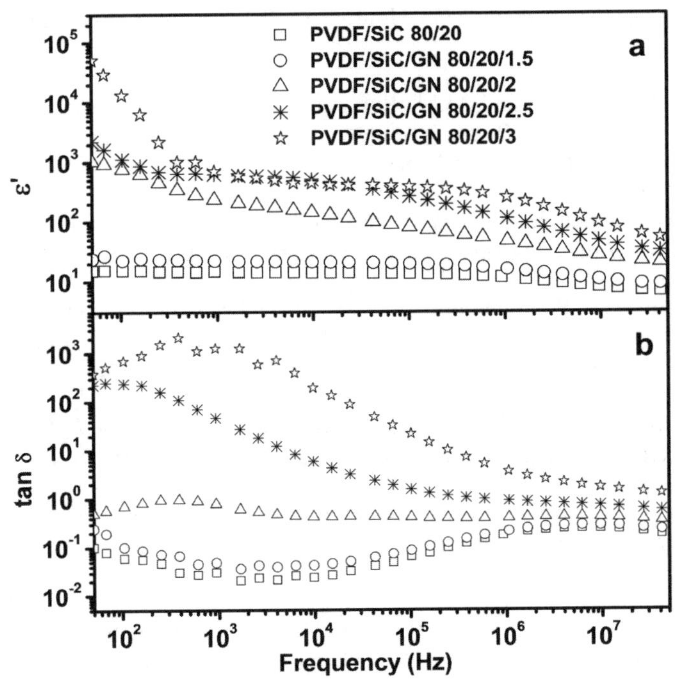

Fig. 6: (a) ε' and (b) tan δ as a function of frequency for PVDF/SiC/GN hybrid

Fig. 6(a)-(b) display the dielectric response of PVDF/SiC/GN hybrid as a function of frequency. Similarly, the permittivity of PVDF/SiC/GN hybrid (Fig. 6(a)) increases with GN additions, but decreases with frequency. However, the increase in permittivity is significant with very small amount of GN additions. Accordingly, a dramatic increase in dielectric loss (Fig. 6(b)) was observed resulting from the formation of conductive network.

Conclusions

Graphite nanosheets doped PVDF/SiC composites were prepared by means of solution mixing process. The structure, morphology and electrical properties of the resulting composites were investigated. SEM micrographs revealed that solution mixing enables a good dispersion state of SiC and GN in PVDF matrix. Electrical measurements showed that GN doped PVDF/SiC/GN hybrid exhibits superior electrical performance than that of binary system. The dielectric constant of PVDF/SiC/GN (80/20/2) hybrid was ~ 200, being twenty times higher than that of neat PVDF. In addition, the dielectric constant and dielectric loss of the composites show strong frequency dependence.

Acknowledgements

This work was financially supported by the Doctor Science Foundation of Liaocheng University (No. 31805), Shandong Province, PRC and Doctoral Research Foundation from Science and Technology Commission of Shandong Province, PRC (No. 2008BS04025).

References

[1] E. Venkatragavaraj, B. Satish, P.R. Vinod and M.S. Vijaya: *Piezoelectric properties of ferroelectric PZT-polymer composites*. Journal of Physics D: Applied Physics, 34, 4(2001), pp. 487-492.

[2] L. Ramajo, M.S. Castro and M.M. Reboredo: *Effect of silane as coupling agent on the dielectric properties of $BaTiO_3$-epoxy composites*. Composites Part A: Applied Science and Manufacturing, 38, 8(2007), pp. 1852-1859.

[3] Z.M. Dang, Y. Shen and C. W. Nan: *Dielectric behavior of three-phase percolative Ni-$BaTiO_3$/ polyvinylidene fluoride composites*. Appl. Phys. Lett., 81, 25(2002), pp. 4814-4816.

[4] I.A. Tchmutin, A.T. Ponomarenko, V.G. Shevchenko, N.G. Ryvkina, C. Klason and D.H. Mcqueen: *Electrical Transport in 0-3 Epoxy Resin–Barium Titanate–Carbon Black Polymer Composites*. J Polym Sci B: Polym Phys, 36, 11(1998), pp. 1847-1856.

[5] C. Zhang, J.F. Sheng, C.A. Ma and M. Sumita: *Electrical and damping behaviors of CPE/BaTiO3/VGCF composites*. Mater. Lett. 59, 28(2005), pp. 3648-3651.

[6] S.C. Tjong and Z.Y. Ma: *High-temperature creep behaviour of powder-metallurgy aluminium composites reinforced with SiC particles of various sizes*. Compos. Sci. Technol., 59, 7(1999), pp. 1117- 1125.

[7] R.M. Rodgers, H. Mahfuz, V.K. Rangari, N. Chisholm and S. Jeelani: *Infusion of SiC Nanoparticles Into SC-15 Epoxy:An Investigation of Thermal and Mechanical Response*. Macromol. Mater. Eng., 290, 5(2005), pp. 423-429.

[8] P.G. Neudeck: *Progress in Silicon Carbide Semiconductor Electronics Technology*. J. Electron. Mater., 24, 4(1995), pp. 283-288.

[9] A. Kassiba, M. Tabellout, S. Charpentier, N. Herlin and J.R. Emery: *Conduction and dielectric behaviour of SiC nano-sized materials*. Solid State Commun., 115, 7(2000), pp. 389-393.

[10] A. Almasri, Z. Ounaies, Y. S. Kim and J. Grunlan: *Characterization of Solution-Processed Double-Walled Carbon Nanotube/Poly(vinylidene fluoride) Nanocomposites*. Macromol. Mater. Eng., 293, 2(2008), pp. 123-131.

[11] J.W. Kim, W.J. Cho and C.S. Ha: *Morphology, crystalline structure, and properties of poly(vinylidene fluoride)/silica hybrid composites*. Journal of Polymer Science: Part B: Polymer Physics, 40, 1(2002), pp. 19-30.

[12] G.H. Chen, W.G. Weng, D.J. Wu, C.L. Wu, J.R. Lu, P.P. Wang and X.F. Chen: *Preparation and characterization of graphite nanosheets from ultrasonic powdering technique*. Carbon, 42, 4(2004), pp. 753-759.

Advanced Materials Research Vol. 279 (2011) pp 10-15
© (2011) Trans Tech Publications, Switzerland
doi:10.4028/www.scientific.net/AMR.279.10

Effect of Quenching Temperature on Microstructures and Tensile Properties of PM Ti-23Al-17Nb Alloy

LANG Zebao[1, a], ZHANG Xuhu[1, b], WANG Liang[1, c] and JIA Wenjun[1, d]

[1]Aerospace Research Institute of Materials & Processing Technology, Beijing 100076, PR China

[a]lzb0413@sina.com, [b]zxh33605@sina.com, [c]wLiangsh@sohu.com and [d]hbjwj@126.com

Keywords: quenching temperature, tensile property, microstructure, Ti_3Al based alloy

Abstract. After treated with different quenching processes near transition temperature of orthorhombic phase, PM Ti-23Al-17Nb alloys with different microstructures were obtained. Microstructures of the alloy were observed by optical microscope, and tensile properties were also tested. In order to analyze the relationship between microstructures and tensile properties at room temperature, fractures of broken test bars obtained in tensile tests were characterized by SEM. The results show that quenching process below transition temperature of orthorhombic phase has few effects on improvement of size and morphology of crystal grain. However, it may lead to an abnormal growth of partial crystal grain, which results in low tensile strength and poor ductility; Quenching process above transition temperature of orthorhombic phase can cause recrystallization of B_2 phase, which make a remarkably improvement on size and morphology of crystal grain. Moreover, the alloy performs the max tensile strength of 1026 MPa and the max ductility of 5.8% after quenched at the temperature of 1020°C.

Introduction

Ti_3Al based alloy has low density and exhibits excellent high temperature behavior, so it can be used as an ideal structure material in aerospace fields [1]. However, Ti_3Al based alloy is hard and brittle in room temperature. After alloyed and treated with complex Thermal Mechanical Process, Ti_3Al based alloy performs some characteristics of plastic material. Therefore, brittleness at room temperature has brought many obstacles in engineering application of Ti_3Al based alloy.

PM technology can be used to prepare brittle materials like Ti_3Al based alloy, which is difficult to be obtained with traditional processing technology. Compacts of PM Ti_3Al based alloy with full density can be obtained by HIPing pre-alloyed powders at a low temperature. Compared with Ti_3Al based alloy obtained by traditional processing technology, PM Ti_3Al based alloy has a fine, homogenous microstructure and exhibits more reliability [2, 3]. Although as-HIP compact performs a high tensile strength, its ductility is very low [3, 4]. According to the study of S. Li, Y. Cui, and et al, microstructures of Ti_3Al based alloy have a great influence on tensile properties at room temperature [4, 5]. Therefore, processing technology and following heat treatments have been the hot point in the study of Ti_3Al based alloy [6-8].

In this study, PM Ti-23Al-17Nb has been chosen as test material, and treated with different quenching processes near transition temperature of orthorhombic phase (abbreviated as O phase in the following paragraph). In the following study, microstructures of test bars obtained in different quenching processes were observed and tensile properties at room temperature were tested. In order to analyse relationship between microstructures and tensile properties at room temperature, fractography of broken test bars in tensile test were performed.

Experimental

Materials. Materials for experimental were obtained by HIPing pre-alloyed Ti-23Al-17Nb powders at 960°C, 3 hours with a pressure of 150 MPa. Chemical content of as-HIP compact was shown in Table 1. Test bars used in different quenching processes were prepared by linear cutting from as-HIP compact of PM Ti-23Al-17Nb alloy.

Table 1: The chemical composite of Ti-23Al-17Nb(at%) pre-alloyed powders

Element	Nb	Al	Fe	C	O	N	Ti
Content (at%)	0.1682	0.2435	0.0003	0.00005	0.0023	0.0002	base

Microstructures of as-HIP compact were observed and shown in Fig. 1. It can be drawn from Fig. 1 that microstructures of the compact include two phases. The black one is α_2 phase and the other with grey color is O phase. Distribution of phases is inhomogeneous and α_2 phase congregates in some spherical areas.

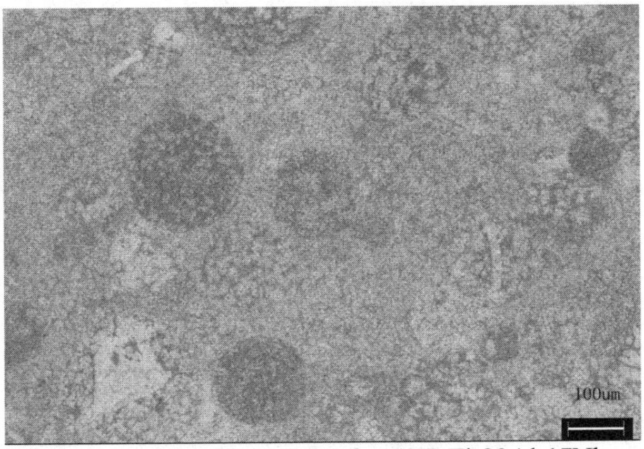

Fig. 1: The microstructure of as-HIP Ti-23Al-17Nb

Quenching process. Differential Thermal Analysis was performed to PM Ti-23Al-17Nb alloy and deduced that transition temperature of O phase was 1008 °C. Different quenching processes near transition temperature of O phase were chosen and shown as Table 2. Microstructures of test bars obtained in different quenching processes were observed and tensile properties at room temperature were test.

Table 2: Different quenching process of PM Ti-23Al-17Nb alloy

No.	Process abbreviation	Detailed
1	HT1	960°C, 45min, oil cooling
2	HT2	980°C, 45min, oil cooling
3	HT3	1000°C, 45min, oil cooling
4	HT4	1020°C, 45min, oil cooling

Results & Discussion

Microstructures of PM Ti-23Al-17Nb alloy obtained in different quenching processes. Optical metallographic photos of PM Ti-23Al-17Nb alloy obtained during different quenching processes are shown in Fig. 2. Microstructures of test bar obtained during HT1 are shown in Fig. 2(a). It can be concluded from Fig. 2(a) that the base of microstructures is α_2 phase with black color. Distribution of α_2 phase and O phase is inhomogeneous. Microstructures of test bar obtained during HT2 are shown in Fig. 2(b). Compared with Fig. 2(a), it can be drawn that content of O phase increases in a large scale. Transformation between α_2 phase and O phase is occurred with the increasing of quenching temperature. Microstructures of test bar obtained during HT3 are shown in Fig. 2(c). Because transition temperature of O phase is about 1008 °C, a few of B_2 phase with white color can be transformed when treated at a temperature of 1000 °C. Size of crystal grain reduces for recrystallization of B_2 phase. Microstructures of test bar obtained during HT4 are shown in Fig. 2(d). Some quantity of white B_2 phases are transformed and exhibited as reticular structures in the area between α_2 phase and O phase. The transformation of B_2 phase reduces size of crystal grains in the alloy.

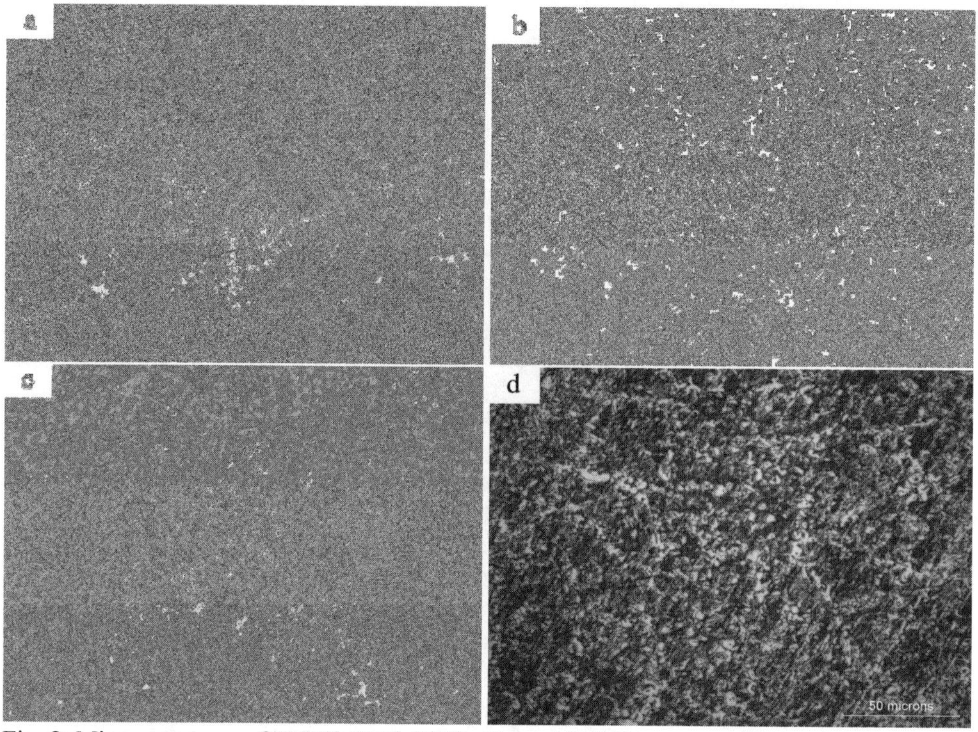

Fig. 2: Microstructures of PM Ti-23Al-17Nb alloy with different quenching processes: (a) HT1; (b) HT2; (c) HT3; (d) HT4;

Tensile properties at room temperature and fractography of PM Ti-23Al-17Nb alloy obtained in different quenching processes. Tensile properties at room temperature of PM Ti-23Al-17Nb obtained during different quenching processes are shown in Table 3. It can be concluded from Table 3 that tensile strength and yield strength of the alloy improve gradually with increasing of quenching temperature from 960 °C to 1020 °C. Tensile strength of the alloy rises from 820 MPa to 1026 MPa and yield strength raises from 736 MPa to 932 MPa. Moreover, ductility of the alloy rises with increasing of quenching temperature. When quenching temperature reaches 1020 °C, ductility of the alloy reaches 5.8%. Therefore, PM Ti-23Al-17Nb alloy performs a better match of tensile strength and ductility after treated by quenching at 1020 °C.

Table 3: Tensile properties of PM Ti-23Al-17Nb at room temperature

Heat treatment	σ0.2 [MPa]	σb [MPa]	δ [%]
1020°C, 45min, oil cooling	962	1026	5.8
1000°C, 45min, oil cooling	891	972	3.2
980°C, 45 min, oil cooling	757	866	1.8
960°C, 45 min, oil cooling	736	820	1.3

In the method of SEM, fracture surfaces of test bar for tensile are observed and shown in Fig. 3-6. Fractographs of test bar obtained during HT1 are shown in Fig. 3. SEM photo shown as Fig. 3(a) is the original area of rupture. It can be drawn from Fig. 3(a) that crystal grain in original area exhibits an abnormally growth, which leads to the formation of crack in the boundary of crystal grain during tensile test. Therefore, fracture of test bar obtained during tensile test is a intergranular one. Fractograph shown in Fig. 3(b) is extension area of fracture in test bar. Distribution and morphology of the crack also performs the characteristics of brittle fracture. Fractographs of test bar obtained during HT2 are shown in Fig. 4. Distribution of crack in original area shown in Fig. (a) is trangranular. It can be determined that fracture of test bar obtained during HT2 is a cleavage one. Extension area of the fracture shown in Fig. 4(b) is a typical "river morphology". Therefore, it can be confirmed that fracture of test bar is a cleavage one. Fractographs of test bar during HT3 are shown in Fig. 5. From Original area of fracture shown in Fig. 5(a), it can be concluded that crack has formed in the area near white B_2 phase. Extension area of frature shown in Fig. 5(b) is characterized by many discrete, short,curved and teared line. Therefore, it can be determined that fracture of test bar is a quasi cleavage one. Fractographs of test bar during HT4 are shown in Fig. 6. Original area of fracture shown in Fig. 6(a) is characterized with many cracked fiber bundles. Extension area of fracture shown in Fig. 6(b) performs with many small pits. Therefore, it can make a conclusion that fracture of test bar obtained during HT4 is a plastic one.

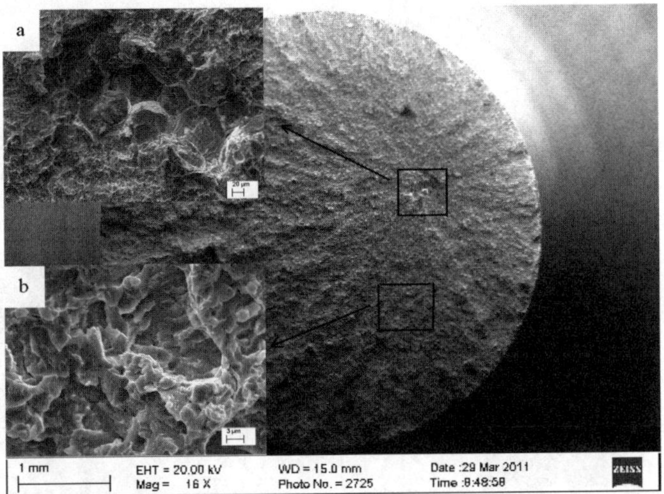

Fig. 3: Fractography of test bar obtained during HT1, (a) original area; (b) expansion area

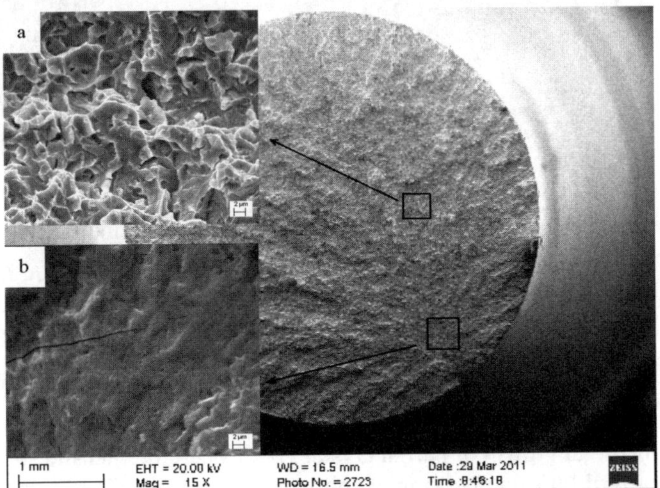

Fig. 4: Fractography of test bar obtained during HT2, (a) original area; (b) extension area

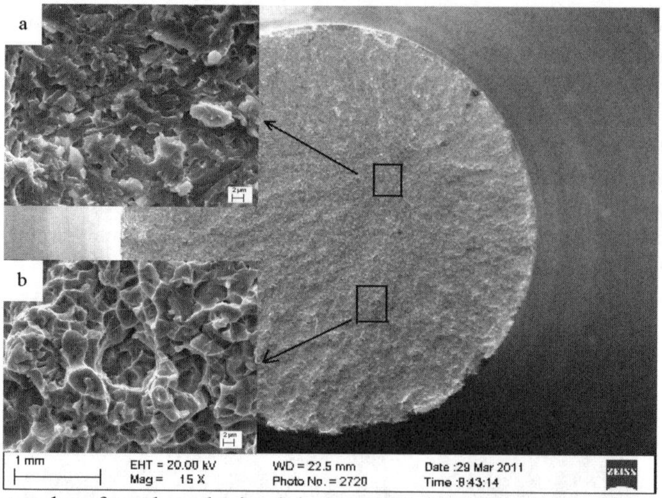

Fig. 5: Fractography of test bar obtained during HT3, (a) original area; (b) extension area

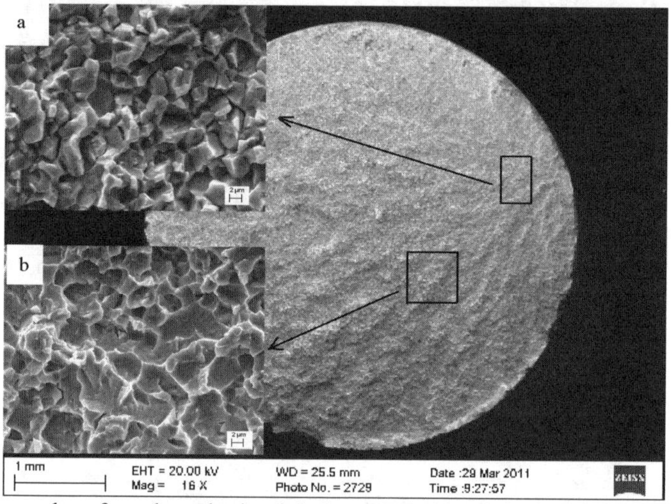

Fig. 6: Fractography of test bar obtained during HT4, (a) original area; (b) extension area

Connection between microstructures and tensile properties at room temperature of PM Ti-23Al-17Nb. Previous studies show that phase equilibrium and transition of PM Ti_3Al is very complex, the morphology, size and content of α_2, O and B_2 phase are sensitive to the heat treatment. Some ideas can be drawn from test and observation referred in this article and list as following.

(1) During HT1, quenching temperature is the same as HIP temperature and below transition temperature of O phase. Such heat treatment as HT1 can not reduce size of crystal grain. However, it may result in an abnormal growth of partial crystal grain. Therefore, tensile strength and ductility of the alloy were improved little and performed poorly. During the tensile test, intergranular fracture of test bar can be occurred.

(2) During HT2, quenching temperature is above HIP temperature and below transition temperature of O phase. Precipitation of O phase can be occurred and result in reduction of crystal size and homogeneousness of O phase. Therefore, tensile strength and ductility of the alloy are improved, but not much. The fracture occurred during tensile test is a cleavage one.

(3) During HT3, quenching temperature is near but below transition temperature of the O phase. Recrystallization of B_2 phase occurred during HT3, but the content of B_2 phase is very little. However,

little changes of microstructures can improve tensile strength and ductility of the alloy in a large scale. During the tensile test, some character of plastic can be observed in fracture of test bar.

(4) During HT4, quenching temperature is near but above transition temperature of O phase, recrystallization of B_2 phase occurred and the content of B_2 phase increased much, which results in reduction of crystal size and homogeneousness of microstructures. Moreover, B_2 phase exhibits as reticular structures, which lead to a remarkable improvement on tensile strength and ductility of the alloy. During tensile test of test bar, plastic fracture can be occurred.

Conclusion

(1) During quenching processes with different temperatures near transition temperature of O phase, tensile properties and microstructures of the obtained alloy are different.

(2) Quenching process above but near transition temperature, recrystallization of B_2 phase occurred and resulted in reduction of crystal size and homogeneousness of microstructures, which lead to a remarkable improvement of tensile strength and ductility of the alloy. When treated with quenching at 1020°C, PM Ti-23Al-17Nb alloy performs a better match of tensile strength and ductility. Tensile strength at room temperature of the alloy can reached 1026 MPa and ductility of the alloy can reach 5.8%.

(3) Quenching process below transition temperature of O phase can not improve tensile strength and ductility of PM Ti-23Al-17Nb alloy obviously. However, it may result in an abnormal growth of partial crystal grain, which leads to a low tensile strength and a poor ductility.

References

[1] H. Jiang: *Development of new materials for aerospace structures abroad.* Aerospace Materials & Technology, 4(1998), pp. 1-8.

[2] T. Lan, C. Han, B. Zhang, M. Li, Y. Zhao, H. Qu: *Investigation of PM Ti-14Al-21Nb.* Rare Metal Materials and Engineering, 21, 3(1992), pp. 53-58.

[3] C. F. Yolton, J. P. Beckman: *PM processing and properties of the ordered orthorhombic alloy Ti-22at.%Al-23at.%Nb.* Materials Science Engineering A, 192/193(1995), pp. 597-603.

[4] Z. Lang, X. Zhang: *PM Processing and Properties of Ti-23Al-17Nb.* Chinese Journal of Rare Metals, 32, 6(2008), pp. 709-713.

[5] T. Lan, B. Zhang, X. Liu and Y. Zhao: *Forming of PM Ti-14Al-21Nb alloy by HIP.* PM Technology, 11, 3(1993), pp. 184-188.

[6] S. Li, Y. Mao, J. Zhang, J. Li, Y. Cheng, Z. Zhong: *Effect of microstructure on tensile properties and fracture behavior of intermetallic alloys.* Trans. Nonferrous Met. Soc. China, 12, 4(2002), pp. 582-586.

[7] Y. Zhang, J. Liu, J Zhang: *Microstructure transition and tensile properties of Ti-22Al-25Nb intermetallic alloy forged in β-phase zone.* Nonferrous Met. Soc. China, 18, 1(2008), pp. 30-35.

[8] Y. Wu, D. Yang: *An investigation of microstructure of Ti_3Al-Nb alloy.* Material Science and Technology, 4, 4(1996), pp. 48-52.

Advanced Materials Research Vol. 279 (2011) pp 16-21
© (2011) Trans Tech Publications, Switzerland
doi:10.4028/www.scientific.net/AMR.279.16

Influence of Melt Holding Time on Microstructure and Mechanical Properties of Ductile Iron

Zhe Liu[1, a], Weiping Chen[1,b] and Yu Deng[1,c]

[1]State Engineering Research Center for Metallic Materials Net-shape Processing, School of Mechanical and Automotive Engineering, South China University of Technology, Guangzhou 510640, China

[a]liuzhe2012@hotmail.com, [b]mewpchen@scut.edu.cn, [c]714732095@qq.com

Key words: Ductile iron, Cooling rate, Antimony, Microstructure

Abstract. Six groups of castings which solidified at different cooling rates with or without addition of 0.03%Sb were fabricated. The effects of the melt holding time, cooling rate, and Sb on the microstructure of ductile iron were studied. Results show that as for the other three blocks with addition of 0.03%Sb, reducing the length of eutectic plateau could improve the graphite morphology, and sufficient residual contents of spheroidizing elements REE and Magnesium could also lead to small and diffuse globular graphite particles no matter what the cooling rate is. As for the other three blocks with addition of 0.03%Sb, no matter what the cooling rate is, the nodularity index values are all about 90%. The main difference among the three blocks is just the size of nodular graphite. And even if two of the blocks solidified at different cooling rates, they own nearly identical NI value and globular graphite average diameter value, which is because they both possess the tolerated rare-earth elements/antimony ratio for spheroidal graphite formation.

Introduction

Graphite morphology and matrix structure in ductile iron can be changed by altering the cooling rate, the amount of spheroidizing elements or minor elements such as antimony and bismuth. The graphite morphology determines the mechanical properties of ductile iron[1-4].

Formation of chunky graphite in the products of heavy section ductile iron castings, due to the low cooling speed and long solidification time, could severely reduce the properties of castings[5-10]. In order to reduce the chunky graphite formation, chills and anti-spheroidizing elements are normally used, which sometimes lowers degenerate graphite level. However, currently a clear understanding of chunky graphite appearance and a safe mastering of metal preparation to avoid it are not yet available. This is partly because there are too many influencing factors on the microstructures and properties of the castings, and partly because of the limitations of experiment, which make the research results not the same.

Solidification of heavy section castings is characterized by very slow cooling rates and lengthy eutectic plateau[11-16]. Accordingly, in the present work, six groups of iron melt with or without addition of Sb remained liquid in different periods of time. The effects of cooling rate and Sb on the microstructure of ductile iron were studied. The results are discussed in view of previous similar reports, and both similarities and discrepancies are enlightened.

Experimental Material and Procedure

The melting process was carried out in a medium-frequency induction furnace of 30kg in capacity. The charge materials used were pig iron and steel scrap. The process for nodulization and inoculation were performed on the melts at a melt temperature of 1480℃. The nodulization process was performed by adding 0.57% nodulizer using the tundish method, and 0.24% inoculant was added into the ladle. After being treated, the molten metal at the temperature of 1250℃were poured into a crucible, which was insulated in 1500℃ knockdown multifunctional resistance furnace, as shown in

Fig. 1. The temperature was accurately controlled by dipping thermocouple into the molten metal. Nominal compositions of six blocks are 3.6%C, 2.2%Si, 0.42%Mn, 0.47%Cu, <0.02%P, <0.01%S. Chemical compositions of nodularizer and inoculant are shown in Table 1.

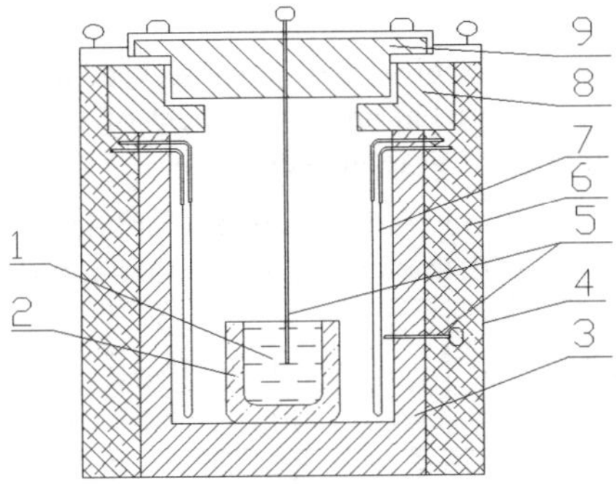

1 Iron melt, 2 Crucible, 3 Brasque, 4 Furnace shell, 5 Thermocouple(Pt-Rh/Pt) recorder, 6 Heat resistant fibre, 7 U-shape silicone molybdenum rods heating elements, 8 Ring inside cover, 9 Furnace cover

Fig. 1: Experimental apparatus for simulative solidification

Table 1: Chemical compositions of nodularizer and inoculant [wt%]

Element	Si	Mg	Ca	Ba	Al	REE[a]	Fe
Nodulizer	42.6	7.05				2.62	Bal.
Inoculant	70.2		1.84	4.22	1.25		Bal.

[a]REE contains 50%Y and 50%Ce, respectively.

In order to study the cooling rate and Sb on the graphite and matrix microstructure of ductile iron, six groups of castings were prepared as shown in Table 2. Iron melts solidified in the furnace according to the schedule in Fig. 2. The microstructure observation was conducted by LEICA DMI5000M intelligent optical microscope. Tensile testing was conducted with a universal testing apparatus (CMT5105, SANS Co. Ltd., China). Brinell hardness measurement was carried out using a Brinell Hardness Tester (HB-3000B, HUAYIN Co.Ltd, China). Compositions for the six groups of fracture specimens were detected by the chemical analysis method, apart from the Mg, Y, Ce and Sb contents, which were measured using the ICP-MS technique.

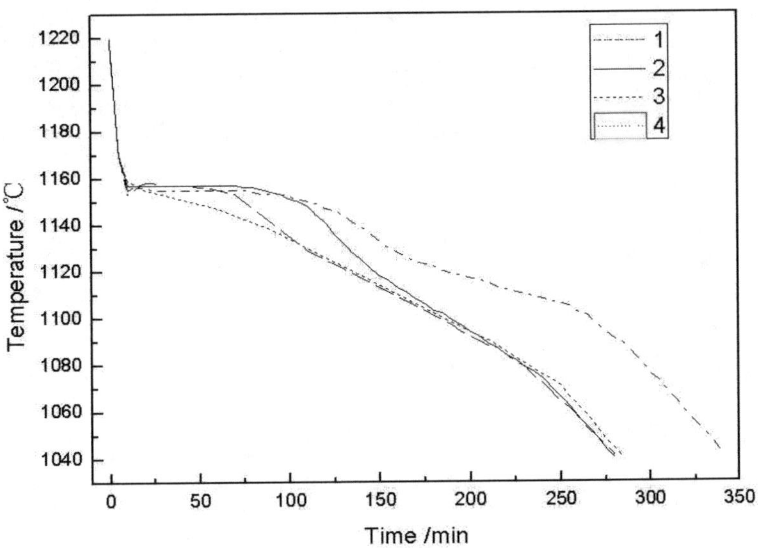

Fig. 2: Cooling curves of iron melts

Table 2: Code of Casting Trials: Indication of cooling curve (1, 2, 3, 4, 5) and addition of 0.03%Sb

Trial	E-1	E-2	E-3	F-1	F-2	F-3
Block	1	3	4	1Sb	2Sb	3Sb

Results

Chemical analyses and mechanical performance testing for six groups of fracture specimens were carried out. In Table 3, a nodularity index NI was also evaluated at an enlargement of 100X as the ratio of the number of spheroidal particles to the total amount of graphite particles.

Table 3: REE, magnesium and antimony content, REE/antimony ratio, globular graphite particles count[N_P], nodularity index[NI], globular graphite average diameter and mechanical properties

Block	REE_{res}	Mg_{res}	Sb	REE/Sb	N_P [mm^{-2}]	NI [%]	D [μ]	σ_b [MPa]	δ [%]	HBW
1	0.025	0.05	—	—	—	30.5	84.6	406	2.4	182
3	0.035	0.076	—	—	—	61.6	56.8	417	4.28	183
4	0.026	0.045	—	—	—	85.2	90.8	517	4.0	220
1Sb	0.026	0.059	0.026	1	24.5	89.6	53.4	708	2.96	234
2Sb	0.029	0.054	0.02	1.45	63.5	88.2	38.2	480	3.98	218
3Sb	0.027	0.054	0.026	1.03	59.2	86.6	34.4	496	4.8	225

Fig. 3 shows the graphite morphology and matrix structure of six groups of specimens. By combining with the relative values shown in Table 3, it's clear that the smaller the globular graphite average diameter, the higher the N_P values are, and the higher the percentage of ferrite content in matrix structure is. Among the former three blocks, the morphology graphite becomes better as thelength of eutectic plateau increases. When 0.03%Sb was added, the latter three blocks owns great graphite nodularity and nodule count.

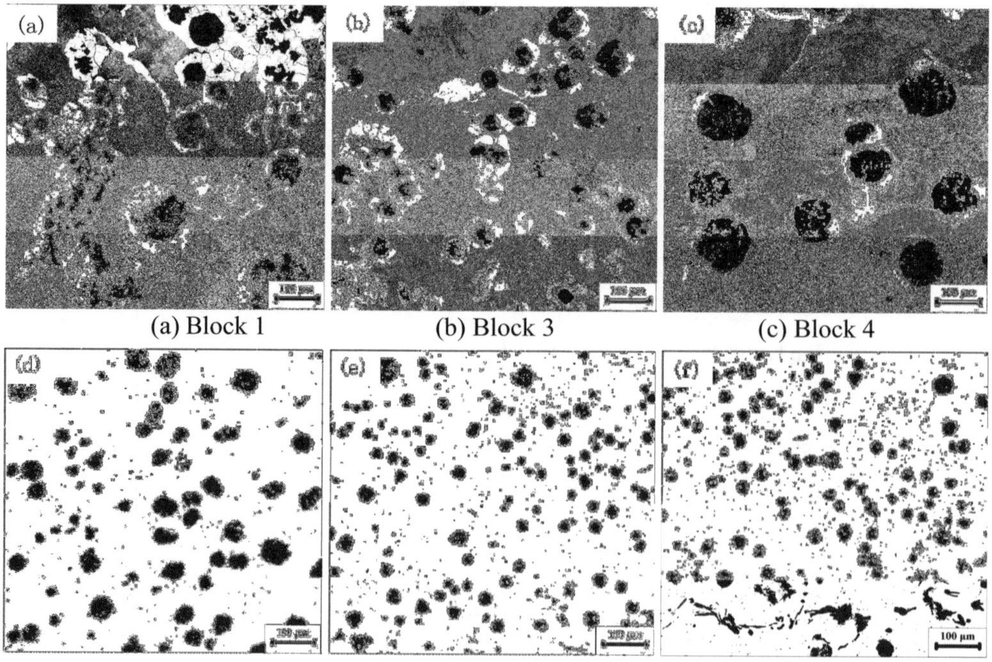

(a) Block 1 (b) Block 3 (c) Block 4

(d) Block 1Sb (e) Block 2Sb (f) Block 3Sb

Fig. 3: Microstructures of the blocks [a, b, c-etched; d, e, f-unetched]

Discussions

Influence of cooling rate without addition of Sb. Melt holding time and cooling rate have a great effect on the microstructure of ductile iron, while the chemical composition and inoculation practice also influence the graphite morphology of ductile iron [17,18].

As for the three blocks without addition of 0.03%Sb, the NI values increases as the length of eutectic plateau decreases. Block 3 possesses the smallest size and largest amount of globular graphite, which is because it possesses the higher residual contents of spheroidizing elements REE and magnesium than block 1 and block 3, so the effective nucleus are sufficient for the nodular graphite formation. A higher content of ferrite exists around chunky graphite, which is due to the chunky graphite causing the existence of carbon poor region around itself.

It was calculated that the percentage of ferrite content in block 1, 3, 4 is 30%, 20% and 5%, respectively. Thus, the tensile strength and hardness values both increase as the length of eutectic plateau decreases, because the percentage of pearlite content is the critical factor for the tensile strength and hardness values. Although block 3 has relatively low value of NI, it possesses the smallest size of globular graphite, which makes it show the highest elongation value.

Influence of cooling rate with addition of Sb. It is common that the NI values increases as the length of eutectic plateau decreases. While as for the latter three blocks, the results are a little different as 0.03%Sb was added. Previous studies show that appropriate addition of Sb is beneficial for the graphite morphology and mechanical properties of heavy section ductile iron. Cerium/antimony ratios in the range of 0.4-0.9 have been found to be tolerated[19].

As for the other three blocks with addition of 0.03%Sb, although the cooling rates and lengths of the eutectic plateau are different, the NI values are all about 90%, because Sb reacts with REE to form high melting point compounds, which can act as nucleus of globular graphite, moreover, the concentration of Sb around graphite can impede the graphite deterioration. The size of nodular graphite of block 1Sb is larger than block 2Sb and block 3Sb, because it remained liquid in the longest period of time. Though the eutectic plateau of block 2Sb is longer than block 3Sb, they own nearly identical NI values and globular graphite average diameter values. It is supposed that the tolerated REE/Sb ratio for spheroidal graphite formation decrease as the length of eutectic plateau decrease. In present study, the REE/Sb ratio of 1.45 for block 2Sb and 1 for block 3Sb are both tolerated, in which case complete neutralization of Sb by REE was achieved. In addition, the antimony addition for block 3Sb is a little excess, which caused the formation of spiky graphite, as shown in Fig. 4.

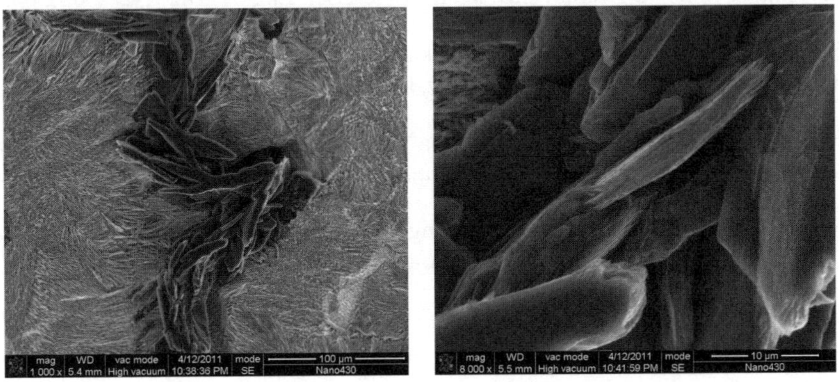

(a) SEM, 1000X (b) SEM, 8000X

Fig. 4: Spiky graphite in block 3Sb

As the nodule count and graphite nodularity of block 2Sb and 3Sb is better than block 1Sb, the elongation values of block 2Sb and 3Sb are much higher, but the tensile strength and hardness values are relatively lower than block 1Sb, since the matrix ferrite content increases as the nodule count increases.

Conclusions

Six groups of small castings which solidified at different cooling rates with or without addition of 0.03%Sb were fabricated. Based on the experimental results, the following conclusions were determined.

As for the three blocks without addition of 0.03%Sb, reducing the length of eutectic plateau could improve the graphite morphology, and sufficient residual contents of spheroidizing elements REE and Magnesium could also result in small and diffuse globular graphite particles no matter what the cooling rate is.

As for the other three blocks with addition of 0.03%Sb, no matter what the cooling rate is, the nodularity index values are all about 90%. The size of nodular graphite of block 1Sb is larger than block 2Sb and block 3Sb. Even if the eutectic plateau length of block 2Sb is longer than block 3Sb, block 2Sb and block 3Sb own nearly identical NI value and globular graphite average diameter value, because these two blocks possess the tolerated REE/Sb ratio for spheroidal graphite formation. It is supposed that the tolerated REE/Sb ratio decreases as the length of eutectic plateau decreases.

Acknowledgements

This work was financially supported by the Significant Science and Technology Project of Guangdong-Energy conservation and emission reduction and renewable energy(2008A080800022).

References

[1] K.F.Nilsson, V.Vokal: *Analysis of ductile cast iron tensile tests to relate ductility variation to casting defects and material microstructure*. Mater. Sci. Eng. A, Vol. 502(2009), pp. 54-63.

[2] C.H.Hsu, S.C.Lee, Y.H.Shy, W.T.Chiou: *Relationship between dynamic and static toughness of flake and compacted graphite cast irons*. Mater. Sci. Eng. A, 282, 1-2(2000), pp. 115-122.

[3] S.Jung, T.Ishikawa, S.Sekizuka, H.Nakae: *Effects of sulfur on interfacial energy between Fe-C melt and graphite*. J.Mater.Sci., 40, 9-10(2005), pp. 2227-2231.

[4] M.Tsujikawa, K.Nagamine, A.Ikenaga, M.Hino: *Influence of graphite morphology on dry sliding wear of flake graphite cast irons*. Int.J.Cast Met.Res. Vol. 21(2008), pp. 76-80.

[5] Wang Liping, Guo Erjun, Chen Hongbin: *Effects of light and heavy rare earths on anti-degradation of nodular cast iron*. Journal of rare earths Vol. 24 (2006), pp. 103-107.

[6] H.W.Hoover: *A Literature Survey on Degenerate Graphite in Heavy Section Ductile Iron*. AFS Trans. Vol. 94(1986), pp. 601-608.

[7] I.Asenjo, P.Larranaga, J.Sertucha, R.Suarez, J.M.Gomez, I.Ferrer, J.Lacaze: *Effect of mould inoculation on the formation of chunky graphite in heavy-section spheroidal graphite cast iron parts*. Int.J.Cast Met. Res., 20, 6(2007), pp. 319-324.

[8] C.Labrecque, M.Gagné.: *Review ductile iron: fifty years of continuous development Can.* Metall. Quart., 37, 5(1998), pp. 343-378.

[9] Z.Zhang, H.M.Flower, Y.Niu: *Classification of degenerate graphite and its formation processes in heavy section ductile iron*. Mater.Sci.Technol. Vol. 5(1989), pp. 657-664.

[10] H.Itofuji, H.Uchikawa: *Formation Mechanism of Chunky Graphite in Heavy-Section Ductile Cast Irons*. AFS Trans. Vol. 98(1990), pp. 429-448.

[11] C.R. Loper, R.W. Heine, R.W. Reesman, and B.H. Shah: *Thermal Analysis of Ductile Iron*. AFS Trans. Vol. 75(1967), pp. 541-547.

[12] P.K. Basutkar and C.R. Loper: *Correlation of Microstructures with Thermal Analysis in Heavy Section Ductile Iron Castings*. AFS Trans. Vol. 79(1971), pp. 169-75.

[13] P.K. Basutkar and C.R. Loper: *Effects of Cerium Additions on the Thermal Analysis of Heavy Section Ductile Iron*. AFS Trans. Vol. 79(1971), pp. 176-185.

[14] M. Wessén, I.L. Svensson and R. Aagaard: *Influence of antimony on microstructure and mechanical properties in thick-walled ductile iron castings*. Int.J.Cast Met.Res. Vol. 16(2003), pp. 119-124.

[15] Z. Ignaszak: *Nodular graphite degeneration in the post-inoculated ductile iron*. Int.J.Cast Met.Res. Vol. 16(2003), pp. 93-97.

[16] R. Källbom, K. Hamberg, M. Wessén and L.E. Björkegren: *On the solidification sequence of ductile iron castings containing chunky graphite*. Mater.Sci.Eng. Vols. A413-A414(2005), pp. 346-351.

[17] A.Javaid, C.R. Loper: *Production of Heavy-Section Ductile Cast Iron*. AFS Trans. Vol.103(1995), pp. 135-150.

[18] M.Bazdar, H.R.Abbasi, A.H.Yaghtin, J.Rassizadehghani: *Effect of sulfur on graphite aspect ratio and tensile properties in compacted graphite irons*. J. Mater.Process.Technol., 209, 4(2009), pp. 1701-1705.

[19] Pan E.N., Chen C.Y.: *Effects of Bi and Sb on Graphite Structure of Heavy-Section Ductile Cast Iron*. AFS transactions Vol. 104(1996), pp. 845-858.

Advanced Materials Research Vol. 279 (2011) pp 22-27
© (2011) Trans Tech Publications, Switzerland
doi:10.4028/www.scientific.net/AMR.279.22

Tribological Behaviors of Self-lubricating Ceramic tool Composite in Various Conditions

CAO Tong-kun[1,a], GE Jun-peng[1,b] and YANG Qi-long[1,c]

[1]College of Electromechanical Engineering, Qingdao University of Science and Technology, Qingdao 266061, Shandong, China

[a]caotk@163.com, [b]gejunpeng0515@126.com, [c]yql851225@163.com

Key words: ceramic composite, self-lubricating, friction and wear behavior

Abstract. Al_2O_3/TiC ceramic composites with the addition of CaF_2 solid lubrication were produced by hot pressing. The effects of various conditions on tribological behaviors of Al_2O_3/TiC/CaF_2 have been studied. The sliding wear tests against cemented carbide and 45 steel were performed on the ceramic composites using ring-block method. Results showed that with the increase of load and sliding speed, the friction coefficient decreased. However, the wear rate increased slightly with the increase of load and decreased with the increase of sliding speed. At low speed, the abrasion resistance of Al_2O_3/TiC/CaF_2 coupled with cemented carbide was better than with 45 steel, while the abrasion resistance showed little difference at high speed. The friction coefficients were little different when coupled with cemented carbide and 45 steel respectively. During the sliding process, at the initial stage, the friction coefficient was waved because the self-lubricating film was not formed on the worn surface; after sliding 628m, the friction coefficient was stable because the self-lubricating film was formed on the worn surface.

Introduction

With the development of the cutting, the traditional methods can not satisfy lubricating requirements for the high speeding cutting and dry cutting, which can produce much hot in process. And a lot of energy and resources were consumed in the manufacture, using and discharge of cutting fluid [1-3]. At present, some researches have been done for dry cutting in domestic and overseas [4, 5], for example, developing new cutting tools with better performance for dry cutting [6-9]. But the friction condition of dry cutting is very harsh, so cutting temperature is extreme high and the tool's thermal wear is severe. These all lead to low life of tools.

In recent years, the concept of self-lubricating tools was proposed [10-12]. A layer of solid lubricant film can be formed on the cutting tool surface in the cutting process. The solid lubricant film has good high-temperature tribology properties. It can not only adapt to high working temperature and a wide range of friction speed, but also bear high load. So the self-lubricating tools can be used in some special environmental conditions. Since the application of self-lubricating tools can reduce the friction and wear, additional lubricating system is not needed.

The concept of "self-lubricating tool" was just proposed in recent years and the study of the self-lubricating tool technology and related basic theories was just begun. Therefore, further studies are needed to solve existing problems in this field.

Experimental Procedure

Friction and wear tests were conduced with a MRH-3 high-speed ring–block tribometer. The block specimen (12×16×5mm) was made of ceramic materials having a polished surface with a surface roughness of 0.08μm. The ring specimen (Ø50×Ø35×15mm) was made of cemented carbide (YG8) with a hardness of HRA89, and 45# hardened steel with a hardness of HRA55. The ring surface was polished to produce a final surface roughness of 0.1μm. Both the block and the ring were rinsed with hexane, and then ultrasonically cleaned in fresh hexane, followed by ultrasonic cleaning with acetone. Each test was run over a period of 10min. The friction coefficient was calculated by dividing the measured tangential force by the applied normal force. At the end of each test, the width of the wear scar on the block specimens was measured with a microscope, and then the wear volume loss *V* of the block specimen was calculated. The mechanical properties of the Al_2O_3/TiC/CaF_2 are shown in Table 1.

Table 1: Mechanical properties of Al_2O_3/TiC/ CaF_2 composites

Material	Volume ratio of Al_2O_3/TiC	Mass fraction of CaF_2 [wt %]	Bending strength [MPa]	*HV* [GPa]
AF1	1:1.2		589	15.4
AF2	1:1.8	10	678	15.5
AF3	1:3		613	16
AF4	1:5.7		516	15.8

Results and Discussion

The Effect of Loads on Tribological Behaviors. Fig. 1(a) and (b) illustrate effect of load on the friction coefficient and wear rate of the self-lubricating ceramic materials with cemented carbide ring. It is indicated that the friction coefficient showed a downward trend with an increase of load. While, the wear rate increased with an increase of load.

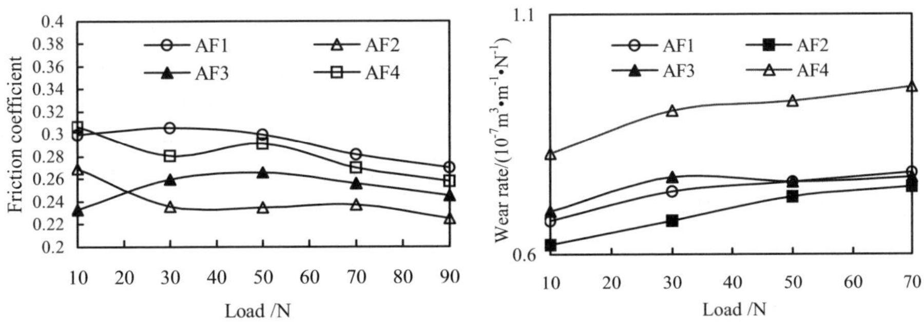

(a) Friction coefficient (b) Wear rate

Fig. 1: Effects of load on the friction coefficient and wear rate of Al_2O_3/TiC/ CaF_2 (400r/min)

Fig. 2(a) shows the SEM photograph of worn surface of Al_2O_3/TiC/CaF_2 self-lubricating ceramic materials under low load (10N). There were numerous scratches and pits on the wear surface, not forming solid lubricating film. Slight surface damage can be observed in the form of a scratched appearance. This suggests that the primary wear mechanism of this composite is abrasive wear. It is because the friction surface temperature is not high under low load and particles of CaF_2 are brittle.

Thus the lubricating film wasn't formed on the surface that leads to high friction coefficient.

Figure 2(b) shows the SEM photograph of worn surface of Al_2O_3/TiC/CaF_2 self-lubricating ceramic materials under high load (70N). In comparison with Figure 2(a), it exhibited a relatively smooth surface, both mechanical plowing grooves and scratches could not be observed, and there was no distinct crack on the wear track. This suggests that a thin solid lubricant film was formed on the wear surface and the thin film had lower shear strength due to its softening under friction heat. This leads to a lower friction coefficient of Al_2O_3/TiC/CaF_2 with high load. At the same time, the solid lubricating film reduced the effect of the load on the wear rate, and the wear rate increased slightly with increasing load.

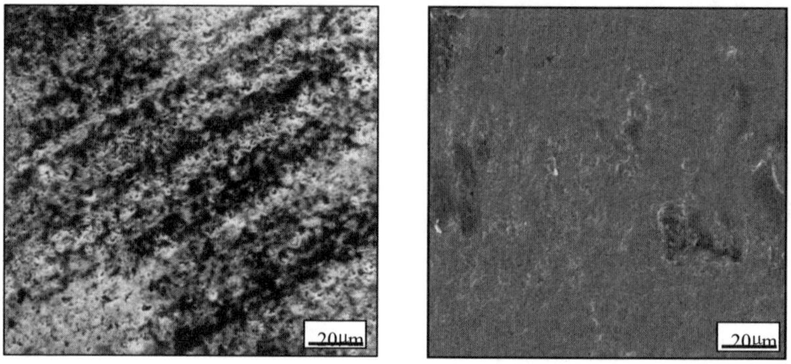

(a) 10N (b)70N

Fig. 2: SEM micrograph of the worn surface of Al_2O_3/TiC/ CaF_2 (400r/min)

The Effect of Friction Speed on Tribological Behaviors. Fig. 3(a) and (b) show the influence of speed on the friction coefficient and wear rate of Al_2O_3/TiC/ CaF_2 with cemented carbide ring. It can be seen that the friction coefficient and wear rate decreased with the speed increasing on the whole. When the speed was 300r/min, the wear rate was the maximum.

Fig. 4(a) and (b) show the SEM photographs of worn surface of Al_2O_3/TiC/CaF_2 self-lubricating ceramic materials with cemented carbide ring. It can be seen that at low speed (300r/min), there were several pieces of solid lubrication film and some shallow mechanical plowing grooves on the wear surface. Because the friction surface temperature was not high at low speed, CaF_2 particles in Al_2O_3/TiC/CaF_2 materials were in a brittle state, and were not easily smeared on the worn surface to form lubricating film. While at high speed, the friction surface temperature is high, and CaF_2 particles softened. Then the particles were easily smeared on the worn surface to form whole lubricating film. Therefore, the friction coefficient and wear rate were lower at high speeds.

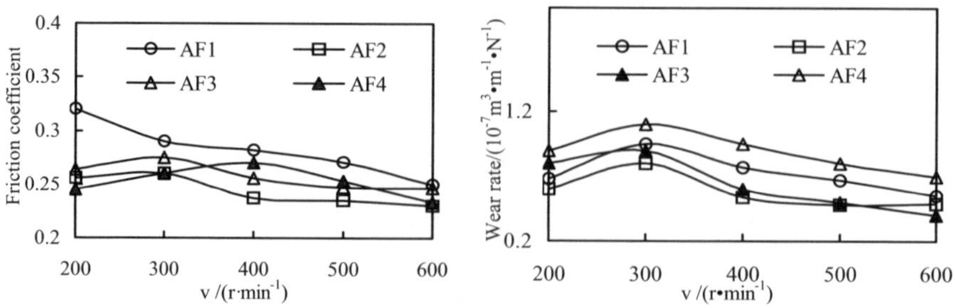

(a) Friction coefficient (b) Wear rate

Fig. 3: Effects of speed on the friction coefficient and wear rate of Al_2O_3/TiC/ CaF_2 (Load 70N)

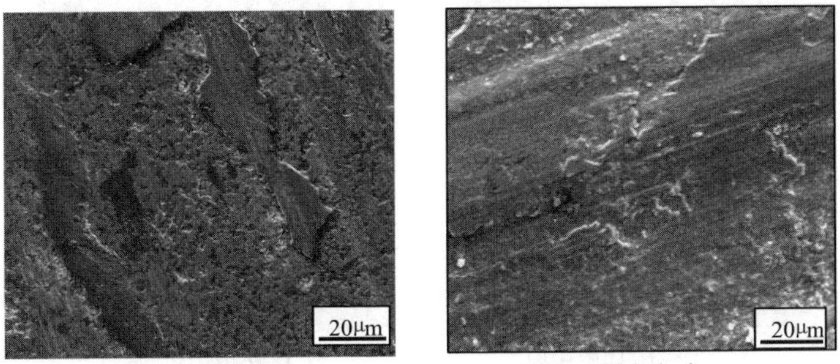

(a) 300r/min (b) 600r/min

Fig. 4: SEM micrograph of the worn surface of Al_2O_3/TiC/ CaF_2 (70N)

The Effect of Pairs on Tribological Behaviors. Fig. 5(a) and (b) show the effect of pairs on the friction coefficient and wear rate of the self-lubricating ceramic materials. It can be seen that the friction coefficients varied slightly when the self-lubricating ceramic materials paired with hardened steel and cemented carbide respectively. Fig. 6 (a) and (b) show the SEM photographs of worn surface of Al_2O_3/TiC/CaF_2 self-lubricating ceramic materials with steel ring. By Fig. 2(b) and 6(b), it can be seen that under higher load, a solid lubricating film was formed on worn surface with cemented carbide and steel respectively, which played a good effect of reducing friction, so the friction coefficient varied slightly.

At the speed of 400r/min, the wear rate with cemented carbide was lower than that with steel. At the speed of 600r/min, the wear rate with cemented carbide and steel varied slightly. By Fig. 4(a) and 6(a), it can be seen that at the low speed the area of the solid lubricating film formed on worn surface with cemented carbide was bigger than that with steel; and at high speed, the areas of the solid lubricating film with cemented carbide and steel varied slightly. So at the low speed, the wear resistance with cemented carbide was better than that with steel. While at the high speed, the wear resistance was almost same.

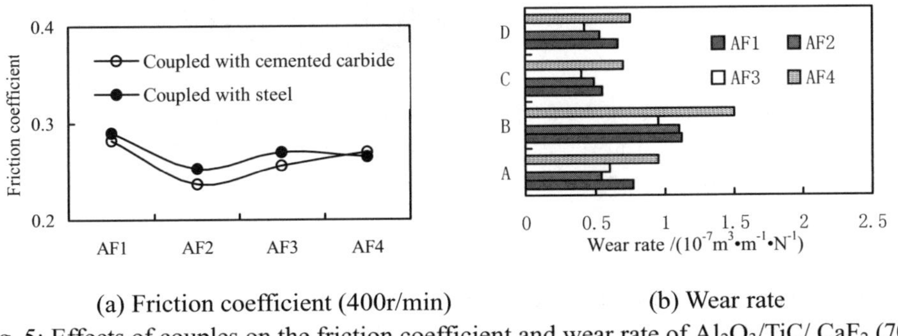

(a) Friction coefficient (400r/min) (b) Wear rate

Fig. 5: Effects of couples on the friction coefficient and wear rate of Al_2O_3/TiC/ CaF_2 (70N)

A Coupled with cemented carbide (400r/min) B Coupled with steel(400r/min)

C Coupled with cemented carbide (600r/min) D Coupled with steel(600r/min)

The Effect of Sliding Distance Tribological Behaviors. The effect of sliding distance on friction coefficient was shown in Fig. 7. It is can be seen that, when the load was 50N and speed was 400r/min, the friction coefficient showed a slight downward trend with an increase of distance. During the sliding process, at the initial stage, the friction coefficient was waved; after sliding 628m, the friction coefficient was stable.

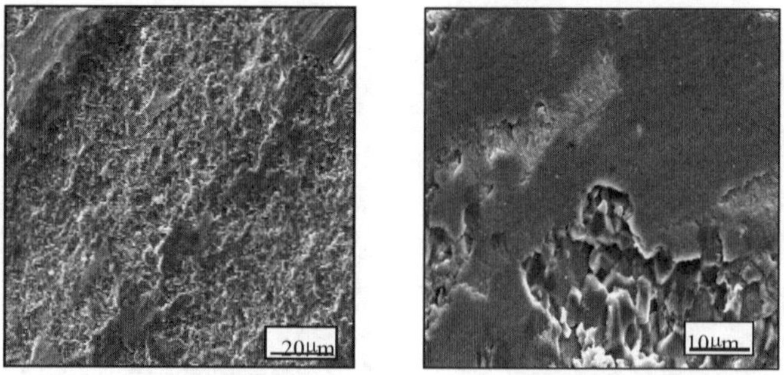

(a) 300r/min (b) 600r/min

Fig. 6: SEM micrograph of the worn surface of Al_2O_3/TiC/ CaF_2 coupled with steel (70N)

Fig. 8(a) and (b) show the SEM photographs of worn surface of Al_2O_3/TiC/ CaF_2 with cemented carbide in different distance. By Fig. 8 (a), it can be seen that there ere only some white powder debris on the worn surface after sliding 188m and a lubricating film didn't exist on there. However, it can be seen from the Fig. 8 (b) that a lubricating film was basically formed on the worn surface after sliding 628m, which played a role in reducing friction. And in the sliding process, the generation of friction heat will slowly rise the friction temperature. The increasing temperature will contribute to reduce the friction coefficient and make the friction coefficient decreased slightly with the growth of friction distance.

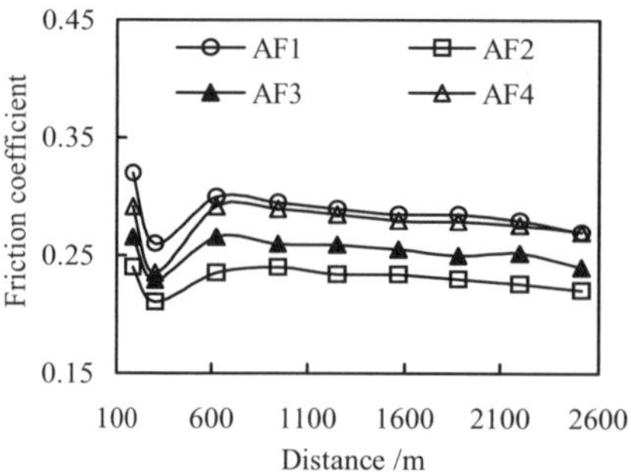

Fig. 7: Effects of sliding distance on the friction coefficient of Al_2O_3/TiC/CaF_2 (Load50N,Speed400r/min)

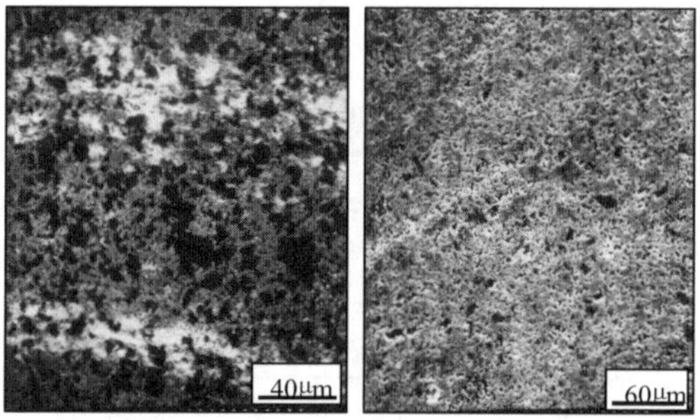

(a)sliding 188m (b) sliding 628m

Fig. 8: SEM micrograph of the worn surface of Al_2O_3/TiC/CaF_2 (Load50N, Speed 400r/min)

Conclusions

(1) With the increase of load and sliding speed, the friction coefficient of Al_2O_3/TiC/ CaF_2 decreased. However, the wear rate increases slightly with the increase of load and decreased with the increase of sliding speed. This is because CaF_2 will change from brittle state to plasticity state at 500℃. Along with the increase of speed and load, friction temperature increased and CaF_2 particles softened, which contribute to forming solid lubricant films easily.

(2) At low speed, the abrasion resistance with cemented carbide was better than with 45 steel, while the abrasion resistance showed little difference at high speed. The friction coefficients were little different when coupled with cemented carbide and 45 steel respectively.

(3) During the sliding process, at the initial stage, the friction coefficient was waved because that the self-lubricating film was not formed on the worn surface; after sliding 628m, the friction coefficient was stable because the self-lubricating film has been formed on the worn surface.

Acknowledgements

This work was supported by the Scientific Research Foundation for the Excellent Young Scientists of Shandong Province (BS2010ZZ003).

References

[1] Sun Dayong: *Advanced Manufacturing Technology.* China Machine Press, Beijing, (2000).

[2] Wang Xibin: *Green Machining Technology in Metal Cutting.* Chinese Journal of Mechanical Engineering Vol. 36 (2000), p. 6.

[3] Jia Xiaoming, Wang Baozhong, Feng Xijing: *Analysis of Green Machining Technology.* Lubrication Engineering Vol.6 (2002), p. 83.

[4] Liu Zhifeng, Zhang Chonggao, Ren Jialong: *Dry Machining Technology and Application.* China Machine Press, Beijing, (2005).

[5] Wang Xi, Xiao Yi, Dong Hai: *Monitoring and Analysis of High Speed Dry Machining Process.* China Mechanical Engineering Vol. 20 (2009), p. 394.

[6] Deng Jianxin, Liu Jianhua, Zhao Jinlong: *Friction and wear behaviors of the PVD ZrN coated carbide in sliding wear tests and in machining processes.* Wear Vol. 264 (2008), p. 298.

[7] V Fox, A Jones, NM Renevier: *Hard lubricating coatings for cutting and forming tools and mechanical components.* Surface and Coatings Technology Vol. 125 (2000), p. 347.

[8] Yu Qixun, Zhang Jingying: *Development and Application of Tool Coating Technology.* Aeronautical Manufacturing Technology Vol. 7 (2007), p. 1.

[9] Wang X, Xing Y M, Ma S L: *Microstructure and Mechanical Properties of MoS_2/titanium Composite Coatings with Different Titanium Content.* Surface and Coatings Technology Vol.201 (2007), p. 5290.

[10] Deng Jianxin, Ge Peiqi, Ai Xing: *Development and Perspective of Lubricating Techniques in Metal Cutting.* Tribology Vol.23 (2003), p. 546.

[11] Deng Jianxin, Cao Tongkun, Sun Junlong: *Self-lubricating Behaviors of Al_2O_3/TiB_2 Ceramic Tools in Dry High-speed Machining of Hardened Steel.* Journal of the European Ceramic Society Vol.25 (2005) p. 1073.

[12] Cao Tongkun, Deng Jianxin: *Self-lubricating Mechanism of A Self-lubricating Ceramic Cutting Tool.* The Chinese Ceramic Society Vol.34 (2006), p.1232.

Advanced Materials Research Vol. 279 (2011) pp 28-32
© (2011) Trans Tech Publications, Switzerland
doi:10.4028/www.scientific.net/AMR.279.28

The Investigation of Creep Behavior for NiAl-28Cr-5.5Mo-0.5Hf-0.02wt.%P Alloy at High Temperature

ZHANG Guangye[a], YE Dongwen[b], WANG Jinlin[c], CHEN Youming[d], LIU Longfei[e], GUO Yuanjun[f]

Hunan Principal Key Laboratory of Heath Maintenance for Mechanical Equipment, Hunan University of Science and Technology, Xiangtan 411201 China

[a]zhangguangye7411@163.com, [b]yedongwen@163.com, [c]jlwang-5@163.com, [d]chenym410@live.cn, [e]wangerce@yahoo.com.cn, [f]yjguo@hnust.edu.cn

Keywords: Intermetallics, NiAl-based alloy, Creep

Abstract. The Microstructure and creep behavior for NiAl-28Cr-5.5Mo-0.5Hf-0.02wt.%P alloy at high temperature have been investigated in this paper. The results reveal that the high temperature creep behavior of the NiAl-28Cr-5.5Mo-0.5Hf-0.02wt.%P alloy is characterized by transient primary creep and dominant steady-state creep as well as ternary creep behavior. The primary creep can be described by Garofalo equation and the steady-state creep can be depicted by Dorn equation. The creep mechanisms are viscous glide of dislocations at lower and middle testing temperatures and dislocation climb at higher temperature. No change of the microstructure for the testing alloy indicates that the creep fracture is controlled by the formation and propagation of cavities and cracks, and the creep fracture behavior obeys Monk man-Grant relationship.

Introduction

NiAl-based intermetallic alloys are potential materials for high temperature structural applications. However, both limited fracture toughness at room temperature and lack of creep strength at high temperature impede their practiced use. DS NiAl/Cr(Mo) eutectic alloy, whose microstructure was characterized by a lamellar Cr(Mo) phase embedded in NiAl matrix, have an average value of 17-24.2MPa.$m^{1/2}$ at room temperature [1-3]. The mechanical properties of intermetallics have revealed that single-, duplex and multi-phase ordered intermetallics can be super plastically deformed under appropriate temperature, initial strain rate and grain sizes [4]. The cast NiAl/Cr(Mo) alloys make the matrix NiAl phase grow with in-*situ* and coherent or semi-coherent precipitation Cr(Mo) strengthening phase[5-8]. The proper additions of nonmetal elements, especially P which worsen the mechanical properties of Fe-based alloys, are found to be effective in improving the high temperature strength, deformation properties and creep resistance in superalloys [9-10]. The various mechanisms by which phosphorus influences the microstructure and mechanical properties of superalloy are arisen from the role of phosphorus P segregation at the grain boundaries which improves the interfacial coherence. Therefore the microstructure and creep behavior of Ni-33Al-28Cr-5.5Mo-0.5Hf-0.02P eutectic alloy are reported.

Experimental Procedures

The nominal Ni-33Al-28Cr-5.5Mo-0.5Hf (%, atom fraction)-0.02P (%, weight fraction) alloy, by using high-pure metal elements and transient alloy Ni-P as starting materials, was induction melted under vacuum conditions and dropped cast into cylindrical chill mold. The samples were hot isostatic pressed (HIP) at 1573 K /200 MPa for 4h. The flat tensile specimens, having a gauge section 2 mm×2.5 mm×16 mm, were electro-discharge machined from the rods. Tensile creep tests were performed on a RCL-3 testing machine. The specimens for optical microscopy were etched in a solution of 30ml HNO_3+ 30ml HCl +40ml CH_3COOH. Scanning electron microscopy (SEM) analysis was performed on JSM6301F. The foil for TEM observation was examined in a Philips TEM-420 analytical electron microscope operated at 150kV.

Results

Microstructure. The SEM micrograph of microstructure for as-cast and HIP treated NiAl-28Cr-5.5Mo-0.5Hf-0.02P alloy is shown in Fig. 1. The alloy as-cast was composed of three phases: (i) the lamellar Cr(Mo) phase, (ii) the NiAl matrix, and (iii) the discontinuously distributed white phase(Fig. 1(a)). The microstructure of this alloy was polycrystalline. Each grain possessed the same crystallographic relation but differed in rotation about the growth axis. The white phase located in the vicinity of the grain boundaries and rarely observed in the center of each grain. Compared to the as cast alloy, the white phase in the testing alloy after the HIP treated located in the vicinity of the grain boundaries was refined and distributed more uniform than that of as cast alloy (shown in Fig. 1 (b)). Fig. 2 shows the characteristic structure of the NiAl-28Cr-5.5Mo-0.5Hf-0.02P alloy after HIP treated. And two kinds size NiAl particles precipitate in Cr(Mo) phase, which the smaller size of NiAl particles coherently precipitated in Cr(Mo) phase, and the larger size of NiAl particles semi-coherently precipitated in Cr(Mo) phase (shown in Fig. 2(a)). Fig. 2(b) shows the Cr(Mo) particles coherently precipitate in NiAl matrix phase.

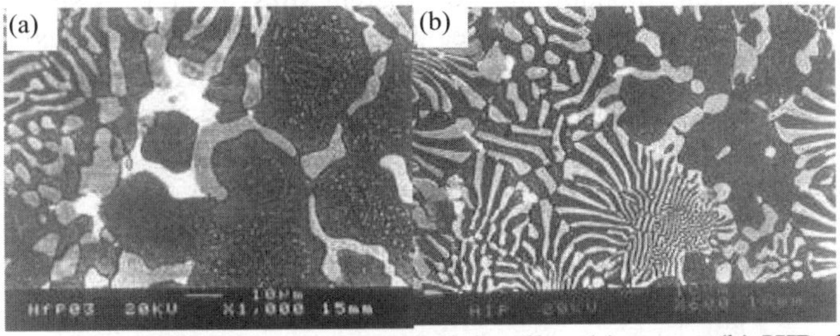

Fig. 1: The microstructures of the alloy (a)as-cast (b) HIPed

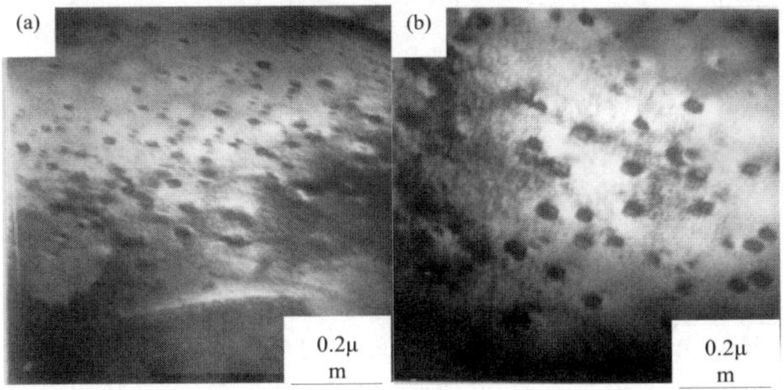

Fig. 2: The TEM microstructures of (a)NiAl particles precipitated in Cr(Mo) phase (b) Cr(Mo) particles precipitated in NiAl phase

Creep Behavior.

Primary Creep Behavior. There is a number of equations describing creep strain vs time in published reports, and Garfalo's and Li's equations are the two most frequently applied to describe the primary creep behavior. The time-dependence of the primary creep strain rate at various applied stress and 1273K are shown in Fig. 3. It is evident that the Li's equation gives less satisfactory representation of primary creep than Garofalo relation. This is not in agreement with resent observations for multiphase Ni-based superalloy IN738LC [11]. In superalloy, the more satisfactory representation of primary creep is Li's equation.

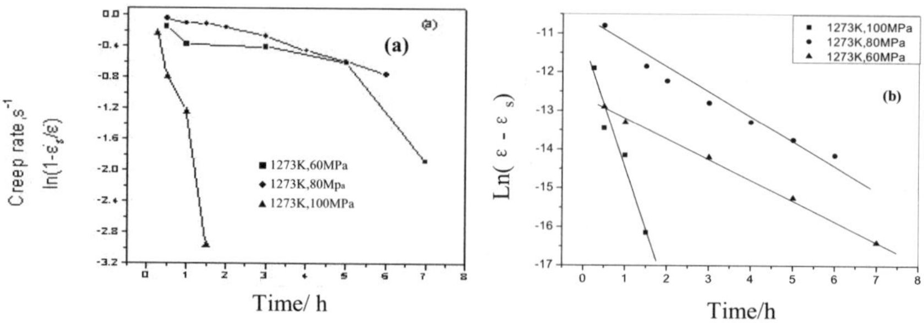

Fig. 3: Dependence of the primary creep rate on the time (a) Li's equation ;(b)Garofalo's equation

Steady-ate Creep. To estimate the tress exponents and activation energy for creep, the creep data from are plotted as strain rate versus stress on double logarithmic scales and are given in Fig. 4(a). The slopes of these plots corresponding to the apparent stress exponents shows a relatively well defined transition from a lower temperature region (1223 K) having stress exponent of 3.42 to a higher temperature region (1323 K) having stress exponent of 5.38. The stress exponent measured for lower temperature is in very good agreement with viscous glide creep. The values of the stress exponents measured at the higher temperature are consistent with a so-called dislocation climb process. With the changing of the temperature, creep experiments were carried out at a modulus-normalized stress lever. The result is presented in Fig. 4(b). The value of activation energy from the above plot locates in the range of 295-428kJ/mol, which was in agreement with that of 308 kJ/mol for Ni diffusion in NiAl at high stress region.

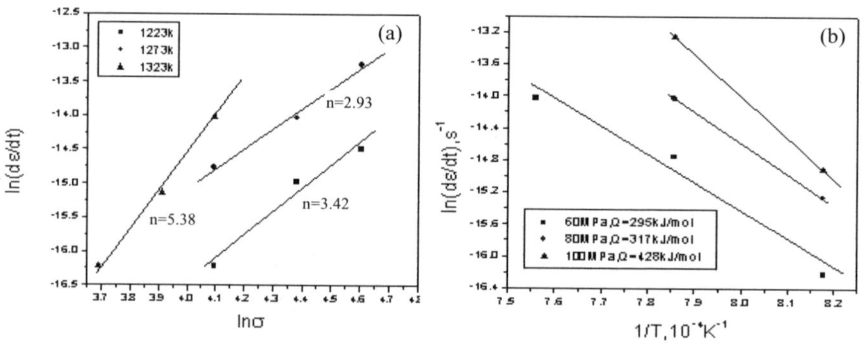

Fig. 4: Dependence of the steady state creep rate on (a) the applied stress (b) testing temperature

Ternary Creep Behavior. Under the testing condition, the values of t_f/t_t fall into the range from 1.15 to 2.8 for the testing alloy. The ternary creep strain and time are of 40-90% and 13-64% respectively. For pure metal and other alloy, the values of t_f/t_t is 1.5, which are structurally stable during creep, and this value being independent of stress and temperature over a wide range. Conversely, when ternary creep is associated with microstructure changes, no matter grain growth or over-aging, large the values of t_f/t_t have been obtained. Larger t_f/t_t only means the creep failure controlled by the formation and extension of cavity or crack in the testing alloy. Crept fracture fellows the relation of Monkman-Grant relation, and the former are given by: $\ln t_f + 0.775\ln\dot{\varepsilon} = 1.104$.

Discussion

As pointed out above, the initial analysis of the creep data as depicted in Fig. 4(a) suggests that the tested alloy exhibits two types of creep behavior. Based on the values of the stress exponents reported, it would conclude that the creep deformation process may be controlled by some form of dislocation climb process for the higher temperature region (Region I). At lower temperature region the viscous glide motion of dislocations controls creep deformation process (Region II). The trends obtained in the present test are in good agreement with the findings of Zhang and Guo et al [12], who studied the steady-state creep behavior of NiAl-25Cr. This alloy tested within a temperature regime corresponding to complete simultaneous deformation of second phase particles along with the NiAl matrix during the creep process exhibits a transition from viscous glide controlled to climb controlled creep when testing was performed at low or intermediate stress levels. The second phase particles deforming along with the NiAl matrix is the α-Cr. However, the authors reported that their activation energy values for region I (climb) and region II (viscous glide) were higher than the values for Ni bulk-diffusion and Al substitutional diffusion, respectively. The apparent activation energy for creep at the tested conditions was 428 kJ/mol, which was much higher than that of 308 kJ/mol for Ni diffusion in NiAl. Therefore, it was concluded that a straightforward analysis of the creep behavior is insufficient and other factors must be involved in the creep deformation process.

The dislocation motion was structured by the second phase particulates (α-Cr(Mo) and rich-Hf phases, which precipitated in the NiAl matrix phase) within the grains dominates(Fig. 5(a)). Once these dislocations were against the force of stricture or the second phase particulates engaged in the deformation, these dislocations would become entangled and formed dislocation network (shown in Fig. 5(b)). The observed homogeneous dislocation tangles is the typical character of viscous glide creep. These features are consistent with the apparent stress exponents measured at this stress level. These dislocation wall structures would evolute to the subgrain boundaries with the increasing of crept testing temperature or time as shown in Fig. 5(c). The presence of dislocation wall structure also correlates well with the apparent stress exponents measured at this stress level (Fig. 4(a)), suggesting the occurrence of dislocation climb. In addition, the formation of subgrain during high-temperature-deformation in NiAl-based alloy has been observed by many researchers [5,8]. Although the creep behavior of the present NiAl-28Cr-5.5Mo-0.5Hf-0.02 (%, weight fraction)P alloy is complicated by additional factors as described above, the TEM images presented here are clearly consistent with the two regions of creep behavior observed.

Conclusions

(1) For primary creep of NiAl-28Cr-5.5Mo-0.5Hf-0.02wt%P alloy, the Li's equation gives less satisfactory representation than that of Garofalo's relation.

(2) The apparent stress exponents are of 3.42 at lower temperature region and 5.38 at higher temperature region. The value of activation energy locates in the range of 295-428kJ/mol at different stress lever.

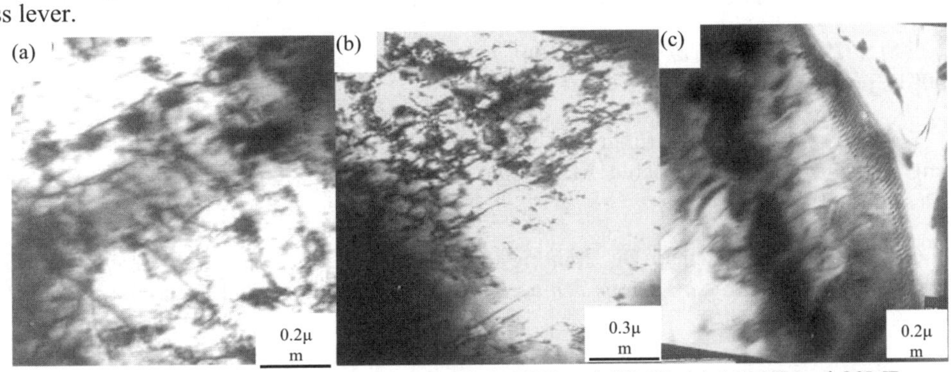

Fig. 5: TEM images of sample crept (a,b) 1273K and 80MPa(c) 1323K and 80MPa

(3) The creep behavior at the temperature range 1223-1323 K reveals two distinct regions of creep behavior present in this material. At higher temperature, the creep deformation was controlled by dislocation climb. At lower temperature, the creep deformation was controlled by viscous glide of dislocations

Acknowledgements

The authors would like to acknowledge National Natural Science Foundation of Hunan Province (contract No.08TP2042-3) and Aid program for Science and Technology Innovative Research Team in Higher Educational Institutions of Hunan Province for their financial supports.

References

[1] Miracle D B: *The physical and mechanical properties of NiAl.* Acta Metallurgica et Materialia, 41, 3(1993), pp. 649-684.

[2] Darolia R: *NiAl Alloys for High-Temperature Structural Applications*. JOM, 43, 3(1991), pp. 44-49.

[3] Du xinghao, Zhang guangye, Guo jianting: *Microstructure and High-Temperature Creep Behavior of NiAl-25at.%Cr Intermetallic Compound.* MATERIALS SCIENCE FORUM, Vol. 475/479 (2005), pp. 771-774.

[4] Nicholls JR, Stephenson DJ: *High temperature coatings for gas turbines*. Intermetallic compounds, Principles and practice, 1995: 489.

[5] S. Suh, M. Dollar and P. Nash.: *Creep in mechanically alloyed NiAl.* Materials Science and Engineering A, Vol.192-193, (1995), pp. 691-697.

[6] Qi, Y. H. Guo, J. T. Cui, C. Y.: *Tensile creep behaviour of NiAl-Cr(Zr) multiphase intermetallic alloy*. Materials Science and Technology, 19, 3(2003), pp. 399-402.

[7] Parviz Yavari, Farghalli A. Mohamed, and Terence G. Langdona: *Creep and substructure formation in an Al-5% Mg solid solution alloy*. Acta Metallurgica, 29, 8(1981), pp. 1495-1507.

[8] V. Krishnamachari, M.R. Notis: *High temperature deformation of polycrystalline NiO and CoO.* Acta Metallurgica, 25, 11(1977), pp. 1307-1313.

[9] D.L.Yaney and W.D.Nix: *Mechanisms of elevated-temperature deformation in the B2 aluminides NiAl and CoAl.* Journal of Materials Sciences, 23, 9(1988), pp. 3088-3098.

[10] Li M S: *The Corrosion of Metals at High Temperatures* . Metallurgical Industry Press, Beijing, (2001), pp. 263.

[11] W.H. Kowalski, Mikrostrukur and meschanische Eignschaften intermetalliischer NiAl-Cr Leigierungen, Fortschrittberichte VDI Reihe 5: Grund-und Werkstoffe Nr 342. VDI Verlag, Düsseldorf, 1993.

[12] Guangye Zhang, Xinghao Du, Jianting Guo, Hengqiang Ye: *BDT and Creep Behaviors of NiAl-25 at. Pct Cr Alloy at Various Temperatures*. J.Mater Science and Technology, 21, 5(2005), pp. 641-646.

Advanced Materials Research Vol. 279 (2011) pp 33-38
© (2011) Trans Tech Publications, Switzerland
doi:10.4028/www.scientific.net/AMR.279.33

CuCr1 alloy Surface Hardening via Ultrasonic assisted Electro-spark Deposition

Yong Xu[1, a], Z.N. Guo[2, b], Guan Wang[3, c] and Y.J. Zhang[4, d]

[1, 2, 3, 4]Faculty of Electromechanical Engineering, Guangdong University of Technology, China

[a]xuyong2927@163.com, [b]znguo@gdut.edu.cn, [c]gxw290@gmail.com, [d]yjzhang@gdut.edu.cn

Keywords: Ultrasonic-aided electro-spark deposition technology (UESD), Resistance Spot Welding (RSW), Electrode, WC, Coating, SEM

Abstract. Resistance Spot Welding (RSW) is a widely used technology. And the rapid wear is the main reason for the short life of RSW electrodes. To improve electrode life during RSW, a novel ultrasonic-aided electro-spark deposition technology (UESD) and device are proposed in this paper. The WC metallurgical bond coating was fabricated on the surface of CuCr1 electrodes by UESD, and some experimental results were also analyzed. Moreover, the surface morphology of WC coating was studied by SEM and the quality of coating wear resistance was analyzed through experimental method. This study provides a novel way to extend the life of common moulds and RSW electrodes.

Introduction

Spot welding is the most widely used type of resistance weld. And RSW is a process in which two pieces of metal are joined by passing current between electrodes positioned on opposite sides of the pieces to be welded. There is no arc with this process, electric current pass through the electrodes and then the material is welded until a suitable amount of material has been heated to an acceptable temperature in order to bond the material to be welded [1,2]. High welding currents, surface roughness and electrode forces directly result shortening the electrode life. To improve electrode life, many methods had been used in the past few years, such as electro-spark deposition (ESD) and laser treatment.

The ESD process is mainly used for repairing damaged parts, tools or equipment, or to extend tool and equipment lifespan by coating the surface with a stronger, tougher, and more durable layer [3]. Electro-spark deposition (ESD) becomes one of the most popular mold repairing method, and is already widely used in Asia and some European industries [4]. Because it can obtains metallurgical bond coating which is more durable than mechanical bond coating [5], it is a lower heat input process due to duty cycles as low as 1% [6]. However, when it was used as a method to extend RSW electrode lifespan, cracks formed early in the deposition process acted as crack nuclei, resulting in numerous stress relief cracks were visible within the coating [7]. A Japanese researcher [8] reported that ultrasonic resonance makes coating grain refinement and promotes the formation of coating. In this paper, a novel UESD technology was used to fabricate WC metallurgical bonds coating on the surface of CuCr1 electrodes.

UESD device and principle

Fig. 1 is the installation drawing and schematic diagram of the UESD device in the process of deposition. As the figure shows, the UESD device is mainly composed of a protecting gas supply system, power control box and UESD integrated part. Gas system provides argon gas to preventing electrode and coating surface oxidating. Power control box contains an ESD pulse power, an ultrasonicator, and a miniature motor. The ESD pulse power has a power of 1 KW with discharge frequency ranging from 40 Hz to 21,000 Hz, and continuously regulating. The ultrasonicator has an output power of 80 W with a maximum 40 KHz vibration frequency. And the rotation speed of the miniature motor ranges from 0 to 2000 r/min. In addition, soft shaft was used for transmitting torque. Adopting this pattern can reduce the weight of UESD integrated part, and it is very helpful for hand-held processing.

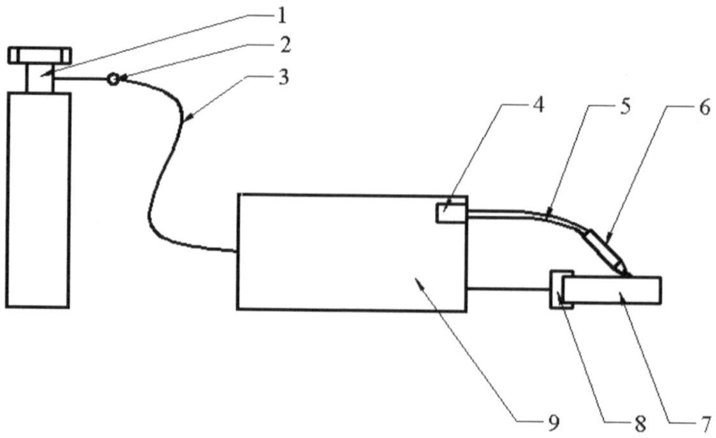

1-Argon gas tank; 2-Reducing valve; 3-Gas tube; 4-Miniature motor;
5-Soft shaft; 6-UESD integrated part; 7- Electrode; 8-The cathode fixture; 9- power control box
Fig. 1: UESD device installation drawing

The UESD integration part is the core of the UESD device, which have both ultrasound and EDM properties. Fig. 2 is the configuration of the UESD integration part, which shows that soft shaft transmission torque to indirect drive the ultrasound spindle. Taper connection was used between ultrasound spindle and ultrasonic horn, and insulated ring was used between both. At the small ends of the ultrasonic horn, the drill chuck was installed via cone connection to clamp the electrode. The hollow high-speed slip ring can realize the transformation from static to rotate, in this case, playing a role in transporting electrical signal to ultrasonic transducer and electrode. The OD of the UESD integration part is just 45mm because of using ultra-thin bearing to connect the shell and ultrasound spindle, and it also ensure the hand-held feasibility.

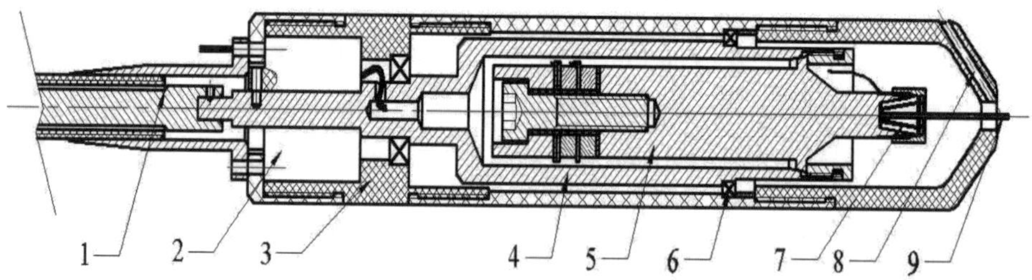

1- Soft shaft, 2-Hollow high speed slip ring, 3-Shell, 4-Ultrasound spindle,
5-Ultrasonic oscillator, 6-Ultra-thin bearing, 7- Chuck, 8-Protection gas through-hole, 9-Electrode
Fig. 2: The configuration of UESD integration part

Machining mechanism is pivotal part in a process, and it is also a theoretical foundation of a process. Therefore, the mechanisms of the ultrasound-assisted electro-spark deposition and ultrasonic polishing process were analyzed. Short durations of electrical pulses ranging from milliseconds to microseconds are used to deposit the electrode onto the substrate. The introduction of ultrasonic is used to effectively improve the gap discharge state, reduce the "anode adhesion" phenomenon and promote the nucleation rate.

Experimental methods

The substrate used in the present study is a precipitation-strengthened and work-hardened Class 2 copper alloy with the C18200 ASTM specification (CuCr1). The Cu substrate was cut into 20 × 10 × 0.6 mm rectangular specimens by a wire electrical discharge machine (WEDM). The material used as UESD electrode is WC rod (2mm in diameter, 50mm in length). Table 1 shows the chemical composition of the electrode and substrate materials, respectively.

Table 1: The chemical composition (wt-%) of electrode and substrate materials

Elements of substrate	Cr	Fe	Cu
Contents [wt-%]	0.85	0.2	Remiander
Elements of electrode	C	CeO_2	W
Contents [wt-%]	0.005	1.85	Remiander

All substrates and electrodes are washed with anhydrous acetone and then rough grinding with 200 grit sand paper, fine grinding with 400 grit sand paper, and finally washed by anhydrous acetone again to obtain a thin and uniform surface oxide layer.

The experiment was conducted using a pulse of 90V, a pulse frequency of 100 Hz, deposition time of 4 min/cm2. During deposition the electrode was oriented preferentially at an angle of 30–90° to the work-piece and rotated at a speed of 800 rpm, and the shielding argon gas was delivered through a tube and flowed at a rate of 4 l/min. The revolving electrode was in traveling contact with the work-piece surface and the electrode material was transferred through the arc of every miniature lightning strike. With a succession of pulse discharge deposition, the deposited layer with a desirable thickness was obtained. The surface morphology and elements content of coating were examined by using scanning electron microscope (SEM).

Results and discussion

ESD coating studies. In the deposition process, the pulse energy released in a very short moment, then forming a small high pressure area, and micro pool was obtained on the surface of substrate. High pressure, rotary motion and ultrasonic sound pressure act as the drive force for the form of pool and splash. When molten droplets were deposited onto the cold copper alloy substrate, they rapidly solidified to form a coating due to the much greater thermal sink of the substrate. The phenomenon of electro eroding happens in three different states which are gas-state, liquid-state and solid-state. And only liquid-state materials contribute to increasing coating thickness. According to the single pulse discharge energy formulate [9],

$$W=\frac{1}{2}C\cdot U^{2}$$
$$P=\frac{1}{2}C\cdot U^{2}\cdot \mathrm{f} \qquad (1)$$

Where, W-single pulse discharge energy, C-loop capacity, U-voltage of pulse power supply, P-discharge power and f- discharge frequency. The more discharge energy, the greater the area of corrosion pits. Japanese scholars concluded a empirical formula for the volume of single pulse corrosion pits through the experimental method [10]. As follows:

$$V=K_{R}\tau^{1.1}I_{P}^{1.4} \qquad (2)$$

Where, V-pit volume, τ - pulse width (μs), I_p -peak current, K_R -a constant related to the nature of materials. Parameter determines the diameter of the discharge pits, and pit depth depends on the material properties of the substrate.

The topographicals of as-coated WC coating are shown in Fig. 3. The typical 'splash' appearance is a characteristic feature of the ESD technique. As shown in the picture A, the center of discharge point is micro-depression area, and around the depression bath is splashed electrode material. What's more, a white halo phenomenon occurs around the outer ring of deposition pool. This phenomenon is most likely caused by differences of thermal physical properties or the cathode cleaning.

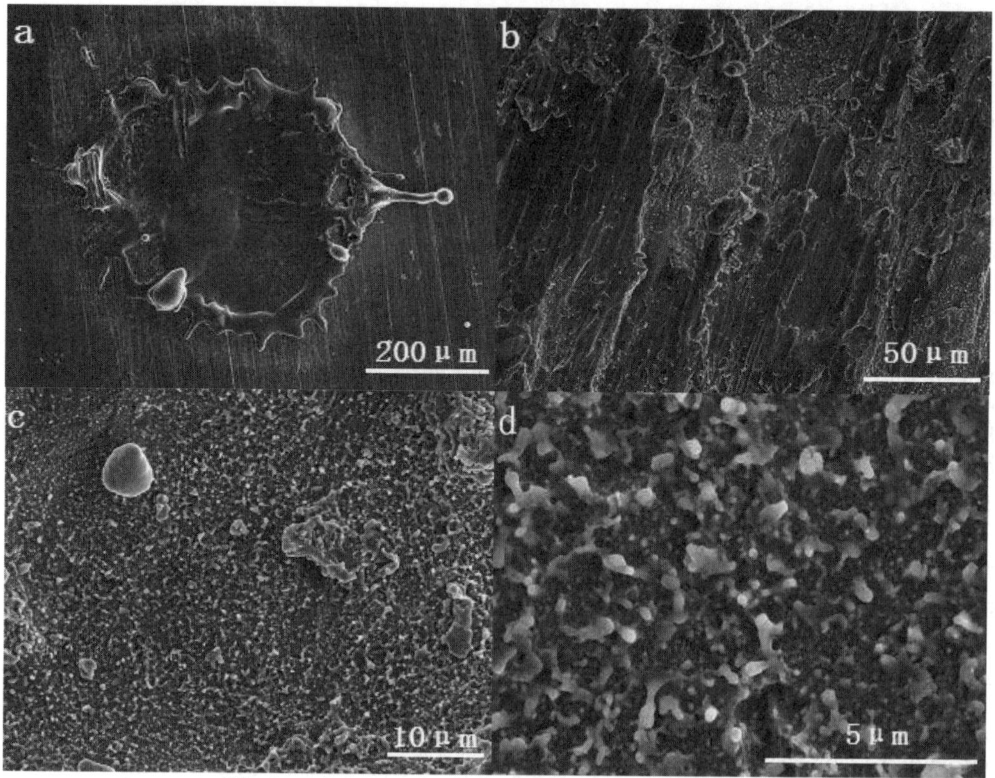

a. Single pulse deposit pit; b. Surface morphology×500;
c. Surface morphology×2000; d. Surface morphology×10000
Fig. 3: Surface morphology of the WC coating on CuCr1alloy

The surface morphology in Picture C was formed by repeated superposition of fine grains. Figure D obviously shows that the coating was formed by the superposition of 0.5-1μm size of fine crystal grains. The emergence of this organizational structure, to a certain extent, direct proof that the introduction of high-frequency ultrasound can improve the crystal nucleation rate, effectively refine grains, and even change the crystal phase.

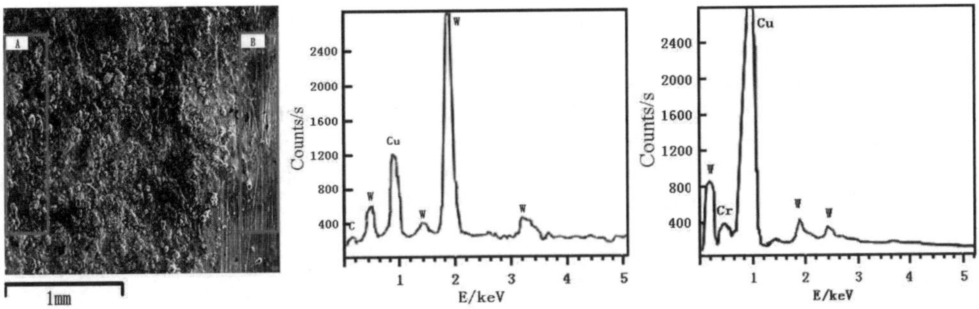

a.Sampling locations b. EDS of area A c. EDS of area B

Fig. 4: EDS of WC coating surface and substrate surface

To recognize the content of various elements in the coating, the energy spectrum of coating and substrate was obtained under the scanning electron microscope, as shown in Fig. 4. Picture A shows that tungsten is the main ingredient in the deposition coating, while a small amount of Cu element also appeared in the coating. To a certain degree, it indicates that the coating is a metallurgical bond one.

Coating performance analysis. The quality of wear resistance is an important indicator of coating performance. Therefore, it is necessary to test coating wear resistance through experimental method. Wear test conditions as follows: attach 180 grits sand paper to the Grinding wheel with rotational speed of 400 r / min , take raw and processed materials as the experimental samples, impose 980N friction load on sample to impel it close to the sandpaper, and without lubrication. Friction experiment are divided into five groups, and each test time is 1 minute increment. All tests are conducted at room temperature conditions. All experimental samples were washed by acetone, and the loss weight of samples were determined by a analytical balance. Each experiment was repeated five times, and the value of average loss weight as a drawing basis. And the curve of lost weight vs. friction time as shown below.

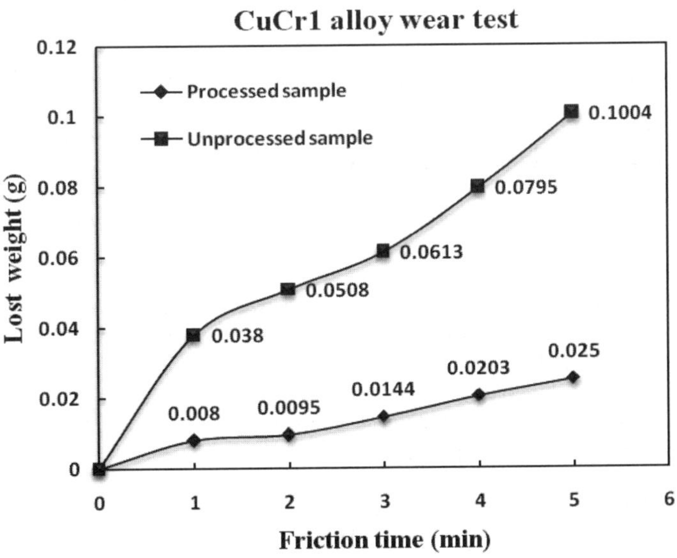

Fig. 5: CuCr1 alloy wear test

It can be seen from Fig. 5, under the same conditions, the wear of two samples are increased with the extension of friction time. However, the loss weight of untreated CuCr1 alloy sample is far greater than the treated one. The trend of curve about processed sample is flat, but unprocessed one is not. In the curve of untreated sample, the slope of the curve is increase, which means the incremental wear is more and more within the same time. In other words, via the UESD process treatment, the wear resistance of CuCr1 surface is superior to untreated one. Thus, the application of UESD technology in CuCr1 alloy electrode surface hardening is expect to be promotion.

Conclusions

In this paper, a novel surface strengthening technology was proposed. What's more, a self-developed UESD machine with high frequency ultrasonic longitudinal vibrating, rotating electrode and high spark energy was introduced.

Via UESD process, WC coating was obtained onto the surface of RSW copper alloy electrode. The topographical and EDS of as-coated WC coating show that WC ingredient permeated into CuCr1 texture, and formed metallurgical bond. In addition, the friction and wear experiments show that the wear resistance of deposition coating is more excellent than raw material. In a certain extent, this study provides a novel way to extend the life of common molds and RSW electrodes.

References

[1] Del Vecchio, E. J.: *Resistance Welding Manual Third Edition: Volume 1*. Resistance Welder Manufacturers' Association, (1956).

[2] Jefferson T.: *The Welding Encyclopedia: Eleventh Edition*. The Welding Engineer Publishing Co., (1943).

[3] Reynolds J. L., Holdren R. L., and Brown L. E.: *Electro-Spark Deposition*. Advanced Materials and Processes, 161, 3(2003).

[4] Amy Harmon Krtanjek, Timonthy J. Langan: *www.moldmaking technology.com,* 2006.

[5] Johnson R. N., Sheldon, G. L.: *Advances in the electro-spark deposition coating process*. Journal of Vacuum Science and Technology, 4, 6(1986), pp. 2740-2746.

[6] Johnson R. N.: *Electro-Spark Deposition: Principles and Applications*. Society of Vacuum Coaters 45th Annual Technical Conference, (2002), pp. 87-92.

[7] Chen, Z. and Y. Zhou: *Surface modification of resistance welding electrode by electro-spark deposited composite coatings: Part I. Coating characterization*. Surface & Coatings Technology, 201, 3 (2006), pp. 1503-1510.

[8] Nishimura, G., Yanagishima, K., Shima, T.: *Ultrasonic Electro-Spark Machining (Part I)*. J. Fac. Eng., Univ. of Tokyo, 25, 1(1957), pp.12-16.

[9] KENNETH PAUL KEES: *Hard-facing with electro-spark deposition*. Thesis of Washington State University College of Engineering, 1983, pp.11-12.

[10] Yuanlun Shuai, Wenxue Yu translated: *Discharge machining principles*. National Defense Industry Press, Beijing, (1983).

Advanced Materials Research Vol. 279 (2011) pp 39-43

doi:10.4028/www.scientific.net/AMR.279.39

In-situ SEM investigation of tensile fracture process of Fe-25Cr-35Ni based superalloy

Jingfeng Guo[1, a], Rengen Xu[2, b], Li Zhao[3, c], Wenzhu Shao[4, d], Lianjie Qin[5, f]

[1, 2, 5]School of Environmental and Materials Engineering, Yantai University, Yantai 264005,China

[3]No. 802, Institute of Shanghai Academy of spaceflight Technology, Shanghai 200090, China

[4]School of Materials Science and Engineering, Harbin institute of technology, Harbin150001, China

[a]gjf100gjf@126.com, [b] xuyoungcn@yahoo.com.cn, [c] xuyoungcn@163.com, [d]wzshao@hit.edu.cn,[f]lianjieqin@126.com

Keywords: Fe-25Cr-35Ni based superalloy, in situ tensile test, micro-crack

Abstract. Initiation and propagation of micro cracks in the Fe-25Cr-35Ni based superalloy were observed and investigated through in situ tensile test in SEM using a single-edge notched specimen. The results show that the micro crack sources may easily occur with one of those, the second phase particles and matrix interface, and the stress concentration region. The micro-cracks which propagated at first for those that are perpendicular to the tensile load would be growing up and connected mutual with the increase of external stress until forming the main crack. While the length of the mail crack enlarged enough in the specimen, it become instable for the main cracks and the fracture occurred rapidly.

Introduction

The Fe-25Cr-35Ni based superalloy is a stable austenitic alloy with high resistance to thermal cycling and thermal stress performance, whose resistance of oxidation and corrosion is also very good [1]. In the petrochemical industry, it has been widely used as high-temperature petrochemical plant parts, such as the reformer tube, radiant tube and ethylene cracking furnace tubes, etc., ethylene steam cracking furnace tube, hydrogen, ammonia and methanol catalytic converter tubes and other heat furnace, pyrolysis furnace and other components, which is also successfully applied to the steel industry, metallurgy and heat radiant tube, bottom rollers, and all kinds of casting parts, so it is the most widely used alloy in heat-resistant casting alloys [2]. For the structural materials, it is necessary to understand and study the formation and propagation of micro-cracks, which have a very significant impact on the static mechanical properties at the tensile stress [3]. This paper describes an in situ SEM dynamic tensile study of the micro cracks formation and propagation in the Fe-25Cr-35Ni based superalloy.

Experimental Procedure

The experimental material is melted by induction furnace and ESR (electroslag remelting), and then forged. The chemical compositions (wt%) of the Fe-25Cr-35Ni based superalloy using in this investigation are as follows: C:0.02~0.10%, Ni: 35~39%, Cr: 23~27%, Nb+Ta: 0.4~0.9%, N: 0.15~0.3%, Fe:≥31%. In situ tensile tests were carried out in a S-570 SEM equipped with a tensile device at room temperature. The sample size is shown in Fig. 1. In order to observe the crack initiation point easily, we choose a single-edge notched specimen. Prior to the tensile test, the specimen was mechanically polished and etched electrolytically with a solution consisting of 13% HF, 7% HNO_3 and 80% HCl for SEM examination. During the test, the load direction of the specimen is shown as the white arrow in the figure.

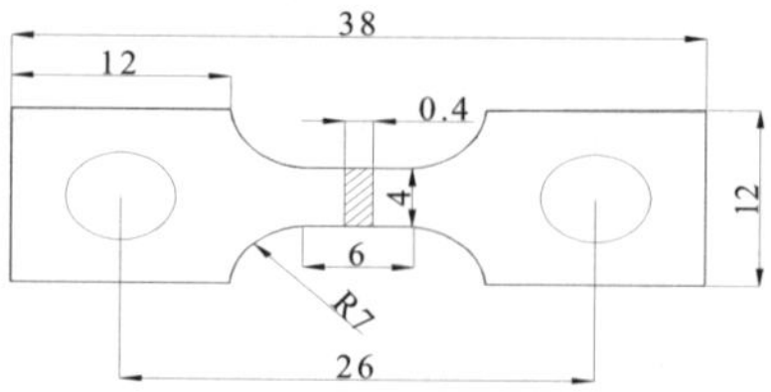

Fig. 1: Dimensions of specimen for dynamic tensile test on SEM (in mm)

Results and discussion

Crack Initiation. The U-shaped notch of the Fe-25Cr-35Ni based superalloy specimen is shown in Fig. 2. The dynamic tension of the Fe-25Cr-35Ni based superalloy is shown in Fig. 3. There was no damage on the specimen at initial load (Fig. 3a); the slip line appeared near the notch of the specimen with the load increasing slowly (Fig. 3b), while the plastic deformation and slip line didn't appear in the other parts of the specimen (Fig. 3c). There are many micro crack nucleation sites shown in Fig. 3. Micro cracks appear mainly in the following regions: the stress concentration region (Fig. 4, A), the second phase particles (or empty) and matrix interface (Fig. 4, B). There are more micro-cracks appearing in the second regions than the stress concentration regions. The second phase particles (or empty) and matrix interface are the main source of the micro crack and have formed a large number of micro-cracks. Because of the tensile stress, the cracks propagate along the direction is perpendicular to the load. The micro cracks reduced the strength and shortened the fracture life of the specimen. The micro cracks didn't appear in the grain and the twin boundaries, and it suggests that the strength of the grain and the twin boundaries are larger than the second phase particles and matrix interface.

Fig. 2: The U-shaped notch of the Fe-25Cr-35Ni based superalloy specimen

The in situ observation of the Fe-25Cr-35Ni based superalloy tensile test showed that the micro cracks mainly appear in the notch tensile stress concentration region and the second phase and matrix interface regions. The initiation sites of micro-cracks are in the region where the slip line is blocked, which shows that the initiation of crack is based on the dislocation pileup mechanism [4]. Because of the difference elastic modulus of matrix and the second phase, when the deformation was caused by the tensile stress, the matrix and the second phase split open and formed micro cracks. There are many micro cracks initiated in the second phase particles and matrix interface, but only a few of these micro cracks can grow up in the stress concentration region and extend into the main crack [5].

Crack Propagation. The tensile stress and the plastic shear strain make the crack propagating, and usually the main crack propagates in the plastic deformation region. For the grains of different orientation, the deformation and deformation resistance are different. In the loading process, the most favorable orientation and possible micro-stress concentration grain is bound to slip at first. With the uniformity of the microstructure decreasing, the selectivity of the initial plastic deformation becomes more obvious [6]. The main crack formed in the dynamic tensile process of the Fe-25Cr-35Ni based superalloy is shown in Fig. 5. The micro-crack which is formed by stress concentration will become the main crack. The propagation process of cracks in the Fe-25Cr-35Ni based superalloy is shown in Fig. 6. Not all micro cracks can be able to propagate to large cracks, and some micro-cracks stop to grow up because of the resistance, while others continue to propagate (Fig. 6 b, c). The crack propagated in the plastic deformation region, when the crack encountered the second phase particles, bypass them and continue to propagate. When the crack formed by the second phase encountered the growing main crack, it became part of the main crack (Fig. 6 e, f). When the main crack is blocked, the micro cracks formed by the holes and the second phase particles continue to grow up and be linked into the main crack, and then become part of the main crack, so the main crack can't continue to propagate in the plastic deformation region (Fig. 6 g, h).

Fig. 3: SEM images for the Fe-25Cr-35Ni based superalloy during dynamic tensile process (a) initial load stage, (b) plastic deformation region, (c) the other parts of the specimen

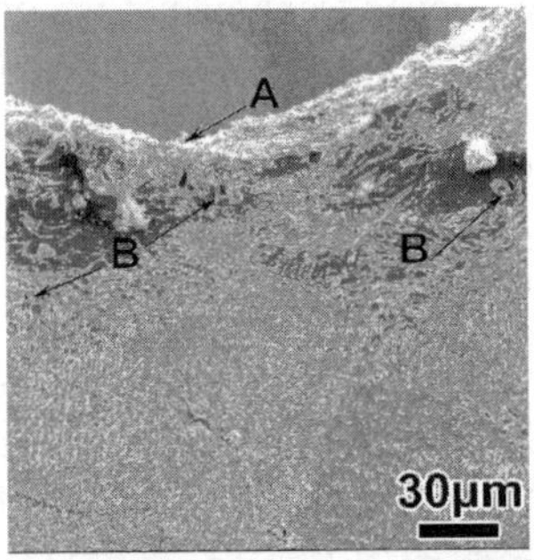

Fig. 4: SEM images of crack nucleation sites in the Fe-25Cr-35Ni based superalloy

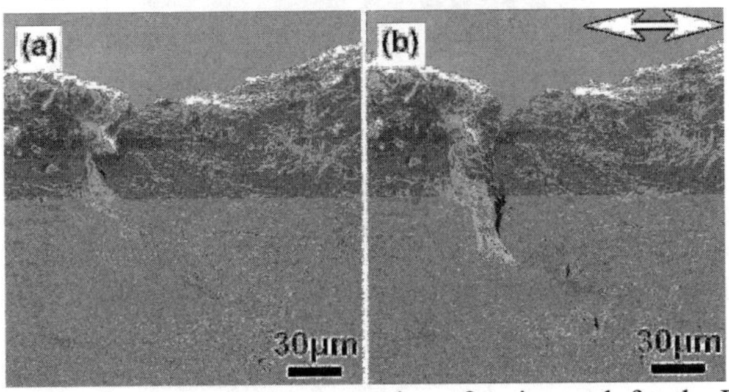

Fig. 5: SEM images of the formation and propagation of main crack for the Fe-25Cr-35Ni based superalloy: (a) in initial stages of crack forming stages, (b) the main crack after further propagation

When the main crack propagates to a certain degree, during loading further the main crack will propagate rapidly, and the specimen will break immediately. So the specimen achieve to the instable propagation stage. The micro cracks formed by the second-phase particles and holes reduce the specimen cross-sectional area, promote the propagation of the main crack and accelerate the fracture. If reduce the amount of the second phase particles and defect in the Fe-25Cr-35Ni based superalloy, it can significantly increase the fracture strength.

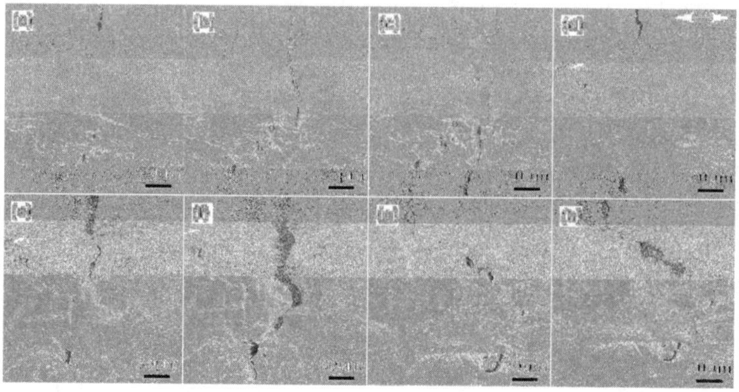

Fig. 6: Propagation process of the cracks from (a) to (h) for the Fe-25Cr-35Ni based superalloy

Fracture Analysis. The dynamic tensile fracture of the Fe-25Cr-35Ni biased superalloy is shown in Fig. 7. The fracture is the dimple pattern, which belongs to the typical ductile fracture. The nucleation of micro porous is along the boundary of the second phase particles, and then the micro porous propagate to dimple-shaped cracks [7]. These holes at the bottom of the dimple were brought about by the second phase particles which are flopped from the bottom of the dimples.

For the fracture behavior of the Fe-25Cr-35Ni based superalloy in situ observation, the fracture process includes: the alloy elastic deformation, the formation of cracks, the large cracks coalescence and breakage stage. At the tensile load, the crack propagates in the plastic deformation region. The micro cracks are formed by the holes, the second phase particles and other defects continue to grow up and be linked into the main crack, and then become part of the main crack. When the main crack is blocked, the micro cracks in front of the main crack which are formed by the holes and the second phase particles continue to grow up and be linked into the main crack, making the main crack propagating forward. When the main crack propagates to a certain degree, the specimen will break rapidly. In addition, a lot of micro cracks stop propagating after the main crack formed.

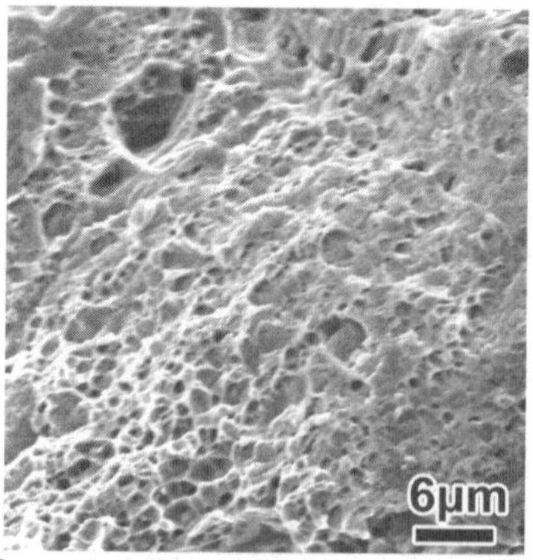

Fig. 7: The fracture of the Fe-25Cr-35Ni biased superalloy

Conclusions

From in-situ SEM tensile studies of the Fe-25Cr-35Ni based superalloy, it was found that:

(1) There are two regions of the micro cracks initiation of the Fe-25Cr-35Ni based superalloy appear: the stress concentration region, the second phase particles (or empty) and matrix interface.

(2) With the increase of load, the micro-crack in the favorable position will continue to propagate and connect with the other micro cracks around, and then become the main crack.

(3) After the main crack formed, many other micro cracks stop propagating, and only the micro cracks near the main crack still propagate, which brings about the main crack continue to propagate. At the tensile load, the crack propagates in the plastic deformation region of the matrix. When the main crack propagates to a certain degree, the specimen will break rapidly. The fracture is the dimple pattern, which belongs to the typical ductile fracture.

References

[1] Xu Zili, Liu Shaoyan: *The Comparison of oxidation resistance about HP-Nb, HP-NbW and HK-40 alloy.* J. Chemical Machinery Vol.1 (1994), pp. 16-20.

[2] Delaunayd, Huntzam: *Influence of structural parameters on oxidation of austenitic Fe-Ni-Cr-Al alloys.* J. Mater Sci. Vol. 89 (1983), pp. 189-194.

[3] Jin Yunxue, Li Jungang: *Crack Forming and Propagation Mechanisms of TiC/Ti Composites during Dynamic Tension.* J. Rare Metal Materials and Engineering Vol. 36 (2007),pp. 764-768.

[4] Shen Xianpu, Lei Tingquan, LiuJianzhuang: *The SEM in situ observation of tensile fracture process in plain carbon dual-phase steels.* J. Acta Metallurgica Sinica.Vol.21(1985), pp. 228-233.

[5] Zeng Qiang, Hou Shu-e,FanYing-wei: *Observation on tensile process of IC6 superalloy in situ.* J. Journal of aeronautical materials.Vol.23 (2003), pp. 143-146.

[6] Luo Xinmin, Chen Kangmin, Luo Gang, Fan Yinhe: *Deformation and fracture behaviors of super alloy welded joint up on scanning electron microscope in-situ tension.* J. Transactions of the China Welding Institution, Vol.26 (2005), pp. 60-64.

[7] Xiong Ronggang, Fu Renyu, Su Yu, Li Lin: *Fracture mechanism of TWIP steel under dynamic tensile condition.* J. Shanghai Metals, Vol.30 (2008), pp. 12-15.

Advanced Materials Research Vol. 279 (2011) pp 44-48
© (2011) Trans Tech Publications, Switzerland
doi:10.4028/www.scientific.net/AMR.279.44

Synthesis of Single-phased BaTi2O5 Powders by Arc-melting

Jian Peng[a], Chuanbin Wang, Ling Li, Qiang Shen and LianMeng Zhang

State Key Lab of Advanced Technology for Materials Synthesis and Processing, Wuhan University of Technology, Wuhan 430070, China

[a]pjdustin@qq.com

Keywords: $BaTi_2O_5$ powders, arc-melting, molar ratio, single-phased

Abstract. Arc-melting was employed to synthesize $BaTi_2O_5$ powders by fast melting the reagent mixture of $BaCO_3$ and TiO_2 with different molar ratios (0.90~1.05:2). The influence of molar ratios of $BaCO_3$ to TiO_2 on the phase of the powders derived from arc-melting was investigated by X-ray diffraction and Raman spectra. When the molar ratio was larger than 0.95:2, a little amount of Ba-rich compound Ba_2TiO_4 was produced besides the main phase of $BaTi_2O_5$. Ti-rich compound $Ba_6Ti_{17}O_{40}$, $BaTi_2O_9$ and TiO_2 began to appear at the ratios ranging from 0.90:2 to 0.95:2. Single-phased $BaTi_2O_5$ powders were obtained by arc-melting the reagent mixture with the molar ratio of 0.95:2.

Introduction

In the BaO-TiO_2 system binary phase-diagram, there are many compounds such as $BaTiO_3$, $BaTi_2O_5$, $Ba_6Ti_{17}O_{40}$, $Ba_4Ti_{13}O_{30}$, $BaTi_4O_9$, $Ba_2Ti_9O_{20}$, $BaTi_5O_{11}$, etc. Among them $BaTi_2O_5$ was found recently to be a promising ferroelectric material available at high temperatures, which showed significant permittivity parallel to the crystallographic *b*-axis near the Curie temperature (~477°C) [1-4] and could be used as a new lead-free ferroelectric material for high density dynamic random access memories (DRAMs) or non-volatile ferroelectric random access memories (FRAMs). Up to now, various preparation methods for $BaTi_2O_5$ [5-9] such as sol-gel, hydrothermal reaction, containerless processing, molten salt and arc-melting, have been developed. Arc-melting is a more convenient method by which $BaTi_2O_5$ powders can be synthesized in short time from the reagent mixture of $BaCO_3$ and TiO_2 with a stoichiometric composition (molar ratio of 1:2). Although the arc-melted $BaTi_2O_5$ powders were thought to be single-phased [9], other impurities besides $BaTi_2O_5$ were still detected in our recent study, implying that the composition of the reagent mixture should be adjusted in order to obtain $BaTi_2O_5$ powders in pure phase.

In the present study, the preparation of single-phased $BaTi_2O_5$ by arc-melting was carried out. The influence of molar ratio of $BaCO_3$ to TiO_2 on the phases of the powders derived from arc-melting was mainly investigated.

Experimental

Micropowders of $BaCO_3$ (99% in purity) and TiO_2 (99% in purity with rutile phase) were used as the starting materials. The reagents were weighed at the molar ratio of 0.90~1.05:2, respectively, and mixed homogeneously. The mixtures were pressed into discs (20 mm in diameter, 5 mm in thickness) and then melted in an arc-melting apparatus (Diavac ACM-S01) which was evacuated to 10^{-3} Pa followed by filling Ar gas of 2.5 kPa. The discs were arc-melted successively for four times and then quenched on a water-cooled copper plate to form button-like pellets. Finally, $BaTi_2O_5$ powders were obtained after crushing these pellets. The crystal structure of the samples was identified by X-ray diffractometer (XRD, Rigaku Ultima III) with Cu kα radiation and Raman shift from a He-Ne laser (633 nm) at a laser power of 3 mW.

Results and Discussion

The X-ray diffraction (XRD) patterns of the powders derived from the reagent mixture of $BaCO_3$ and TiO_2 (in molar ratio of 1:2) are shown in Fig. 1(a). The standard patterns of $BaTi_2O_5$ (JCPDS 34-0133) are also presented in Fig. 1(b) for comparison. From the patterns, no apparent amorphous peaks were observed, indicating the good crystallization of the as-synthesized powders. Both the peak position and relative intensity of the powders are in well consistent with those of JCPDS 34-0133, except the unidentified peak at 2θ =31.53°. In previous investigation reported [9], button-like bulk derived from arc-melting was cut along the plane parallel with copper plate, and the cut bulk showed a significant (010) orientation. However, other peaks with much smaller intensity were not well indentified, including the unidentified peak mentioned above. Since the locations of the two stronger peaks (211) and (031) of Ba_2TiO_4 (JCPDS 34-1481) almost coincide with the peaks of (-203) and (003) of $BaTi_2O_5$ (JCPDS 34-0133), and the unidentified peak in Fig. 1(b) also locates at nearly the same position of the third strongest peak (131) of Ba_2TiO_4 (JCPDS 34-1481), the small amounts of impurity in the powders was supposed to be Ba_2TiO_4 tentatively.

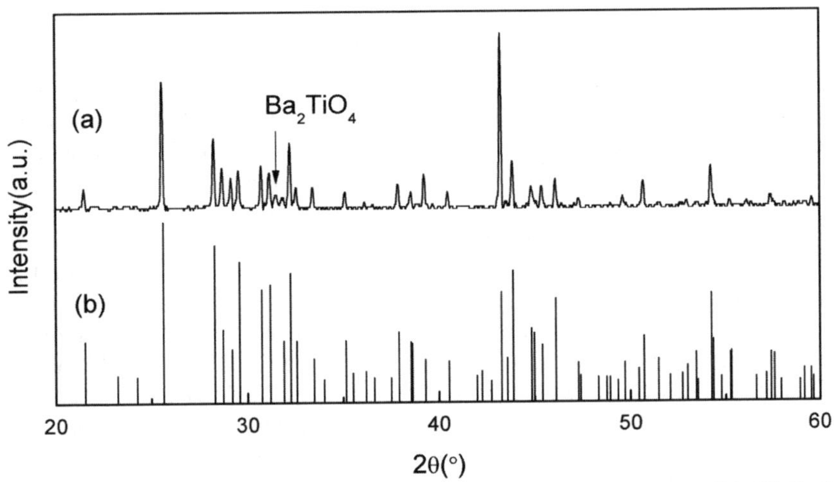

Fig. 1: XRD patterns of (a) arc-melted powders and (b) standard patterns of $BaTi_2O_5$ (JCPDS 34-0133)

Since it is impossible to indentify the phase from only one peak in XRD patterns, Raman spectrum was introduced in this study subsequently. Raman spectrum can be useful as a supplementary method to indentify the crystal phase, owing to its independence of orientation. The Raman spectrum of the arc-melted powders is shown in Fig. 2. Two stronger bands at 340, 589cm^{-1} and other band at 216, 241, 278, 308, 370, 411, 437, 484, 529, 644, 704, 780 and 880 cm^{-1} can be indexed to a pure $BaTi_2O_5$ phase [8]. And the relatively weak band at 760cm^{-1} can be indexed to Ba_2TiO_4 as 761cm^{-1} is the strongest band of Ba_2TiO_4[10]. In summation, the small amounts of impurity should be Ba_2TiO_4.

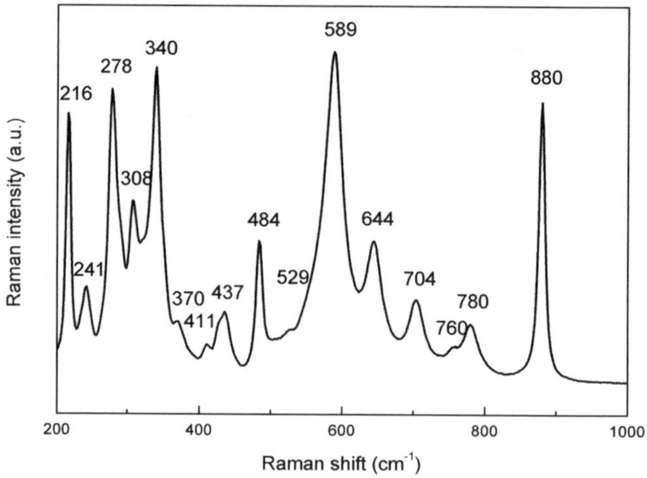

Fig. 2: Raman spectrum of arc-melted powders

Further investigations about the influence of molar ratios of $BaCO_3$ to TiO_2 on the phase of the arc-melted powders were carried out. Fig. 3 and Fig. 4 show the XRD patterns and Raman spectra of the powders derived from $BaCO_3$ and TiO_2 with molar ratios of 0.90~1.05:2, respectively. As shown in Fig. 3, no characteristic reflection peaks indexed to other BaO-TiO_2 compounds, besides $BaTi_2O_5$ were detected in XRD patterns of the powders derived from $BaCO_3$ and TiO_2 with the ratio of 0.95:2, indicating single-phased $BaTi_2O_5$ was obtained by arc-melting. When the molar ratio was larger than 0.95:2, Ba-rich compound Ba_2TiO_4 was produced besides the main phase of $BaTi_2O_5$. Ti-rich compound $Ba_6Ti_{17}O_{40}$, $BaTi_2O_9$ and TiO_2 began to appear at the ratios ranging from 0.90:2 to 0.95:2. Raman spectrum in Fig. 4 can also prove the analysis above. At the ratios ranging from 0.90:2 to 0.92:2, bands indexed to $Ba_6Ti_{17}O_{40}$ (140, 200, 258, 365, 552, 624 cm^{-1}) [11], $Ba_2Ti_9O_{20}$ (151, 802 cm^{-1}) [12] and TiO_2 (448 cm^{-1}) [13] were detected. And Ba_2TiO_4 related band (760 cm^{-1}) was indentified at the ratios ranging from 1.00:2 to 1.05:2. A summary about the phases of the arc-melted powders derived from the reagent mixture of $BaCO_3$ and TiO_2 with different molar ratios (0.90~1.05:2) is given in Table 1.

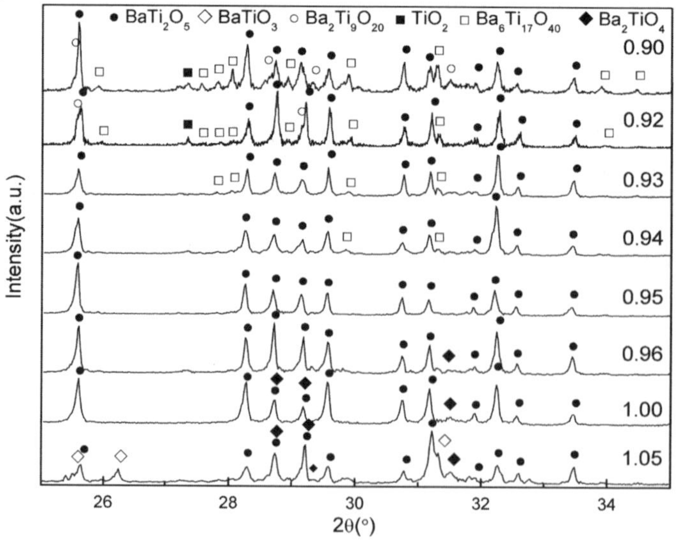

Fig. 3: XRD patterns of arc-melted powders derived from the reagent mixture of $BaCO_3$ and TiO_2 with different molar ratios

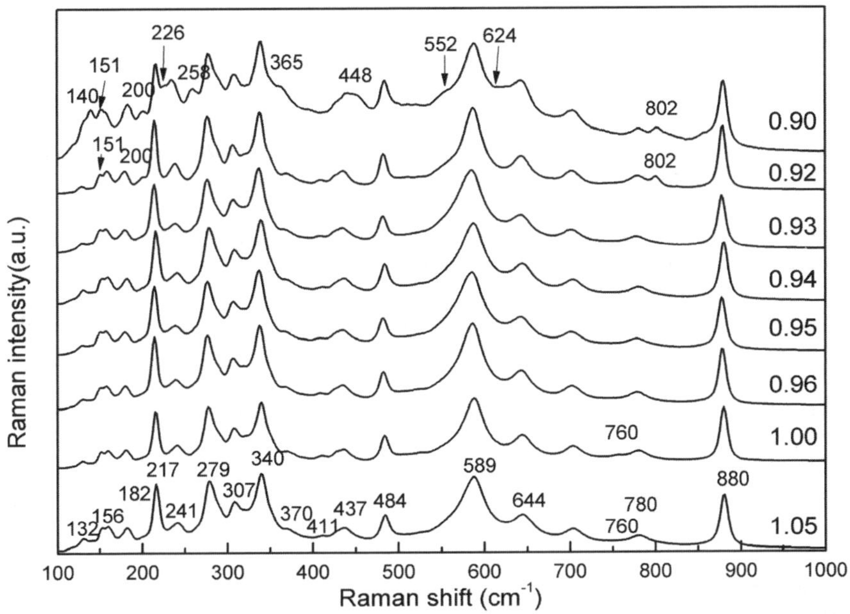

Fig. 4: Raman spectra of arc-melted powders derived from the reagent mixture of $BaCO_3$ and TiO_2 with different molar ratios

Table 1: Phases of arc-melted powders derived from the reagent mixture of $BaCO_3$ and TiO_2 with different molar ratios

Molar ratio of $BaCO_3$ to TiO_2	Phases
0.90:2	$BaTi_2O_5+Ba_6Ti_{17}O_{40}+Ba_2Ti_9O_{20}+TiO_2$
0.92:2	$BaTi_2O_5+Ba_6Ti_{17}O_{40}+Ba_2Ti_9O_{20}+TiO_2$
0.93:2	$BaTi_2O_5+Ba_6Ti_{17}O_{40}$
0.94:2	$BaTi_2O_5+Ba_6Ti_{17}O_{40}$
0.95:2	$BaTi_2O_5$
0.96:2	$BaTi_2O_5+Ba_2TiO_4$
1.00:2	$BaTi_2O_5+Ba_2TiO_4$
1.05:2	$BaTi_2O_5+Ba_2TiO_4+BaTiO_3$

Summary

$BaTi_2O_5$ powders were synthesized by arc-melting the reagent mixture of $BaCO_3$ and TiO_2 with different molar ratios (0.90~1.05:2). For the arc-melted powders derived from the $BaCO_3$ and TiO_2 mixture with a stoichiometric composition (1:2), a little amount of impurity of Ba_2TiO_4 was still detected, whereas single-phased $BaTi_2O_5$ powders could be obtained at the molar ratio of 0.95:2. When the molar ratio was larger than 0.95:2, Ba-rich compound Ba_2TiO_4 was produced besides the main phase of $BaTi_2O_5$. Ti-rich compound $Ba_6Ti_{17}O_{40}$, $BaTi_2O_9$ and TiO_2 appeared at the ratios ranging from 0.90:2 to 0.95:2.

Acknowledgement

This work was financially supported by International Science and Technology Cooperation Project (2009DFB50470), National Natural Science Foundation of China (50802071) and International Science and Technology Cooperation Project of Hubei Province (2010BFA017).

References

[1]T. Akashi, H. Iwata, and T. Goto: *Preparation of $BaTi_2O_5$ single crystal by a floating zone method.* Mater. Trans., 44, 4(2003), pp. 802-804.

[2] T. Akashi, H. Iwata, and T. Goto: *Dielectric property of single crystalline $BaTi_2O_5$ prepared by a floating zone method.* Mater. Trans., 44, 8(2003), pp. 1644-1648.

[3] Y. Akishige: *Synthesis and physical properties of single crystals and ceramics of new ferroelectric $BaTi_2O_5$*. Jpn. J. Appl. Phys., 44, 9B(2005), pp. 7144-7147.

[4] Y. Akishige, K. Fukano, and H. Shigematsu: *New ferroelectric $BaTi_2O_5$*. Jpn. J. Appl. Phys., 42, 8A(2003), pp. L946-L948.

[5] S. Tangjuank, T Tunkasiri: *Sol-gel synthesis and characterization of $BaTi_2O_5$ powders*. Appl. Phys., A 81, 5(2005), pp. 1105-1107.

[6] L. Wang, G. Li, Z. Zhang: *Synthesis of $BaTi_2O_5$ nanobelts*. Materials Research Bulletin, 41, 4(2006), pp. 842-846.

[7] Y.B. Xu, G.H. Huang, H. Long: *Sol-gel synthesis of $BaTi_2O_5$*. Materials Letters, 57, 22-23(2003), pp. 3570-3573.

[8] Z. Deng, Y. Dai, W. Chen, and X.M. Pei: *Synthesis and characterization of single-crystalline $BaTi_2O_5$ nanowires*. J. Phys. Chem., C 114, 4(2010), pp. 1748-1751.

[9] R. Tu, T. Goto: *Dielectric properties of poly- and single-crystalline $BaTi_2O_5$*. Materials Transactions, 47, 12(2006), pp. 2898-2903.

[10]F. Wijzen, J. Rocha, A. Esculcas, B. Gilbert, P. Tarte, A. Rulmont: *Structural studies of dilute lacunar solid solutions of PO_4^{3-} ions in β-barium orthotitanate, β-Ba_2TiO_4*. Spectrochimica Acta, Part A 55, 2(1999), pp. 325-332.

[11]J. Javadpour and N.G. Eror: *Raman spectroscopy of higher Titanate phases in the $BaTi0_3$-$Ti0_2$ system*. J. Am. Ceram. Soc., 71, 4(1988), pp. 206-213.

[12]W. Li, N. Wang, W.J. Wu, Y.L. Wang, M.Y. Zhao: *Synthesis of $Ba_2Ti_9O_{20}$ powder via precipitation–mixing route*. Materials Letters, 59, 2-3(2005), pp. 162-165.

[13]P. Xiao, S.B. Zheng, J.L. You, G.C. Jiang, H. Chen, H. Zeng, *Structure and Raman spectra of Titanium oxides*, Spectroscopy and spectral analysis, 27, 5(2007), pp. 936-939.

Advanced Materials Research Vol. 279 (2011) pp 49-53
© (2011) Trans Tech Publications, Switzerland
doi:10.4028/www.scientific.net/AMR.279.49

Isothermal Dielectric Relaxations of Poly(vinylidene fluoride) Filled with Silicon Carbide Nanoparticles

Yuchao Li[1, a], Xiangcai Ge[1, b], Sie Chin Tjong[2, c]

[1]Department of Materials Science and Engineering, Liaocheng University, Shandong, PRC

[2]Department of Physics and Materials Science, City University of Hong Kong, Kowloon, HK

[a]cityu.lyc@gmail.com, [b]gexiangcai@lcu.edu.cn, [c]aptjong@cityu.edu.hk

Keywords: Silicon carbide, Dielectric spectroscopy, PVDF, Nanocomposites

Abstract. The dielectric relaxation behavior of poly (vinylidene fluoride) based composites filled with beta silicon carbide nanoparticles were investigated over a wide frequency range and temperature intervals. The composites exhibited dielectric relaxations in the tested frequency range and the relaxations of composites can be well described via the modulus formalism of dielectric spectroscopy. Further, activation energy determined from the isothermal dielectric relaxations tended to decrease with increasing SiC indicating the promotion of SiC to the dipole relaxations of PVDF.

Introduction

Nanocomposites of 0-3 connectivity, consisting of polymer matrix loaded with various ceramics, have attracted a considerable interest in the past few decades for their potential applications in many electronic devices and smart sensors [1-3]. Ferroelectric ceramics such as barium titanate ($BaTiO_3$), lead titanate ($PbTiO_3$) and lead zirconium titanate (PZT) filled with a varity of polymers are examples of such system. The properties of these composites can be tailored to specified requirements by combining the excellent dielectric, piezoelectric and pyroelectric activities of ceramics and the flexibility of polymers.

Semiconductive silicon carbide (SiC), a very hard and strong ceramic material, is commonly used to improve the mechanical and thermal properties of metallic materials [4]. Moreover, SiC-based electronic devices are presently being developed for use in severe conditions [5]. The application of SiC technology has led to a rapid development in electronic devices, such as light-emitting diodes, sensors, transistors, transducers and so on [6-8]. However, few attentions are paid on SiC filled polymers. Herein, SiC was successfully introduced into PVDF matrix via solution casting technique. The electrical properties of the resulting composites was characterised via the dielectric relaxation spectroscopy (dielectric analysis, DEA).

Experimental

Commercial grade of PVDF (Kynar 740, 1.78 g/cm^3) was supplied by Atofina Chemicals Inc. (USA). Beta-SiC (45-55 nm) particles were purchased from Nano-structured Amorphous Materials Inc, USA. Both were used as received.

The PVDF/SiC nanocompostites were fabricated via solution casting technique. Briefly, PVDF was previously dissolved in the N, N-Dimethylformamide (DMF) solvent in a flask. SiC with contents varying from 0 to 50 wt% was added to the PVDF/DMF solutions. After mechanical stirring 2h and ultrasonic oscillation 8h, the solvent was then evaporated completely under the temperature of 70 °C and the resulting PVDF/SiC nanocomposites were obtained. Finally, the composites were compression molded under 230 °C for characterization.

SEM images were recorded in a JEOL JSM-820 scanning electron microscope. A thin gold layer was sputter-coated on each specimen prior to SEM observation. The isothermal dielectric relaxation behavior of PVDF/SiC composites is measured with an AC impedance analyzer (Agilent model 4294) over a wide frequency range from 50 to 5×10^7 Hz under various temperatures.

Results and Discussion

Fig. 1(a)-(b) illustrate SEM images of representative PVDF/SiC 10 wt% and PVDF/SiC 30 wt% nanocomposites. Solution enables homogeneous dispersion of SiC nanoparticles in the PVDF matrix. No pores were found in the composites after compression molding.

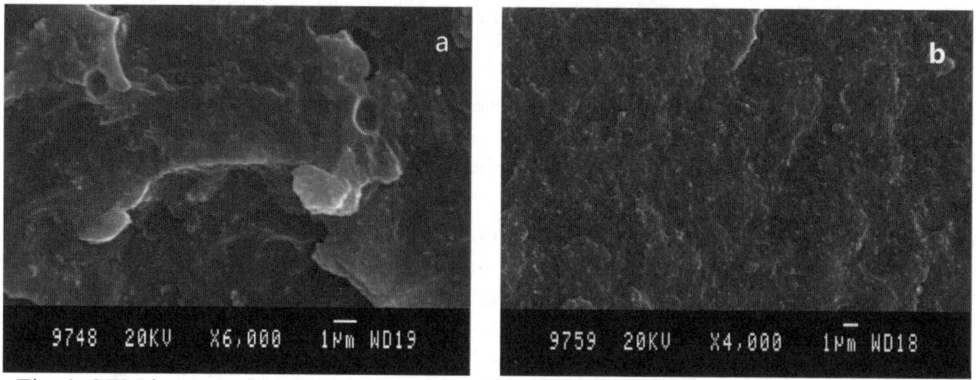

Fig. 1: SEM images of (a) PVDF/SiC 10 wt% and (b) PVDF/SiC 30 wt% nanocomposites

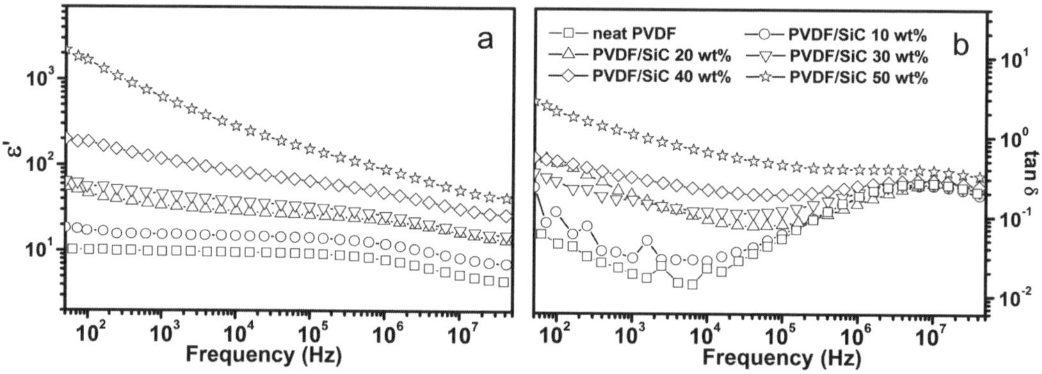

Fig. 2: Dielectric constant of PVDF/SiC nanocomposites as a function of SiC concentrations

Dielectric spectroscopy is a powerful tool in studying relaxation phenomena in polymers and composites. It may provide much information about the location and activation energy of relaxation transitions, the dipole moment of the subunits involved, concentration and mobility of charge carriers, and so forth. Fig. 2(a)-(b) depict the frequency dependence of dielectric constant and loss tangent under various SiC concentrations. As expected, the dielectric constant (Fig. 2a) increases with increasing β-SiC contents while it tends to decrease with increasing frequency due to the mismatch of dipole relaxation with applied frequencies. The loss tangent (Fig. 2b) increases with increasing silicon carbide particularly at low frequencies and a broad relaxation peak was observed in the vicinity of 10^7 Hz in the dielectric spectrum. At 1 kHz, the dielectric constant of PVDF/SiC 40 wt% nanocomposite reaches as high as 116 (tanδ = 0.3), being ten times higher than that of neat PVDF.

Dielectric spectroscopy makes a detailed analysis possible of the kinetics by which the different relaxation phenomena are governed. The isothermal dielectric spectrum of PVDF/SiC nanocomposites (Fig. 3) has been examined in which the electric modulus formalism has been applied to the analysis of data. The modulus formalism, firstly introduced by McCrum et al. [9], is usually used to investigate the dielectric relaxations of composites. The complex modulus of dielectric behavior can be expressed via Eq. 1:

$$M^* = \frac{1}{\varepsilon^*} = \frac{1}{\varepsilon' - j\varepsilon''} = \frac{\varepsilon'}{\varepsilon'^2 + \varepsilon''^2} + j\frac{\varepsilon''}{\varepsilon'^2 + \varepsilon''^2} = M' + jM''. \quad (1)$$

where *M'* and *M''* are the real and imaginary part of dielectric modulus, respectively.

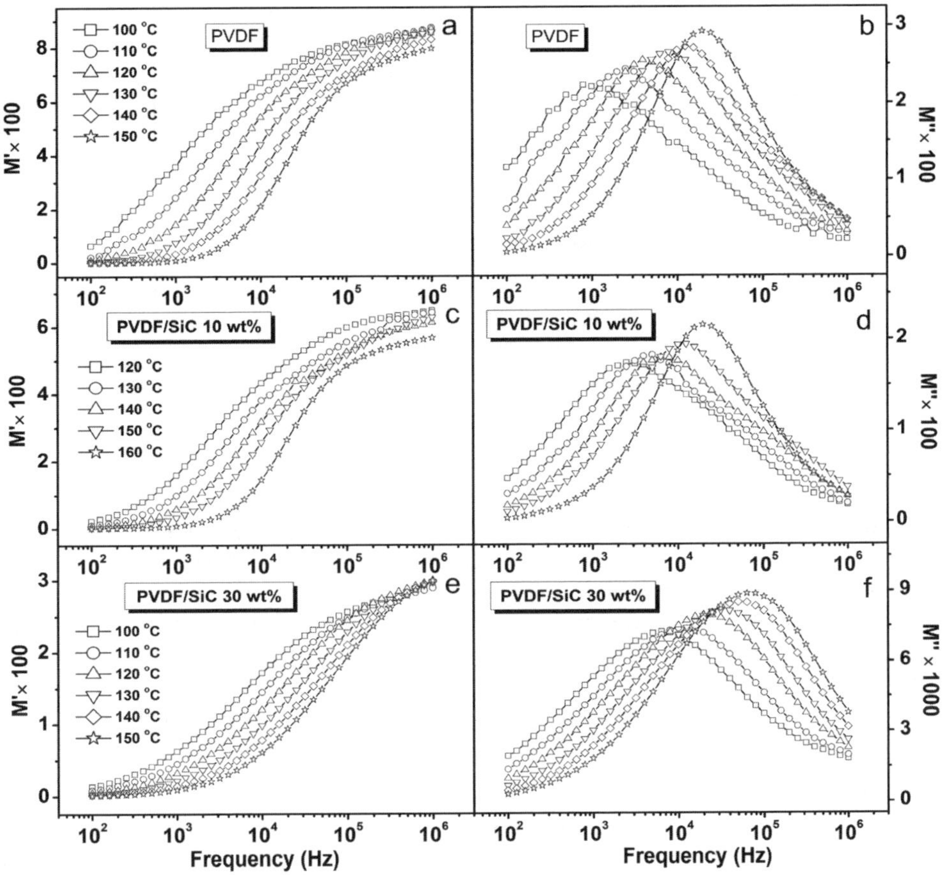

Fig. 3: (*a, c, e*) M' and (*b, d, f*) M'' as a function of frequency at different temperatures

Fig. 3(a)-(f) display the isothermal relaxations of PVDF/SiC systems. Distinct relaxations can be observed from the modulus expressions of dielectric spectroscopy. It is seen that the relaxation peaks of PVDF and PVDF/SiC nanocomposites shift to higher frequency with increasing temperature. Further, the whole spectrum shifts to higher frequencies with the addition of SiC as well. The isothermal dielectric relaxations of composites can be described via the Arrhenius-type equation where the maximum peak frequency displays a power relationship with temperature (Eq. 2):

$$f = f_0 \exp(-\frac{E_b}{RT}). \quad (2)$$

where *f* is the maximum peak frequency, T is the applied temperature and E_b the activation energy.

Fig. 4 shows the activation energy determined from the isothermal relaxations of PVDF/SiC nanocomposites and the inset of Fig. 4 illustrates the Arrhenius plots of the above studied samples. The data obey Eq. 2 well and the activation energy tends to decrease with increasing SiC contents. The value of E_b is found to decrease with increasing SiC concentrations. The activation energy is related to the dipolar reorientations of PVDF matirix. Therefore, the addition of SiC promotes the relaxation of PVDF dipoles.

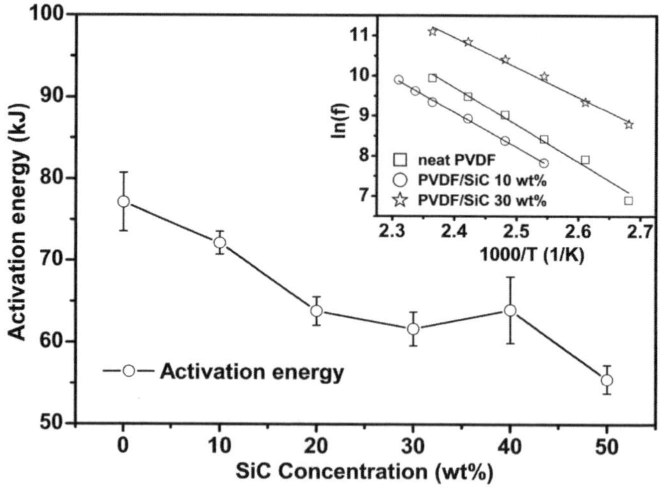

Fig. 4: Activation energy determined from isothermal dielectric relaxation of PVDF and PVDF/SiC nanocomposites.

Conclusions

PVDF/SiC nanocomposites were fabricated via solution mixing and subsequent compression molding process. Solution enables homogeneous dispersion of SiC in PVDF matrix. The dielectric properties of PVDF/SiC nanocomposites were evaluated at various frequency and temperature ranges. The results showed that the permittivity increases with increasing SiC content. The modulus formalism of dielectric spectroscopy can be well used to describe the dielectric relaxation process of composites. Further, activation energy determined from the isothermal dielectric relaxations tended to decrease with increasing SiC indicating the promotion of SiC to the dipole relaxations of PVDF.

Acknowledgements

This work was financially supported by the Doctor Science Foundation of Liaocheng University (No. 31805), Shandong Province, PRC and Doctoral Research Foundation from Science and Technology Commission of Shandong Province, PRC (No. 2008BS04025).

References

[1] Y. Kobayashi, T. Tanase, T. Miwa and M. Konno: *Fabrication and dielectric properties of the BaTiO3-polymer nano-composite thin films*. J. Eur. Ceram. Soc. Vol. 28 (2008), p. 117.

[2] E. Venkatragavaraj, B. Satish, P. R. Vinod and M. S. Vijaya: *Piezoelectric properties of ferroelectric PZT-polymer composites*. J. phys. D Vol. 34 (2001), p. 487.

[3] N.G. Devaraju, E.S. Kim and B.I. Lee: T*he synthesis and dielectric study of* $BaTiO_3$*/polyimide nanocomposite films*. Microelectron. Eng. Vol. 82 (2005): p. 71.

[4] S.C. Tjong and Z. Y. Ma: *High-temperature creep behaviour of powder-metallurgy aluminium composites reinforced with SiC particles of various sizes.* Compos. Sci. Technol. Vol. 59 (1999), p. 1117.

[5] P.G. Neudeck: *Progress in Silicon Carbide Semiconductor Electronics Technology*. J. Electron. Mater. Vol. 24 (1995), p. 283.

[6] D.M. Brown, E. Downey, S. Emily and D. Schneider: *SiC flame sensors for gas turbine control systems*. Solid-State Electron Vol. 42 (1998), p. 755.

[7] S. Yoshida, H. Daimon, E. Sakuma and S. Misawa: *Schottky-barrier field-effect transistors of 3C-SiC*. J. Appl. Phys. Vol. 60 (1986), p. 2989.

[8] A.D. Kurtz and A.A. Ned: *High temperature pressure transducer fabricated from beta silicon carbide*. U.S. patent 6327911, (2001).

[9] N.G. McCrum, B.E. Read and G. Williams: *Anelastic and Dielectric Effects in Polymeric Solids*, Dover Publications, Wiely: London 1967.

Advanced Materials Research Vol. 279 (2011) pp 54-60
© (2011) Trans Tech Publications, Switzerland
doi:10.4028/www.scientific.net/AMR.279.54

Microstructure and Wear Behavior of Si3N4/TiC Nanocomposite Ceramic Tool Material

Zhijie Lü[1, 2, a], Jun Zhao[3] and Weimin Wang[1]

[1]School of Materials Science and Engineering, Shandong University, Jinan, 250061, CHN

[2]School of Mechanical and Electronic Engineering, Shandong Jianzhu University, Jinan, 250101, CHN

[3]Key Laboratory of High Efficiency and Clean Mechanical Manufacture of MOE, School of Mechanical Engineering, Shandong University, Jinan, 250061, CHN

[a]zhijie@sdu.edu.cn

Key words: Si_3N_4/TiC, Nanocomposite, Microstructure, Cutting Tool Material, Wear Mechanism

Abstract. Si_3N_4/TiC nanocomposite cutting tool materials were fabricated by adding Si_3N_4 and Al_2O_3 nanocomposites with Al_2O_3 and Y_2O_3 as additives. The microstructures of materials were analyzed by scanning electron microscopy (SEM) and transmission electron microscopy (TEM), and then wear resistance of Si_3N_4/TiC nanocomposite cutting tool materials was studied when machining cast iron and quenched steel in continuous dry turning experiments. Micrographs of the worn tools materials were observed with SEM. Results showed that TiC nanoparticles distribute in the matrix grains and lead to the crack deflection as well as crack pinning. Due to TiC nanoparticles and β-Si_3N_4 quasiwhiskers, the crack trajectories exhibited crack deflection, rod-like grain bridging and pull-out. Wear mode of Si_3N_4/TiC inserts is mainly abrasive wear with slighter adhesive wear when machining cast iron. The dominant wear patterns during the cutting of quenched steel were adhesion, abrasion, as well as oxidation and diffusion under high temperature.

Introduction

The advancement of ceramic cutting tools has promoted machining technology to a new level. Today, there are two basic types (Al_2O_3-based and Si_3N_4-based) and 40 varieties of ceramic tool materials covering more than 200 different brands [1-3]. It is on special cutting conditions and difficult-to-cut materials that the potential of ceramics tools could be exhibited [4, 5]. It can be inferred that ceramic tool materials are the ones that have great potential and broad prospects in the future despite their unpopularity at present. A challenging research subject is how to increase their strength and toughness as well as the working reliability of ceramic tool materials.

Compared to Al_2O_3-based ceramic materials, Si_3N_4-based ceramic materials exhibit more excellent mechanical properties, such as low coefficient of thermal expansion, high chemical reliability, creep resistance, flexural strength and thermal-shock resistance. Due to its covalent nature of bonding, it's difficult to get densified. Simultaneously, pure Si_3N_4 ceramics have low hardness and bad wear resistance. Therefore it's imperative to improve their mechanical properties with different toughening and strengthening mechanisms [6, 7].

Ceramic nanocomposites are materials in which a reinforcing ceramic phase of nanometric dimensions is added to a ceramic matrix. Nanocomposites are one of the particle-dispersed materials, but the strengthening and toughening mechanisms are quite different, where the strength of nanocomposites was improved significantly and also moderate enhancement in toughness was found[8, 9] .During the present work, by adding nanoparticle, Si_3N_4/TiC ceramic cutting tool was fabricated with high flexural strength and high fracture toughness.

Today commercial ceramic inserts are used for cast iron and superalloy machining. These inserts are also proposed for machining carbon steels in order to increase productivity [10, 11]. They stand for their high material hardness/hot hardness at elevated temperature, wear resistance and chemical stability [12, 13]. During machining, a contact pair between inserts and work piece/chip is formed and is subjected to conditions promoting intense abrasion, adhesion, diffusion, plastic deformation over a portion of ceramic inserts. These results in different forms of tool wear like flank wear, crater wear, notch wear *etc*. Basically, the tool wear observed in ceramic cutting tool can be attributed to two main categories: wear activated mechanically and wear activated chemically. Usually the wear in a ceramic tool is a complex phenomenon and can relate to several wear modes/mechanisms.

In this paper, mechanical properties and microstructure of Si_3N_4/TiC cutting tool materials were studied. Some experimental results on the cutting performance of the developed ceramic tool materials are reported and wear mechanisms are analyzed with microstructure observation.

Experiment

Material Processing. High-purity α-Si_3N_4 (99.7%) with mean particle size of 0.5 μm and 130 nm was used in this investigation, and TiC (99%) powder used had an average particle size of 50 nm. Al_2O_3 and Y_2O_3 were used as additives. The weight fractions of various constituents are listed in Table 1.

Table 1: Constituents of nanocomposite [wt %]

Constituents	Si_3N_4 [0.5 μm]	Si_3N_4 [130 nm]	TiC [50 nm]	Y_2O_3+ Al_2O_3	Mo+ Ni
Weight percentage	55	10	28	5	2

PEG (polyethylene glycol) was used as dispersant. According to the optimum parameters, the starting powders were mixed for 12 h using deionized water in a polyethylene jar. After ball milling, the slurry was vacuum-dried and passed through a 120 μm sieve. Finally, the mixed powder was hot-pressed in nitrogen atmosphere in a BN-coated graphite mold at 1750 ℃ for 60 min under 30 MPa to obtain circular disks of approximate 50 mm in diameter by 4 mm in thickness.

Mechanical Tests and Machining Parameters. The mechanical properties were measured on prismatic bars with a nominal rectangular section of 4.0 mm× 3.0 mm× 40 mm. Subsequently, they were polished with 9, 3 and 1μm diamond paste to achieve a mirror finish and to eliminate any residual stresses generated during machining. The tensile edges were beveled to reduce the effect of edge flaws.

The flexure strength was measured by three-point bend tests. The tensile surface of the bars was perpendicular to the hot-pressed direction. The tests were carried out in electromechanical testing machine at a cross-head speed of 0.5 mm/min.

For the microstructural analysis, the polished specimens were etched in melting NaOH for 60 s to reveal the grain boundaries. The polished and etched surfaces were sputter-deposited with a thin Au-Pd coating for 20 min.

Fractured surfaces of specimens were observed by scanning electron microscopy (SEM). Specimens for transmission electron microscopy (TEM) were prepared by following the conventional preparation method: grinding, dimpling, ion beam thinning and finally C-coating.

The insert mounted on a commercial tool holder has the following geometry: rake angle $\gamma_o = -5^0$; clearance angle $\alpha_o = 5^0$; inclination angle $\lambda_o = -5^0$; main deflection angle $\kappa_r = 45^0$;

The cutting tests were carried out by using CA6140 lathe. Each ceramic insert was tested three times in continuous dry turning. The workpieces and machining cutting parameters are listed in Table 2.

Table 2: Machining parameters

Workpiece materials	Cutting speed v [m/min]	Depth of cut a_p [mm]	Feed rate f [mm/r]
Cast iron(HB 320)	217	0.5	0.1
T10A (HRC 61~62)	156	0.2	0.1

Results and Discussion

Mechanical Properties and Microstructure. The results of the mechanical tests, including the flexural strength, fracture toughness, vickers hardness were 920Mpa, 7.2Mpa•m$^{1/2}$ and 16.95GPa respectively. Major phases in the sample were β-Si_3N_4 and TiC, and analysis showed that the secondary (grain boundary) phase in this sample was $Si_{1.8}Al_{0.2}O_{1.2}N_{1.8}$ [14].

The micrographs of a fracture surface and an indentation crack (Fig. 1) show that the fracture mode of this nanocomposite is mainly transgranular (Fig. 1a), which is favorable for the increment of toughness. Meanwhile, grains bridging and pull-out mechanisms can be found in Fig. 1(b). The combined effects of multi-reinforcing-toughening mechanisms lead to the higher toughness of the nanocomposite, compared with Al_2O_3-based ceramics.

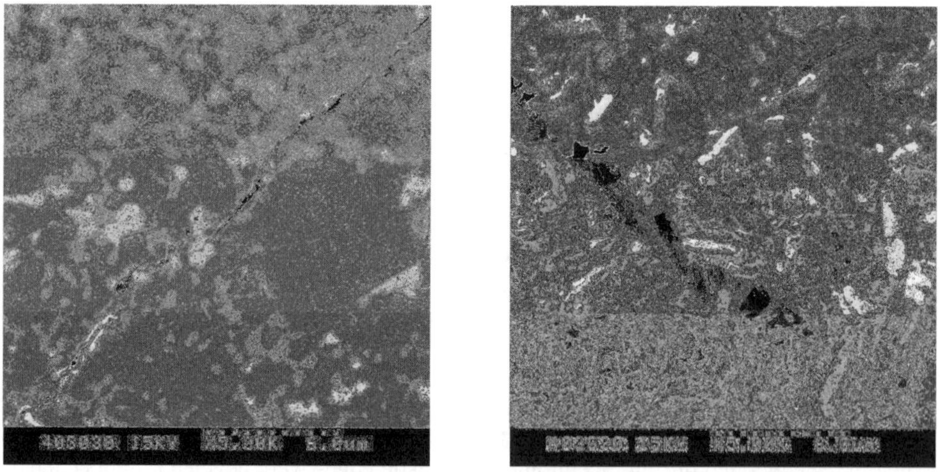

(a) Cross section been polished (b) Cross section been etched

Fig. 1: micrographs of indentation crack

Fig. 2 shows SEM micrographs of the etched surface of the Si_3N_4/TiC nanocomposites specimens. All materials consist chiefly of elongated β-Si_3N_4 grains. Voids in the structure are the result of dissolution of matrix phase by excess chemical etching which was conducted to observe three-dimensional microstructures of elongated grains.

The TEM micrograph in Fig. 3 shows the general microstructure of specimen, showing an interlocking structure. TEM observations show that the TiC particles can be found both inside the grain and on the grain boundaries. According to Niihara classification[15], this material is an "intra/inter" nanocomposite.

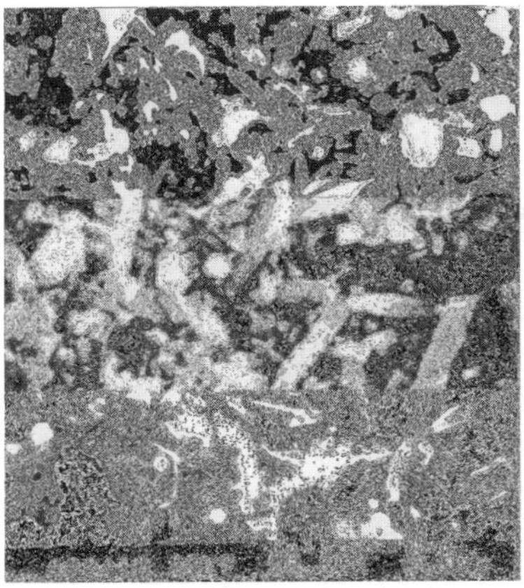

Fig. 2: SEM micrograph of etched surface

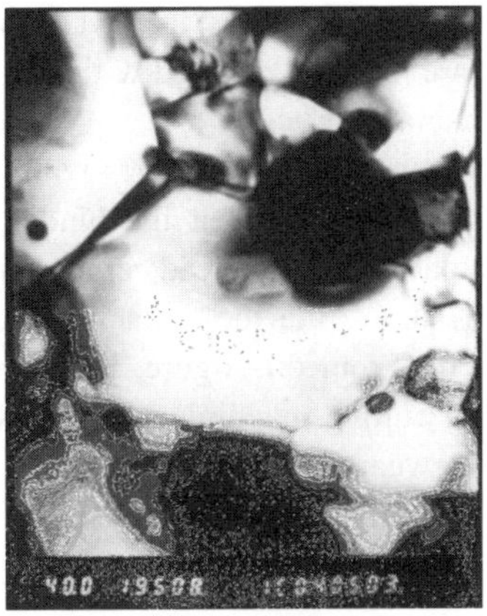

Fig. 3: TEM micrograph of nanocomposite

At some of the grain boundaries and multiple grain junctions, an intergranular phase is observed. The origin of this phase comes from crystallization of additive phases and a little powder impurities. The toughening of Si_3N_4/TiC nanocomposites is an integrated result which is combined by strengthening of main boundaries. The main boundaries possess much strong bonding based on crack pinning around intergranular nanoparticles. Toughness action of intergraular nanoparticles is due to formation of residual stress fields which is favorable to cracks transgranular propagation, formation of microcracks and sub-grain boundaries.

Wear Morphology of Si_3N_4/TiC Inserts. The wear morphologies of ceramic inserts are shown in Fig. 4 when continuously dry turning cast iron (Fig. 4a_1~c_1) and quenched steel (Fig. 4a_2~c_2) respectively.

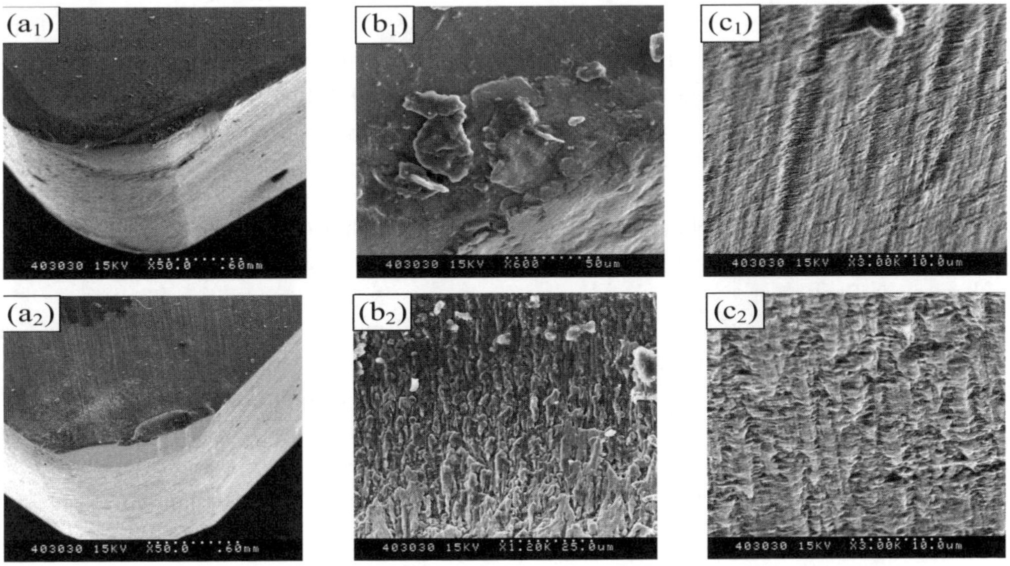

Fig. 4: Wear morphologies of inserts

Fig. 4(a_1) shows the wear region of ceramic inserts on machining cast iron, including rake and flank wear after machining length 600 m. The widest of the wear land is in the vicinity of the tool nose which is different from conventional cutting where tool flank wear occurs at the point further from tool nose.

Fig. 4(b_1) shows that rake face was bench slope. The bottom of the crake and the zone near their inner edge are partially covered with iron chips pasted with the ceramic, and many of these metallic particles are cracked.

Fig. 4(c_1) reveals some grooves with abrasive traces, being typical abrasive wear. The cause of the appearance may be that the binder of the tool is abraded by hard carbide particles of the workpiece material, which lead ceramic insert grains to be detached from the bond.

Fig. 4(a_2) shows the tool wear region of insert on machining quenched steel, after machining length 500 m. From this figure, it can be seen that the tool crater wear is different from that in conventional cutting machining, which was formed at some distance from the cutting edge, while the rake face wear was adjacent to the cutting edge. This mode of tool wear is mainly due to higher cutting temperature (900-1000℃) on the rake face. The hardness of the tool materials decreases at such high cutting temperature, which aggravates the adhesion, abrasion, plastic deformation and oxidation.

Fig. 4(b_2) shows that the ceramic surface exhibits plastic deformation provoked by the chip flow. In the upper part of the picture the effect of adhesive wear is evident. In the image of the crater bottom, plastic deformation is more evident. Because of TiC inclusions, the stress induced dislocation motion of Si_3N_4 should be pinned, and therefore increase the tool resistance to the wear mechanism related to plastic deformation.

It is noticed that the bottom of the crater and the zone near their inner edge are partially covered with iron chips pasted with the ceramic because of adhesion, and many of these metallic particles are cracked. The unusual brittle behavior of the metal can be attributed to the high cooling rate of the thin steel chips. Under these conditions, the steel T10A containing 1.0 wt.% carbon transforms to the martensitic structure and cracks under the influence of tensile stresses generated by the difference in thermal expansion coefficient between the metal and the ceramic .

The crater wear of inserts is obvious and adhesive wear is relatively serious even at low cutting speed. For flank wear, there are not only a large number of ploughed grooves of different depths, but also a small amount of adhesives on the flank face (Fig.4c_2). With the increase of cutting speed, cutting temperature increases consequently. As a result, the adhesive is intensified both in the crater area and flank face, coupled with oxidation wear.

The worn flank of the insert has the same aspect as the crater bottom (Fig. 4c_2). Flank wear is generally attributed to the rubbing of the tool along the machined surface and high temperature causing abrasive and/or adhesive wear. Thus abrasion, diffusion and adhesion are the main wear mechanisms in flank wear.

According to Ref. [16, 17], Si_3N_4/TiC ceramic material may be oxidized to produce SiO_2 and TiO_2 under high temperature, meanwhile, the SiO_2/TiO_2 grains were molten for many factors. The chemical dissolution of Si_3N_4/TiC grains in wear chips should be more severe considering cutting speed and higher toughness of quenched steel. Adhesion, abrasion, oxidation and element diffusion are not isolated, and they are relative to each other. The variation of cutting parameters, which may affect the wear mechanisms, will accelerate the wear of the ceramic cutting tool.

Conclusion

(1) Si_3N_4/TiC nanocomposites were fabricated successfully by hot-pressing with oxide additives. The study shows that TiC nanoparticles distribute in the matrix and at the grain boundaries. The results indicate that the microstructure of the composite materials is characterized by the mixing of intergranular and intragranular phases. According to the fracture pattern, low contents of nano-particles can refine matrix grains and result in the crack deflection and propagation, and it is also effective to pin cracks. The structure of the ceramic material is framework structures which inlay each other.

(2) The wear morphologies Si_3N_4/TiC nanocomposite ceramic cutting tool on machining tests of cast iron and quenched steel under different cutting conditions are investigated. Wear mode is mainly abrasive wear with slighter adhesive wear when machining cast iron. The dominant wear patterns of Si_3N_4/TiC inserts during the cutting of quenched steel were adhesion, abrasion, as well as oxidation and diffusion under high temperature.

Acknowledgement

This work was supported by the National Basic Research Program of China (2009CB724402). The National Natural Science Foundation of China (50875156). China Postdoctoral Science Foundation (20090451306). ShanDong Postdoctoral Innovative Projects (200902017).

References

[1] Y.H. Zhou, J. Zhao, X. Ai: *Cutting performance and wear mechanisms of an Al_2O_3-Based micro-nano-composite ceramic tool*. Key Eng. Mater. Vol.443 (2010), pp. 244-249.

[2] J. Zhao, X.L. Yuan, Y.H. Zhou: *Cutting performance and failure mechanisms of an Al_2O_3/WC/TiC micro-nano-composite ceramic tool.* Int. J. Refract. Met. Hard Mater. Vol.28 (2010), pp. 330-337.

[3] B. Zou, C.Z. Huang, H.L. Liu, *et al.*: *Cutting Performance and Wear Mechanism of Si3N4-Based Nanocomposite Ceramic Tool.* Key Eng. Mater. Vol.443 (2010), pp. 324-329.

[4] G.M. Zheng, J. Zhao, X.Y. Song, *et al.*: *Ultra High Speed Turning of Inconel 718 with Sialon Ceramic Tools.* Adv. Mater. Res. Vol.126 (2010), pp. 653-657.

[5] S. Sun, M. Brandt, M. Dargusch: *Thermally enhanced machining of hard-to-machine materials-A review.* Int. J. Mach. Tools Manuf. Vol.50 (2010), pp. 663-680.

[6] J.X. Deng, Z.X. Duan, D.L. Yuan, *et al.*: *Fabrication and performance of Al_2O_3/(W, Ti) C+ Al_2O_3/TiC multilayered ceramic cutting tools.* Mater. Sci. Eng. A. Vol.527 (2010), pp. 1039-1047.

[7] Y. Kang, S. Kang: *WC-reinforced (Ti, W)(CN).* J. Eur. Ceram. Soc. Vol.30 (2010), pp. 793-798.

[8] S.M. Choi, H. Awaji: *Nanocomposites-a new material design concept.* Sci. Technol. Adv. Mater. Vol.6 (2005), pp. 2-10.

[9] R. Gadow, F. Kern, A. Killinger: *Manufacturing technologies for nanocomposite ceramic structural materials and coatings.* Mater. Sci. Eng. B. Vol.148 (2008), pp.58-64.

[10]J. Souza, M. Nono, M. Ribeiro, *et al.*: *Cutting forces in turning of gray cast iron using silicon nitride based cutting tool.* Mater. Des. Vol.30 (2009), pp. 2715-2720.

[11]Y. Qiao, X. Ai, Z.Q. Liu, *et al.*: *Machinability Investigation in High Speed Turning of Powder Metallurgy Nickel-Based Superalloy with Sialon Ceramic Inserts.* Adv. Mater. Res. Vol.139 (2010), pp. 805-808.

[12]G.A. Ibrahim, C. Haron, C. Hassan, *et al.*: *Tool Wear Mechanism in Continuous Cutting of Difficult-to-Cut Material under Dry Machining.* Adv. Mater. Res. Vol.126 (2010), pp. 195-201.

[13]J. Zhao, X. Yuan, Y. Zhou: *Cutting performance and failure mechanisms of an Al2O3/WC/TiC micro-nano-composite ceramic tool.* Int. J. Refract. Met. Hard Mater. Vol.28 (2010), pp. 330-337.

[14]Z.J. Lu, X. Ai, J. Zhao: *Mechanical properties and microstructure of* Si_3N_4*-TiC nanocomposites.* J. Mater. Sci. Technol. Vol.21 (2005), pp. 899-902.

[15]K. Niihara: *New design concept of structural ceramics-ceramic nanocompsites.* J. Ceram. Soc. Jpn. Vol.99 (1991), pp. 974-982.

[16]Z.J. Lv, J. Zhao, X. Ai: *Oxidation behavior of Si3N4/TiC nanocomposite ceramic tool materials.* J. Chin. Ceram. Soc. Vol.36 (2008), pp. 210-214.

[17]X. Zhao, J. Liu, B. Zhu, *et al.*: *Wear simulation of Si3N4 cutting tool material on a pin-on-disc tester.* Ceram. Int. Vol.23 (1997), pp. 483-488.

Advanced Materials Research Vol. 279 (2011) pp 61-65

doi:10.4028/www.scientific.net/AMR.279.61

Effect of laser ablation on $Ba_{0.5}Sr_{0.5}Co_{0.8}Fe_{0.2}O_{3-\delta}$ membrane for methane conversion

Zigang Shen [1, a], Xiaorui Guo [2,b], Wuqing Zhang [1,c] and Lijun Pan [3,d]

[1]Department of Physics, Zhengzhou Normal University, Zhengzhou 450044, China

[2]Zhengzhou University of Light Industry, Zhengzhou 450002, China

[3]School of Physical Engineering, Zhengzhou University, Zhengzhou 450052, China

[a]liebao100@sina.com, [b]rabbits2004@sina.com, [c]zhangwuqing@126.com, [d]lijunpan515@yahoo.com

Keywords: Partial oxidation of methane, Membrane reactor, Oxygen permeation

Abstract. A cross-shaped pattern was formed on the surface of $Ba_{0.5}Sr_{0.5}Co_{0.8}Fe_{0.2}O_{3-\delta}$ oxygen permeation membrane by laser ablation in order to increase the specific surface area of the membrane. The membrane was used to assemble a reactor for partial oxidation of methane (POM) to syngas in the present of Ni/ZrO_2 catalyst. The CH_4 conversion and CO selectivity of the membrane reactor can reach 98.8% and 91.5%, respectively, and the oxygen permeation flux through the membrane was 11.0 ml/cm^2 min at 850°C. The effect of space velocity (SV) on CH_4 conversion rate and CO selectivity in such reactor was discussed.

Introduction

During the past decades, natural gas has attracted a considerable and increasing attention because natural gas reserves have exceeded ever-diminishing liquid reserves. Extensive efforts have focused on direct and indirect conversion of methane, the most abundant component of natural gas, to value-added products, particularly towards easily transportable fuels [1-3]. Direct conversion involves oxidative coupling of methane to ethylene, and the elective oxidation of methane to methanol and formaldehyde. But such technique has not been applied to industry because the reaction products are easily to be further oxidized to CO_2 and H_2O, which leads to the low selectivity of the aimed products. Indirect conversion involves oxidation of methane to form syngas ($CO+H_2$) by either steam reforming or partial oxidation, then using syngas to produce liquids fuel and methanol by Fishcer-Tropsch technology [4]. Although partial oxidation of methane (POM) reduces the production cost and simplifies the operation process, downstream processing requirements cannot tolerate nitrogen and thus pure oxygen is required. To lower the cost of producing pure oxygen, using oxygen permeation dense ceramic membranes to assemble POM reactor is regarded as a promising way in resent years, which can supply the required pure oxygen continuously.

In the present time, it is believed that the mixed conducting materials with peroviskite structure may be the most promising materials for POM reactors. Jin et al. and Balachandran et al. have investigated POM to syngas by using tubular $La_{0.2}Sr_{0.8}Co_{0.2}Fe_{0.8}O_{3-\delta}$ (LSCF) and $SrCo_{0.8}Fe_{0.2}O_{3-\delta}$ (SCF) membrane reactors, respectively[5,6]. They found that the membranes broke into several pieces after the introduction of methane to membrane reactor at 850°C. Shao et al. firstly reported a new mixed-conducting material $Ba_{0.5}Sr_{0.5}Co_{0.8}Fe_{0.2}O_{3-\delta}$ (BSCF), which showed high oxygen permeability (reached about 1.4ml/cm^{2}min for a 1.80mm thickness membrane at 950°C) and improved stability [7]. POM to syngas using LiLaNiO/γ-Al_2O_3 as catalyst in a disk-type BSCF membrane reactor was also investigated [8]. Wang et al. investigated the performance of the BSCF tubular membrane reactor for oxidative coupling of methane to C_2 [9]. Recently, to further increase the oxygen permeation flux through BSCF membrane, a pattern was ablated on the surface of BSCF membrane by laser beam and the oxygen permeation flux through BSCF membrane after laser ablation (BSCF-LA) increased under air/helium oxygen gradient due to the increase in the specific surface area of the membrane [10]. The purpose of present study is to investigate the performance of BSCF-LA membrane reactor for POM reaction.

Experimental

BSCF samples were prepared by the common solid state reaction method [10]. Fig. 1 shows a sketch of BSCF membrane reactor for POM. BSCF disc was tightened on an open-ended alumina tube by ceramic glass glue which can provide a perfect sealing and almost no air leak [11]. Ni/ZrO_2 was used as catalyst because Ni/ZrO_2 showed high carbon deposit-resisting ability and stability in the carbon dioxide reforming of methane to syngas [12].

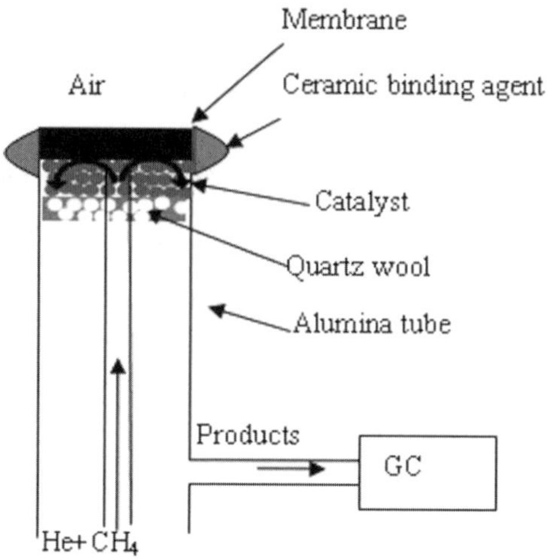

Fig. 1: The configuration of membrane reactor for POM

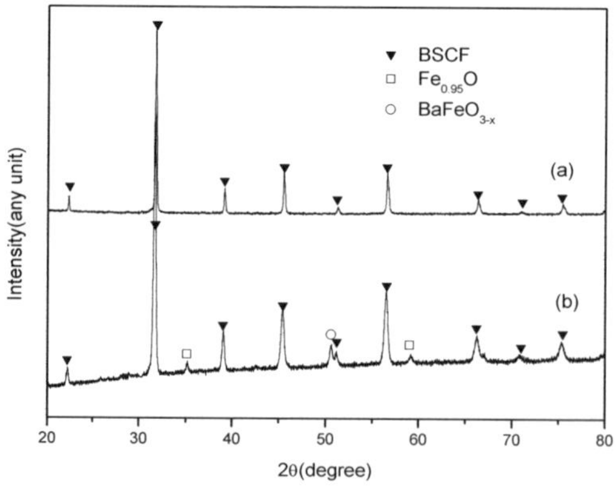

Fig. 2: XRD patterns of (a) BSCF powder and (b) BSCF membrane surface after laser ablation.

Results and discussion

Fig. 2 shows the X-ray diffraction patterns of the BSCF powder and BSCF membrane surface after laser ablation. After laser ablation, additional diffraction peaks appeared at 2θ =35.12°, 51.10° and 58.98°, which can be ascribed to the characteristic diffraction peak of $BaFeO_{3-x}$ and $Fe_{0.95}O$, respectively. These impurities could be generated in the laser ablation process. But the main structure of the samples was still BSCF phase, which indicated that the laser ablation had no serious effect on the perovskite phase. Similar result can be found in others work [13]. Fig. 3(a) shows the cross section of a stripe. The width of the stripe was about 155μm and the depth was about 24μm. Fig. 3(b) shows a top view of the surface with cross shaped pattern. Many pores were formed due to the material melting and shrinking in the laser ablation process, which had been discussed by Linde et al. [14]. Since the profile of the stripes was arc form and many larger pores were formed, it could increase the specific surface of the membrane effectively, which is beneficial for enhancing the oxygen permeation flux through membrane.

Fig. 3: SEM image of BSCF membrane. (a) The cross section of a stripe, (b) The top view of cross-shaped pattern

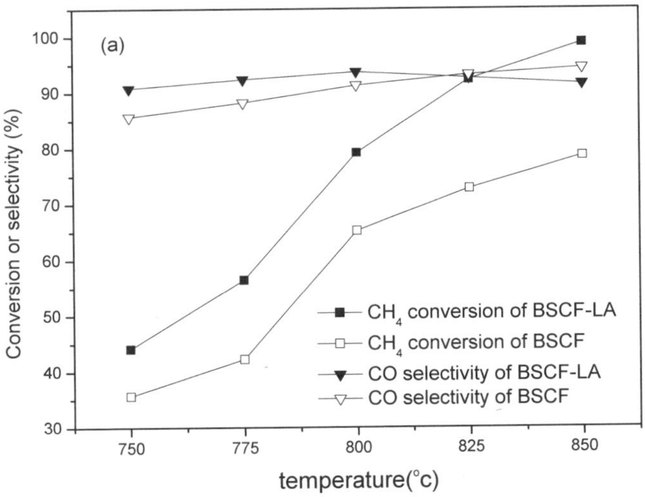

Fig. 4: (a) The temperature dependence of methane conversion and CO selectivity of BSCF membrane reactor and BSCF-LA membrane reactor with 50% helium diluted methane at the flow rate of 40ml/min, respectively.

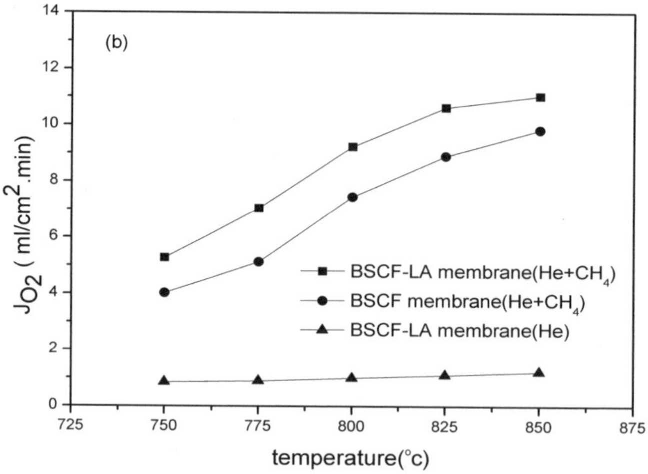

Fig. 4: (b) The dependence of oxygen permeation flux on temperature and atmosphere.

The temperature dependence of the catalytic performance of the BSCF and BSCF-LA membrane reactors are shown in Fig. 4. For BSCF membrane reactor, 78.6% CH_4 conversion, 94.3% CO selectivity and 9.84 ml/cm^2 min oxygen permeation flux were achieved at 850°C. While, for BSCF-LA membrane reactor, 98.8% CH_4 conversion, 91.5% CO selectivity and 11.0 ml/cm^2 min oxygen permeation flux were achieved at 850°C. Shao et al also studied the performance of BSCF membrane reactor for POM using LiLaNiO/γ-Al_2O_3 catalyst with 10wt.% nickel loading [8]. Their result is similar to ours.

It can be seen in Fig. 4 that the oxygen permeation fluxes and the methane conversion rates of both BSCF and BSCF-LA membrane reactors increased with the increase of temperature. The increase of the oxygen permeation flux is due to the increase of the oxygen ion diffusion and the surface exchange kinetics of the membrane with temperature. And the increase of the oxygen permeation flux will result in the increase of the methane conversion. The CO selectivity rates of both BSCF and BSCF-LA membrane reactors increased with the increase of temperature in the temperature range from 750°C to 800°C. From 800°C to 850°C, the CO selectivity rates of BSCF-LA membrane reactor decreased slightly, while BSCF membrane reactor increased. The CO selectivity is related to not only the temperature but also the CH_4/O_2 ratio. A higher temperature is favorable to endothermic reforming reaction and to increasing CO selectivity. However, a stoichiometric ration of 2 is ideal for the partial oxidation reaction ($CH_4 + 0.5O_2 = CO + H_2$). CO selectivity will decease if the CH_4/O_2 ratio deceases ($CH_4/O_2 \leq 2$), because some of the CO formed by reforming reaction ($CH_4 + H_2O = CO + 3H_2$, $CH_4 + CO_2 = 2CO + 2H_2$) could be oxidized with oxygen on the membrane surface to produce CO_2 ($CO + O_2 = CO_2$) [15]. For BSCF-LA membrane reactor, the CH_4/O_2 ratio was below 2.0 in the temperature range from 800°C to 850°C since its oxygen flux was larger than the BSCF membrane, therefore, its CO selectivity deceased.

For practical application, the value of oxygen permeation flux is a key parameter for a membrane reactor. Fig. 4(b) presents the temperature dependence of the oxygen permeation fluxes through BSCF and BSCF-LA membranes under different atmosphere. The oxygen permeation fluxes through BSCF and BSCF-LA membranes under CH_4/He atmosphere were 8 and 10 times higher than that through BSCF-LA membrane under pure He/air at 850°C, respectively. This can be explained as follows. Methane consumes oxygen resulting in a very lower oxygen partial pressure at the permeation side of membrane reactors, which in turn leads to a higher oxygen partial gradient between the two sides of membranes and significantly increases the oxygen permeation flux through membrane.

Conclusion

Laser ablation was used to increase specific surface area of the BSCF membrane. The performance of BSCF-LA membrane reactor with Ni/ZrO_2 catalyst for partial oxidation of methane was investigated in the temperature range from 750ºC to 850ºC. The CH_4 conversion and CO selectivity of BSCF-LA membrane reactor can reach 98.8% and 91.5%, respectively, and the oxygen permeation flux through BSCF-LA membrane was 11.0 ml/cm^2 min at 850ºC, which was modestly higher than that through BSCF membrane.

References

[1] H. J.M. Bouwmeester: *Dense ceramic membranes for methane conversion.* Catal. Today, Vol.82 (2003), pp.141-150.

[2] A. Tavolaro and P.Tavolaro: *The preparation of transition metal-containing mordenite catalytic tubular composite membranes.* Catal.Commun. Vol.10 (2008), pp.586-591.

[3] A. Thursfield and I. S. Metcalfe: *The use of dense mixed ionic and electronic conducting membranes for chemical production.* J. Mater. Chem. Vol.14(2004), pp.2475-2485.

[4] G.Henrici-Olive and S.Olive: *The Fischer-Tropsch Synthesis: Molecular Weight Distribution of Primary Products and Reaction Mechanism.* Angew.Chem.Int.Ed.Engl. Vol.15 (1976), pp.136-141.

[5] W. Jin et al.: *Tubular lanthanum cobaltite perovskite-type membrane reactors for partial oxidation of methane to syngas.* J. Membr. Sci. Vol. 166 (2000), pp.13-22.

[6] U. Balachandran and J.T. Dusek: *Dense ceramic membranes for partial oxidation of methane to syngas.* Appl. Catal. A Vol.133 (1995), pp.19-29.

[7] Z.P.Shao et al.:*Investigation on the permeation behavior and stability of a $Ba_{0.5}Sr_{0.5}Co_{0.8}Fe_{0.2}O_{3-\delta}$ oxygen membrane.* J. Membr.Sci. Vol. 172 (2000), pp.177-188.

[8] Z.P. Shao et al.: *Performance of a mixed-conducting ceramic membrane reactor with high oxygen permeability for methane conversion.* J. Membr. Sci. Vol. 183 (2001), pp.181-192.

[9] H.H. Wang, Y. Cong and W.S.Yang: *Oxidative coupling of methane in $Ba_{0.5}Sr_{0.5}Co_{0.8}Fe_{0.2}O_{3-\delta}$ tubular membrane reactors.* Catal. Today, Vol. 104 (2005), pp.160-167.

[10] Z.G. Shen, P.X. Lu and X. Hu: *Improving the oxygen permeability of $Ba_{0.5}Sr_{0.5}Co_{0.8}Fe_{0.2}O_{3-\delta}$ membranes by laser ablation.* J. Inorg. Mater. Vol. 2 (2010), pp.221-224.

[11] X. Qi, F. T. Akin and Y. S. Lin: *Ceramic–glass composite high temperature seals for dense ionic-conducting ceramic membranes.* J. Membr. Sci. Vol. 193 (2001), pp.185-193.

[12] J.M. Wei et al.: *Highly active and stable Ni/ZrO2 catalyst for syngas production by CO2 reforming of methane.* Appl. Catal. A, Vol.196 (2000) pp.167-172.

[13] F. Meriche et al.: *Micro structuring of $LiNbO_3$ by using nanosecond pulsed laser ablation.* Appl. Surf. Sci. Vol.254(2007), pp.1327.

[14] D. Von Der Linde and K. Sokolowski-Tinten : *The physical mechanisms of short-pulse laser ablation.* Appl.Surf.Sci. Vol.154-155 (2000), pp.1-10.

[15] H.H. Wang et al.: *Partial oxidation of methane to syngas in a perovskite hollow fiber membrane reactor.* Catal. Commun. Vol.7 (2006), pp.907-912.

Advanced Materials Research Vol. 279 (2011) pp 66-70
© (2011) Trans Tech Publications, Switzerland
doi:10.4028/www.scientific.net/AMR.279.66

A Study on the Phase Transformation Behavior and Characteristics of $BaTi_{1-x}(Li_{2/3}Nb_{2/3})_xO_3$ Solid Solution

Peng Wu[a], Jingji Zhang[b], Lina Sun[c], Jingtao Liu[d], Huang Zhou[e], Shengyong Jin[f] and Jiangying Wang[g]

College of Materials Science and Engineering, China Jiliang University, Hangzhou, China 310018

[a]wupeng100200@yahoo.cn, [b]zjjtongji@gmail.com, [c]sunlina0708@163.com, [d]we134652@163.com, [e]zzhh429@163.com, [f]1248079714@qq.com, [g]wjyliu@163.com

Keywords: Phase transformation, Dielectric properties, Ferroelectricity

Abstract. Microstructure, phase transformation behavior and electrical properties of $BaTi_{1-x}(Li_{2/3}Nb_{2/3})_xO_3$ with x=0.01-0.40 ceramics have been investigated. Only perovskite phase is detected for all the compositions. The room temperature state of $BaTi_{1-x}(Li_{2/3}Nb_{2/3})_xO_3$ ceramics experiences a transformation from a ferroelectric phase to a paraelectric phase with increasing x. Meanwhile, the permittivity of $BaTi_{1-x}(Li_{2/3}Nb_{2/3})_xO_3$ ceramics reduces, while the $Q\times f$ value increases. A degree of frequency dispersion of dielectric peak is obviously increased as x increases from 0.01 to 0.10. It is of great interest that a dielectric abnormity represented by a broad dielectric peak at 200-400K is observed for the composition with x=0.40.

Introduction

Since the discovery of $BaTiO_3$ in the early 1940s, there has been a continuous development of using it in new industrial and commercial applications. Because $BaTiO_3$ is the most easily compatible host compound for different dopants, the occupation by both isovalent and aliovalent cation at equivalent sites in $BaTiO_3$ can tailor the electrical properties to meet various device applications [1-3]. Buscaglia et al. [4] reported that aliovalent cation substitution in perovskite lattice served as donors or acceptors which can affect the electrical properties greatly. It is well known that donor dopants can induce cationic defects while occupying A or B-site of perovskite lattice [5-6]. Yanxia Li et al. [7] reported that Mg-doped $Ba_{1+k}(Ti_{1-x}Ca_x)O_{3-x+k}$ ceramics had the poorest resistivity properties compared to Ca- or Mg-doped $BaTiO_3$. Isovalent cation substitution for Ba or Ti, such as $(Ba,Sr)TiO_3$, $Ba(Zr,Ti)O_3$ and $Ba(Sn,Ti)O_3$ ceramics, have been studied extensively [8-10]. However, the investigation on complex cation $(Li_{2/3}Nb_{2/3})^{4+}$ modified $BaTiO_3$ ceramics has not been reported.

In this paper, we report a study on the microstructure, phase transformation behavior and electrical properties of $BaTi_{1-x}(Li_{2/3}Nb_{2/3})_xO_3$ ceramics with x up to 0.40.

Experimental procedure

$BaTi_{1-x}(Li_{2/3}Nb_{2/3})_xO_3$ ceramics were prepared through the conventional solid-state reactions. High-purity $BaCO_3$ (99.0%), TiO_2 (98.0%), Li_2CO_3 (98.0%) and Nb_2O_5 (99.5%) powder were used as starting materials. Mixtures based on the composition of $BaTi_{1-x}(Li_{2/3}Nb_{2/3})_xO_3$ with x=0.01, 0.05, 0.10, 0.20, 0.30 and 0.40 were ball-milled with zirconia media in ethanol for 24h and then dried at 110°C for 12h. After drying, the powders were calcined at 1150°C for 4h and then remilled for 24h. The calcined powders, mixed with 8wt% polyvinyl alcohol (PVA), were pressed into pellets at 100MPa. The green pellets were kept at 550°C for 6h to remove the solvent as well as the binder and then sintered at 1450°C/4h for x=0.01-0.10, 1400°C/4h for x=0.20, 1350°C/4h for x=0.30 and 1300°C/4h for x=0.40 in air, respectively.

Phase compositions of the ceramics were investigated by means of X-ray diffraction (XRD, Bruker D8 Advanced, Germany) with Cu K_α radiation. Permittivity and loss tangent as a function of temperature were measured using a HP 4284A precision LCR meter (Agilent, Palo Alto, CA) at frequencies from 1kHz to 1MHz in the temperature range of 125~450K. The field-induced hysteresis loops and current of $BaTi_{1-x}(Li_{2/3}Nb_{2/3})_xO_3$ ceramics were measured by a Sawyer-Tower circuit. Dielectric permittivity and loss at microwave frequencies were measured by the Hakki-Coleman dielectric resonator method using a network analyzer (AV 3629A) in combination with a resonating cavity [11].

Results and discussion

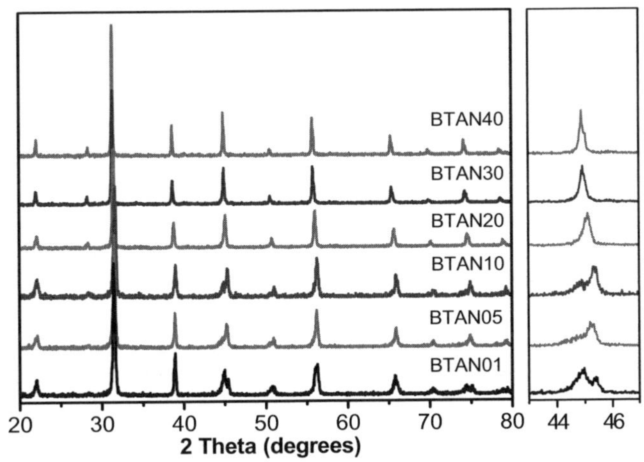

Fig. 1: XRD patterns of $BaTi_{1-x}(Li_{2/3}Nb_{2/3})_xO_3$ ceramics.

XRD patterns of the $BaTi_{1-x}(Li_{2/3}Nb_{2/3})_xO_3$ with x=0.01, 0.05, 0.10, 0.20, 0.30 and 0.40 ceramics are given in Fig. 1. Only perovskite phase is detected for all the compositions. The evolution of the diffraction peak in the ranges of 2θ from 43 to 47 degree with x, shown in Fig. 1, indicates that complex cation $(Li_{2/3}Nb_{2/3})^{4+}$ substitution has the tendency to make $BaTiO_3$ shift from tetragonal to cubic symmetry. It should be noticed that the diffraction peak shifts toward low angle with increasing x, which can be readily ascribed to the differences in ion radius of Li^+, Nb^{5+} and Ti^{4+}. The ionic radius of Li^+ (0.760 Å) and Nb^{5+} (0.640 Å) in 6-fold coordination is larger than that of Ti^{4+} (0.605 Å) [12], which leads to expansion of the crystal cells of $BaTiO_3$.

Temperature dependence of permittivity and loss tangent for all samples measured at different frequencies, varying from 1kHz to 1MHz, is displayed in Fig. 2. The dielectric anomalous peaks of cubic-tetragonal phase transition (paraelectric-ferroelectric transition) for $BaTi_{1-x}(Li_{2/3}Nb_{2/3})_xO_3$ ceramics are markedly suppressed and shifted toward low temperature with increasing x. Meanwhile, the other dielectric peak of tetragonal-orthorhombic phase transition disappears. This could result from the pinching of phase transition [13]. As x increases up to 0.20, the temperature at the permittivity maxima T_m reduces and it is below 150K. The reduction in T_m is due to the deterioration of the ferroelectric long-range order [13]. On the other hand, the internal stress may cause the shift of T_m. The internal stress increases as the grains size decreases (not shown here) and leads to the shift of T_m toward low temperature [14].

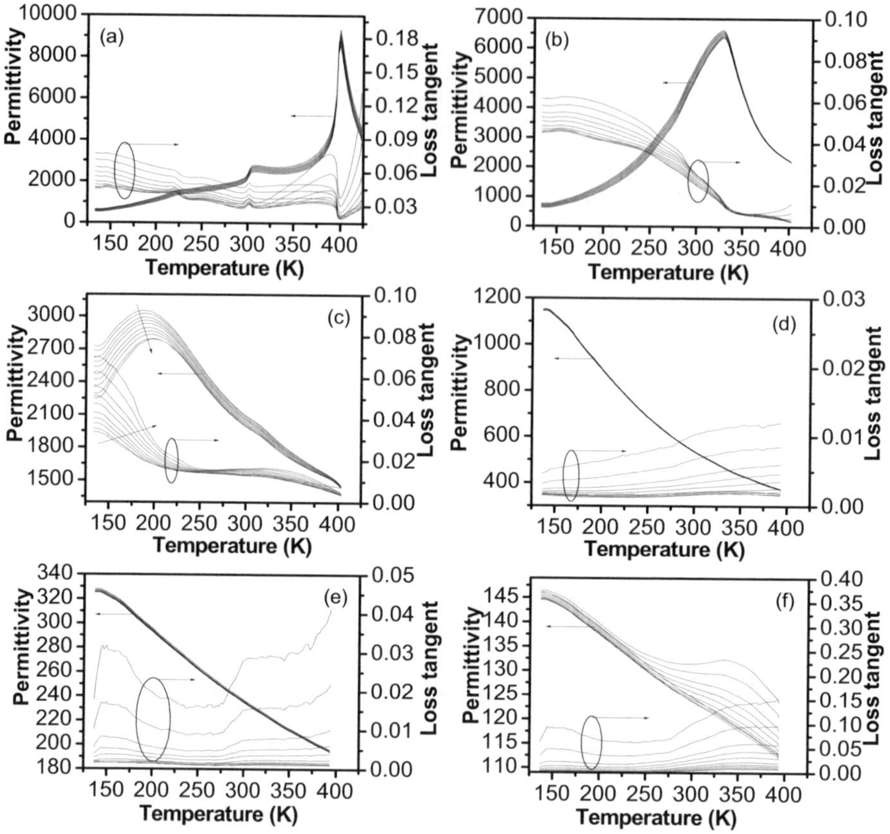

Fig. 2: Temperature dependences of the permittivity and loss tangent for $BaTi_{1-x}(Li_{2/3}Nb_{2/3})_xO_3$ ceramics: (a) x=0.01, (b) x=0.05, (c) x=0.10, (d) x=0.20, (e) x=0.30 and (f) x=0.40, respectively.

As shown in Fig. 2, the spectral features closely depend on the composition. For x=0.01, there is a strong and sharp dielectric peak at 397K with no frequency dispersion, similar to that observed in pure $BaTiO_3$. With increasing x, the sharp dielectric peak gradually becomes a diffuse peak with a frequency dispersion feature. A small shift of T_m toward high temperature with increasing frequency is observed for x=0.05 and 0.10. A degree of frequency dispersion of permittivity below T_m is obviously enhanced as x increases. This behavior may be ascribed to the complex cation $(Li_{2/3}Nb_{2/3})^{4+}$ substitution for Ti^{4+} leading to a disordered distribution of different ions on the equivalent crystallographic sites, which is similar to the results reported by Yu and Ye [13]. For the composition with x≥0.20, the dielectric peaks are not visible. Interestingly, for the composition with x=0.40, a dielectric abnormity represented by a broad dielectric peak at 200-400K is observed. This result is very similar to the dielectric abnormity of complex perovskite $Ba(Fe_{1/2}Nb_{1/2})O_3$ ceramics [15], which originates from an oxygen defect. It is very important to note that the dielectric abnormity in loss peak begins to appear for the composition with x=0.20.

Fig. 3 exhibits the polarization-field (*P-E*) loops of $BaTi_{1-x}(Li_{2/3}Nb_{2/3})_xO_3$ ceramics measured at 1kHz and room temperature. The *P-E* loops for x=0.01 and 0.05 show the ferroelectric character.

The saturated polarization, the remnant polarization and the switching field for x=0.01 are 11.3μC/cm^2, 4.3μC/cm^2 and 3.0kV/cm, respectively. The saturated polarization, the remnant polarization and the switching field markedly reduce with increasing x. The very slim hysteresis loops are observed for 0.10≤x≤0.30, which denotes the paraelectric nature at room temperature for the composition with 0.10≤x≤0.30. Especially, the linear *P-E* behavior of the composition with x=0.20 and 0.30 implies the quantum paraelectric nature. The nonlinear *P-E* behavior is visible for x=0.40, which is probably related to an oxygen defect. These results are in good agreement with the XRD analyses.

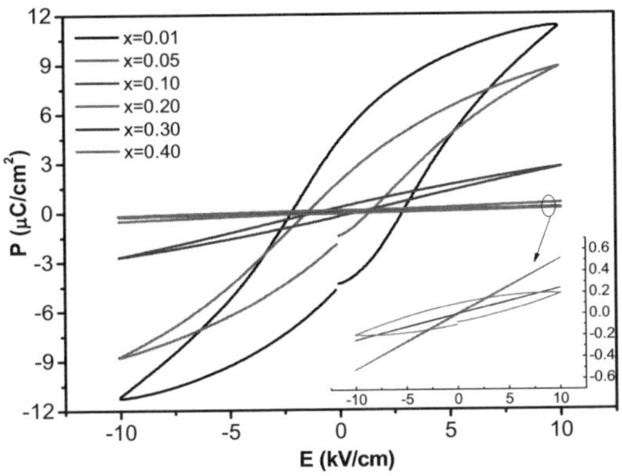

Fig. 3: *P-E* hysteresis loops of $BaTi_{1-x}(Li_{2/3}Nb_{2/3})_xO_3$ ceramics.

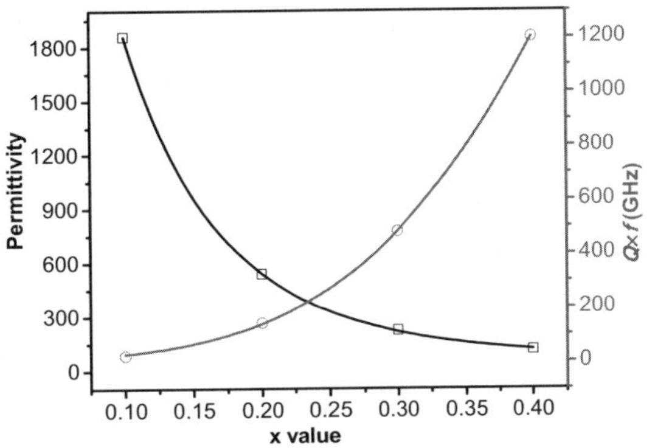

Fig. 4: Permittivity and $Q\times f$ value of $BaTi_{1-x}(Li_{2/3}Nb_{2/3})_xO_3$ ceramics with x.

Microwave dielectric parameters of the compositions with x=0.10-0.40 are measured at room temperature and permittivity and $Q\times f$ value of $BaTi_{1-x}(Li_{2/3}Nb_{2/3})_xO_3$ ceramics with x are shown in Fig. 4. It is noticeable that the permittivity of $BaTi_{1-x}(Li_{2/3}Nb_{2/3})_xO_3$ ceramics initially markedly reduces and then slightly decreases with increasing x. Comparatively, the $Q\times f$ value initially slightly increases and then obviously increases. The evident change in permittivity and $Q\times f$ value for the compositions with high $(Li_{2/3}Nb_{2/3})^{4+}$ levels is probably attributed to the close packed oxygen octahedral resulting from the substitution of Ti^{4+} by larger Li^+ and Nb^{5+} ions.

Conclusion

Microstructural evolution and macroscopic dielectric analysis have been performed for the $BaTi_{1-x}(Li_{2/3}Nb_{2/3})_xO_3$ compositional sequence. Only perovskite phase is detected for all the compositions and the diffraction peak is shifted toward low angle with increasing x. The dielectric peaks of ceramics are suppressed, broadened and shifted to low temperature with increasing the $(Li_{2/3}Nb_{2/3})^{4+}$ concentration. Meanwhile, the permittivity of $BaTi_{1-x}(Li_{2/3}Nb_{2/3})_xO_3$ ceramics reduces, while the $Q\times f$ value increases. A degree of frequency dispersion of dielectric peak is obviously increased as x increases from 0.01 to 0.10. However, a dielectric abnormity represented by a broad dielectric peak at 200-400K is observed for the composition with x=0.40.

Acknowledgements

This research was supported by the Research Fund for higher Education of Zhejiang Province (No. 01101158) and the undergraduate innovative project of Zhejiang Province (No. 2010R409016).

References

[1] D.Y.Lu, X.Y. Sun, M. Toda: *A novel high-k 'Y5V' barium titanate ceramics co-doped with lanthanum and cerium.* Journal of Physics Chemistry Solids Vol.68 (2007), pp. 650-664.

[2] S.Rout, E. Sinha, S. Panigrahi: *Dielectric properties and diffuse phase transition in* $Ba_{1-x}Mg_xTi_{0.6}Zr_{0.4}O_3$ *solid solutions.* Materals Chemistry Physics Vol.101 (2007), pp. 428-432.

[3] P. Zheng, J.L. Zhang, S.F. Shao, Y.Q. Tan, C.L. Wang: *Piezoelectric properties and stabilities of CuO-modified* $Ba(Ti,Zr)O_3$ *ceramics.* Applied Physics Letters Vol.94 (2009), pp. 032902.

[4] M.T. Buscaglia, V. Buscaglia, M. Viviani, P. Nanni, M. Hanuskova: *Influence of foreign ions on the crystal structure of* $BaTiO_3$. Journal of the European Ceramic Society Vol.20 (2000), pp. 1997-2007.

[5] J. Ravez, A. Simon: *Spontaneous transition from relaxor to ferroelectric state in new lead-free perovskite ceramics.* Ferroelectrics Vol.240 (2000), pp. 1579-1586.

[6] Z. Li, H. Fan: *Relaxation associated with the synergetic oxygen vacancies and electrons in* $(Ba_{1-x}Bi_x)_{0.9}Sr_{0.1}TiO_{3-\delta}$ *ceramics.* Journal of Applied Physics Vol.108 (2010), pp. 034103-7.

[7] Y. Li, X. Yao, L. Zhang: *Dielectric properties and microstructure of magnesium-doped* $Ba_{1+k}(Ti_{1-x}Ca_x)O_{3-x+k}$ *ceramics.* Ceramics International Vol.30 (2004), pp. 1283-1287.

[8] J.J. Zhang, J.W. Zhai, X.J. Chou, J. Shao, X. Lu, X. Yao: *Microwave and infrared dielectric response of tunable* $Ba_{1-x}Sr_xTiO_3$ *ceramics.* Acta Materialia Vol.57 (2009), pp. 4491-4499.

[9] T. Teranishi, T. Hoshina, H. Takeda, T. Tsurumi: *Ultrawideband dielectric spectroscopy of* $BaZr_xTi_{1-x}O_3$ *ceramics and its microscopic mechanism of polarization.* Journal of Applied Physics Vol.105 (2009), pp.114102-4.

[10] X. Wang, B. Li: *Dielectric audio-frequency dispersion in* $Ba(Ti_{1-x}Sn_x)O_3$ *ferroelectrics.* Solid State Communication Vol.149 (2009), pp. 537-539.

[11] B.W. Hakki, P.D. Coleman: *A dielectric resonator method of measuring inductive capacities in the millimeter range.* IRE Transactions on Microwave Theory Techniques Vol.8 (1960), p. 402.

[12] R.D. Shannon: *Revised effective ionic radii and systematic studies of interatomic distances in halides and chalcogenides.* Acta Crystallographic A Vol.32 (1976), pp. 751-767.

[13] H. Yu, Z.G. Ye: *Dielectric properties and relaxor behavior of a new* $(1-x)BaTiO_3$-$xBiAlO_3$ *solid solution.* Journal of Applied Physics Vol.103 (2008), pp. 034114-5.

[14] Y. Chen, X.L. Dong, R.H. Liang, J.T. Li, Y.L. Wang: *Dielectric properties of* $Ba_{0.6}Sr_{0.4}TiO_3/Mg_2SiO_4/MgO$ *composite ceramics.* Journal of Applied Physics Vol.98 (2005), pp. 064107-5.

[15] Z. Wang, X.M. Chen. L. Ni, X.Q. Liu: *Dielectric abnormities of complex perovskite* $Ba(Fe_{1/2}Nb_{1/2})O_3$ *ceramics over broad temperature and frequency range.* Applied Physics Letters Vol.90 (2007), pp. 022904-3.

Advanced Materials Research Vol. 279 (2011) pp 71-76
© (2011) Trans Tech Publications, Switzerland
doi:10.4028/www.scientific.net/AMR.279.71

A Comparison between B_4C-ZrB_2-Al Composite and B_4C-Al Composite on Microstructure and Mechanical Properties

LÜ Peng

Department of science research, The Chinese People's Armed Police Forces Academy, Hebei 065000, China

Penglv1982@163.com

Keywords: B_4C-Al, B_4C-ZrB_2-Al, vacuum infiltration, microstructure, mechanical properties

Abstract. B_4C-Al and B_4C-ZrB_2-Al composites were fabricated by infiltrating aluminum into porous B_4C and B_4C-ZrB_2 preforms in vacuum. The effect of ZrB_2 addition on the microstucture and mechanical properties of the B_4C-Al composites were investigated. The flexural strength and the fracture toughness of composite improved greatly as the result of ZrB_2 addition. The ZrB_2 addition inhibited the reaction between B_4C and Al. The infiltrated aluminum addition was the leading reason for the fracture toughness improvement of the composites. Inter/transgranular fracture mode with many tear ridges and dimples was showing in the fracture surface of the B_4C-ZrB_2-Al composite. The relationships between the microstructures and the mechanical properties of the B_4C-ZrB_2-Al composites were discussed.

Introduction

Boron carbide (B_4C) ceramics possess excellent physical and mechanical properties of high melting point and hardness, good impact and wear resistance, excellent resistance to chemical agents as well as high capability for neutron absorption [1-5]. The low density of B_4C and its high Young's modulus recommend this material for the construction of light-weight armor such as military helicopter and similar aero-applications [6-7]. Design and processing of new materials with improved ballistic resistant properties is becoming one of the challenging tasks of modern engineering. The B_4C-Al composite has also attracted many interests due to the low density and good ductility of Al. Jung *et al.* [8] have successfully fabricated a B_4C-Al composite having enhanced fracture toughness by infiltrating aluminum into B_4C preform in vacuum. However, much less is known about the effect of ZrB_2 addition on the microstructures and mechanical properties of B_4C-Al composites. In the presented study, the B_4C-Al and B_4C-ZrB_2-Al composites were prepared by infiltration technique. The comparison between B_4C-ZrB_2-Al composite and B_4C-Al composite on microstructure and mechanical properties was presented in detail and the toughening mechanism and fracture mode of the B_4C-ZrB_2-Al composite were also analyzed.

Experimental

B_4C (B_4C >98 wt.%, Dalian Juxing Superhard Material Ltd., China), TiO_2 (TiO_2 >99 wt.%, Shenyang Guangda Lighting Ltd., China), and aluminum alloy (trademark 5083, Suzhou Yongxie Aluminum Co. Ltd., China) were used as the starting materials. Two types of B_4C powders were used with different mean sizes (d_{50}), 20 and 1 μm, respectively. The mass ratio of B_4C (d_{50} = 20 μm) to B_4C (d_{50} = 1 μm) was 4:1. The mean size of ZrO_2 powder was 1 μm. The B_4C and ZrO_2 powders were mixed using phenolic resin-acetone solution as adhesive and ethanol as media. ZrB_2 was obtained by the reaction of B_4C, ZrO_2, and C shown as follows:

$$B_4C + 2ZrO_2 + 3C = 2ZrB_2 + 4CO\uparrow \qquad (1)$$

where C was supplied from the carbonization of the phenolic resin. The raw powders were weighed considering that the ZrB_2 content in the B_4C-ZrB_2 preforms were 35 wt.%. The mixed powders were pressed into plate about ϕ40 mm × 10 mm in size and then dried at 373 K for 43.2 ks. Porous B_4C

preform was prepared at 2273 K for 1.8 ks in in vacuum, and porous B_4C-ZrB_2 preform was *in situ* synthesized first heated at 1673 K for 3.6 ks in vacuum and then at 2223 K for 1.8 ks in an Ar atmosphere, both using pressureless sintering. B_4C-Al and B_4C-ZrB_2-Al composites were fabricated by infiltrating aluminum into the porous B_4C and B_4C-ZrB_2 preforms in vacuum at 1373 K for 7.2 ks.

The crystal phases were identified by X-ray diffraction (Cu K_α). The microstructure was observed by optical microscopy (OM), scanning electron microscopy (SEM), transmission electron microscopy (TEM) and confocal laser scanning microscope (CLSM). The phase compositions were identified by energy-dispersive spectroscopy (EDS) analysis. The hardness (HRA) was evaluated by Rockwell hardness meter. The flexural strength and the fracture toughness were evaluated using WE-10A hydraulic universal testing machine, respectively. The flexural strength was measured by three-point bending tests (specimen size = 40 mm × 4 mm × 3 mm, bend span = 30 mm and load speed = 0.05 mm/min). The fracture toughness was measured by single edge notched beam method (specimen size = 40 mm × 5 mm × 2.5 mm, notch width = 0.2 mm, notch depth = 2.5 mm, bend span = 20 mm and load speed = 0.05 mm/min).

Results and discussion

Fig. 1 shows XRD patterns of B_4C-Al composite (Fig. 1(a)) and B_4C-ZrB_2-Al composite (Fig. 1(b)). As shown in Fig. 1(a), there were B_4C, AlB_2, Al_3BC and Al phases in the B_4C-Al composite. B_4C, ZrB_2 and Al were identified in the B_4C-ZrB_2-Al composite and no peaks of AlB_2 and Al_3BC phases appeared shown in Fig. 1(b). It was assumed that the ZrB_2 addition could completely inhibit the reaction between B_4C and Al during aluminum infiltration.

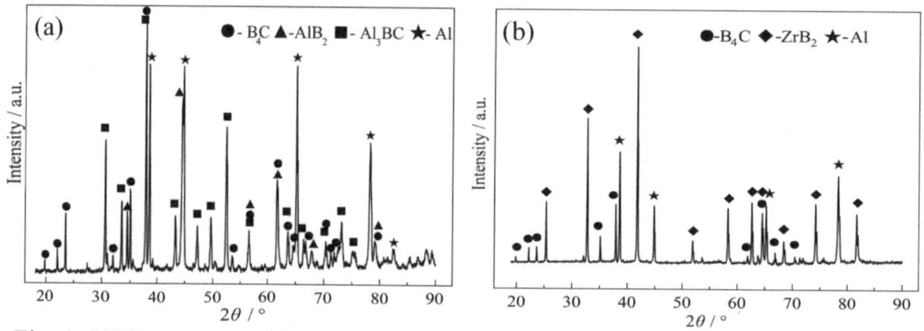

Fig. 1: XRD patterns of the B_4C-Al composite (a) and B_4C-ZrB_2-Al composite (b)

Fig. 2(a) and Fig. 2(b) show the SEM micrographs of the B_4C and B_4C-ZrB_2 preforms. The large gray phases were identified as B_4C and the small white phases were ZrB_2 based on EDS analysis. It was obvious that the size of B_4C became small as the result of the ZrB_2 addition and the ZrB_2 particles dispersed homogenously in B_4C matrix showing in Fig. 2(b). It was assumed that the B_4C particle refinement was probably caused by the *in situ* reaction shown in Eq. 1. The partial edges of the B_4C particles next to ZrO_2 were consumed to in situ synthesize ZrB_2. Srivatsan *et al.* [9] has reported that the formation of TiB_2 effectively inhibited the B_4C grain growth. And the *in situ* synthesize ZrB_2 could be considered to having the same effect as TiB_2. Fig. 2(c) and Fig. 2(d) depict the optical micrographs of the B_4C-Al and B_4C-ZrB_2-Al composites. It was obvious that the amount of Al phase was much more and the Al phase showed a continuous distribution existing in the skeleton structure of B_4C matrix showing in Fig. 2(d) comparing with Fig. 2(c).

Table 1 shows the mechanical properties of the B_4C-Al and B_4C-ZrB_2-Al composites. The hardness had almost no change with the ZrB_2 addition. On the other hand, the flexural strength increased from 362 to 522 MPa and the fracture toughness increased from 5.8 to 8.6 MPa·m$^{1/2}$ with the ZrB_2 addition as shown in Table 2. The ZrB_2 addition and smaller grain size were considered the major influencing factors on the increase of the flexural strength. The fracture toughness was improved significantly due to the B_4C grain refinement and more ductile aluminum existing. It was confirmed that more infiltrated aluminum was available to provide a higher toughness to the composite.

Fig. 2: SEM images of the performs of B_4C (a) and B_4C-ZrB_2 (b) and OM images of the composites of B_4C-Al (c) and B_4C-ZrB_2-Al (d)

Table 1: Mechanical properties of the B_4C-Al and B_4C-ZrB_2-Al composites

Composite	Hardness [HRA]	Flexural strength / [MPa]	Fracture toughness / [$MPa{\cdot}m^{1/2}$]
B_4C-Al	82.2 ± 0.37	362 ± 6.1	5.8 ± 0.12
B_4C-ZrB_2-Al	82.2 ± 0.58	522 ± 7.6	8.6 ± 0.08

The effect of the ZrB_2 addition on the fracture toughness of the B_4C-Al composites can be confirmed by the fracture feature of the composites shown in Fig. 3. The B_4C-Al composite showed transgranular fracture mode and there were a small amount of tear ridges as shown in Fig. 3(a). B_4C-ZrB_2-Al composite showed inter/transgranular fracture mode as shown in Fig. 3(b). The amount of tear ridges and dimples observed on the fracture surface of the composites increased greatly with the ZrB_2 addition, which could effectively prove the fracture toughness improvement. The formation of the tear ridges and the dimples were due to the infiltrated aluminum addition. The crack bridging action caused by the ductile metals was schematically explained shown in Fig. 3(c) [10]. When the crack propagated in the ceramic matrix composite, the ductile metal phase passed by the crack was opened until fractured. This process would consume considerable energy and toughen the composite with the metal plastic deformation. When the energy of the crack was exhausted, the crack propagation would be stopped. The contribution of the metal phase to the toughness could be calculated by following:

$$\Delta G = f\int_0^{u^*} \sigma_u du \quad (2)$$

where f, u, u^* and σ_u were the area percentage of metal phase in the crack surface, crack opening displacement, critical crack opening displacement on the tensile failure of metal phase and nominal stress, respectively. Fig. 3(d) was the amplified photograph of the marked region in Fig. 3(b). As shown in Fig. 3(d), the crack propagated into ceramic particle and formed the transgranular fracture mode. When the crack kept propagating into the ductile Al phase, the crack was finally closed owing to the crack bridging action and the shielding effect on the crack tip caused by the ductile aluminum addition [11-14].

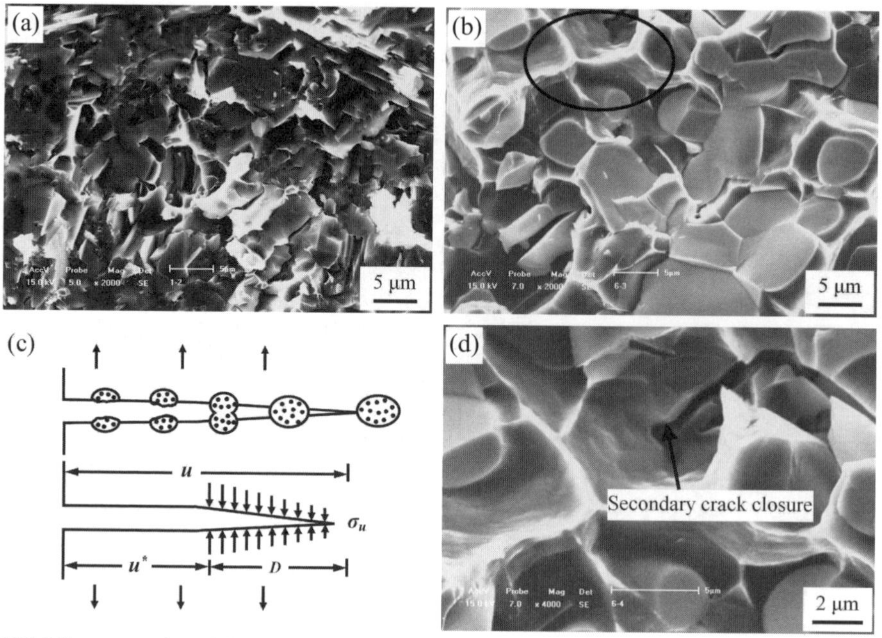

Fig. 3: SEM fractographs of the composites and the schematic diagram of crack bridging action by ductile metal: (a) B_4C-Al; (b) B_4C-ZrB_2-Al; (c) schematic diagram of crack bridging mechanism; (d) amplification of the marked region in Fig. 5(b)

Fig. 4 shows the TEM images of the twins in the B_4C matrix of the B_4C-ZrB_2-Al composite: bright-field image (Fig. 4(a)) and the corresponding electron diffraction pattern (Fig. 4(b)). The TEM images showed a number of striations, which were confirmed to be twins with twin plane $(01\bar{1}\bar{1})$ as shown in Fig. 4. The stress concentration might be avoided on the junctions between twins and the interfaces between twin and matrix. The twins in the B_4C matrix as a kind of subgrain boundary could be effective barriers for crack propagation, which might improve the fracture toughness of the composites [15]. In addition, the thermoelastic stress was usually generated when the material was subjected to temperature change during specimen preparation due to the difference in thermal expansion expansion coefficients between B_4C (α = 4.5 × 10^{-6} K^{-1}) and ZrB_2 (α = 6.88 × 10^{-6} K^{-1})[16].

Fig. 4: TEM images showing twins in the B_4C matrix of the B_4C-ZrB_2-Al composite: (a) bright-field image; (b) electron diffraction pattern

When crack propagated into the stress field, the crack deflection and divarication occurred accompanied with energy absorption, which also enhanced the fracture toughness of the composite.

Furthermore, the fracture surfaces were observed by confocal laser scanning microscope and the fracture characteristics were simulated showing in Fig. 5. The B_4C-Al composite showed the brittle fracture mode with the smooth facture surface, the short crack propagation path and the small energy consumption as shown in Fig. 5(a) and (b). And in the fracture surface of B_4C-ZrB_2-Al composite showing in Fig. 5(c) and (d), the rougher surface, the longer crack propagation path and the more energy consumption indicated that the fracture toughness of the B_4C-ZrB_2-Al composite was much more than that of the B_4C-Al composite.

Fig. 5: CLSM fractographs of the B_4C-Al (a) and B_4C-ZrB_2-Al (c) and the corresponding numerical simulations of the B_4C-Al (b) and B_4C-ZrB_2-Al (d)

Conclusions

Owing to the addition of ZrB_2 (35 wt.%), the reaction between B_4C and Al was inhibited completely and the B_4C-ZrB_2-Al composite only consisted of B_4C, ZrB_2 and Al. The hardness (HRA) of the composite had almost no change, the flexural strength increased from 362 to 522 MPa, and the fracture toughness increased from 5.8 to 8.6 $MPa \cdot m^{1/2}$ as the result of ZrB_2 addition. The infiltrated aluminum addition was the leading reason for the fracture toughness improvement of the composites. The ZrB_2 addition made more aluminum existing after infiltration process. Besides that, the improvement of fracture toughness was well related to the B_4C grain refinement, the difference in thermal expansion between the B_4C matrix and the ZrB_2 particle, and the existence of the twins in B_4C matrix. The composite showed inter/transgranular fracture mode, and the amount of the tear ridges and dimples on the fracture surface of the composite increased with the addition of ZrB_2.

References

[1] G. Fanchini, V. Gupta, A.B. Mann, M. Chhowalla: *In situ monitoring of structural changes in boron carbide under electric fields*. J. Am. Ceram. Soc., 91, 8 (2008), pp. 2666-2669.

[2] S. Hayun, D. Rittel, N. Frage, M.P. Dariel: *Static and dynamic mechanical properties of infiltrated B_4C-Si composites.* Mater. Sci. Eng. A 487, 1-2 (2008), pp. 405-409.

[3] T.K. Roy, C. Subramanian, A.K. Suri: *Pressureless sintering of boron carbide*. Ceram. Int. 32, 3 (2006), pp. 227-233.

[4] H. Lee, R.F. Speyer, W.S. Hackenberger: *Sintering of boron carbide heat-treated with hydrogen*. J. Am. Ceram. Soc. 85, 8 (2002), pp. 2131-2133.

[5] M.W. Chen, J.W. McCauley, J.C. LaSalvia, K.J. Hemker: *Microstructural characterization of commercial hot-pressed boron carbide ceramics*. J. Am. Ceram. Soc. 88, 7 (2005), pp.1935-1942.

[6] E.M. Heian, S.K. Khalsa, J.W. Lee, Z.A. Munir: *Synthesis of dense, high-defect-concentration B_4C through mechanical activation and field-assisted combustion*. J. Am. Ceram. Soc. 87, 5 (2004), pp. 779-783.

[7] L.S. Sigl, H.J. Kleebe: *Microcracking in B_4C-TiB_2 composites*. J. Am. Ceram. Soc., 78, 9 (1995), pp.2374-2380.

[8] J. Jung, S. Kang: *Advances in manufacturing boron carbide-aluminum composites*. J. Am. Ceram. Soc. 87, 1 (2004), pp. 47-54.

[9] T.S. Srivatsan, G. Guruprasad, D. Black, R. Radhakrishnan, T.S. Sudarshan: *Influence of TiB_2 content on microstructure and hardness of TiB_2-B_4C composite*. Powder Technol. 159, 3 (2005), pp.161-167.

[10] M.F. Ashby, F.J. Blunt, M. Bannister: *Flow characteristics of highly constrained metal wires.* Acta Metall., 37, 7(1989), pp. 1847-1857.

[11] J.F. Bartolome, M. Diaz, J.S. Moya: *Influence of the metal particle size on the crack growth resistance in mullite-molybdenum composites*. J. Am. Ceram. Soc. 85, 11 (2002), pp. 2778-2784.

[12] S.I. Cha, S.H. Hon, G.H. Ha, B.K. Kim: *Mechanical properties of WC-10Co cemented carbides sintered from nanocrystalline spray conversion processed powders*. Int. J. Refract. Met. Hard Mater. 19, 4-6 (2001), pp. 397-403.

[13] S. Dubey, T.S. Srivatsan, W.O. Soboyejo: *Fatigue crack propagation and fracture characteristics of in-situ titanium-matrix composites*. Int. J. Fatigue, 22, 2 (2000), pp. 161-174.

[14] J.A. Yeomans: *Ductile particle ceramic matrix composites—Scientific curiosities or engineering materials*. J. Eur. Ceram. Soc. 28, 7 (2008), pp. 1543-1550.

[15] P. Van Houtte, L. Delannay, S.E. Schoenfeld: *The work of crack formation as measure of toughness of hard metals*. Int. J. Plasticity Vol. 7(1983), pp. 107-110.

[16] R. Liu, H. Ru, Y. Zhao, D. Tang: *In situ synthesis of B_4C ceramics toughened by ZrB_2 particles*. Chin. J. Mater. Res. 20, 6(2006), pp. 611-616.

Advanced Materials Research Vol. 279 (2011) pp 77-82
© (2011) Trans Tech Publications, Switzerland
doi:10.4028/www.scientific.net/AMR.279.77

Synthesis and electrochemical properties of $Li_4Ti_5O_{12}$/CuO anode material for Li-ion batteries

Ji-Ping Zhu[1, a], Guang Yang[1, b], Jun-jie Zhao[1, c], Qing-song Wang[1, d], and Hong-wei Yang[1, e]

[1]School of Materials Science and Engineering, Hefei University of Technology, Hefei 230009, P. R. China

[a]jpzh@ustc.edu.cn, [b]damatuan0226@163.com, [c]jjzhao216@163.com, [d]qshfut@126.com, [e]hongwei.7620@yahoo.com.cn

Key words: CuO coating, Anode material, High performance, $Li_4Ti_5O_{12}$, Li-ion batteries

Abstract. Spinel-type electrode material $Li_4Ti_5O_{12}$/CuO was synthesized by a two-step solid state reaction method. Cupric acetate ($Cu(Ac)_2$) was used as raw material mixed into $Li_4Ti_5O_{12}$ to coat the surface of $Li_4Ti_5O_{12}$ with CuO in order to improve the electrochemical and cycle performances. The samples were characterized by XRD and the grain size analysis was researched by laser particle-size analyzer. The curves of charge/discharge and cycles $Li_4Ti_5O_{12}$/CuO were researched. The results showed that the electrochemical properties of the material were improved. Among the three samples investigated, it was found that the $Li_4Ti_5O_{12}$/CuO (coating amount was 4wt%) possesses the best electrochemical performance, and the first discharge specific capacity was 224, 173, 165 and 162mAh/g at 0.5C, 1C, 5C and 10C rate, respectively.

Introduction

Li-ion batteries have been widely used in different areas owing to their excellent performances[1-5]. At present, the anode materials for those batteries are various kinds of carbon materials, such as graphite, coke and MCMB (Mesocarbon microbeads). However, the batteries with carbon materials may easy to give rise to explode when over-charged[6]. Under high power charge pulses, the carbon/graphite electrode can be polarized to the extent that highly reactive lithium metal is deposited on the surface of the electrode particles. If the thermal management system is inadequate (or fails), this lithium will create a hazard[7]. The capacity of carbon materials will reduce greatly during the Li-ion insertion/extraction process. Due to those reasons the carbon materials are not right choices for high performance batteries.

Presently, spinel-type $Li_4Ti_5O_{12}$ as anode materials of lithium ion batteries has attracted increasingly attentions owing largely to its zero-strain characteristic during Li-ion insertion/extraction process when compared with other materials[8-10]. The spinel-type $Li_4Ti_5O_{12}$ has been found to show following advantages: (1) relative low cost; (2) easy to prepare; (3) excellent cycling performance; (4) long life and (5) a very flat charge-discharge voltage plateau. However, the bare $Li_4Ti_5O_{12}$ is isolating, and it exhibits a very poor electronic and ionic conductivity. And the isolation of the material hindered its application in practice[6]. In order to overcome this obstacle, researchers are trying to enhance its electrochemical performance by coating or doping with other materials[10-12].

Copper is a material with high conductivity, coating $Li_4Ti_5O_{12}$ with copper would increase its conductivity, and copper doping can increase the carrier concentration of $Li_4Ti_5O_{12}$ material, so the conductivity would be improved to some extent[13,14]. Here, cupric acetate was used as raw material to coat the $Li_4Ti_5O_{12}$ material with CuO and the final product was $Li_4Ti_5O_{12}$/CuO. The electrochemical properties can be improved to some extent.

Experimental

The $Li_4Ti_5O_{12}$/CuO materials were obtained through a two-step solid state reaction method. Lithium carbonate (Li_2CO_3) and anatase-type titania (TiO_2) were used as starting materials, the ratio of the two was the atomic ratio Li:Ti=4.2:5, according to the formula of $Li_4Ti_5O_{12}$, and the weight of Li_2CO_3 was 8% excessive approximately for Li_2CO_3 would be evaporated during sintering. The materials were ball-milled for 8h in ethanol and then the mixture was dried at 80℃ for 12h. The dried powders were then pre-sintered in a tube furnace at 750°C for 6h and then sintered at 850°C for 18h in atmosphere. The sintered mixture was white powder which was bare $Li_4Ti_5O_{12}$, and the powder was quartered. One of them was used as pure $Li_4Ti_5O_{12}$, and the other three quarters were marked with A, B and C, respectively. $Cu(Ac)_2$ was mixed into sample A, B and C, and the amount of $Cu(Ac)_2$ were 3%, 4% and 5% of the weight of the $Li_4Ti_5O_{12}$ powder, respectively. The powder was mixed by ball-milling in ethanol for 8h, and then those mixtures were dried and sintered at 500°C for 5h in air to obtain the $Li_4Ti_5O_{12}$/CuO (sample A, B and C). Those final samples were analyzed by X-ray diffraction (Philips X Pert Pro Super X ray diffraction) using Cu Kα radiation. Particle size of the powder were characterized by Laser particle size analyzer (Mastersizer 2000, Malvern UK companies). The charge/discharge capacities and the cycling behaviors were tested by Neware battery test system (Neware BTS 2300, Shenzhen).

The sintered powder, acetylene black and PVDF were mixed into a slurry according to the weight ratio of 8:1:1 and pressed into films, and then dried in vacuum at 120°C for 12h. The experimental half cells were assembled in Ar-filled glove box with Celgard 2400 microporous polyethylene membrane as the separator, metallic lithium foil as the counter electrode and 1M $LiPF_6$ in the mixture of ethyl carbonate (EC) and dimethyl carbonate (DMC) (1:1 in vol. ratio) as the electrolyte. The specific capacity and current density was based on the mass of active material, which was typically 4-8mg per electrode. The cells were galvanostatically cycled on a multi-channel battery cycler (Neware BTS 2300, Shenzhen) in the voltage range of 0.8-2.5V with current rates from 0.5C to 10C rate.

Results and Discussion

Fig. 1 is the X-ray diffraction of pristine $Li_4Ti_5O_{12}$ and $Li_4Ti_5O_{12}$/CuO(sample A, B and C). The inset of the figure is the enlargement of the XRD pattern of sample C. All the patterns showed spinel structure (JCPDS card number: 49-0207). However, there were obvious differences between the coated materials and pristine $Li_4Ti_5O_{12}$. As shown in the upper right in the figure, some peaks of CuO were found in the curves of the samples (JCPDS card number: 78-0428). No evidences could be found in the figure supported that there were other impurities existed, from which can be supposed that the $Cu(Ac)_2$ was decomposed into CuO completely.

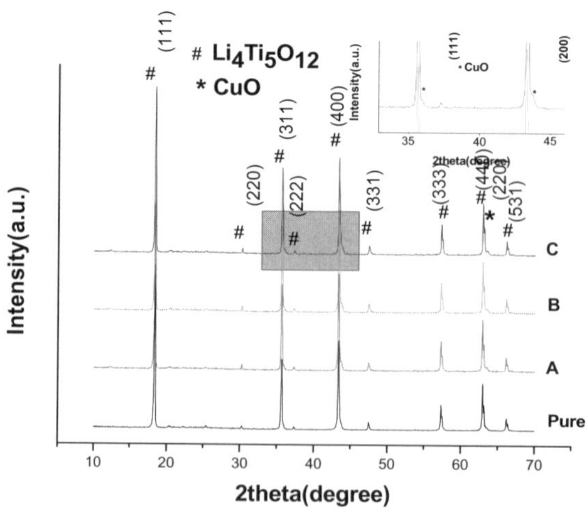

Fig. 1: X-ray diffraction of pure $Li_4Ti_5O_{12}$ and $Li_4Ti_5O_{12}$/CuO (sample A, B and C).

The particle size distribution of pristine $Li_4Ti_5O_{12}$ powder and the $Li_4Ti_5O_{12}$/CuO(sample A, B and C) are shown in Fig. 2. As can be seen, sample B and C displayed poor size distribution, and both of them had two peaks. The result made us believe the polydispersity of the particle size due to the unstability of the process of ball-milling during the sample preparation. The average particle size of the pristine $Li_4Ti_5O_{12}$ was 0.852μm and the $Li_4Ti_5O_{12}$/CuO (A, B and C) were 1.021, 3.554 and 3.468μm, respectively. Obviously, sample A possessed a very narrow particle size distribution with one peak.

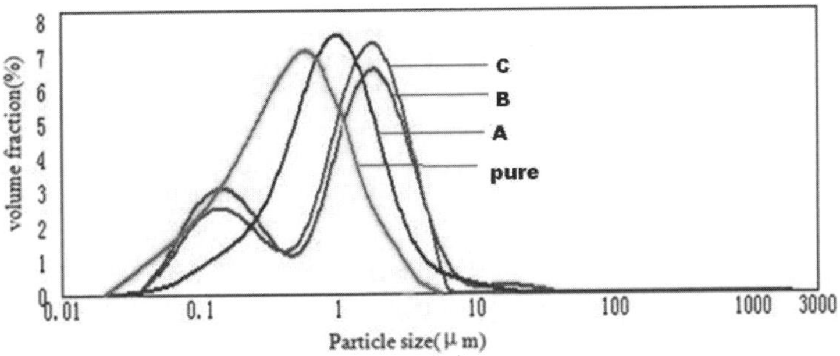

Fig. 2: The particle size distribution of pure $Li_4Ti_5O_{12}$ and the $Li_4Ti_5O_{12}$/CuO (sample A, B and C)

The SEM image of $Li_4Ti_5O_{12}$/CuO (Fig. 3) shows that the particle size is about 0.3–1.2μm. The sample shows irregular particle shape and relatively small particle size, but plenty of the particles are aggregated. It is apparent that $Li_4Ti_5O_{12}$/CuO exhibits regular uniform particles with good size. There was no second phase could be found in the image which implied that the CuO was coated homogeneously.

Fig. 3: The SEM image of sample B

The effects of CuO coating on the electrochemical performance of $Li_4Ti_5O_{12}$ were investigated by charge/discharge measurement under constant current densities. Fig. 4 displayed the first charge/discharge capacity of all three samples at the current 0.5C rate between 0.8 and 2.5V. It can be seen the first discharge capacity of the $Li_4Ti_5O_{12}$/CuO (sample A, B and C) were 168, 224 and 172mAh/g, respectively. It is obvious that sample B possessed the highest discharge capacity. All of the three rate capacities of the samples were higher than the practical capacity of pristine $Li_4Ti_5O_{12}$ which is 160mAh/g. That means the method of coated CuO to $Li_4Ti_5O_{12}$ is effective, and the amount of coating plays a key role during the process of synthesis. We can arrive at a conclusion that the sample coated with 4wt% $Cu(Ac)_2$ (sample B) has optimal electrochemistry performance. The cycles and discharge special capacity of the sample B was investigated in following parts.

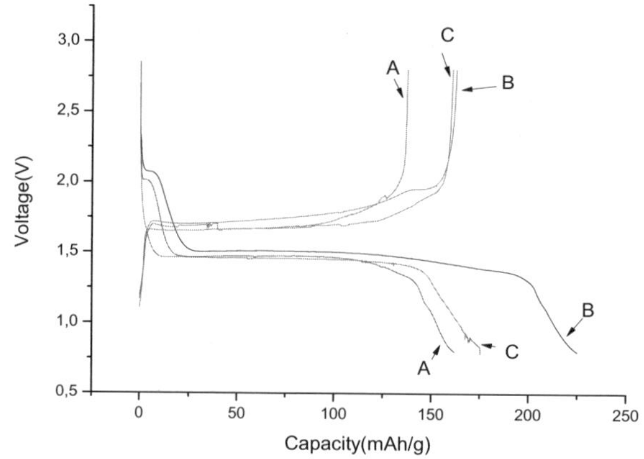

Fig. 4: The charge/discharge capacity of $Li_4Ti_5O_{12}/CuO$

According to the above part, sample B possessed the highest capacity. In order to investigate the rate performance of this material, the sample was tested at 1C, 5C and 10C current, respectively. And the results were revealed by Fig. 5(a). The discharge capacity at 0.5C, 1C, 5C and 10C were 224, 173, 165 and 162mAh/g, respectively. However, the capacity of pristine $Li_4Ti_5O_{12}$ at same condition was 160, 150, 107 and 54mAh/g, respectively[15]. Apparently, the electrochemical performance of $Li_4Ti_5O_{12}$ was highly improved by the coating with CuO, especially the performance at high rate. Fig. 5(b) is the cycle performance of the sample with 4wt% $Cu(Ac)_2$ and also tested by the battery cycler. The cell was charge/discharge for ten cycles at 1C and 5C, respectively. The reversible capacities were approximately 169.6mAh/g after cycled at 1C after 10 cycles and 156.8mAh/g at 5C. The conservation ratio of capacity was 97.8% and 98.1% compared to the corresponding first cycle, respectively. And the conservation ratio even raise as the rate of discharge increased. According to the curves, the zero-strain characteristic of $Li_4Ti_5O_{12}$ was not harmed after the coating.

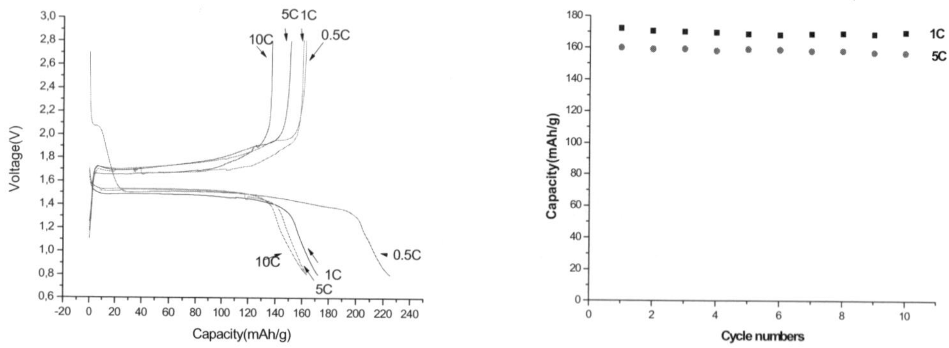

(a) (b)

Fig. 5: (a)Curves of capacity of $Li_4Ti_5O_{12}/CuO$ (sample B) cycled at different rates (b)The cycle performance of $Li_4Ti_5O_{12}/CuO$ (sample B) at 1C and 5C

Based on the plenty results mentioned above, the specific capacity, high rate performance and cycling stability were improved. The method of coating with CuO was effectively. As the amount of $Cu(Ac)_2$ added to the $Li_4Ti_5O_{12}$ raised, the rate capacity improved. However, the capacity declined when the amount exceeded 4wt%. In Wang's study[14], $Li_4Ti_5Cu_xO_{12+x}$ ($0 \leq x \leq 1.67$) was synthesized by a solid-state reaction method, and they found the samples ($0<x<1.67$) were composed of two spinel phases $Li_4Ti_5O_{12}$ and $Li_4CuTi_3O_8$, which were similar in the lattice parameter and believed to intergrow with each other. The $Li_4CuTi_3O_8$ was improved the specific capacity largely. However, due to its structure, the irreversible capacity increased as the amount of Cu raised. The $Li_4Ti_5Cu_{0.15}O_{12.15}$ was found as the optimal nominal composition. In comparison to Wang's investigation, our samples were obtained through two-step synthesis. The bare $Li_4Ti_5O_{12}$ was obtained in first step, and then $Cu(Ac)_2$ was mixed with the bare $Li_4Ti_5O_{12}$ to get $Li_4Ti_5O_{12}$/CuO. The temperature was relative low during the second step of synthesis, and there were no other substances existed in the final production except $Li_4Ti_5O_{12}$/CuO. Thereby, the zero-strain characteristic during lithium insertion/extraction process of pristine $Li_4Ti_5O_{12}$ was conserved, and the cycleability of our $Li_4Ti_5O_{12}$/CuO was as excellent as bare $Li_4Ti_5O_{12}$.

The XRD shows there were CuO peaks, but the image of $Li_4Ti_5O_{12}$/CuO shows the sample was single phase. We can suppose that $Cu(Ac)_2$ was well distributed when mixed into the $Li_4Ti_5O_{12}$, and decomposed into CuO. CuO is a kind of semi-conductor, and the conductivity is 0.1-1S/m. Therefore, the conductivity of $Li_4Ti_5O_{12}$/CuO is much higher than bare $Li_4Ti_5O_{12}$. Due to the reason, the electrochemical properties were improved to some extent.

Conclusion

Composite anode material $Li_4Ti_5O_{12}$/CuO was obtained by a two-step solid state reaction. $Cu(Ac)_2$ was mixed into pristine $Li_4Ti_5O_{12}$ powder and decomposed into CuO on the surface of $Li_4Ti_5O_{12}$ particle. X-ray distribution shows the $Li_4Ti_5O_{12}$/CuO have a spinel structure and some peaks CuO appear in the patterns. The coating did improve the electrochemical performance of the samples, and the amount of coating plays a significant role during the synthesis. Compared the data of the three samples, we come to a conclusion that 4wt% was the optimal amount of coating.

Acknowledgements

We thank for the financial supports from the National Natural Science Foundations of China (No.20771032), Anhui Province (Nos. 070414200, 090414179), Key Project of Natural Science Research of Universities in Anhui Province (KJ2010A261), Changzhou science and technology plan (Ce20100054) and national innovation experiment program for university students (No. 091035910).

References

[1] Byoungwoo Kang, Gerbrand Ceder: *Battery materials for ultrafast charging and discharging.* Nature Vol. 458, pp. 190-193.

[2] Yun Jung Lee, Hyunjung Yi, Woo-Jae Kim, et al.: *Fabricating Genetically Engineered High-Power Lithium-Ion Batteries Using Multiple Virus Genes*. Science, 324, 5930(2009), pp. 1051-1055.

[3] John B. Goodenough, Youngsik Kim: *Challenges for Rechargeable Li Batteries*. Chem. Mater., 22, 3(2010), pp. 587-603.

[4] Wei-Ming Zhang, Jin-Song Hu, Yu-Guo Guo, et al.: *Tin-Nanoparticles Encapsulated in Elastic Hollow Carbon Spheres for High-Performance Anode Material in Lithium-Ion Batteries.* Advanced Materials, 20, 6(2008), pp. 1160-1165.

[5] D Djian, F Alloin, S Martinet and H Lignier: *Macroporous poly(vinylidene fluoride) membrane as a separator for lithium-ion batteries with high charge rate capacity.* Journal of Power Sources, 187, 2(2009), pp. 575-580.

[6] Shahua Huang, Zhaoyin Wen, Xiujian Zhu and Zuxiang Lin: *Effects of dopant on the electrochemical performance of Li4Ti5O12 as electrode material for lithium ion batteries.* Journal of Power Sources, 165, 1(2007), pp. 408-412.

[7] A. N. Jansen, A. J. Kahaian, K. D. Kepler, et al.: *Development of a high-power lithium-ion battery*. Journal of Power Sources Vol. 81-82 (1999), pp. 902-905.

[8] C. M. Julien, M. Massot and K. Zaghib: *Structural studies of* $Li_{4/3}Me_{5/3}O_4$*(Me = Ti, Mn) electrode materials: local structure and electrochemical aspects*. Journal of Power Sources, 136, 1(2004), pp. 72-79.

[9] F.O. Ernst, H.K. Kammler, A. Roessler, et al.: *Electrochemically active flame-made nanosized spinels: LiMn2O4, Li4Ti5O12 and LiFe5O8*. Materials Chemistry and Physics, 101, 2-3(2007), pp. 372-378.

[10] Rui Cai, Xing Yu, Xiaoqin Liu and Zongping Shao: $Li_4Ti_5O_{12}$*/Sn composite anodes for lithium-ion batteries: Synthesis and electrochemical performance*. Journal of Power Sources, 195, 24(2010), pp. 8244-8250.

[11] Hidetoshi Utsunomiya, Tsuyoshi Nakajima, Yoshimi Ohzawa, et al.: *Influence of conductive additives and surface fluorination on the charge/discharge behavior of lithium titanate (*$Li_{4/3}Ti_{5/3}O_4$*)*. Journal of Power Sources, 195, 19(2010), pp. 6805-6810.

[12] Jizhang Chen, Li Yang, Shaohua Fang and Yufeng Tang: *Synthesis of sawtooth-like* $Li_4Ti_5O_{12}$ *nanosheets as anode materials for Li-ion batteries*. Electrochimica Acta, 55, 22(2010), pp. 6596-6600.

[13] Shahua Huang, Zhaoyin Wen, Bin Lin, et al.: *The high-rate performance of the newly designed* $Li_4Ti_5O_{12}$*/Cu composite anode for lithium ion batteries*. Journal of Alloys and Compounds, 457, 1-2(2008), 400-403.

[14] Da Wang, Hua-Yun Xu, Man Gu and Chun-Hua Chen: $Li_2CuTi_3O_8$–$Li_4Ti_5O_{12}$ double spinel anode material with improved rate performance for Li-ion batteries. Electrochemistry Communications, 11, 1(2009), pp. 50-53.

[15] Hiromasa Shiiba, Masanobu Nakayamab and Masayuki Nogami: *Ionic conductivity of lithium in spinel-type* $Li_{4/3}Ti_{5/3}O_4$–$LiMg_{1/2}Ti_{3/2}O_4$ *solid-solution system*. Solid State Ionics, 181, 21-22(2010), pp. 994-1001.

Advanced Materials Research Vol. 279 (2011) pp 83-87
© (2011) Trans Tech Publications, Switzerland
doi:10.4028/www.scientific.net/AMR.279.83

Visible-light Driven $LaNiO_3$ Nanosized Photocatalysts Prepared by a Sol-gel Process

Peisong Tang[a], He Sun[b], Feng Cao[c], Jintian Yang[d], Shengliang Ni[e] and Haifeng Chen[f]

Department of Chemistry, Huzhou Teachers College, Huzhou 313000, P.R. China

[a]tangps@hutc.zj.cn, [b]sunhepeter@hotmail.com, [c]caofeng@hutc.zj.cn, [d]yjt@hutc.zj.cn, [e]nisl@hutc.zj.cn, [f]headder@hutc.zj.cn

Keywords: $LaNiO_3$, visible-light, photocatalysis, sol-gel

Abstract. The $LaNiO_3$ nanoparticles were prepared by a sol-gel process. The $LaNiO_3$ nanoparticles were characterized by powder X-ray diffraction (XRD), scanning electron microscopy (SEM) and UV-Vis diffuse reflectance spectroscopy (DRS). XRD and SEM demonstrate the successful synthesis of single phase perovskite $LaNiO_3$ and an average grain size of 80 nm in diameter. It was found that the as-prepared $LaNiO_3$ shows strong visible-light absorption with absorption onset of 545 nm, indicating a narrow optical band gap of 2.28 eV. Consequently, $LaNiO_3$ nanoparticles show high visible-light photocatalytic activity for decomposition of methyl orange in comparison with the commercial Degussa P25. The photocatalytic experiment shows the high photocatalytic activity for the decomposition of methyl orange under visible-light irradiation, which is attributed to the strong visible-light absorption.

Introduction

Photocatalysis using solar irradiation is highly expected to be an efficiency and ideal green technology for removal of contaminants in air and river water [1]. In the past decades, titanium dioxide (TiO_2) has been paid extensive attention because of its high catalytic activity, stability, and low-cost, etc [2]. However, TiO_2 with the band gap of 3.2 eV (anatase) becomes active only under ultraviolet (UV) light radiation, which only occupies 5% of the total solar energy reaching the earth, and limits the photocatalytic efficiency in the practical application [3]. Currently, much effort has been devoted to develop low-cost, effective and simple photocatalytic materials that are active under visible-light [4]. Zou et al. [5], Kudo et al. [6], and our group [7] recently reported several visible-light responsive photocatalysts. A photocatalytic material with high visible-light activity is eagerly desired.

Zhao et al. [8] synthesized $LaNiO_3$ powders by chemical coprecipitation method and first found the photodegradation percentage of azo blue after 80 min on the best sample under UV light irradiation reached to 90%. Zhu et al. [9] studied the photocatalytic activity of $LaNiO_3$ for simulated wastewater. They prepared the powder by solid state reaction method and the results revealed that 80% methyl orange could be degraded under UV light irradiation for 160 min. Recently, $LaNiO_3$ was found to bear a much narrow band gap and showed efficient visible-light photocatalytic activity[10]. Consequently, $LaNiO_3$ is expected to be an efficient visible-light photocatalyst. Herein, single phase perovskite $LaNiO_3$ was synthesized using $La(NO_3)_3{\cdot}6H_2O$ and $Ni(NO_3)_2{\cdot}6H_2O$ as starting materials by a sol-gel process. The photocatalytic activity of the as-prepared $LaNiO_3$ was investigated by decomposition of methyl orange under visible-light.

Experimental section

Synthesis of nanocrytalline $LaNiO_3$. Analytical-grade lanthanum nitrate ($La(NO_3)_3{\cdot}6H_2O$), nickel nitrate ($Ni(NO_3)_2{\cdot}6H_2O$), citric acid ($C_6H_8O_7{\cdot}H_2O$) and chemical-grade methyl orange (MO) were purchased from Shanghai Jingchun Chemical Ltd. Nanocrystalline $LaNiO_3$ was synthesized by a sol-gel process. Typically, lanthanum nitrate ($La(NO_3)_3{\cdot}6H_2O$), nickel nitrate ($Ni(NO_3)_2{\cdot}6H_2O$) in

1:1 molar ratio were dissolved in 100 mL deionized water to form a transparent solution. Citric acid in 1:5 molar ratio with respect to the metal nitrates was added to the solution as a coordinate reagent. The pH value of the solution was adjusted to 7 by adding ammonia. The mixture was held at 30 °C for 2 hours under magnetic stirring to produce the sol, which was dried at 100 °C for 24 hours to form the dried gel powders. The resultant powder was calcined at 700 °C for 4 hours, and was ground and stored at room temperature for further use.

Characterization. Powder X-ray diffraction (XRD) was performed with a XD-6 X-ray diffraction using Cu Kα radiation (λ=0.15406 nm) operated at 36 kV and 20 mA. The morphology of the samples was observed using a Hitachi S-4800 field emission scanning electron microscope (SEM) with an accelerating voltage of 5 kV. UV-visible diffuse reflectance spectra (DRS) and UV-visible absorption spectra were taken with a Hitachi UV-4100 spectrometer.

Photocatalysis measurement. The photocatalytic activity of $LaNiO_3$ nanocrystals was evaluated by decomposition of MO in a self-assembled apparatus in which a metal halogen lamp (150 W) equipped with a JB400 filter used as the irradiation source. Typically, 20 mg $LaNiO_3$ powders were added to a 10 mL MO aqueous solution (10 mg/L), and then magnetically stirred for 30 min in the darkness to get the adsorption equilibrium followed by visible-light irradiation. After a given time, the solution was centrifuged, and the upper transparent solution was subjected to the UV-visible absorption spectra measurement. The concentration of the residual MO was evaluated by the absorbance at 464 nm. The commercial photocatalyst P25 (Degussa, Germany) was used as control.

Results and Discussion

Fig. 1 shows the XRD pattern of $LaNiO_3$ nanoparticles. All peaks can be indexed to the typical perovskite structure (JCPDS card #: 33-0711). No impurities are observed, which indicates the successful synthesis of monophasic $LaNiO_3$ by the sol-gel process. The sharp and narrow peaks suggest the high crystallization of the as-prepared $LaNiO_3$ indicating the successful synthesis of $LaNiO_3$. The crystallite size was calculated to be 80 nm, according to the scherrer's equation:

$$D = \frac{0.89\lambda}{\beta \cos\theta} \tag{1}$$

based on the (110) peak [11]. The excellent single perovskite $LaNiO_3$ nanoparticles are obtained. Furthermore, the SEM image shows the significant aggregation of $LaNiO_3$, but a careful observation will find that these aggregates consist of particles ranging from 70 nm to 90 nm, and the average crystallite size was about 80 nm consistent with the result from XRD (Fig. 2).

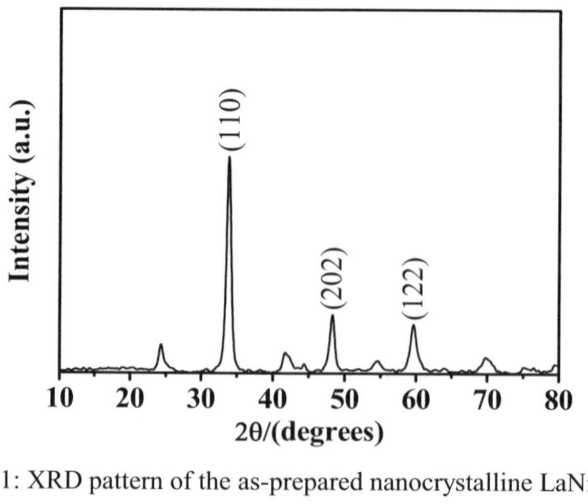

Fig. 1: XRD pattern of the as-prepared nanocrystalline $LaNiO_3$

Fig. 2: SEM image of the as-prepared nanocrystalline $LaNiO_3$

The DRS spectroscopy was usually used to reveal the electronic state of a photocatalyst material, and is very useful for photocatalysis. The diffuse reflectance spectra were measured through transformation of Kubelka- Munk equation from the transmission spectra of sample. We investigated the UV-visible DRS spectrum of the as-prepared $LaNiO_3$, as shown in Fig. 3. The as-prepared $LaNiO_3$ shows strong visible-light absorption with the absorption onset of 545 nm, which indicates the optical band gap of 2.28 eV. Therefore, the prepared $LaNiO_3$ nanoparticles are expected to be excited by the visible-light irradiation, and to result in the photogenerated electron-hole pairs.

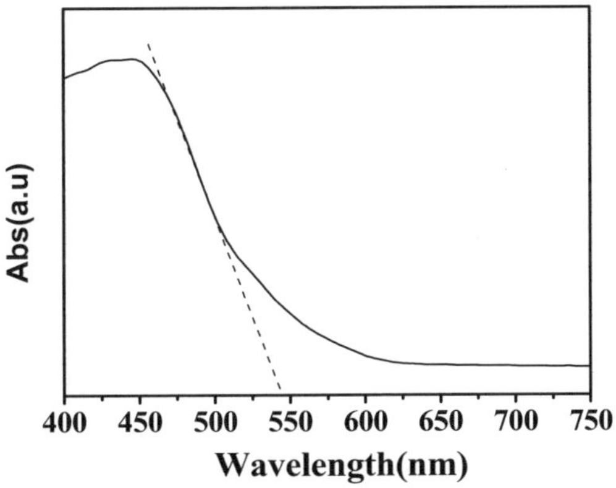

Fig. 3: Diffuse reflectance spectrum of the as-prepared nanocrystalline $LaNiO_3$

Organic dyes exist in the river due to the release from dye industry, which imposes environmental risk to human beings. Therefore, removal of these organic contaminants will benefit the green environment around us, as well as the purification of water sources. Herein, the photocatalytic activity of the as-prepared $LaNiO_3$ was investigated by decomposition of MO under visible-light, and the commercial P25 photocatalyst (Degussa, Germany) was used for the control sample. As seen from Fig. 4, MO is very stable under visible-light, and there was no observable decomposition of MO even after 90 min visible-light exposure. After visible-light exposure for 90 min, MO was decomposed by 20 % in the presence of P25. Furthermore, in the case of $LaNiO_3$, MO was decomposited by 70% within 90 min, which is obviously higher active than P25. As shown Fig. 3, $LaNiO_3$ has a narrow optical band gap and strong visible-light absorption. The high visible-light photocatalytic activity $LaNiO_3$ is mainly attributed to the strong visible-light absorption. Specifically, the visible-light irradiation can excite the $LaNiO_3$, and result in the efficient separation of electron-hole pair, which contributes to the oxidation decomposition of H_2O and organic compounds.

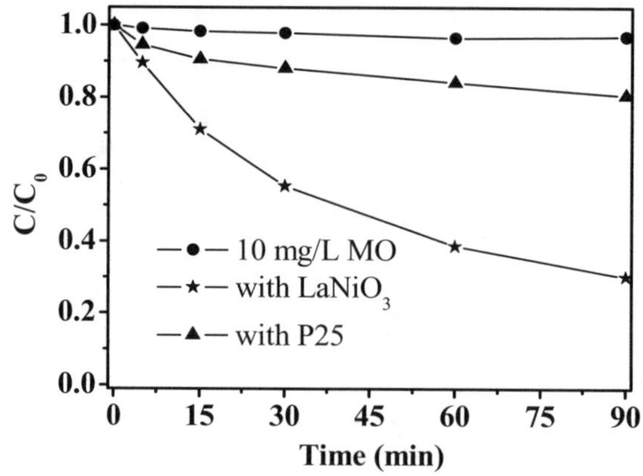

Fig. 4: Plot of photocatalytic decomposition efficiency of methyl orange as a function of irradiation time in the presence of nanocrystalline $LaNiO_3$

Conclusions

The single phase perovskite $LaNiO_3$ was prepared using $La(NO_3)_3{\cdot}6H_2O$ and $Ni(NO_3)_2{\cdot}6H_2O$ as starting materials via a sol-gel process. The optical absorption property and the visible-light photocatalytic activity were investigated. It was found that $LaNiO_3$ has a narrow band gap of 2.28 eV and strong visible-light absorption. Furthermore, $LaNiO_3$ shows high visible-light photocatalytic activity for decomposition of methyl orange. To sum up, $LaNiO_3$ is expected to be one kind of promising photocatalytic materials under solar irradiation.

Acknowledgements

The authors acknowledge the financial supports from Zhejiang Province Natural Science Foundation of China (No.Y4100471 and Y4110641), Zhejiang Province Education Department Project of China (No.Y200903866 and ZC200805662), Science and Technology Planning Project of Zhejiang Province (2008F70042).

References

[1] M. R. Hoffmann, S. T. Martin, W. Choi, and D. W. Bahnemann: *Environmental Applications of Semiconductor Photocatalysis*. Chem. Rev. Vol. 95 (1995), pp. 69-96.

[2] Y. H. Ng, I. V. Lightcap, K. Goodwin, M. Matsumura, and P. V. Kamat: *To What Extent Do Graphene Scaffolds Improve the Photovoltaic and Photocatalytic Response of* TiO_2 *Nanostructured Films*. J. Phys. Chem. Lett. 1, 15 (2010), pp. 2222-2227.

[3] S. Sakthivel and H. Kisch: *Daylight photocatalysis by carbon -modified titanium dioxide.* Angew. Chem. Int. Ed. Vol. 42(2003), pp. 4908-4911.

[4] J. Gole, J. Stout, C. Burda, Y. Lou and X. Chen: *Highly Efficient Formation of Visible Light Tunable* $TiO_{2-x}N_x$ *Photocatalysts and Their Transformation at the Nanoscale.* J. Phys. Chem. B, 108, 4 (2004), pp. 1230-1240.

[5] J. Tang, Z. Zou, M. Katagiri, T. Kako, J. Ye: *Photocatalytic degradation of MB on* MIn_2O_4 *(M = alkali earthmetal) under visible light: effects of crystal and electronic structure on the photocatalytic activity.* Catal. Today Vol. 93–95(2004), pp. 885-889.

[6] J. Yu, Y. Zhang, A. Kudo: *Synthesis and photocatalytic performances of $BiVO_4$ by ammonia co-precipitation process*. J. Solid State Chem. 182, 2(2009), pp. 223-228.

[7] P. Tang, H. Chen, F. Cao, G. Pan, K. Wang, M. Xu, Y. Tong: *Nanoparticulate SnS as an efficient photocatalyst under visible-light irradiation.* Mater. Lett. Vol. 65(2011), pp. 450-452.

[8] X. Zhao, D. Chen, X. Lou, Q. Cheng, and Y. Lu: *Study on Photocatalytic Property of Perovskite Composite Oxide $LaNiO_3$*. J. Henan Nor. Univ. 33, 2(2005), pp. 69-72.

[9] J. Zhu, N. Zhang, J. Zhong, S. Tong: *Preparation of La-Ni mixed oxides by solid phase method and their photocatalytic activity.* Industr. Catal. 14, 3(2006) , pp. 48-51.

[10] Y. Li , S. Yao, W. Wen, L. Xue, Y.Yan: *Sol-gel Combustion Synthesis and Visible-Light Driven Photocatalytic Property of Perovskite $LaNiO_3$*. J. Alloys Compd. Vol. 491 (2010) , pp. 560-564.

[11] A. Zachariah, K. V. Baiju, S. Shukla, K. S. Deepa, J. James, and K. G. K. Warrier: *Synergistic Effect in Photocatalysis As Observed for Mixed-Phase Nanocrystalline Titania Processed via Sol-gel Solvent Mixing and Calcination.* J. Phys. Chem. C, Vol. 112(2008), pp. 11345-11356.

Advanced Materials Research Vol. 279 (2011) pp 88-92
© (2011) Trans Tech Publications, Switzerland
doi:10.4028/www.scientific.net/AMR.279.88

The field emission of globe-like diamond microcrystalline aggregate films

Jin-hai Gao [1,a], Wu-qing Zhang[1,b] and Zhen Li[1,c]

[1]Department of Physics, Zhengzhou Normal University, Zhengzhou 450044, China

[a]gjhzjgxx@sina.com, [b]zhangwuqin@126.com, [c]lizhenzhengyi@163.com

Keywords: The diamond microcrystalline aggregates, Field emission, Chemical vapor deposition

Abstract: The globe-like diamond microcrystalline aggregates films were fabricated by microwave plasma chemical vapor deposition method. The field emission properties and emission stability of the films were tested using a diode structure in vacuum. It was found that the globe-like diamond microcrystalline aggregates films exhibited good electron emission properties and stability. The turn-on field of 0. 55 V /μm and the current density of 11mA/cm^2 at the electric fields of 2.73V/μm were obtained. At the successive operator circles, the turn-on field tends to stabilize at 1. 08V /μm and the current density of 6.6 mA/cm^2 is obtained.

Introduction

Field emission displays have emerged as a leading front runner for low-power panel application because of very thin profile, high production efficiency, fast response, high brightness, wide operating temperature range, excellent viewing angles, excellent picture quality at lower cost and possible expansion of size[1]. Carbon based materials such as diamond[2], diamond-like carbon[3], amorphous carbon (a-C)[4], and carbon nanotubes (CNTs)[5] were widely investigated to be cold cathode because of their low or negative electron affinity. However, there were a limited number of reports which deal with big current density and stability of the field emission of globe-like diamond microcrystalline aggregate films, which were grown in the high methane concentration (H2/CH4=100/10 SCCM) using a microwave plasma chemical vapor deposition (MPCVD) system in our laboratory. The characteristic of field emission of the globe-like diamond microcrystalline aggregate films is reported in this paper.

Experimental

The ceramic wafer was used as the substrate. For getting a good cohesion between carbon film and ceramic substrate, Ti film was sputtered onto the substrate as a buffer layer (the thickness is about 200 nm). The globe-like diamond microcrystalline aggregate films were deposited on the Ti buffer layer using MPCVD system. Prior to the carbon films growth, the polished substrates were ultrasonically rinsed in acetone and de-ionized water. The source gas was a mixture of H2 and CH4 with flow rates of 100 and 10 SCCM, respectively. During the deposition process, the total pressure of 6.0 kPa, the substrate temperature of 800 oC, and the microwave power of 1700 W were kept for 150 mins. The samples were characterized using both Raman spectroscopy and x-ray diffraction spectrum (XRD). Raman spectrometer equipped with a Renishaw, Renishaw-2000 spectrometer and a YAG laser. The Raman scattering was obtained by the excitation of 514.5 nm wavelength of laser beam. The XRD pattern was measured using RANalytical, X'Pert PRO spectrometer with X-ray source of Cu target. Scanning electron microscopy (SEM) was utilized to image surface topography, which was a field emission scanning electron microscope of model JSM-6700F/INCA-ENERGY and operated at voltage of 10 kV. The macroscopic field-electron-emission current-voltage curve was measured by diode structure in a vacuum of below 5×10-5 Pa. The distance between the sample cathode and a phosphor-coated indium-tin oxide glass anode was about 275μm.

Results and discussion

Fig. 1(a) and (b) show the SEM photograph of the fabricated globe-like diamond microcrystalline aggregates in different amplification. The density of the globe-like diamond microcrystalline aggregates is about 10^5 /cm^2, and the average size of globe-like diamond aggregate is 10μm, which contained plenty of well-shaped faceted diamond .

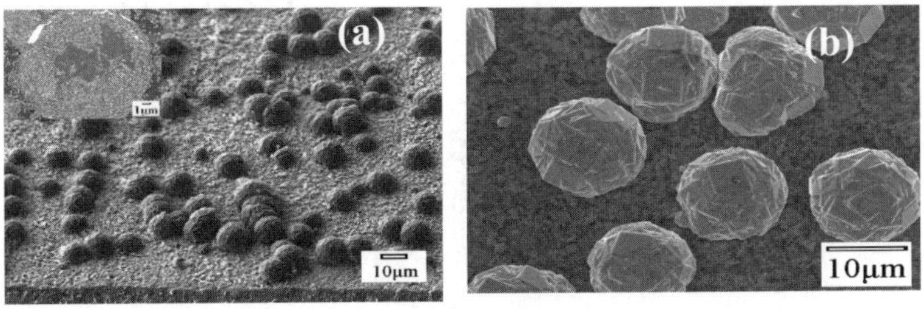

Fig. 1: SEM photograph of the fabricated globe-like diamond microcrystalline aggregates: (a) side face; (b) face. The inset in Fig. 1(a) indicates amplification of a globe-like diamond microcrystalline aggregate.

The Raman spectrums of the globe-like microcrystalline diamond aggregates and substrate surface between the aggregates are shown in Fig. 2(a) and Fig. 2(b), respectively. The peak at 1332.1 cm-1, which is shown in Fig. 2(a), is due to diamond grains Raman scattering and declare that the globe-like microcrystalline diamond aggregates are composed of diamond crystal. Fig. 2(b) is typical Raman spectrum of the a-C[6], which indicates that the globe-like microcrystalline diamond aggregates grow up on the a-C surface formed on the deposition process.

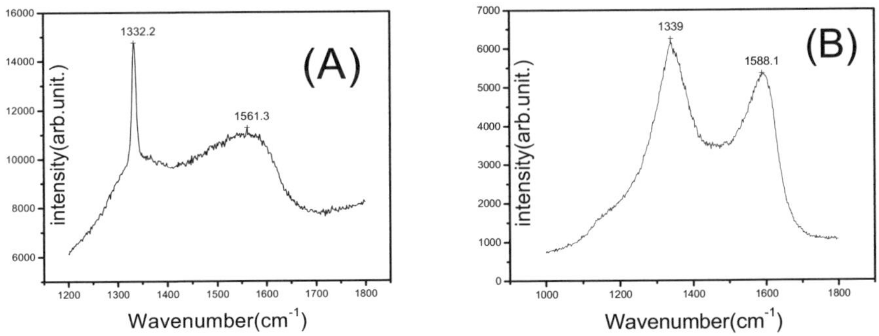

Fig. 2: Raman spectrums of the carbon films: (a) globe-like microcrystalline diamond aggregates; (b) between the aggregates.

A peak at 2θ of 440 corresponding to diamond d-value of 2.0536nm has been observed in XRD spectrum of the sample, which also leads to the conclusion that the obtained film consists of crystalline diamond structure.

The field emission properties of the globe-like microcrystalline diamond aggregates have been tested. In the first operation, the turn-on field is a very low value of 0.55V/μm and the current density reaches 11mA/cm2 at the electric fields of 2.73V/μm (curve 1). At the successive operation circles, the turn-on field tends to stabilize at 1.08V/μm and the current density reaches 6.6mA/cm2 at the electric fields of 2.73V/μm (curve 3) shown in Fig. 3. The globe-like microcrystalline diamond aggregates exhibit good electron field emission and field emission stability after few operation circles. The inset in Fig. 3 is the Fowler–Nordheim (F-N) plots corresponding to the curves in Fig. 3.

These plots of the first circle electric field and the third electric field operation in the electric field are almost straight line, which indicates that the mechanism of the electron emission from the diamond microcrystalline aggregate is linked to the tunnel effect.

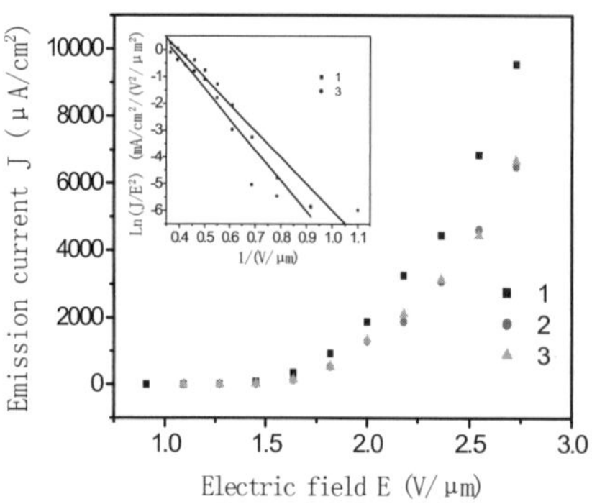

Fig. 3: J-E curves and the corresponding FN plot of the globe-like microcrystalline diamond aggregates

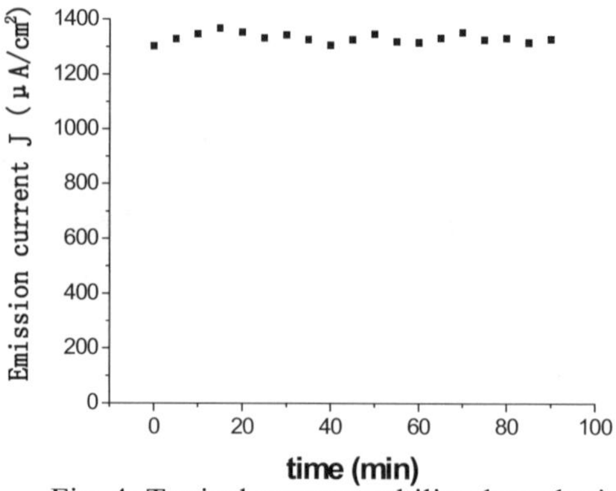

Fig. 4: Typical current stability data obtained at electric field 2.2V/μm.

The experimental results of the field emission for diamond films, which were reported by some authors in recent years, are shown in Table 1. The field emission of the globe-like microcrystalline diamond aggregates have superior to other type of diamond. The globe-like diamond microcrystalline aggregates exhibiting good electron field emission would be attributed to possible mechanism of the diamond edge emission [7] and good electron channels from substrate to surface by the grain boundary [8].

Table 1: Field emission for the six samples of diamond in recent years.

author	fabricate	turn-on	current density	Reference
Ying-Chieh Chen	MPECVD porous alumina patterned nanodiamond films	10 V/μm	3.8 mA/cm² (37 V/μm)	Diamond & Related Materials 15 (2010) 324
Yu-Fen Tzeng	MPECVD ultrananocrystalline diamond films	6.4 V/μm	6mA/cm² (13.6V/μm)	Diamond & Related Materials online 25 October 2007
F. Y. Xie	dip-coating technique nanodiamond films	>2 V/μm	1mA/cm² (7.4 V/μm)	J. Vac. Sci. Technol. B 26 (1) Jan/Feb 2008
J.C. Madaleno	MPCVD nanocrystalline diamond films	5.5 V/μm	10mA/cm² (11 V/μm)	APPLIED PHYSICS LETTERS 92 (2008) 023113

It was known that the surfaces of all kinds of carbon films grown by CVD are terminated by hydrogen. The desorption of hydrogen may result in the increase of surface work function of thc film and then increase the barrier to field emission[9], so that the field emission current becomes smaller and smaller untilled stable. After the two operation circles, the field emission current of the globe-like microcrystalline diamond aggregates is stable. The tested stability of the field emission is shown in Fig. 4. The current stability of the samples was monitored continuously for a period of 90mins at 1.35mA/cm2 under electric field 2.2V/μm. The area of the field emission is $0.2926 cm^2$. The fluctuation in emission current for samples is within 2.6%. The electron emission stability is superior to the CNT [10] and a-C [11].

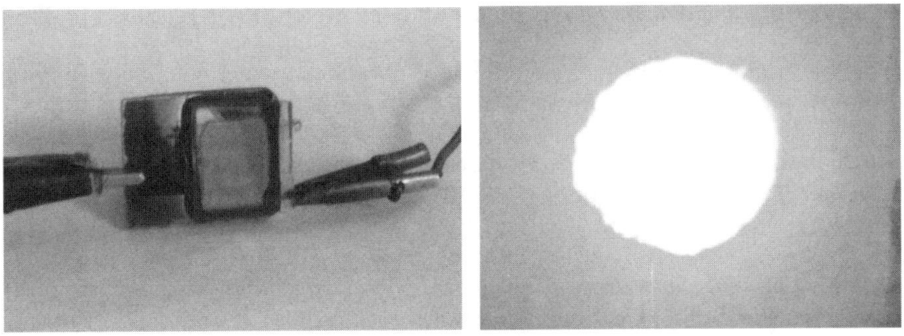

Fig. 5: The globe-like diamond microcrystalline aggregates cathode fluorescent device of (a), the fluorescent image of (b).

A fluorescent device with the globe-like diamond microcrystalline aggregates cathode was fabricated shown in Fig. 5(a). The fluorescent image within a radius of 0.5cm is shown in Fig. 5(b). The fluorescent intensity of the device is more than 104cd/m2.

Summary

We have successfully synthesized the globe-like diamond microcrystalline aggregates using a MPCVD system. The globe-like diamond microcrystalline aggregates exhibit good electron emission properties and stability. The good emission properties of the diamond aggregates are considered to be combined results of diamond grains and the lots of the grain boundaries in the diamonds microcrystalline aggregates. The stability would be possible attributed to the stable diamond character.

Acknowledgement

This work was supported by a fund of Science and Technology Foundation of Education Ministry of China (No.205091).

References

[1] Teplier F, et al.: *Fabrication of high performance low temperature poly-silicon backplanes on metal foil for flexible active-matrix organic light emission diode displays.* Thin Solid Films, 515, 19(2007), pp. 7428-7432.

[2] Jinhai Gao, Limin Zhao, et al.: *Field emission properties of the globe-like diamond microcrystalline aggregate films grown by MPCVD.* Physica B: Condensed Matter, 405, 1(2010), pp. 318-321.

[3] S. G. Wang, Qing Zhang, et al.: *Electron field emission enhancement effects of nano-diamond films.* Surface and Coatings Technology, 167, 2-3(2003), pp.143-147.

[4] Chung S J et al.: *Stability of electron emission current in hydrogen-free diamond.* J. Appl. Phys. 82, 8(1997), pp.4047-4050.

[5] Jing Lv, Yongxiang Zhao and Hailong Zhao: *Influence of pretreatment conditions of Ti film on field emission properties of carbon nanotubes films.* Synthetic Metals, 160, 19-20(2003), pp. 2051-2054.

[6] C.L. Dong, et al.: *Effect of surface treatments on the electronic properties of ultra-nanocrystalline diamond films.* Diamond Relat. Mater., 17, 7-10(2008), pp.1150-1153.

[7] D. Pradhan et al.: *Low temperature growth of ultrananocrystalline diamond film and its field emission properties.* Diamond Relat. Mater., 15, 11-12(2006), pp. 2001-2005.

[8] B. K. Choi, et al. : *Thermionic electron emission from chemical vapor deposition diamond by nanosecond laser heating.* Journal of Vacuum Science & Technology B, 27, 2(2009), pp.557-561.

[9] E. Ozturk, et al.: *The dependence of the intersubband transitions in square and graded QWs on intense laser field.* Solid State Communications, 132, 7(2004), pp.497-502.

[10] Bin Zhao, et al: *Tunable field emission properties of carbon nanotube arrays by engineering Fe catalyst.* Materials Letters, 63, 29(2009), pp. 2556-2559.

[11] Qi Yang, et al.: *Field-effect transistor based on a combination of nanometer film and undoped semiconductor* . Applied Surface Science, 253, 8(2007), pp.3927-3929.

Advanced Materials Research Vol. 279 (2011) pp 93-96
doi:10.4028/www.scientific.net/AMR.279.93

Synthesis and Gas Sensitivity of SnO_2 nanoparticles

NI Zifeng[1, a], CHEN Guomei[2, b], YING Pengzhan[3, c]

[1]School of Mechanical Engineering, Jiangnan University, Wuxi, Jiangsu, China, 214122

[2]Wuxi Yongda Sewage Treatment Co., Ltd. Wuxi, Jiangsu, China, 214122

[3]School of Materials Science and Engineering, China University of Mining and Technology, Xuzhou, Jiangsu, China, 221008

[a]nizf@jiangnan.edu.cn, [b]cgm0208@163.com, [c]ypz3889@cumt.edu.cn

Keywords: Tin dioxide (SnO_2), Nanoparticles, Gas sensitivity

Abstract. Tin dioxide (SnO_2) nanoparticles have been synthesized in bulk quantity by thermal evaporation of SnO powder. The x-ray powder diffraction (XRD) analysis indicates that the nanoparticles are the tetragonal rutile structure of SnO_2. Scanning electron microscopy (SEM) analysis shows that the size of the synthesized SnO_2 nanoparticles is relatively homogeneous with diameter of about 10 nm. Gas-sensing components have been manufactured with the SiO_2 nanoparticles. Their performances indicate that it has high sensitivity and selectivity to LPG, and the max sensitivity appears at 280 °C, compared with C_2H_5OH, H_2, CO and CH_4.

Introduction

In the past decades, the semiconductor sensors have become important analytical devices for detecting combustible and toxic gases in domestic and industrial environment [1]. Tin dioxide (SnO_2), which is a typical n-type oxide semiconducting materials, has been studied and used very widely for gas sensing owing to its high sensitivity, low coat, and quick response [2]. SnO_2 gas sensors can detect reducing gases based on a reaction between the semiconductor and the gas ambient which produces conductance changes in the conductivity of the semiconductor [3]. The gas sensing and other properties of the SnO_2 materials are affected by many factors. One such important factor is the microstructure of the SnO_2 materials such as their size and shape [4].

The development of nanotechnology provides excellent sensitive materials for sensors. In recent years, nanocrystalline SnO_2 particles have been used to improve the gas-sensing properties, because the nanocrystalline SnO_2 particles have smaller size, higher surface-to-volume ratio, larger proportion of grain boundaries on the crystal surface compared with conventional materials. These may possibly affect the interaction between oxygen and the SnO_2 surface, inducing changes in the resistivity of the SnO_2 [5]. SnO_2 nanoparticles can be synthesized by using different methods, such as: sol-gel[6], hydrothermal[7], hydrolytic[8], polymeric precursor[9], ion implantation[10], carbothermal reduction[11], thermal evaporation[12] and surfactant mediated[13].

In this paper, we report the preparation and configuration of the SnO_2 nanoparticles. The gas sensitivity of the component made by SnO_2 nanoparticles is also studied.

Experimental Details

The SnO_2 nanoparticles were prepared by thermal evaporation of SnO powder (2g, 200 mesh, 99.9% purity). Firstly, the pure SnO powder was put into an alumina boat. This boat was then placed in the middle of a quartz tube and inserted into a horizontal tube furnace. The reaction chamber was heated to 800°C rapidly (about 20 min) from room temperature and held at this temperature for 2h, and then cooled to room temperature. During the experiment, Ar was kept flowing through the tube at a rate of 80 sccm (standard cubic centimeters per minute), and the flow pressure in the alumina tube was kept at about 1 atm. After reaction, it would be observed that white powder-like products have been formed in high yield on the walls of the alumina boat.

The morphology, size and crystal structure of the synthesized products were characterized by using x-ray powder diffraction (XRD) (Philips PW 1710 with Cu K_α radiation), scanning electron microscopy (SEM, JEOL JSM-6300). The gas sensitivity measurement was taken by the gas sensitivity testing system (HW-30A).

Results and discussion

The XRD pattern (Fig. 1) from the as-synthesized bulk samples reveals the crystal structure and phase purity of the products. All the diffraction peaks can be indexed to the tetragonal rutile structure of SnO_2, which are consistent with the standard values for bulk SnO_2. No impurities, such as unreacted SnO, were detected.

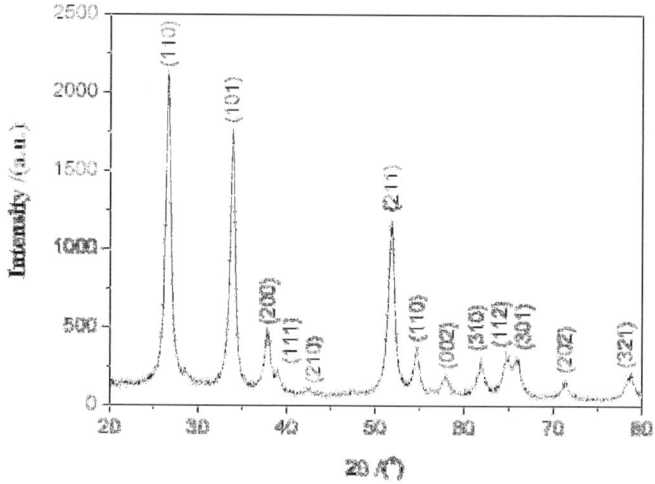

Fig. 1: The XRD pattern recorded from the synthesized SiO_2 nanoparticles

A low-magnification SEM image (Fig. 2 (a)) reveals the as-synthesized products which consist of a large quantity of nanoparticles. Detailed SEM observation shows that the size was relatively homogeneous with diameter of about 10 nm (Fig. 2 (a)).

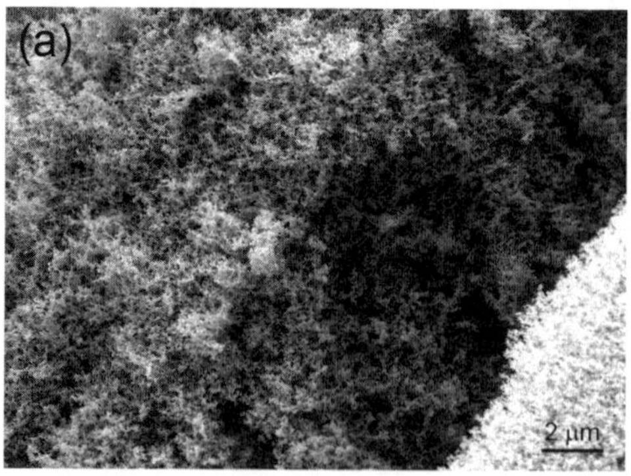

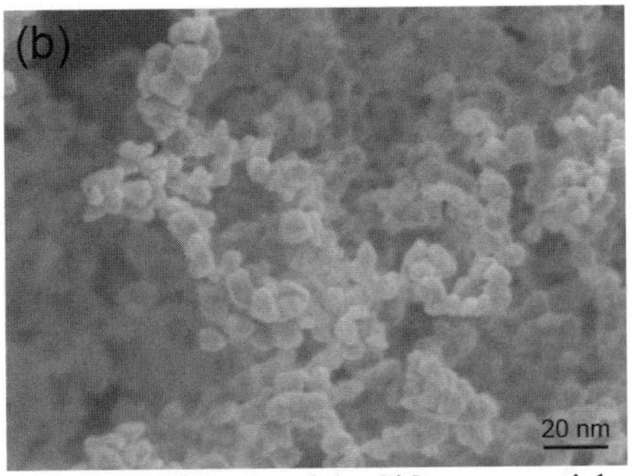

Fig. 2: SEM images of the SiO_2 nanoparticles

Gas-sensing components have been manufactured with prepared SnO_2 nanoparticles. Fig.3 shows the relation between gas sensitivity and working temperature of the components to different gases. The gas concentration used in this test is 1.0×10^{-3}. The component is more sensitive to LPG, compared with C_2H_5OH, H_2, CO and CH_4, and the max sensitivity appears at 280 °C.

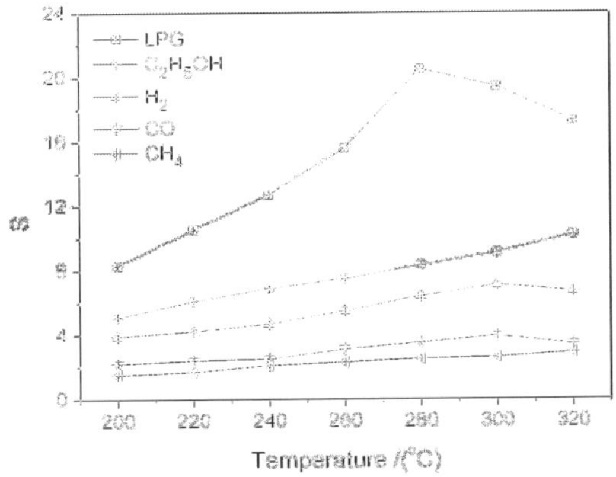

Fig. 3: The relationship between sensitivity and temperature of the SiO_2 nanoparticles

Fig.4 shows the relation between gas sensitivity and gas concentration of the components at 280 °C. The value of gas sensitivity increases with the increase of gas concentration, but different gas has different range. The component shows a high selectivity to LPG compared with other gases.

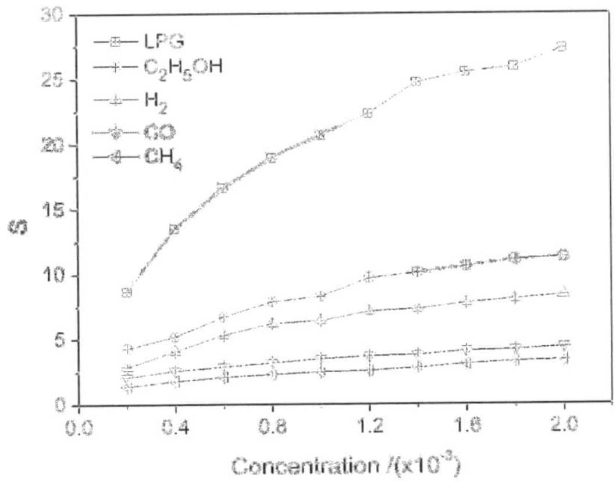

Fig. 4: The relationship between gas sensitivity and gas concentration of the SiO_2 nanoparticles

Conclusions

Tin dioxide (SnO_2) nanoparticles have been synthesized by thermal evaporation of SnO powder. The morphology, size, crystal structure and gas sensitivity of the synthesized SnO_2 nanoparticles were investigated. It was found that:

(1) The size of the synthesized SnO_2 nanoparticles was relatively homogeneous with diameter of about 10 nm.

(2) Gas-sensing components have been manufactured with the SiO_2 nanoparticles. Their performances indicate that it has high sensitivity and selectivity to LPG compared with C_2H_5OH, H_2, CO and CH_4.

Acknowledgements

This research work was supported by the Fundamental Research Funds for the Central Universities (JUSRP11115).

References

[1] Y. Shimizu, E.D. Bartolomeo, E. Traversa, et al.: *Effect of surface modification on No.2 sensing properties of SnO_2 varistor-type sensors*. Sens. Actuators B, Vol. 60 (1999), p. 118.

[2] M. Nitta, M. Haradome: *Semiconductor gas sensors*. Journal of Applied physics, Vol. 48 (1979), p. 977.

[3] W. Liu, X. Cao, Y. Zhu, et al.: *The effect of dopants on the electronic structure of SnO_2 thin film*. Actuators B Chem. Vol. 66 (2000), p. 219.

[4] S.G. Ansari, P. Boroojerdian, S.R. Sainkar, et al.: *Grain size effects on H_2 gas sensitivity of thick film resistor using SnO_2 nanoparticles*. Thin Solid Films, Vol. 295 (1997), p. 271.

[5] R. Ramamoorthy, M.K. Kennedy, H. Nienhaus, et al.: *Surface oxidation of monodisperse SnO_x nanoparticles*. Sensors and Actuators B, Vol. 88 (2003), p. 281.

[6] J.R. Zhang, L. Gao: *Synthesis and characterization of nanocrystalline tin oxide by sol-gel method*. J. Solid State Chem. Vol. 177 (2004), p. 1425.

[7] H.C. Chiu, C.S. Yeh: *Hydrothermal synthesis of SnO_2 nanoparticles and their gas-sensing of alcohol*. J. Phys. Chem. C, Vol. 111(2007), p.7256.

[8] Z.X. Deng, C. Wang, Y.D. Li: *New hydrolytic process for producing zirconium dioxide, tin dioxide, and titanium dioxide nanoparticles*. J. Am. Ceram. Soc. Vol. 85 (2002), p. 2837.

[9] E.R. Leite, I.T Weber, E. Longo, et al.: *A new method to control particles size and particles size distribution of SnO_2 nanoparticles for gas sensor applications*. Adv. Mater. Vol. 12(2000), p. 965.

[10] M.A. Taglienta, V. Bello, G. Pellegrini, et al.: *Synthesis and characterization of SnO_2 nanoparticles embedded in silica by ion implantation*. Nuclear Instruments and Methods in Physics Research B, Vol. 268(2010), p. 3063.

[11] E.R. Leite, J.W. Gomez, M.M. Oliveira, et al.: *Synthesis of SnO_2 nanoribbons by a carbothermal reduction process*. J. Nanosci. Nanotechnol. Vol. 2(2002), p. 125.

[12] S.H. Luo, Q. Wan, W.L. Liu, et al.: *Photoluminescence properties of SnO_2 nanowhiskers grown by thermal evaporation*. Prog. Solid State Chem. Vol. 33(2005), p. 287.

[13] J. Pal, P. Chauhan: *Structural and optical characterization of tin dioxide nanoparticles prepared by a surfactant mediated method*. Mater. Charact. Vol. 60(2009), p. 1512.

Advanced Materials Research Vol. 279 (2011) pp 97-100
© (2011) Trans Tech Publications, Switzerland
doi:10.4028/www.scientific.net/AMR.279.97

Preparation and Properties of Cement Bricks Based Paraffin for Thermal Energy Storage by Phase Change

Haifeng Chen[a], Peisong Tang[b], Feng Cao[c] and Minhong Xu[d]

School of Life Science, Huzhou Teachers College, Zhejiang, China, 313000

[a]headder@hutc.zj.cn, [b]tangps@hutc.zj.cn, [c]caofeng@hutc.zj.cn, [d]minhongxu@hutc.zj.cn

Keywords: expanded perlite, paraffin, phase change material, cement bricks

Abstract. Using paraffin as phase change material, expanded perlite (EP) as porous materials, the EP/paraffin was prepared by vacuum impregnation treatment and tested by scanning electron microscopy (SEM). Adding other building materials, cement bricks based paraffin (CBBP) for thermal energy storage were molded and tested by means of water absorption, compressive strength and actual working of thermal energy storage. The testing results showed that paraffin was absorbed into the holes and cracks of EP, and CBBP had not linear change of various properties versus amount of cement or paraffin. Its mechanical strength was full compliance with national standard GB 21144-2007-T. At last, by phase change material exothermic or endothermic, CBBP effectively regulated outdoor surface temperature in summer: in temperature rising stage, CBBP was 4.63 °C lower than general cement bricks (GCB, no paraffin); in the cooling stage, CBBP was 2.30 °C higher than GCB.

Introduction

Phase Change Materials (PCMs) used in building materials could reduce the indoor temperature fluctuations due to environmental temperature because of its high-thermal energy storage and isothermal behavior[1-3], while reducing building energy consumption. Thus, many researchers had focused on finding and researching the PCMs which had the high latent heat, stable performance, low cost and easy preparation. Currently used PCMs had inorganic and organic categories. Inorganic PCMs in general not only had corrosive, and they had the shortcomings of supercooling and phase separation in the phase transition process; most of the organic PCMs had characterized by poor heat-transfer performance and low storage utilization. In addition, cost and technology were important reasons for limitting large-scale promotion of PCMs. Among the discussed PCMs, paraffin had desirable properties, such as high heat-fusion, little or no supercooling, thermal and chemical stability [3].

In this experiment, we used paraffin as phase change material, expanded perlite (EP) as porous materials, to prepare the porous materials filled with the PCM by vacuum impregnation treatment[4]; then adding sand, cement, reinforcing concrete agent and water, cement bricks based paraffin (CBBP) for thermal energy storage were molded under forming press 50 MPa. Most of the current literature focused on the performance of the PCMs or modified PCMs, but the actual performance of the application materials can not be equated to the final products or devices. So the bricks were tested on its performance of water absorption (%), compressive strength (MPa), density (g·cm-3) and energy-saving with testing device [5] in the actual working environment. Through this work, we tried to prepare phase change thermal insulation cement bricks that significantly reduced outdoor surface temperature during the day in summer.

Experimental

Materials. Cement was obtained from Zhengzhou Dengfeng Smelting Materials Co., Ltd. (Henan, China); Paraffin was used as phase change material; Sands (average particle size 20 ±7.0 mm) were provided by Zhejiang Yangtze River Delta Building Materials Co., Ltd.(Zhejiang, China); reinforcing concrete agent was supplied by Shangdong Zhongsen Building Materials Technology Co.,

Ltd.(Shangdong, China); Expanded perlite (average particle size: 2.5 ± 1.0 mm, Ture dengsity: 1257 Kg m^{-3}, Bulk density 70 Kg m^{-3}, Porosity : 0.77; average pore size : 20 ± 10 um) was obtained from Shenzhen Fanghao Industrial Co., Ltd.

Preparation of the Cement Bricks Based Paraffin. Firstly, the expanded perlite filled with paraffin by vacuum impregnation treatment was prepared in an electric furnace; the EP/ paraffin was taken out from the furnace and dried. Secondly, sands, cement and reinforcing concrete agent were mixed for 2 h. Then water was added while continue stirring, and the EP/paraffin was added. Thirdly, after the slurry was fully stirred, the slurry was poured into pre-made mold (20 cm × 10 cm ×5 cm) under forming press 50 MPa. At last, under natural conditions the CBBP were obtained after 2 weeks curing time. In addition, by the same method in accordance with the above, cement bricks with different contents of paraffin and cement were prepared for comparison test, and general cement bricks (GCB) had no paraffin.

Scanningelectronmicroscopy(SEM). SEM analyses were performed at room temperature on a Hitachi S4800 Field-Emission apparatus. Accelerating voltages was 5.0 KV.

Actual working of thermal energy storage. The CBBP and GCB were laid on the square divided into two areas. square surface temperatures and air temperature were recorded by artificial intelligence temperature controller(AT4310, Changzhou Applent Instruments Co., Ltd. Jiangshu, China) which was connected to a computer. Temperature rising stage: detection of temperature from sunrise until the afternoon; cooling stages: detection of temperature from sunset until the evening.

Results and discussion

Macro-and microstructure. Porous structure of EP was an ideal vehicle for the phase change material. As shown in Fig. 1(a), EP inside had many holes which were ranging from 5-30 um, and holes were connected. Paraffin was adsorbed in the holes of different sizes within EP by vacuum impregnation treatment. A large number of small cracks were distributed in the EP, and the cracks had a certain adsorption capacity of paraffin. As shown in Fig. 2(a), amount and size of holes both reduced, indicating that paraffin into the holes and cracks of EP after vacuum impregnation.

Fig. 1: SEM micrographs of expanded perlite by different conditions.(a: before vacuum impregnation; b: after vacuum impregnation)

The effect of paraffin. Taking sands 880 g, EP 500 g, cement 250 g, reinforcing concrete agent 5 g, water 120 g and different amounts of paraffin (90 g, 130 g, 150 g, 170 g, 200 g, 240 g) adsorbed by EP, some CBBP were prepared. As shown in Fig. 2(a), because the absorption of paraffin was a gradual process from the outside, EP's surface adsorption rate was gradually increased; after the internal pores and cracks of EP were filled of paraffin, the adsorption rate was gradually decreased. As shown in Fig. 2(b), adding the organic paraffin was negatively affected on the mechanical performance of cement bricks[3]. But while the adsorption of paraffin on EP reached a certain value (the amounts of paraffin > 150 g), the compressive strength of CBBP was gradually increased. This was consistent with the phenomenon of the decline in water absoption and the basic stability of density, as shown in Fig. 2(a).

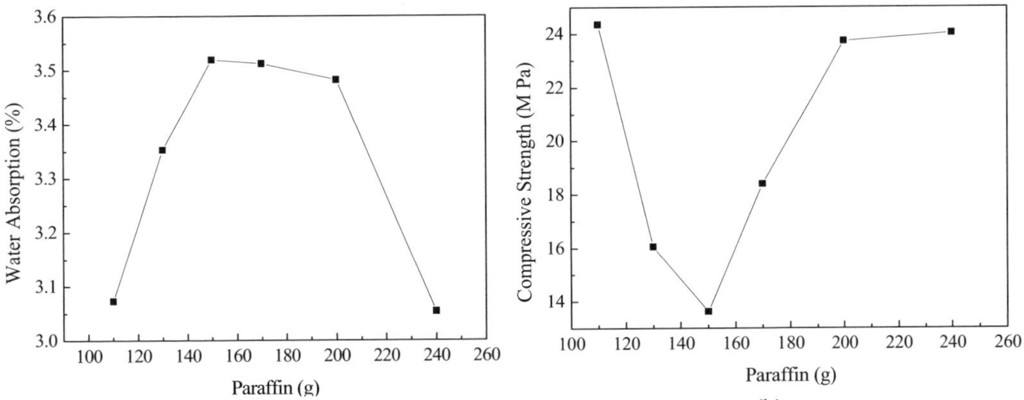

Fig. 2: The effect of different amounts paraffin on the properties of CBBP.(a: water absoption; b: compressive strength)

The effect of cement. Taking sands 880 g, EP 500 g, paraffin 250 g, reinforcing concrete agent 5 g, water 120 g and different amounts of cement (150 g, 200 g, 250 g, 300 g, 350 g, 400 g), some CBBP were prepared. Fig. 3(a) showed the water absoption of CBBP changes with the amount of cement. As coagulant, cement played the role of network connectivity of various components in the bricks. Taking into account the cost of cement, water absorption was not a major factor. The effect of cement was more important on the mechanical properties, as shown in Fig. 3(b). Fig. 3(b) showed 250 g of the cement content is an important critical point. When CBBP had 300 g cement, the value of compressive strength was about 3 times than 250 g.

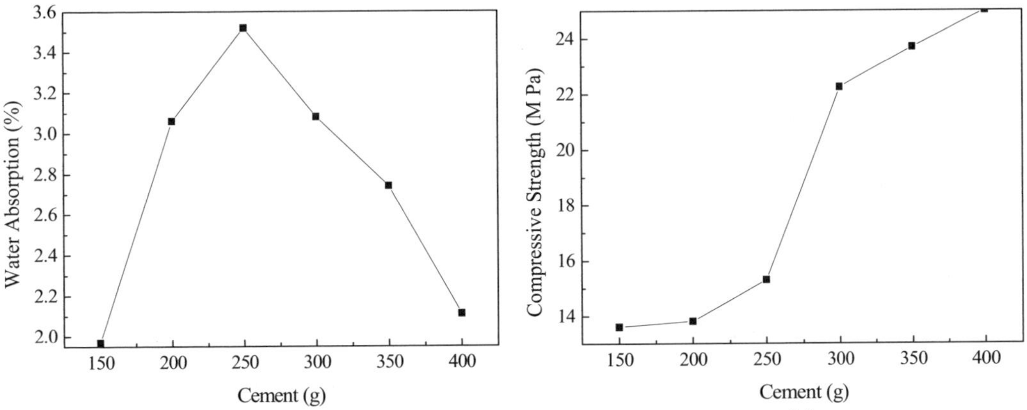

Fig. 3: The effect of different amounts cement on the properties of CBBP.(a: water absoption; b: compressive strength)

Actual working of thermal energy storage. Fig. 4 showed that three curves of environment (curve a), general cement bricks (curve b) and CBBP (curve c) on temperature in the day (1) and evening (2). As shown in Fig. 4(1), while reaching phase transition point of paraffin CBBP absorbed heat in temperature rising stage, CBBP had a lower heating rate than GCB, and there was temperature difference $\Delta T_H = 4.63$ °C. As shown in Fig. 4(2), when the temperature began to drop and is below the phase transition point, CBBP released energy in cooling stages, and it was temperature difference $\Delta T_L = 2.30$ °C. ΔT_H and ΔT_L could be used to measure the phase change thermal energy storage effect: the greater the temperature difference, the better thermal energy storage.

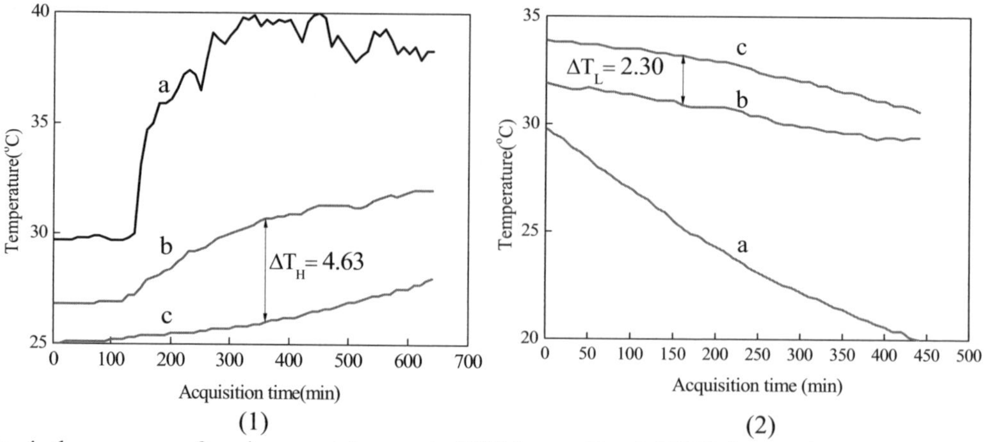

Fig. 4: three curves of environment (curve a), GCB (curve b)and CBBP (curve c) on temperature in the day (1) and evening (2).

Conclusion

Paraffin was absorbed into the holes and cracks of expanded perlite by vacuum impregnation treatment. The feasibility of this approach was better, but it could not prevent leaking of paraffin. In short, adsorption rate of paraffin needs to be improved. Paraffin and cement have much effect on CBBP. The formula was reasonable adjusted on actual situation, including mass of paraffin > 150 g and mass of cement > 250 g. By phase change material exothermic or endothermic, CBBP could regulate outdoor surface temperature in summer.

Acknowledgements

This work was financially supported by Science and Technology Planning Project of Zhejiang Province (2008F70042), the Zhejiang Natural Science Foundation (NoY4100471& Y4110641), Zhejiang Municipal Education Commission (Y200908366) and Huzhou Natural Science Foundation (2010YZ01).

References

[1]P. Zhang, Y. Hua, L. Song, J.X. Ni, W.Y. Xing and J. Wang: *Effect of expanded graphite on properties of high-density polyethylene/paraffin composite with intumescent flame retardant as a shape-stabilized phase changematerial*. Sol. Energy Mater. Sol. Cells. Vol. 94(2010), pp.360-365

[2]L. Bayés-García, L. Ventolà, R. Cordobilla, R. Benages, T. Calvet, M.A. Cuevas-Diarte: *Phase Change Materials(PCM) microcapsules with different shell compositions: Preparation, characterization and thermal stability*. Sol. Energy Mater. Sol. Cells. Vol. 94 (2010), pp.1235-1240.

[3]Atul Sharma, V.V. Tyagi, C.R. Chen, D. Buddhi: *Review on thermal energy storage with phase change materials and applications, Renewable and Sustainable Energy Reviews*. Renewable and Sustainable Energy Reviews. Vol. 13 (2009), pp. 318-345.

[4]Takahiro Nomura, Noriyuki Okinaka and Tomohiro Akiyama: *Impregnation of porous material with phase change material for thermal energy storage*. Mater. Chem. Phys. Vol. 115 (2009), pp.846-850.

[5]Z.Y. Wang, X.T. Lin, F. Chen, X.S. Wang, H. Ruan, X.P. Fan and G.D. Qian: *Preparation and Evaluation of phase change energy storage building material*. New Building Materials. Vol. 11 (2006), pp. 35-37 (In Chinese).

Advanced Materials Research Vol. 279 (2011) pp 101-105
© (2011) Trans Tech Publications, Switzerland
doi:10.4028/www.scientific.net/AMR.279.101

Chemical Composition and Structure of the Boride Inclusions in Synthetic Diamonds Grown under HPHT

Jianhong Gong[1,2,a], Shuxia Lin[1,b] and Jun Gao[1,c]

[1]School of Mechanical, Electrical & Information Engineering of Shandong University at weihai, Weihai, 264209

[2]Post-doctoral station, Mechanical Engineering of Shandong University, Jinan, 250061

[a]gongjh@sdu.edu.cn, [b]linshuxia@sdu.edu.cn, [c]shdgj@sdu.edu.cn

Key words: diamond, boride, structure, chemical composition, HPHT

Abstract. Boride inclusions in the synthetic diamond single crystals grown from Fe-Ni-C-B system under high temperature and high pressure were studied in the present paper. Both chemical composition and structure of the inclusions incorporated into the diamond during the process of diamond growth were successfully determined by transmission electron microscopy (TEM) and scanning electron microscopy (SEM). It was found that the inclusions related to boron trapped in the diamond consisted of f.c.c. $(FeNi)_{23}(CB)_6$, $(FeNi)_3(CB)$, Fe_2B, FeB, hexagonal Ni_3B, and B_4C.

Introduction

Boron-doped diamond can be considered the ideal covalent semiconducting material because of its large bandgap, high electron and hole mobilities [1-5] as well as its exceptional thermal conductivity [1], which make diamond a suitable material for use in high-power high-temperature and high-speed electronic devices serving in extreme conditions. So it is significant to explore the boron compounds in diamond. However, earlier studies about boron impurities are almost on the diamond film grown by chemical vapor deposition (CVD) method [6-8]. This paper pays attention to the boron inclusions in the diamond single crystals grown under HPHT from Fe-Ni-C-B solvent catalyst, which are cheap and convenient for industry production. The chemical composition and microstructure of some borides incorporated in the diamond were studied by transmission electron microscopy (TEM) in present paper.

Experiments

The diamond single crystals were synthesized under a high temperature of 1570 K and a high pressure of 5.5 GPa for 12 min in a Fe-Ni-B catalyst–graphite system. The HPHT-grown diamonds were treated in a boiling mixture of sulfuric acid and nitric acid for 1 h to remove the metallic and graphite inclusions on the surface of diamonds. An HPHT as-grown diamond of about 0.8 mm in diameter was first processed by electrolytic etching, then stress was applied to the diamond, after that the diamond was etched again. The HPHT as-grown diamonds were directly thinned by argon–ion beam milling machine until they were suitable for TEM observation. For the studies, a Philips CM-30 transmission electron microscope was used to study dislocations in the HPHT as-grown diamond with an operating voltage of 100 kV. And the energy dispersive X-ray spectrometer (EDS) system was used in combination with selected area electron diffraction (SAD) to examine the composition and the crystal structure of the inclusions in the synthetic diamond grown from the Fe-Ni-C-B system.

Results and Discussions

The diamond crystal synthesized in the presence of an iron–nickel-boron alloy catalyst under high temperature and high pressure had size of 0.6 mm dimensions. The shape of diamond crystals was imperfect hexakisooctahedron and their color was strong blue and opaque. Fig. 1(a) illustrates a diamond crystal with faulted hexakisooctahedron shape. Fig. 1(b) shows the electron diffraction pattern from the [001] zone axis of a diamond, which indicates the hexagonal crystal structure.

Fig. 2(a) shows a TEM image of a kind of metallic inclusion related to boron. The Fe and Ni elements can be detected from the EDS analysis of the metallic inclusion. Further identification by corresponding selected area electron diffraction (SAD) pattern from the [731] zone axis of the metallic inclusion, as shown in Fig. 2(b) indicates that the metallic inclusion is f.c.c. $(Fe, Ni)_{23}(CB)_6$. The formation of $(Fe, Ni)_{23}C_6$ has been reported in [9]. It is inclined to think that $(Fe, Ni)_7C_3$ formed under a pressure ranging from 4.7 to 6GPa during the process of the diamond crystal growth , when iron and nickel reacted with graphite trapped in the diamond, firstly $(Fe, Ni)_7C_3$ appears as a metastable phase (900-1050℃), then it would transform into a more stable phase $(Fe, Ni)_{23}C_6$ and $(Fe, Ni)_3C$ as the temperature decreased [10].

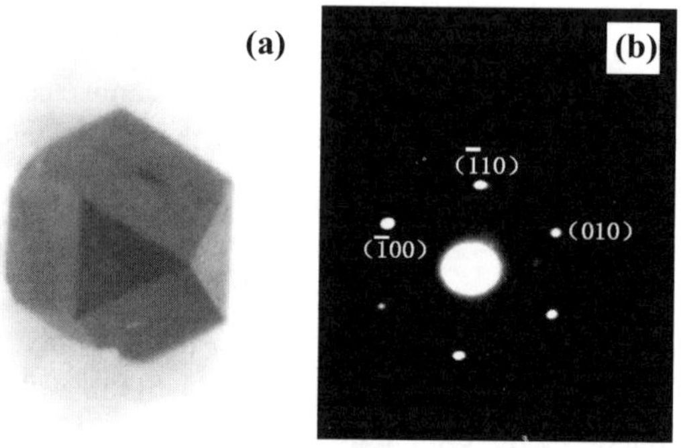

(a) SEMimage of the diamond; (b) SAD pattern from the zone axis of [001]

Fig. 1: A synthetic diamond crystal

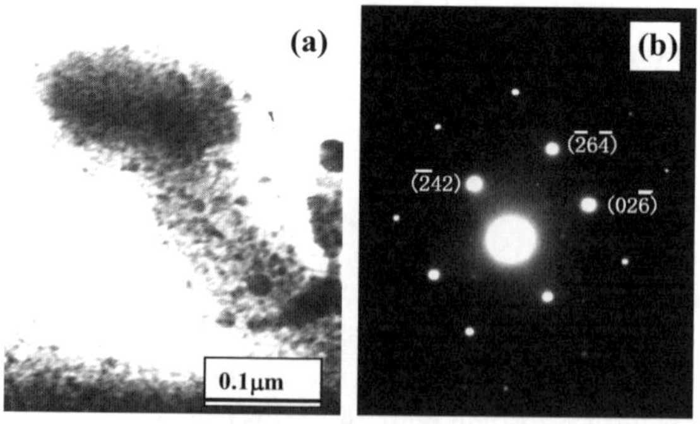

(a) TEM image of $(FeNi)_{23}(CB)_6$; (b) Corresponding SAD pattern from zone axis of [731]

Fig. 2: F.c.c. $(FeNi)_{23}(CB)_6$ inclusion

Fig. 3 illustrates the TEM image of polycrystalline $(Fe, Ni)_3(C,B)$ in the diamond. EDS analysis of this phase suggests the existence of iron and nickel. Fig. 3(b) is the corresponding SAD ring, which indicates that the $(Fe, Ni)_3(C,B)$ is orthorhombic. Pavel ever confirmed the Fe_3C presence as main phase of metallic inclusions in diamond crystals grown in Fe-C system [11] and $(FeNi)_3C$ presence in synthetic diamond crystals grown in Fe-Ni-C systems [12]. The microstructure, characterization, thermal and magnetic properties of the $(Fe, Ni)_3C$ were studied using scanning electron microscopy, X-ray diffraction, magnetic measurement, hardness data, MÖssbauer and reflection optical microscopy [12-14]. For similar atom radius, boron atoms are easy to take place of carbon atoms of carbide and then (Fe, $Ni)_{23}(C,B)_6$ and (Fe, $Ni)_3(C,B)$ are formed. They are remained as inclusions in the diamond crystals during the course of sharp quenching process.

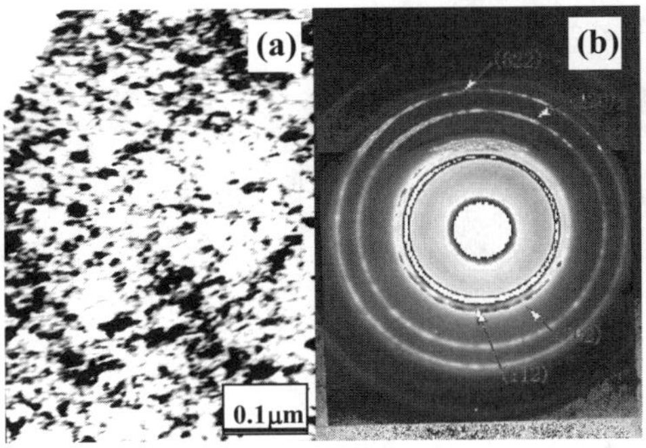

(a) TEM image of $(FeNi)_3$ (CB) (b) Corresponding SAD pattern
Fig. 3: Polycrystalline $(FeNi)_3$ (CB) inclusion

Other inclusions boride trapped in the diamond are shown in Fig. 4-6. Fig. 4 illustrates the TEM image of (Fe, Ni)B. EDS analysis demonstrates the existence of iron and nickel, and further studies of the corresponding SAD pattern of the inclusion from the [211] zone axis indicated that this inclusion is the orthorhombic (Fe, Ni)B phase. Fig. 5(a) is the TEM image of body-centered tetragonal Fe_2B, which is identified by the corresponding electron diffraction pattern from [031] zone axis in Fig. 5(b). The TEM image of Ni_3B is illustrated in Fig. 6(a), and Fig. 6(b) is corresponding electron diffraction pattern of [201] zone axis. These borides are not found in the catalyst before synthesis, which could be formed during the diamond growth under the condition of high temperature and high pressure. The driving force for this inclusion to be incorporated into growing diamond is chemical absorption.

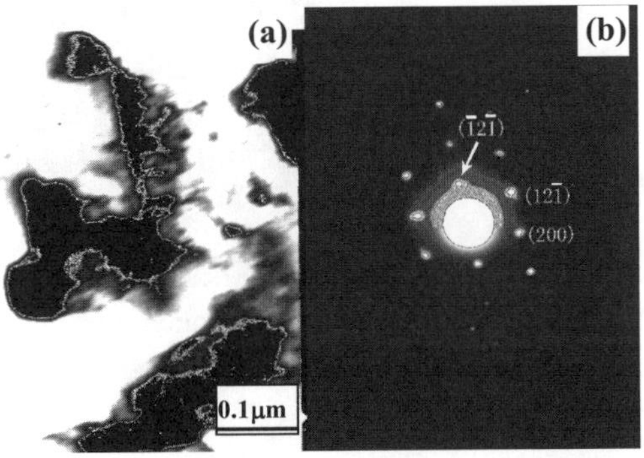

(a) TEM image; (b) Corresponding SAD pattern from the zone axis of [012]
Fig. 4: Orthorhombic (Fe,Ni)B inclusion

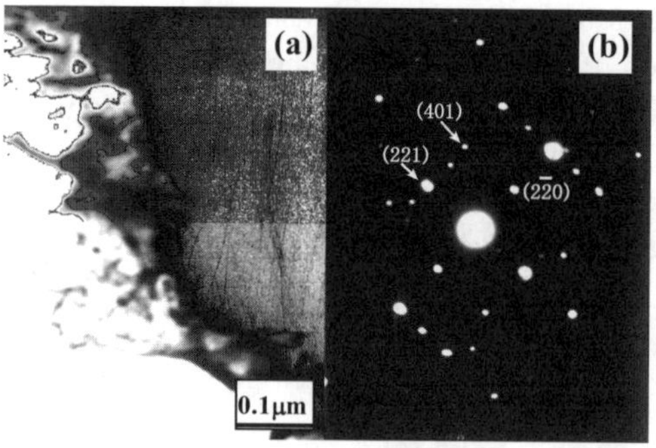

Fig. 5: Tetragonal Fe_2B inclusion (a) TEM image of Fe_2B; (b) Corresponding SAD pattern from the zone axis of [031].

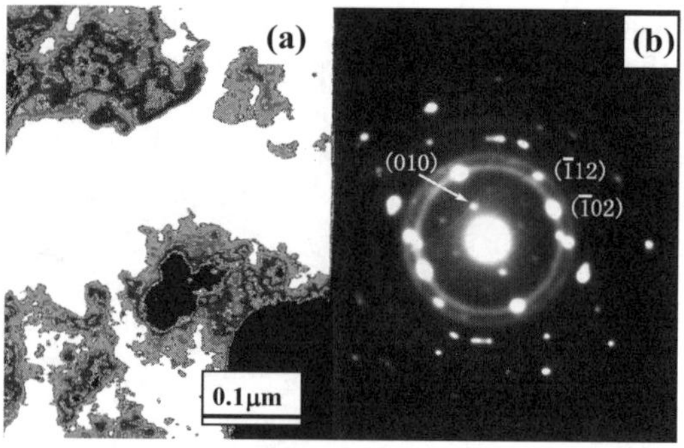

Fig. 6: Orthorhombic Ni_3B inclusion (a) TEM image of Ni_3B; (b) Corresponding SAD pattern from the zone axis of [201].

Fig. 7 shows boron carbide contained in the diamond crystal. It can be identified from the corresponding electron diffraction pattern of [211] zone axis illustrated in Fig. 7(b). The boron atoms diffused into diamond from molten alloy catalyst. And boron diffusion into diamond lattice was studied in [15]. Boron concentration fluctuated under the influence of temperature and pressure and has more concentration than carbon atoms at some local area. Interparticle collision and coalescence of carbon and boron atoms resulted in the formation of B_4C. Within the temperature range 1000-1100°C Fe-B_4C would decompose into a mixture of $Fe_3(C, B)$ crystals and graphite with relatively high microhardness in the place of initial particles of B_4C according to the constitutional diagram of the Fe-B-C [16].

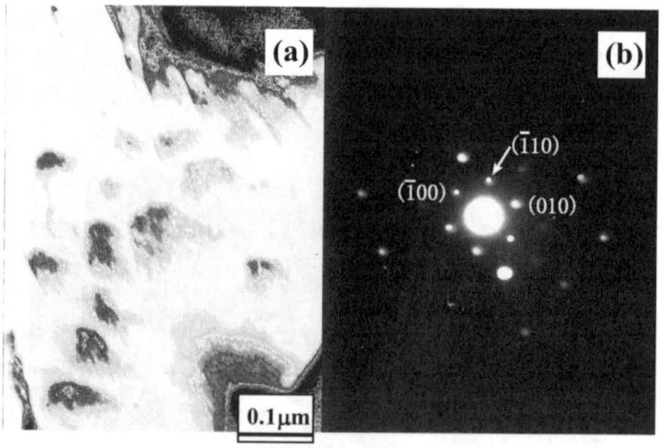

(a) TEM image of B_4C; (b) Corresponding SAD pattern from the zone axis of [211]

Fig. 7: Hexagonal B_4C inclusion

Summary

The diamond crystal was synthesized under a high pressure of 5.3 GPa and a high temperature of 1570 K in the presence of an iron–nickel-boron alloy catalyst. During the process of diamond crystal growth, boron atoms diffused into the diamond lattice with iron and nickel atoms and reacted to form carbides containing boron and boride. TEM and SEM were used to examine both the chemical composition and structure of the inclusions borides incorporated into the diamond successfully and directly. It was shown that the inclusions borides trapped in the diamond consisted of f.c.c. (FeNi)23(CB)6, orthorhombic $(FeNi)_3$ (CB), orthorhombic (Fe, Ni)B, tetragonal Fe_2B, orthorhombic Ni_3B and hexagonal B_4C. It is consumed that the boron could exist as boride in the diamond.

References

[1] Wang M, Li H S, Li M S, et al.: *Effect of boron contained in the catalyst on thermal stability of boron-doped diamond single crystals*. Chinese Journal of High Pressure Physics, 22, 2(2008), pp. 215-219.

[2] Barjon J, Habka N, Mer C, et al.: *Resistivity of boron doped diamond*. Physica status Solidi-rapid Research Letters, 3, 6(2009), pp. 202-204.

[3] Soltani A, Talbi A, Mortet V, et al.: *Diamond and Cubic Boron Nitride: Properties, Growth and Applications*. 2010 Wide Bandgap Cubic Semiconductors: From Growth to Devices, Vol. 1292 (2010), pp.191-196.

[4] Luong JHT, Male KB, Glennon JD: *Boron-doped diamond electrode: synthesis, characterization, functionalization and analytical applications*. Analyst, Vol 134 (2009), pp. 1965-1979.

[5] Achard, J; Silva, F; Issaoui, R, et al.: *Thick boron doped diamond single crystals for high power electronics*. 4th International Conference on New Diamond and Nano Carbons, 2010 Suzhou, China.

[6] Li HD, Zhang T, Li LA, et al.: *Investigation on crystalline structure, boron distribution, and residual stresses in freestanding boron-doped CVD diamond films*. Journal of crystal growth, Vol 12 (2010), pp. 1986-1991.

[7] Show, Yoshiyuki; Matsukawa, Toshikazu; Ito, Hirokazu: *Structural changes in CVD diamond film by boron and nitrogen doping*. Diamond and Related Materials, 9, 3(2000), pp.337-340.

[8] Muramatsu Y, Iihara J, Takebe T, et al.: *Chemical analysis of impurity boron atoms in diamond using soft X-ray emission spectroscopy*. Analytical Sciences, Vol 24 (2008), pp. 831-834.

[9] Yin, Long-Wei, Zou, Zeng-Da, Li, Mu-Sen: *Some inclusions and defects in a synthetic diamond single crystal*. Journal of Crystal Growth, 218, 2(2000), pp.455-458.

[10] Y.A. Kocherzhinskii, O.G. Kulik, V.Z. Turkevich: *Phase equilibria in the Fe-Ni-C and Fe-Co-C systems under high temperatures and high pressures*. High Temp.-High Press. 25 (1993), pp.113.

[11] Pavel, E. Baluta, Gh. Barb, D: *Nature of the metrallic inclusions in syntetic diamond crystals synthesized at approximately 5.5 Gpa in Fe-C system*. Solid State Communications, 76, 4(1990), pp. 531-533.

[12] Pavel, E.; Baluta, Gh.; Barb, D.: *$(Fe_{1-X}Me_x)_3C$ carbide identification in synthetic diamond crystals produced at high pressure and high temperature in Fe-Me-C systems, where Me=Ni, Co*. Materials Letters, 10, 1-2(1990), pp. 62-65.

[13] Pavel, E.; Baluta, Gh.; Barb, D.: *Complex carbides in synthetic diamond crystals produced at approx 5.5 Gpa*. Journal of Materials Science, 28, 6(1993), pp. 1645-1647.

[14] Pavel, E.; Baluta, Gh.; Giurqiu, C.: *Invar property of the metallic inclusions in synthetic diamond crystals grown in the Fe-Ni-C system*. Materials Science & Engineering A: Structural Materials: Properties, Microstructure and Processing, A150, 1(1992), pp. L9-L10.

[15] Sung, T.; Popovici, G.; Prelas, M.A.: *Boron diffusion coefficient in diamond*, Diamonds for Electronic Applications. Nov 27-Dec 1 (1995).

[16] Turov, Yu.V.; Khusid, B.M.; Voroshnin, L.G.: *Structure formation under sintering of powder iron-boron carbide composite*. Poroshkovaya Metallurgiya, n6(1991), pp. 25-31.

Advanced Materials Research Vol. 279 (2011) pp 106-110
© (2011) Trans Tech Publications, Switzerland
doi:10.4028/www.scientific.net/AMR.279.106

Preparation and properties of recycled polypropylene /carbon nanotube composites

Gao Jing Long[1, a], Liu Yan Hui[1, b] and Li Dong Ming[1, c]

[1]College of Material Science and Engineering, Shenyang Ligong University, Shenyang 110168, China

[a]xflgjl@126.com, [b]liouyh@126.com, [c]6333787@qq.com

Keywords: carbon nanotubes, CNTs-reinforced composites, recycled polypropylene

Abstract. Recycled polypropylene (PP)/carbon nanotube (CNTs) composites with different CNTs fraction were prepared by the melting blend method. The effects of CNTs content on the thermal properties and mechanical properties were mainly investigated. The results show that the thermal degradation of the composites shifts towards higher temperatures as the concentration of CNTs is increased. With increasing CNTs content, tensile strength and elongation at break increase firstly and then decrease. When CNTs content is 3 %, tensile strength and elongation at break are 34.71 and 27.00, respectively. Moreover, a unique tensile rupture characteristic was found by SEM observations, which explained the critical broken theory of the PP/CNTs composites.

Introduction

Nanotechnology is assumed to be one of the key technologies in the next century. Especially, the development and improvement of nanocomposites come to the fore. Engineering of some properties can be best accomplished through combining materials at smaller and smaller size scales. With the need for increasingly demanding combinations of material properties, nanocomposites have become an interest area for possible solutions. Nanocomposites are composed of a matrix material and fibers or particles in the nanometer size range. Varying the combinations and architectures of these materials affects the material properties on the nanoscales and macroscales. Thermal, electrical, magnetic, mechanical and tribological properties can be engineered for specific applications[1].

In recent years, nanometer-scale reinforcing filling particles such as clay silicates, calcium carbonates, silicon dioxide and carbon nanotube(CNTs) have attracted much great interest in the field of polymer composite materials[2-4]. Among these filling nanoparticles, CNTs are an especially attractive class of fillers for polymers due to their intriguing mechanical and thermal properties. They represent one of the strongest and toughest materials as known. CNTs long-radius ratio is about 100-1000, and tensile strength is about 45×10^9 Pa in comparison to steel with 2×10^9 Pa[5]. However, CNTs density is only 120-1.8 g/cm^3, about one sixth of the steel[6].

Much work related to CNTs has been done on thermoplastic polymers[7–10], thermosetting polymers[11] and elastomers[12,13], which have successfully proven their applicability to various fields of technical application, especially for load-bearing application. Polypropylene(PP) is one of the most commonly used thermoplastic polymers which has been utilized in automobiles, appliances, construction, house wares, etc. PP have been compounded with carbon nanotube to prepare composites from different point of views and use different methods[14-16]. However, the work done on recycled plastic/CNTs systems is still limited. Nowadays, the disposal of waste plastic materials is giving rise to increase environmental and economic problems around the world. Among these plastic wastes, PP represents the second largest volume segment of the plastic wastes after polyethylene. Thus, used plastics are becoming a potential worldwide source of raw materials.

In this work, recycled polypropylene (RPE) /CNTs composites were prepared by the melting blend method. Influence of CNTs on the mechanical properties of the PP was studied. Microscopic structure of composites and thermal properties were analyzed using scanning electron microscope (SEM) and thermogravimetry (TG), respectively.

Experimental

Materials. RPP pellets were obtained from Yingkou Jiuyun plastic Industries, Ltd. (Yingkou, Liaoning, China). CNTs were from Institute of Metal Research of the Chinese Academy of Sciences. Prior to melt mixing, the CNTs were heated to 450 ℃ for 1 h to eliminate adsorbed species. Absolute ethyl alcohol and acetone were purred from Xinxi Chemical Co. (China).

Preparation of PP/CNTs Composites. Fig. 1 is a schematic showing the manufacturing procedure used for production of the composite material. In steps a and b, CNTs powder was added into the mixture solution of absolute ethyl alcohol and acetone(the ratio of volume,1:1) and was placed under ultrasonic stirring for 3 h. After ultrasonic stirring, the PP powder was casted into the black suspension and was mechanically stired for 3 h, then filtration. With the aim of adequate mixing of the CNTs and PP, the mixture was additionally ground. After dried, the mixture was then hot pressed and cut into small pellets. In step f, high-shear mixing was used to disperse the CNTs homogenously throughout the PP matrix. The mixing was performed by a XSS-300 torque rheometer at 140 ℃ for 20 min with the speed of 65 rpm.

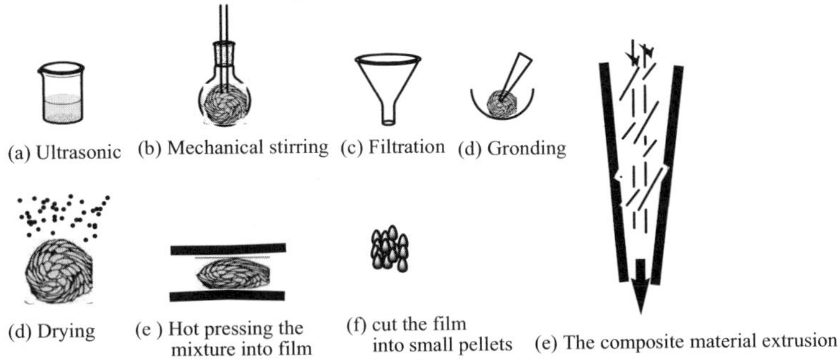

Fig. 1: Procedures for preparing CNTs/PP composite

Characterization. Thermogravimetric analysis was used to study the thermal stability characteristics of resultant composites with a BEIF FRC/T-2 thermogravimetric analyzer, under argon at a scan rate of 10 ℃/min from room temperature to 1000 ℃. Tensile mechanical properties of the composites were measured with a Reger 10kN electronic universal testing machine (Model RGWT, SHENZHEN,China) at room temperature. All the specimens(80×25×1.5mm) were measured for tensile strength according to GB/T 2412-1998. For each treatment level, three replications were conducted. The morphology of the composites was studied by a PhilipXL-20 scanning electron microscope, and the fracture surfaces of the specimens after tensile test were sputter-coated with gold before analysis.

Results and Discussion

Thermal Properties. Fig. 2 shows the temperature dependence of the sample weight, for different loading conentrations of CNTs (ranging from 0% wt. up to 5% wt.CNTs). In order to ensure a good reproducibility of expermental data, all samples had a weight of about 10.0 ± 1.5 mg. To easily observe the general trends and the differences between different TGA data sets, all data sets (excepting the one for pristine polymer) have been shifted upwards along the vertical axis by 5 % relative to the previous data set. The experimental data were not shifted along the horizontal axis. From Fig. 2 it is noticed that the thermal degradation of the composites shifts towards higher temperatures as the content of CNTs is increased. This substantiates the interactions between RPP macromolecular chains and CNTs revealing the formation of an RPP-CNTs interface with enhanced thermal stability. Within the experimental errors, no additional mass loss has been noticed as the temperature of the pp-CNTS composites was raised from 700℃ to 1000 ℃.

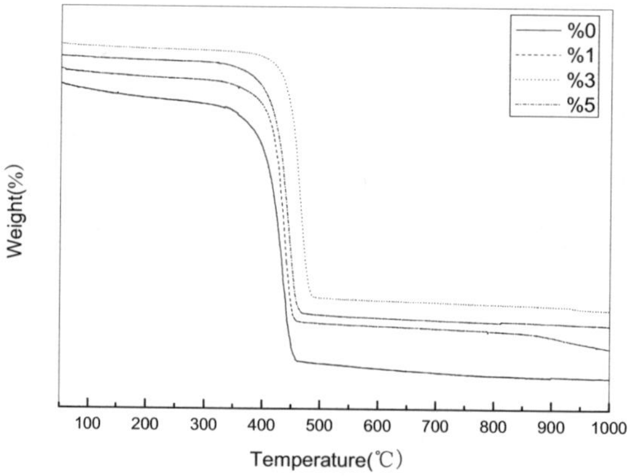

Fig. 2: TGA curves of RPP/CNTs nanocomposites filled with CNTs ranging from 0 % wt. up to 5 % wt.

Mechanical Prosperities. The effects of CNTs content on mechanical properties of composites are listed in Table 1. With increasing CNTs content, tensile strength and elongation at break increase firtly and then decrease. When CNTs content is 3 %, tensile strength and elongation at break increase are 34.71 and 27.00, respectively.

Table 1: Effects of CNTs content on mechanical properties of composites

CNTS content [%]	Tensile strength [Mpa]	Elongation at break [%]
0	18.61	13.00
1	28.59	14.50
3	34.71	27.00
5	23.25	16.00

The improvement in the tensile properties arises from several factors. Firstly, CNTs promote the crystallization of PP and the crystallites strengthen the composites. Secondly, CNTs are well dispersed in the matrix, allowing a more uniform load distribution. Thirdly, interaction between the PP chains and CNTs enable a more efficient load transfer from the matrix to the nanotubes. It was well understood that for fibre-reinforced polymer composites, fibres could toughen polymer matrix through several energy-consuming events, such as fibres fracture, fibres pullout and fibre bridging. In the present work, similar features of carbon nanotubes were also observed. Fig. 3 illustrates a high resolution SEM image of the fracture surfaces of the specimens (wt.3 %) after tensile test, where nanotube fracture (A) and pullout (B) could be recognized. As shown in the SEM micrograph Fig. 3, the nanotubes were broken and being pulled out upon tensile fracture. To discern the nanotube fracture from the nanotube pullout mainly lay on the lengths of left nanotubes, which protrude from the fracture surface. As seen in Fig. 3, a majority of the left nanotubes have the lengths of several hundred nanometers. This value seems to be close to the critical fracture length of nanotubes, ranging from several tens to several hundreds nanometers.

Based on the above discussion, it is noted that for composites containing CNTs, there is a downturn in mechanical properties when the effective CNTs content was increased to 5.0 wt%. The downturn in mechanical properties can be attributed to the presence of small bundles of CNTs in the RPP-CNTs composite. Similar downturns of the mechanical properties of composites containing polymer-grafted CNTs at higher CNTs loadings have been reported [17].

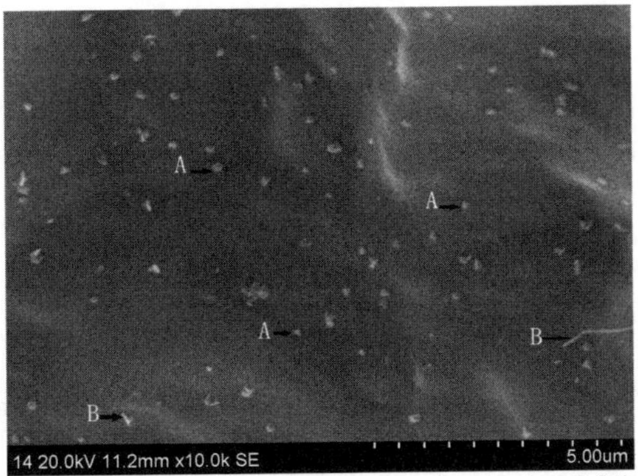

Fig. 3: SEM image of the fracture surfaces of the specimens (wt.3 %) after tensile test

Conclusions

Recycled polypropylene/carbon nanotube composites were prepared by the melting blend method. As a result, these CNTs played an important role of bridging molecular chain–CNTs in PP matrix, resulting in improvements in the thermal and mechanical properties of the RPP/CNTs composites. Nanotube reinforcement enhances the tensile strength and elongation at break when compared to pristine recycled polypropylene. Mechanical properties of the nanocomposites increases with nanotube content and maximum increase was observed for 3 wt.% CNTs filled nanocomposites. Microscopic picture indicates a broken theory of the RPP/CNTs composites.

Acknowledgements

This work was financially supported by the Liaoning Province Office of Education Foundation (2009A621).

References

[1] B.B. Johnson, M.H. Santare and J.E. Novotny: *Wear behavior of Carbon Nanotube/High Density Polyethylene composites* . Mechanics of Materials Vol.41 (2009), pp. 1108.

[2] S.P. Bao, S.C. Tjong: *Mechanical behaviors of polypropylene/carbon nanotube nanocomposites: The effects of loading rate and temperature* . Materials Science and Engineering A Vol. 485 (2008), pp. 508.

[3] W. Bauhofer, J. Z. Kovacs: *A review and analysis of electrical percolation in carbon nanotube polymer composites* . Composites Science and Technology Vol.69 (2009), p. 1486.

[4] Z.W.Jiang , R.J. Song and W.G. Bi: *Polypropylene as a carbon source for the synthesis of multi-walled carbon nanotubes via catalytic combustion* . Carbon Vol.45 (2007), p. 449.

[5] C. Wei, K. Cho and D. Srivastava: *Tensile strength of carbon nanotubes under realistic temperature and strain rate.* Phys. Rev B: Conden. Matter. Mater. Phys Vol. 67(2003), p. 1154.

[6] Y. Mamunya, A. Boudenne and N. Lebovka: *Electrical and thermophysical behaviour of PVC-MWCNT nanocomposites* . Composites Science and Technology Vol.68 (2008), p. 1981.

[7] S. Lee, E. Cho and S. Jeon: *Rheological and electrical properties of polypropylene composites containing functionalized multi-walled carbon nanotubes and compatibilizers* . Carbon Vol.45 (2007), p. 2810.

[8] T. Kashiwagi, J. Fagan and J. F. Douglas: *Relationship between dispersion metric and properties of PMMA/SWNT nanocomposites* . Polymer Vol.48 (2007), p. 4855.

[9] T. Mcnally, P. Pftschke and P. Halley: *Polyethylene multiwalled carbon nanotube composites* . Polymer Vol.46 (2005), p. 8222.

[10] Q. Wang, J. Dai and W. Li: *The effects of CNT alignment on electrical conductivity and mechanical properties of SWNT/epoxy nanocomposites* . Composites Science and Technology Vol.68 (2008), p. 1644.

[11] L. Bokobza: *Multiwall carbon nanotube elastomeric composites: A review* . Polymer Vol.48 (2007), p. 4907.

[12] Y. Li, H.Shimizu: *Toward a Stretchable, Elastic, and Electrically Conductive Nanocomposite: Morphology and Properties of Poly[styrene-b-(ethylene-co-butylene)-b-styrene]/Multiwalled Carbon Nanotube Composites Fabricated by High-Shear Processing*. Macromolecules Vol.42 (2009), p. 2587.

[13] G. Farzi, A. Sohaib and E. Beyou: *Effect of radical grafting of tetramethylpentadecane and polypropylene on carbon nanotubes' dispersibility in various solvents and polypropylene matrix* . Polymer Vol.50 (2009), p. 5901.

[14] S. H. Liao, C. C. Wenga and C. Y. Yena: *Preparation and properties of functionalized multiwalled carbon nanotubes/polypropylene nanocomposite bipolar plates for polymer electrolyte membrane fuel cells* . Journal of Power Sources Vol.195 (2010), p. 263.

[15] I. Alig, D. Lellinger and S. M. Dudkin: *Conductivity spectroscopy on melt processed polypropylene–multiwalled carbon nanotube composites: Recovery after shear and crystallization* . Polymer Vol.48 (2007), p. 1020.

[16] M.Yang, V.Koutsos and M. Zaiser: *Size effect in the tensile fracture of single-walled carbon nanotubes with defects*. Nanotechnology Vol.18 (2007), p. 155708.

Advanced Materials Research Vol. 279 (2011) pp 111-114
© (2011) Trans Tech Publications, Switzerland
doi:10.4028/www.scientific.net/AMR.279.111

Crystal Structure and Resistance-temperature characteristic of the Boron-doped Diamonds Synthesized from Fe-Ni-C-B System

Jianhong Gong[1,2,a], Shuxia Lin[1,b] and Jun Gao[1,c]

[1]School of mechanical, electrical & information engineering of Shandong University at weihai, Weihai, 264209

[2]Post-doctoral station, Mechanical Engineering of Shandong University, Jinan, 250061

[a]gongjh@sdu.edu.cn, [b]linshuxia@sdu.edu.cn, [c]shdgj@sdu.edu.cn

Key words: diamond, boron, resistance, ionization energy

Abstract. In the present paper, boron-doped diamond was synthesized by static pressure method using Fe-Ni-C-B system catalyst, whose resistance-temperature characteristic curve was studied. Experiments results proved that the boron-doped diamond has different ionization energy in different temperature interval and the reasons were analyzed. The maximum operating temperature is about 773K for such boron-doped diamond. The research provides experimental basis for high-temperature semiconductor diamond.

Introduction

Diamond containing boron is a special kind of diamond with many excellent properties such as thermal resistance, high mechanical strength, corrosion resistance and heat dissipation [1-3]. More significantly, it processes some extrcellent electrical properties, such as wide bandgap, high thermal conductivity, high electron saturation velocity and low dielectric constant, which can be used in semiconductor devices, for example, diamond rectifiers, diamond triode transistors, diamond thermometers and so on. It is the most promising electronic materials working under high temperature and poor work environment [4, 5]. Natural boron-doped diamonds are too few to meet the demand of development of modern industry. In this paper, the resistance-temperature characteristic curve of boron-doped diamonds synthesized by Fe-Ni-C-B system catalyst under high temperature and high pressure were studied.

Experimental Methods

The diamond was synthesized by the static pressure catalyst method. As starting materials, the graphite discs (purity 99.9%) were placed alternatively with several discs of Fe-Ni- B alloy. The high-pressure experiment was carried out in a cubic anvil apparatus. This cell assembly was brought up to pressure of 5.3–5.5 GPa by a modified belt-type high pressure apparatus and then heated to a temperature of about 1300°C. After keeping the sample at this temperature for 11 min, the temperature was decreased rapidly by turning off the electric power. The diamond grown in the solvent metal was recovered by dissolving the metals and the residual graphite in a boiling mixture of H_2SO_4 and HNO_3 acids, followed by extensive washing.

Diamond surface morphology was observed by optical microscopy.

Boron-doped diamond was heated by tube furnace, whose resistance-temperature coefficient was measured by EST121 type digital high resistance micro-current measurement. Heating rate was 5 ℃ per minute and the data was recorded every 20 ℃ interval.

Results and Analysises

Fig. 1 is single crystals of normal diamond (Fig. 1a) and boron -doped diamonds (Fig.1b) under the optical microscopy. It shows that the normal diamonds are mostly hexoctahedron, light yellow and transparent. By comparison, the boron-doped diamonds are mostly blue-black octahedron. Studies have proved that high boron concentration would make crystal colors darken , and the boron degree is different on different crystal faces of diamonds [6,7].

Fig. 1: Diamond single crystals a) normal diamonds; (b) Boron-doped diamonds;

The resistivity of the semiconductor is much higher than that of metal. The semiconductor resistivity is usually decreased with increasing temperature; inversely, the metal resistivity increases with warming temperatures. That is, the resistance of semiconductor material has negative temperature coefficient [8].

According to resistivity formula:

$$\rho = \rho_0 e^{\Delta E/2KT} \quad \text{(K: Boltzmann constant)} \tag{1}$$

the resistance and temperature has the relationship:

$$\ln R = \frac{\Delta E}{2KT} + \ln R_0, \tag{2}$$

$$\eta = \frac{\Delta E}{2K} \tag{3}$$

Therefore the ionization energy of semiconductor materials is

$$\Delta E = 2K\eta \tag{4}$$

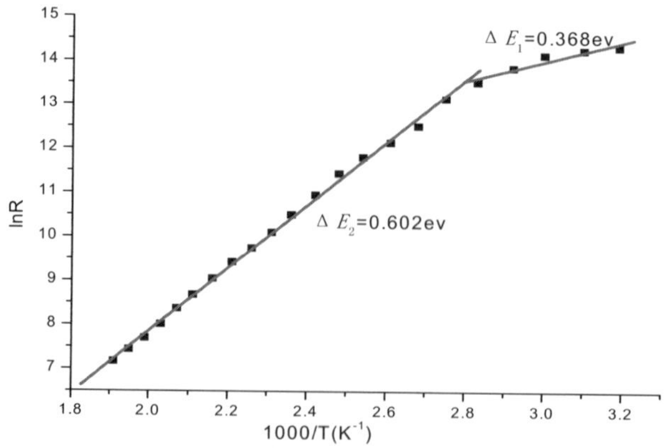

Fig. 2: Plot of the lnR versus 1/T of B-doped diamond

Fig. 2 shows relationship of lnR and 1/T of boron-doped diamond with changing temperature, and the heating range is 293K-570K. As can be seen from the graph, the resistance of boron-doped diamond is approximately $10^5\Omega$ magnitude at room temperature, which was significantly lesser than pure diamond. Resistance is decreased with increasing temperature, which indicates that the diamond has a negative temperature coefficient of resistance, that is, semiconductor behavior.

Fig. 2 also demonstrates two linear regions from 293K to 570K for the resistance-temperature curve: in the lower temperature region (293K ~ 350K), resistance decreases slowly with increasing temperature. Ionization energy of boron-doped diamond in low temperature range can be available as: $\Delta E_1=2\eta_1 K=0.368$ev by curve fitting method of Eq. 4. In this temperature range electrical conductivity is due to boron atoms within the diamond. Boron atoms cause holes by replacing carbon atoms, and acceptor level is formed. So conductance increases are caused by the hole on the shallow acceptor level transiting to the valence band with rising temperature. The experimentally results coincide with study results of boron-doped diamond films by Kiyota (ionization energy for 0.35 ~ 0.39ev[9]).

In higher temperature range (350K ~ 570K), the resistance decreases rapidly with increasing temperature. Depending on linear correlation coefficient of lnR and 1/T and Eq. 4, the ionization energy is $\Delta E_2=2\eta_2 K=0.602$ev.

The reason of higher ionization energy is that more impurity atoms participate in the conduction process with increasing temperature and some impurity atoms are activated to electric conduction, which were not ionized at low temperatures [10]. Additionally, some structural defects of diamond also participate in the electric conduction on continuous levels of diamond central band in the high temperature, which has been verified by elliptical points photometric [11].

To further study the resistance-temperature characteristics of boron-doped diamond, the temperature is heated to 843K and the test curve is shown in Fig. 3.

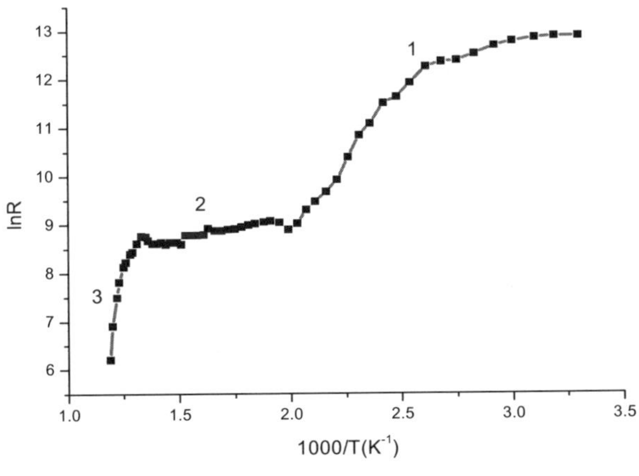

Fig. 3: Plot of the lnR versus 1/T of B-doped diamond

Curve 1 in Fig. 3 indicates that the conductive behavior is similar to that in Fig. 2. Curve 2 shows the resistance is almost constant in range of 523K ~ 773K temperature, and it is because the impurities have ionized almost and the diamond is in semiconductor saturated zone. At this time, lattice vibration scattering becomes the main effect factors on carrier mobility instead of the impurity scattering; namely, carrier concentration is constant essentially with changing temperature. When it is heated to 773K the diamond is in intrinsic ionization zone and the ionization energy is calculated as 3.97ev, which was shown in curve 3 of Fig. 3. The experimental result shows the boron-doped diamond can work at 773K, which is much higher than operating temperature of silicon devices. Therefore, boron-doped diamonds of semiconductor devices made more suitable for high temperature conditions.

Conclusions

(1) The non-boron diamonds are light yellow, while boron-doped diamonds are black blue. The former is a hexoctahedron and the latter is octahedral, and even both types diamond are cubic structure.

(2) In the lower temperature range (293K to 350K), ionization energy of boron-doped diamond is 0.368ev; in higher temperature range (35oK ~ 570K), the ionization energy is 0.602ev; the ionization energy is 3.97ev when the diamond is heated to 773K.

References

[1] Wang Mei, Li H S, Li M S, et al.: *Effect of boron contained in the catalyst on thermal stability of boron-doped diamond single crystals.* Chinese Journal of High Pressure Physics, 22, 2(2008), pp. 215-219.

[2] Gong J H, Li M S, Xu B, et al.: *Influence of boron-doped catalyst on crystal structure and thermal stability of synthetic diamond.* Diamond and Abrasives Engineering. Vol. 1 (2005), pp. 18-21.

[3] Soltani A, Talbi A, Mortet V, et al.: *Diamond and Cubic Boron Nitride: Properties, Growth and Applications.* 2010 Wide Bandgap Cubic Semiconductors: From Growth to Devices, Vol. 1292 (2010), pp.191-196.

[4] Luong JHT, Male KB, Glennon JD: *Boron-doped diamond electrode: synthesis, characterization, functionalization and analytical applications.* Analyst, 134 ,10(2009) pp. 1965-1979.

[5] J. Isberg, J. Hammersberg: *Single crystal diamond for electronic applications.* Diamond and Related Materials, 13, 2 (2004), pp. 320-324

[6] V.D. Blank, M.S. Kuznetsov, S.A. Nosukhin, et al.: *The influence of crystallization temperature and boron concentration in growth environment on its distribution in growth sectors of type IIb diamond.* Diamond & Related Materials, 16 ,4-7(2007), pp. 800-804.

[7] J.Q. Zhang, H.A. Ma, Y.P. Jiang, et al. : *Effects of the additive boron on diamond crystals synthesized in the system of Fe-based alloy and carbon at HPHT.* Diamond & Related Materials, 16, 2(2007), pp. 283-287

[8] Liu E K, Zhu B S, Luo J S: *Semiconductor Physics.* Fourth Edition, National Defence Industry Press, Beijing (1994), pp. 96-102

[9] Kiyota, Hideo; Okushi, Hideyo; Ando, Toshihiro: *Electrical properties of a Schottky barrier formed on a homoepitaxially grown diamond (001) film.* Diamond and Related Materials, 5, 6-8 (1996), pp. 718-722

[10] Deguchi M, Kitabatake M, Hirao T: *Electrical properties of boron-doped diamond films prepared by microwave plasma chemical vapor deposition.* Thin solid films, 281-281, 1-2 (1996), pp. 267-270

[11] Babak Sadigh, Thomas J. Lenosky: *Large enhancement of boron solubility in silicon due to biaxial stress.* Appl. Phys. Lett., 80, 25 (2002), pp. 4738-4740.

Advanced Materials Research Vol. 279 (2011) pp 115-119
© (2011) Trans Tech Publications, Switzerland
doi:10.4028/www.scientific.net/AMR.279.115

Preparation and properties of High-density polyethylene/silica composites

Gao Jing Long[1, a], Liu Yan Hui[1, b] and Wei Shou Dong[1, c]

[1]College of Material Science and Engineering, Shenyang Ligong University, Shenyang 110168, China

[a]xflgjl@126.com, [b]liouyh@126.com, [c]6333787@qq.com

Keywords: organic-inorganic composites, High-density polyethylene, silica, reinforced

Abstract. High-density polyethylene (HDPE)/silica composites with different silicon dioxide fraction were prepared by the melting blend method. The effects of silicon dioxide content on the thermal properties and mechanical properties were mainly investigated. The results show that the thermal degradation of the composites shifts towards higher temperatures as the concentration of silicon dioxide is increased. With increasing silica dioxide content, the tension yield strength of HDPE/silicon composites increased, but its breaking elongation decreased. When silicon dioxide content is 5 %, the tension yield strength and its breaking elongation are 29.33 MPa and 102.82 %, respectively.

Introduction

The research in polymer materials is increasingly focusing on the development of inorganic-polymer composites, which have attracted a great deal of interest from material scientists since their applications have dramatically improved material properties in engineering plastics, enhanced rubber, coatings, and adhesives[1-3]. Compared to pure polymers, polymeric composites are claimed to exhibit markedly improved properties, such as modulus, strength, stiffness, flame retardancy, dimensional stability, electrical conductivity, barrier performance, solvent and heat resistance, wettability and dyeability depending on type and content of nanoparticles used[4,5].

The most commonly used inorganic particles are SiO_2, TiO_2, ZnO, $CaCO_3$, etc. Among particles and these filling inorganic particles, silicon dioxide particles are an especially attractive class of fillers for polymers due to their intriguing properties, such as transparency, surface smoothness, retardancy and cheap. It has been applied in electrical engineering, electronics and consumer goods. If dispersed to the scale of 10–50 nm, it can improve the heat distortion temperature, the modulus and the barrier properties of the matrix. Much work related to silicon dioxide has been done on thermoplastic polymers[6-8], thermosetting polymers[9,10] and elastomers[11-13],which have successfully proved their applicability to various fields of technical application. Many research efforts have been devoted to reduce the high surface energy and the particle interactions. Additionally, filler treatment will lead to a better dispersion due to the more probable interactions of the polymer with the modification than with the inorganic particle surface[3].

High-density polyethylene (HDPE) is the most commonly used thermoplastic polymers. HDPE made by low pressure processed is currently the preferred material for its excellent low temperature resistance and abrasion resistance. However, HDPE has poor mechanical property and low heat distortion temperature, which limits its application to some extent.

The present studies showed that mechanical properties of interface have significant effect on the organic-inorganic composites and their final fracture mechanism. When inorganic filler particles were bonded tightly to the organic polymer matrix obtained composites had excellent comprehensive performance. By exploring the role of interfacial bonding in improving the performances of nanocomposites, this paper focuses on how the compatibilization between filler and matrix affect thermal or mechanical properties of silica-filled polyethylene composites. In order to generate different interfacial interactions, the silicon dioxide particles were treated by chemical grafting with the silane coupling agents. The so formed composites were characterized by FT-IR identify the chemical structure of the obtained material. Thermal properties were investigated by thermogravimetricanalyses (TGA), while mechanical properties were tested with electronic universal testing machine.

Experimental

Materials. HDPE particles were obtained from Liaoyang Petrochemical Company. (Liaoyang, Liaoning, China). Silicon dioxide (industrial-grade, the particle size less 61 μm) used to prepare composites was produced by Tonghua Shuanglong Group Chemical Co., Ltd. Silane coupling agent KH550 was provided by Nanjing Shuguang Chemical Group Co., Ltd. Absolute ethyl alcohol was pured from Xinxi Chemical Co.(china)

Preparation of High-density Polyethylene/silica Composites. The High-density polyethylene/silica composites were prepared by the following process: 10 g silicon dioxide was dispersed in the aqueous solution, in which 20 g of silane coupling agent KH550, 72 mL absolute ethyl alcohol and 8 mL distilled water were added. The suspension was mechanically stirred for 3 h, then filtration. The modified silicon was obtained after the filter cake was dried at 45℃ under vacuum. HDPE and 1,3, 5 phr modified silicon powder were first pre-mixed using a Haibo AXN internal mixer. The manufacturing temperature was kept at 160℃ and the screw speed amounted to 60 rpm for 10 min. After that the blends were compression-moulded into sheet samples (80×5×2mm) at 180±5 ℃ and 100 bar for 5 min in an industrial press.

Characterization. Thermogravimetric analysis was used to study the thermal stability characteristics of resultant composites with a BEIF FRC/T-2 thermogravimetric Analyzer, under argon at a scan rate of 10 ℃/min from room temperature to 800 ℃. Tensile properties of the specimens were tested using a Reger 10kN electronic universal testing machine at a crosshead rate of 100 mm/min at room temperature according to the GB/T 2412-1998 standard. The tensile yield strength and elongation at break could be directly obtained from the stress — strain curves by the provided software. FTIR spectra of the composites was recorded on a microscope FT-IR US Nicolet 380 apparatus in transmittance mode. Thin films, of about 150 mm, were obtained by mould pressing.

Results and Discussion

FT-IR Analysis. Fig. 1 shows FTIR spectra of the HDPE and the HDPE/silica dioxide composites (3 wt. % silicon dioxide). It can be seen that there is a very pronounced band appearing at 1090 cm^{-1}, together with two less pronounced bands at 804 and 471 cm^{-1}, corresponding to the vibration absorption of Si–O– Si groups, indicating that all the samples are composed of silica.

There are peaks at 3600,2880, 1460, 758 cm^{-1} in the LDPE and the composite materials. These peaks are associated with the characteristic vibration of water (OH), carbohydrate (CH), and methyl (CH_3), respectively. These results verify that silica dioxide has been bonded on the HDPE through the polymerization.

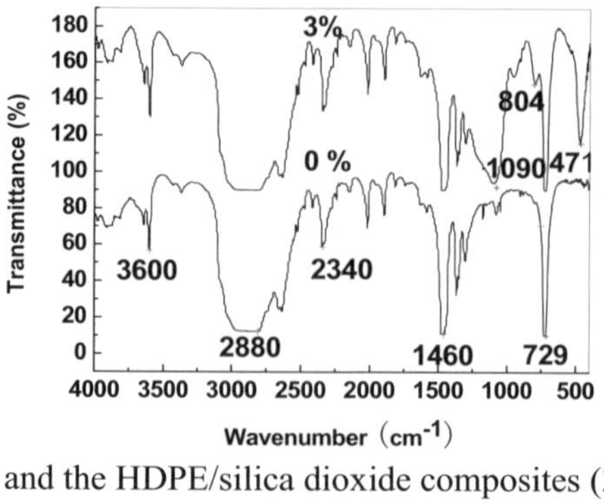

Fig. 1: FT-IR of HDPE and the HDPE/silica dioxide composites (3 wt. % silicon dioxide)

Thermal Properties. Fig. 2 shows the temperature dependence of the sample weight, for different loading conentrations of CNTs (ranging from 0% wt. up to 5% wt.CNTs). In order to ensure a good reproducibility of expermental data all samples had a weight of about 10.0 ± 1.5 mg. To observe the general trends and the differences between different TGA data sets easily, all data sets (excepting the one for pristine polymer) have been shifted upwards along the vertical axis by 5 % relative to the previous data set. The experimental data were not shifted along the horizontal axis. From Fig. 2 it is noticed that the thermal degradation of the composites shifts towards higher temperatures as the content of CNTs is increased. This substantiates the interactions between RPP macromolecular chains and CNTs revealing the formation of an RPP-CNTs interface with enhanced thermal stability. Within the experimental errors, no additional mass loss has been noticed as the temperature of the pp-CNTS composites was raised from 700℃ to 1000℃.

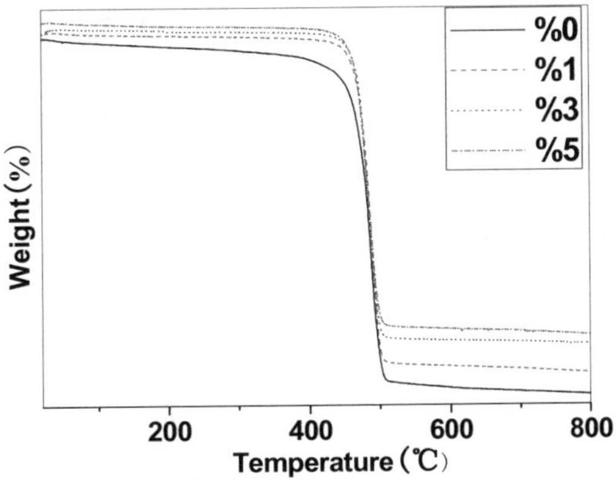

Fig. 2: TGA curves of HDPE/ silica dioxide composites filled with silica dioxide is ranging from 0% wt. up to 5% wt.

Mechanical Properties. The typical tensile stress–strain curves of the composite samples with respect to their silicon dioxide content are shown in Fig. 3, including original molded HDPE for comparison. For HDPE, the typical behavior of a ductile material is observed with a very high elongation at break (1301%). The strain hardening region appears at about 1250% of elongation and after which tensile strength increases almost linearly with strain until fracture eventually occurs. Also in case of the composites with lower loading (1 wt.% silicon dioxide), a ductile behavior is observed with necking, with lower strain at fracture and without strain hardening. The composites bearing higher nanotube content (3 and 5wt.% silicon dioxide) showed a brittle behavior with breaking after the yield point. Tensile properties of the composites are shown in Table 1. It was observed that the Young's modulus and yield stress of the composites are substantially increased relative to the neat polypropylene. The percentage increase of Young's modulus is from 7.24 % to 17.84 % and that of yield stress is from 1.58 % to 7.99 % with an increase of silicon dioxide concentration. In case of strain at fracture, it decreases linearly from 37.97 % to 92.08 %, as shown in Table 1. Moreover, samples of the neat polymer and the composites showed neck formation at the yield stress. However, extent of necking decreases with increase in silicon dioxide content. It is due to the fact that at higher silicon dioxide content the silicon-rich areas become bigger and tend to show characteristics of forming an interconnecting network. When such network-density becomes higher at higher silicon dioxide concentrations they can potentially lead to intense strain localization because of hindered plastic deformation/strain of the matrix. The strain localization around the dense network of silicon dioxide ultimately causes matrix cracking due to severe modulus mismatch between polymer and silicon dioxide and hence ductile yielding behavior is substantially reduced. Therefore, decrease of the ductility at higher silicon dioxide content might be attributed to the presence of silicon dioxide aggregates.

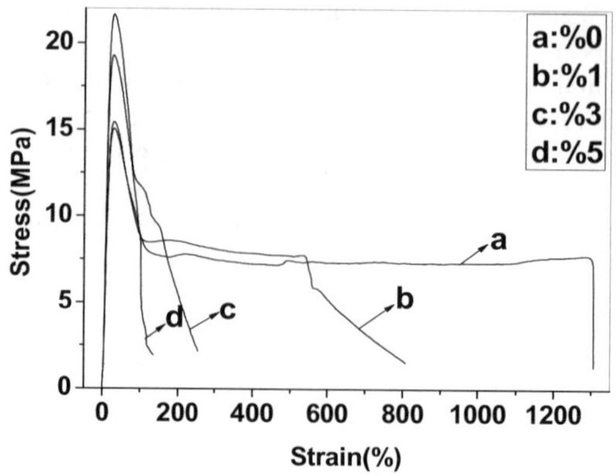

Fig. 3: Tensile behavior: stress–strain curves of HDPE and HDPE/silica dioxide composites

Table 1: Tensile properties of HDPE and HDPE/silica dioxide composites

Silicon dioxide content [%]	Modulus [Mpa]	Yield strength [Mpa]	Strain at break [%]
0	175.52	27.16	1301
1	188.23	27.59	807
3	193.15	28.99	264
5	206.84	29.33	103

Conclusions

HDPE/ silica dioxide composites were prepared by melting the blend method. As a result, these silica dioxide particles played an important role in PP matrix, resulting in improvements in the thermal and mechanical strength properties of the HDPE/ silica dioxide composites. Young's modulus of the composites was increased by the addition of silica dioxide. Silica dioxide reinforcement enhances the yield stress, while reducing the ductility.

Acknowledgements

This work was financially supported by the Liaoning Province Office of Education Foundation (2009A621).

References

[1] M. Bailly , M. Kontopoulou and K. E.Mabrouk: *Effect of polymer/filler interactions on the structure and rheological properties of ethylene-octene copolymer/nanosilica composites* . Polymer Vol.51 (2010), pp. 5506-5515.

[2] T.H.Zhou ,W.H.Ruan andY.L.Mai: *Performance improvement of nano-silica/polypropylene composites through in-situ cross-linking approach* . Composites Science and Technology Vol. 68 (2008), pp. 2858-2863.

[3] O. Aso, J.I. Eguiaza'bal, J. Naza'bal: *The influence of surface modification on the structure and properties of a nanosilica filled thermoplastic elastomer.* Composites Science and Technology Vol.67 (2007), pp.2854-2863.

[4] A.Zhu, A.Cai and W. D Zhou: *Effect of flexibility of grafted polymer on the morphology and property of nanosilica/PVC composites* . Applied Surface Science Vol.254 (2008), p.3745.

[5] N. Moussaif: *Mechanically reinforced biodegradable nanocomposites. A facile synthesis based on PEGylated silica nanoparticles* :Polymer, 51, 26(2010), p.6132.

[6] M. Garcia, G. V.Vliet and S. Jain: *Polypropylene/SiO_2 Nanocomposites with Improved Mechanical Properties*. Reviews on Advanced Materials Science Vol(6), 2004, p.169.

[7] Y.C.Ou, F.Yang and Z.Z. Yu: *A new conception on the toughness of nylon 6/silica nanocomposite prepared via in situ polymerization*. Journal of Polymer Science Part B: Polymer Physics Vol.36, (1998), p.789.

[8] E. Kontou and M .Niaounakis: *Thermo-mechanical properties of LLDPE/SiO2 nanocomposites*. Polymer Vol.47 (2006), p.1267.

[9] F. N. Ahmad, M. Jaafar and S. Palaniandy: *Effect of particle shape of silica mineral on the properties of epoxy composites* . Composites Science and Technology Vol.68 (2008), p. 346.

[10] C.P. Turssi, J.L. Ferracane and K.Vogel: *Filler features and their effects on wear and degree of conversion of particulate dental resin composites* :Biomaterials Vol.26 (2005), p. 4932.

[11] J.S. Parent, A. Liskova andR. Resendes: *Isobutylene-based ionomer composites: siliceous filler reinforcement* :Polymer Vol(45) 2004, p. 8091.

[12] H.Yang, Q.Zhang and M.Guo: *Study on the phase structures and toughening mechanism in PP/EPDM/SiO2 ternary composites* .Polymer Vol.47 (2006), p. 2106.

[13] V.Arrighi, I.J. McEwen and H. Qian: *The glass transition and interfacial layer in styrene-butadiene rubber containing silica nanofiller* :Polymer Vol.44 (2003), p. 6259.

Advanced Materials Research Vol. 279 (2011) pp 120-125

doi:10.4028/www.scientific.net/AMR.279.120

Highly soluble poly (aryl-ether benzoxazole) random copolymers by aromatic nucleophilic polycondensation: Synthesis and properties

Jinhuan Li[1,a], Haiyang Wang[1,b] and Ruihai Wang[1,c]

[1]College of Materials Science and Technology, Nanjing University of Aeronautics and Astronautics, Nanjing 210016, China

[a]jinhuan_nj@nuaa.edu.cn, [b]why.nuaa@163.com, [c]wrh880122@163.com

Keywords: poly(aryl-ether benzoxazole)s, highly soluble polymers, synthesis, polybenzoxazoles, poly(aryl ether)s

Abstract. Aromatic polybenzoxazoles (PBOs) are very promising materials for civil, aerospace and electronics industries due to their outstanding mechanical properties, good thermooxidative stability and chemical resistance. However, the feasibility of the wide applications is hindered due to the difficulty of their processing derived from the very rigid structure. Poly(aryl-ether benzoxazole) random copolymers (PAEXBOx) were prepared by nucleophilic polycondensation. Ary-ether ketone and aryl-ether sulfone segments along with flexible linkages and bulky groups were introduced to provide PAEXBOx with good solubility in organic solvents. PAEXBOx exhibit the improved Tgs, and intermediate properties between pristine PBOs and poly(acryl ether)s (PAEs). The structure and properties of the polymers were systematically characterized.

Introduction

Aromatic polybenzoxazoles (PBOs) are a class of heterocyclic polymers and their general chemical structure is shown in Scheme 1 (a) and (b). They exhibit outstanding thermooxidative stability, high tensile property, superior chemical resistance along with lower water absorption and dielectric constant[1]. Typically, the mechanical properties, thermal resistance and flame retardancy of PBO fibers with the trademark of Zylon® far exceed the first generation superior fibers, such as P-Aramid[2,3]. In view of the properties, PBOs are also promising for polymer films, moulding resins and coating materials required by the aerospace and electronics industries[4-6]. However, they are generally difficult to process because of the high transition temperature and poor solubility in organic solvents. Poly(p-phenylene benzobisoxazole), the polymer used for preparing Zylon® fiber (Scheme 1 (a), Ar = 1,4-phenylene), can only be dissolved in such strong acids as 100 % sulfuric acid, methane sulfuric acid, polyphosphoric acid (PPA) and trifluoroacetic acid[3,7]. Therefore, PBO fibers are synthesized in PPA, followed by spinning the PPA polymer solution in liquid-crystalline state[3,8-10].

In order to find versatile applications of PBOs, we prepared novel PBO copolymers with the structure as shown in Scheme 1 (c). Aryl-ether ketone (AEK) and aryl-ether sulfone (AES) are highly designable structures for providing possible functionality with good processibility. Thus, AES and AEK segments are introduced for the structure modification to prepare poly (aryl-ether benzoxazole)s (PAEXBOx). The intermediate properties between the PBO and the poly(aryl ether ketone) (PAEK)/poly(aryl ether sulfone) (PAES) were obtained. The resulted polymers have the good processibility derived from the PAEK/PAES and excellent tensile properties and enhanced toughness from the PBO.

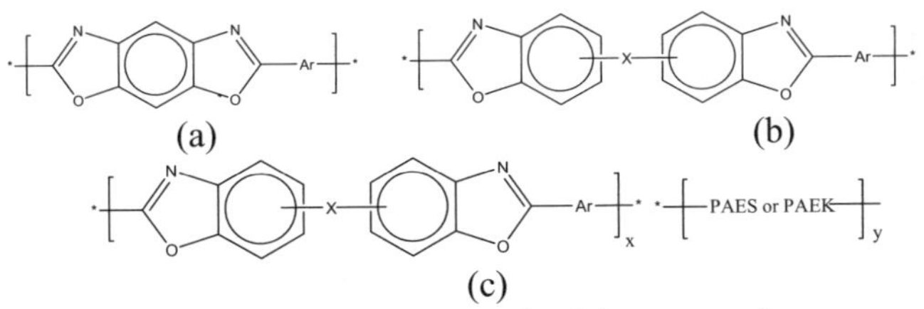

Scheme 1: Structure of polybenzoxazoles

Experimental

Materials. 2,2'-bi[2-(4-fluorophenyl)benzoxazol-5-yl] hexafluoropropane (6FBO) was synthesized in our lab. Other reagents were used as received or purified using the relevant methods.

Polymer Synthesis. PAEXBOx including poly(aryl ether-sulfone benzoxazole)s (PAESBOx) and poly(aryl ether-ketone benzoxazole)s (PAEKBOx) were synthesized as shown in Scheme 2 through changing the ratio of benzoxazole monomer to aryl-sulfone/aryl-ketone monomer. PAEXBOx without ether-sulfone, ether-ketone and benzoxazole segments are also synthesized for comparison, and designated as PEBOx (pristine PBO without ether-sulfone/ether-ketone), PAESx (pristine PAES without ether-ketone/benzoxazole) or PAEKx (pristine PAEK without ether-sulfone /benzoxazole), respectively. The number x in subscript represents the structures 1 or 2 derived from the bisphenols of 6FOH or HF, respectively. The synthesis was carried out in DMAc using K_2CO_3 as a catalyst, and using toluene as an azeotropic agent following the reference [11] procedure.

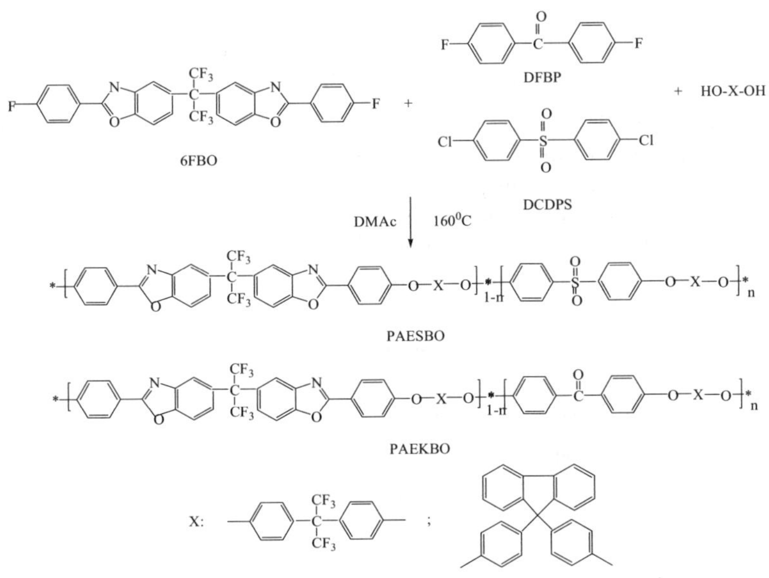

1 2

Scheme 2: Synthesis of PAEXBOx

Results and Discussion

Polymer synthesis. Hexafluoroisopropylidene (6F) groups in 6FBO provide high solubility of the oligomers, preventing them from precipitating out from the solvents during the polymerization (see Scheme 2). The electron-withdrawing effects of 6F and benzoxazole groups determine the high activity of the fluorine substitution. In order to guarantee the high solubility of the final polymers, the diphenols with bulky groups 1 and 2 were used. The polymerization in DMAc proceeded smoothly.

After ~12 h reaction, the polymers reached high molecular weight, which is deduced from high viscosity of the solution. FTIR and ^{1}H NMR (see Figure 1) confirmed the chemical structure of the polymers.

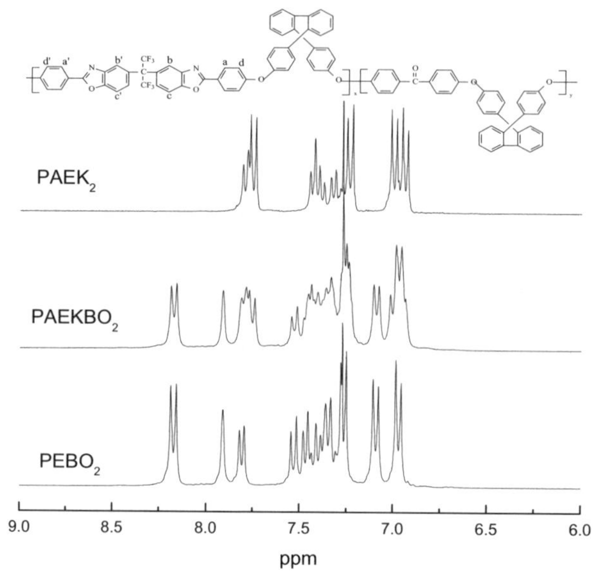

Fig. 1: ^{1}H NMR of $PAEK_2$, $PEBO_2$ and $PAEKBO_2$

Polymer properties. Table 1 summarizes the solubility of PAEXBOx in given solvents. Except $PAES_1$, all the other polymers can't be dissolved in DMSO due to the low solubility of 6FBO in DMSO. $PAES_1$ shows the best solubility and $PAES_2$ the worst. It was noted that $PAES_2$ could be dilutedly dissolved in NMP. It is obvious that 6FOH with 6F groups dedicated more to the solubility improvement than HF. Notably, most copolymers show a good solubility in such commonly used solvents as NMP, DMAc, m-Cresol, THF and $CHCl_3$, leading to the flexibility of PAEXBOx solution processing.

Table 1: Solubility of PAEXBOx.

Polymer	Solubility [a]						
	NMP	DMAc	DMF	DMSO	m-Cresol	THF	$CHCl_3$
$PEBO_1$	+	+(-)	-	-	+	+	+
$PEBO_2$	+	+	-	-	+	+	+
$PAEKBO_1$	+	+	+	-	+	+	+
$PAEKBO_2$	+	+	-	-	+	+	+
$PAESBO_1$	+	+	+	-	+	+	+
$PAESBO_2$	+	+	-	-	+	+	+
$PAEK_1$	+	+	+(-)	-	+	+	+
$PAEK_2$	+	-	-	-	+	+	+
$PAES_1$	+	+	+	+(-)	+	+	+
$PAES_2$	-	-	-	-	+	-	-

+: Soluble; -: insoluble; +(-): turbid at room temperature

Thermal gravimetric analysis (TGA) showed that all the polymers exhibit the 5 % weight loss at the temperature of > 460 °C. The thermal stability of the polymers with the structure 2 is higher than that of the structure 1 polymers because hexafluoroisopropylidene groups provide weak linkages in the polymer backbones. For the structure 1 polymers with the same composition at the low benzoxazole level, the thermal stability of the $PAEKBO_1$ is slightly higher than that of the $PAESBO_1$. Double scan measurements of differential scanning calorimetry (DSC) confirmed that the polymers exhibit the single glass transition temperatures of 169-296 °C which is much higher than that of the conventional poly(ether ether ketone)s (PEEK) (Tg = 143 °C). As shown in Fig. 2, with the increase of benzoxazole, Tgs increase gradually and reach the maximum for $PAEBO_1$ and $PAEBO_2$. The Tgs of ether-ketone polymers as expected is lower than that of ether-sulfone polymers. However, the curve slope rises more steeply. Notably, for the same composition the structure 2 polymers show much higher Tgs than the structure 1 polymers. Specifically, Tgs increase by 60-80 °C for the ether-ketone polymers if the structure is changed from 1 to 2. It is known that the conventional PEEK has a low Tg and thus suffers from creep behavior. Therefore, the synthesis of poly(aryl ether ketone)s (PAEKs) with high Tgs is of particular interest. Undoubtedly, the structure 2 derived from HF is effective to improve the glass transition temperature, and the present PAEKBOx, expecially $PAEKBO_2$ are promising candidates for this purpose. Wide-angle X-ray diffraction (WAXD) measurements as shown in Fig. 3 were performed to clarify the morphology of the polymers, confirming that the present polymers show no evidence of crystalline or liquid crystalline morphologies, and are amorphous structure in nature.

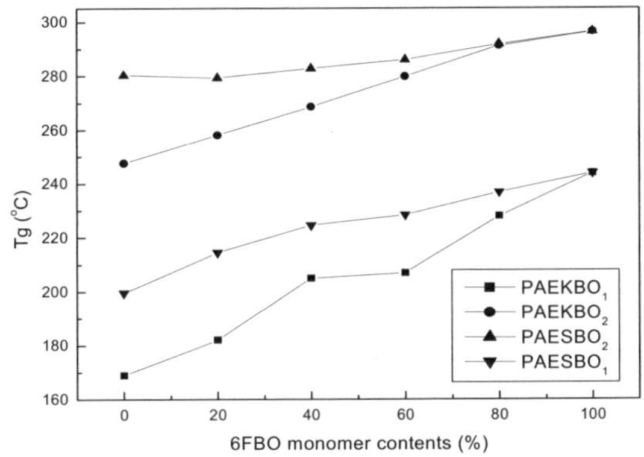

Fig. 2: Tgs of polymers versus benzoxazole composition

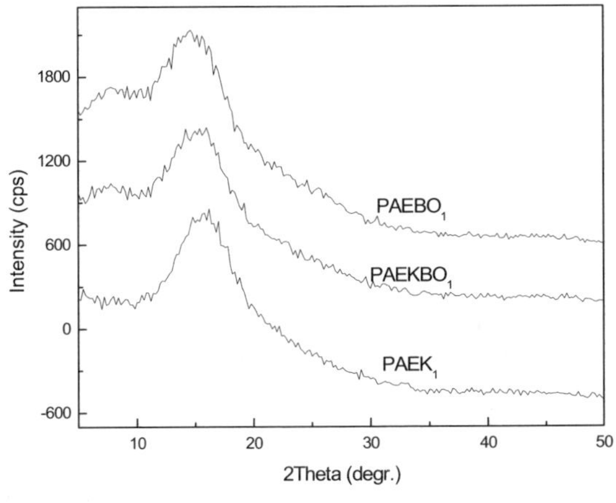

Fig. 3: WAXD of $PAEKBO_1$, $PAEBO_1$ and $PAEK_1$

Typical results of DMA experiments are displayed in Fig. 4. Comparisons of the tensile storage moduli among $PAEK_2$, $PAEKBO_2$ and $PEBO_2$ reveal the impact of the benzoxazole composition on the mechanical behavior of the polymers. The tensile moduli near room temperature are in the range of 1.5-3.5 GPa. The slop difference of the curves at the temperatures near Tg shows the sensitivity of the tensile properties to the polymer composition. For these three compositions, with the Tg increase the negative slope of log modulus versus temperature decreases, showing the enhancement of mechanical properties due to the benzoxazole cooperation. They show a large drop in modulus above Tg, indicating an amorphous melt. This allows these materials to be thermoformed at 50-70 °C above their Tgs. It is known that some PBOs such as poly(p-phenylene benzobisoxazole) don't show any Tgs. However, the PAEKBOx behave in many respects as thermoplastics with characteristics similar to those of PAEK, PAES and poly(ether imide).

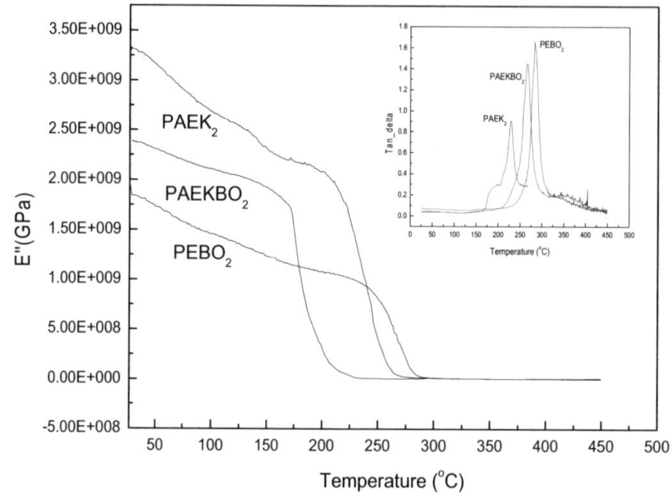

Fig. 4: Storage modulus and tensile loss tangent (top right corner) as a function of temperature for PAEKBOx

Because the PAEKBOx can be dissolved in organic solvents their membranes are fabricated. The mechanical properties of the polymer membranes are comparable to many aromatic polyimides. For PAEKBOx, the monomer HF tends to provide a higher tensile strength and thus a larger elongation with a similar modulus. Compared with PAEKBOx, PAESBOx have a slightly higher strength and modulus. Due to the dedication of HF, the strength of $PAEKBO_2$ and $PAESBO_2$ is larger than that of $PAEKBO_1$ and $PAESBO_1$.

Conclusion

In this contribution, high soluble poly(aryl-ether benzoxazole) copolymers were prepared by aromatic nucleophilic polycondensation. The copolymers could form tough and flexible membranes. The copolymers, especially poly(ether-ketone benzoxazole)s with fluorene groups, show good thermal stability. Compared with the relevant poly(aryl ether)s they exhibit the improved single glass transition temperature largely due to the dedication of benzoxazole structure. The copolymers show amorphous structure in nature. The copolymers have the intermediate properties between pristine PBOs and PAEs, namely behave as PAEs with enhanced performance and as PBOs with good processiblity, offering the prospect of being high performance materials for such application as films, coatings for optical and electronic devices and gas separation membranes.

Acknowledgements

Authors are very grateful for the support of Doctoral Fund of Ministry of Education of China (200802871049).

Reference

[1] J. F. Wolfe: *Encyclopedia of Polymer Science and Technology*. Wiley, New York, (1988).

[2] G. Maier: *Low dielectric constant polymers for microeletronics*. Progress Polymer Science Vol. 26 (2001), pp.3-65.

[3] Y. H. So: *Rigid-rod polymers with enhanced lateral interactions*. Progress Polymer Science Vol. 25 (2000), pp.137-157.

[4] G. Maier, S. Banerjee, J. Haußmann, R. Sezi: *High-temperature polymers for advanced microelectronics*. High Performance Polymer Vol. 13 (2001), pp.107-115.

[5] T. D. Dang, P. T. Mather, M. D. Alexander Jr., C. J. Grayson, M. D. Houtz, R. J. Spry, F. E. Arnold: *Synthesis and characterization of fluorinated benzoxazole polymers with high T_g and low dielectric constant*. Journal of Polymer Science, Part A: Polymer Chemistry Vol. 38 (2000), pp.1991-2003.

[6] S. L. Hsu, W. Chen: *A novel positive photosensitive polybenzoxazole precursor for microelectronic applications*. Polymer Vol. 43 (2002), pp.6743-6750.

[7] X. D. Hu, S. E. Jenkins, B. G. Min, M. B. Polk, S. Kumar: *Rigid-rod polymers: Synthesis, processing, simulation, structure, and properties*. Macromolecular Materials Engineering Vol. 288 (2003), pp.823-843.

[8] J. F. Wolf: *Materials research symposium proceeding*. Materials Research Society, Pittsburgh, (1989).

[9] S. Kumar, T. D. Dang, F. E. Amold, A. R. Bhattacharyya, B. G. Min, X. Zhang, R. A.Vaia, C. Park, W. W. Adams, R. H. Hauge, R. E. Smalley, S. Ramesh, P. A. Willis: *Synthesis, structure and properties of PBO/SWNT composites*. Macromolecules Vol. 35 (2002), pp.9039-9043.

[10] J. Li, X. Chen, X. Li, H. Cao, H. Yu, Y. Huang: *Synthesis, structure and properties of carbon nanotube/poly(p-phenylene benzobisoxazole) composite fibers*. Polymer International Vol. 55 (2006), pp.456-465.

[11] J. G. Hilborn, J. W. Labadie, J. L. Hedrick: *Poly(aryl ether-benzoxazoles)*. Macromolecules Vol. 23 (1990), pp.2854-2861.

Advanced Materials Research Vol. 279 (2011) pp 126-131
© (2011) Trans Tech Publications, Switzerland
doi:10.4028/www.scientific.net/AMR.279.126

The Effect of Microstructure on Properties of Fe-Cr-C-Nb/Ti Hardfacing Alloy

Wenbo Tang[1, a], Yungang Guo[1, b] and Hongrui Wang[1, c]
[1]School of Material Science and Engineering, Zhengzhou University, Zhengzhou 450001, China
[a]twb6751@zzu.edu.cn, [b]gyg0850152@sina.com, [c]lovehray@163.com

Key words: hardfacing, wear resistance, crack resistance, microstructure, carbides

Abstract. Hardfacing is one of the most useful and economical ways to improve the performance of components submitted to severe wear conditions. In this paper, a new kind of alloy called Fe-Cr-C-Nb/Ti alloy system for wear resistant successfully with the shielded metal arc welding (SMAW) method has been studied. The microstructure and wear resistant of hardfacing alloys reinforced with primary carbides were compared in this study. Meanwhile, the average hardness, the abrasion weight loss and microstructure of deposited metal were systematically studied by optical microscopy, scanning electronic microscopy and energy dispersive spectrum analysis. The results showed that the microstructure of the best optimizing hardfacing layer was the mixed martensite and little retained austenite, and NbC/TiC particles distributing dispersively in the matrix. The amount of low-carbon martensite and high-carbon martensite was identical. The alloy system showed high wear resistance due to the formation of dispersed MC type carbides and good toughness due to the exist of low carbon martensite in the matrix. The hardfacing alloy reinforced with complex carbides was also investigated, the microstructure was analyzed and its hardness and wear resistance were evaluated. In conclusion, the distribution, the chemical composition and the amount of the carbides, as well as the matrix microstructure are all factors to influence the crack resistance, hardness and wear resistance of the hardfacing alloys.

Introduction

Hardfacing is a commonly employed method to improve surface properties of agricultural tools, components for mining operation, soil preparation equipments and others [1]. For example, when rolls need to be repaired due to damage during hot rolling of steels, hardfacing the roll surface makes considerable contributions to lower material cost, higher productivity, improved product quality, and longer roll life [2]. In particular, hypereutectic high-chromium white irons and CrC, VC, WC alloys are most widely used in many applications because of their excellent wear resistance and low cost [3]. However, the carbide of Cr ,V, W straightly deposited to base metal have a series of problems such as easily cracking, bad bonding and being prone to disengage[4]. The recent studies on hardfacing alloys have focused on the improvement of the wear resistance, crack resistance, the microstructural renovation and their applications to work services [5~7]. In fact, the wear resistance and crack resistance are two of the most necessary properties of hardfacing layers; therefore, it is very important to realize that hardfacing layer of carbide quantity, morphology and distribution, and the alloy composition, matrix organization is how to affect the wear resistance and crack resistance.

In this study, the new hardfacing alloy is a kind of Fe-based hardfacing alloy reinforced with hard particle of the mixed NbC and TiC. The microstructure of the newly developed Fe–Cr-C–Nb/Ti hardfacing alloys is compared with those of the commercial Fe–Cr–C alloys, and the relationships between the microstructure and wear resistance, crack resistance are all analyzed by several different methods.

The main purpose of the present research is to produce a new electrode using NbC/TiC particles to reinforce the hardfacing metal. In addition, combined with wear-resisting behavior and influence factor, the relationship between the microstructure and wear resistance of the electrode was discussed.

Experimental Details

Materials and Welding Parameters. Using H08A welding core and the coating fluxes of calcium fluoride, calcium carbonate, silicate phosphor, sodium water glasses, the electrode belongs to basic slag system electrode. The hardfacing alloys containing Cr, Nb, Fe and a constant amount of C, Ti, Mn, Mo and Si were elaborated. Additions of these different alloying elements were to gain good mechanical and wear resistance properties.

The substrate material was Q235 steel (ASTM grade G3101,Q235 is the Chinese grade of this steel) with dimensions of 100mm×100mm×20 mm and 400mm×200mm×20 mm. It is a low-carbon steel. A commercial electrode of the Fe–Cr–C was also deposited over the steel substrate by the same welding technique and designated as A. Similarly, the self-made Fe–Cr–C–Nb/Ti electrode is designated as B. The chemical composition of A and B is reported in Table 1.

Table 1: Chemical composition (wt %) of materials used

Number	C	Cr	Nb	Mo	Si	Mn	Ti	Ni	V	P	S
A	2.00	17.4	-	-	4.09	2.80	0.11	0.11	0.06	0.13	0.01
B	0.61	2.37	1.54	0.59	1.29	1.17	0.6	06	-	0.02	0.01
H08A	≤0.10	≤0.20	-	-	≤0.30	0.30~0.5	-	≤0.30	-	≤0.03	≤0.03
Q235	0.15	-	-	-	0.30	0.30	-	-	-	0.050	0.045

The hardfacing alloys were deposited by shielded manual arc welding (SMAW) under direct current with a reverse polarity. The electrodes were dried at 200°C for 2 hours before welding, and the test plate was not preheated. The welding parameters are voltage-30 V, current-140 A, welding speed-150 mm/min. The thickness of each welding bead is up to 3~4 mm. The total thickness is 15mm.

Microscopy and Metallography. Rectangular block shaped samples of dimension 20 mm×10 mm×10 mm were cut from the deposition with electrospark linear cutting machine for observations of the microstructure .The metallographic specimens were ground on SiC abrasive paper of 1000 mesh size. The polished specimens of hardfacing alloys were subsequently etched with the admixture of nitric acid and alcohol (3ml HNO_3, 97ml alcohol). Then, the metallographic specimen was observed by optical microscopy (OM) and scanning electron microscope (SEM) which equipped with energy dispersive spectrum (EDS). In the procedure, the carbide size, the substrate's distribution and the grain size are discussed.

Hardness Test. The macrohardness of the hardfacing layers was measured by the Rockwell hardness (HRC) method. The hardness value is the average of five measurements.

Abrasive-wear Test. Dry sliding wear tests using the pin-on-disc method were performed on 5.5-mm-diameter and 30 mm long cylindrical specimens impinging vertically on a rotating disc coated with the 180 mesh SiC abrasive. The samples were then ground to remove the sharp edges from wear pins before the sliding tests. The specimens were gravity loaded each with a mass of 20N, 30N, 40N. Schematic diagram of wear testing machine is shown in Fig. 1. The parameters of the abrasive wear test are shown in Table 2. All the specimens should be cleaned in Acetone and rinsed with warm air before the tests. The wear resistance was determined by the mass loss results from every 250m of sliding, and each wear test was conducted using a new SiC paper. The weight loss was obtained from the difference in weight of the specimen before and after the sliding test by an electronic balance (accuracy 0.0001 g). Several tests were carried out in order to evaluate the average loss of weight for both types of samples. The relative wear resistance was determined by the following formula [8]:

$$\varepsilon = W_s / W_t \quad (1)$$

where ε is the relative resistance, W_s is the mass loss of unit area in reference sample and W_t is the mass loss of unit area in tested sample. The macrohardness of the hardfacing layers and result of wear resistance test are shown in Table 3.

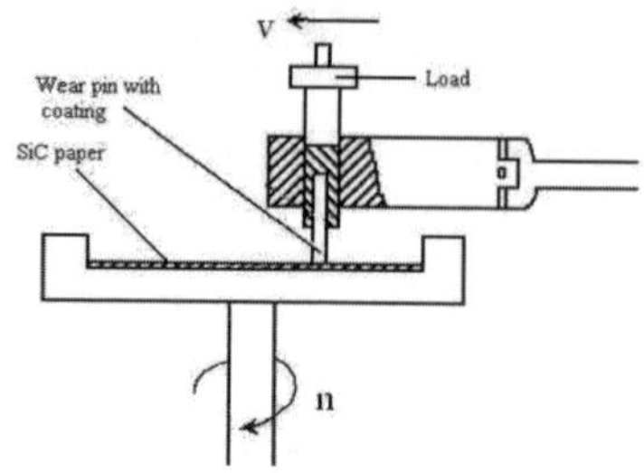

Fig. 1: schematic diagram of wear testing set up. (Modle:ML-100,CHINA)

Table 2: Parameters of abrasive wear test

Diameter of wheel [mm]	Speed of rotation [rpm]	Sliding distance [m]	radial feed [mm/ r]
260	60	250	1

Table 3: The hardness of hardfacing layer and result of wearability test

Number	HRC	Mass loss [g]			Relative
		Load/20 N	Load/30 N	Load/40 N	wear resistance
A	56.4	0.0633	0.0889	0.1189	1
B	59.7	0.0396	0.0523	0.0712	1.7

Results and Discussion

The characterization of microstructure has been studied with optical microscopy after etching with the admixture of nitric acid and alcohol. Typical microstructures of the hardfacing layers are shown in Fig. 2.

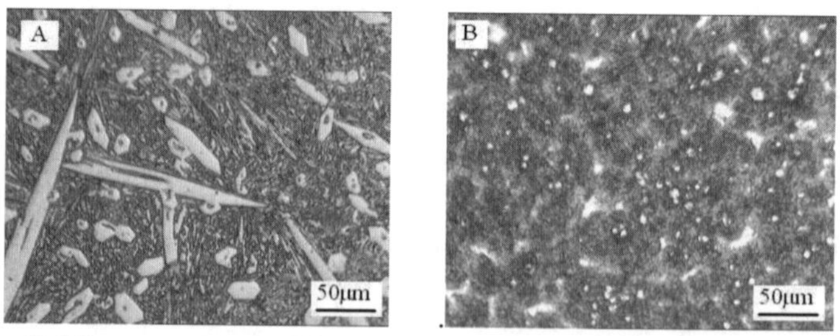

Fig. 2: Microstructures of the Fe-based hardfacing alloys investigated: (A) The Fe–Cr–C alloy, (B) The Fe–Cr–C–Nb/Ti alloy.

Fig. 2(A) and (B) shows homogeneous distribution of fine carbides. However, there are many differences in the pictures. In Fig. 2(A), white blocks distribute on the eutectic carbide and martensite matrix. The white large block phase is mainly M_7C_3 type chromium carbide. Its form is rod-like. The transverse section is hexagonal type. Its hardness value is about 1800 HV [9]. This phase offers hardness and wear resistance to the layers. It makes the average macrohardness of the hardfacing layer reach 56.4 HRC. In addition, in the condition of non-preheating and non-postheating, there are some cracks on the hardfacing layer. Under high load condition, the microcracks easily propagate along the rod type carbide or the matrix to form a long crack. The fracture of the hardfacing layers can be speeded up, even the cracks throughout the transverse section of hardfacing layers. The total crack length was reached up to 910mm in the layers of 100mm×100mm area. All these have further influence on the properties of matrix.

The microstructure of B alloy is that white particles are dispersively distributed in the mixed martensite matrix, and the amount of low-carbon lath martensite and high-carbon needle martensite was identical. In the SEM image of Fig. 3, it can be seen more clearly that there are a number of small white particles with sizes about 5–10μm embedded in the layer. The particles have good strengthening effect to the matrix.

The area of white particles (b area) was detected by spectrum analysis, and the result is shown in Fig. 3(b). It can be seen that Nb and C are rich in the phase. It contains 47.59% niobium and 6.82% carbon (wt. %), indicating that this is the NbC hard particle carbide (2800HV). The black core area of white particles (c area), shown in Fig. 3(c), mainly contains 13.96%Ti (at. %), besides C, N and O elements can be detected, and its exterior area is a Nb-containing carbide. This physical phenomenon is easy to be explained. The titanium carbide is formed primarily from the melt at about 3000°C. Thus, they act as a nucleus for the complex carbides of M_7C_3 type, which can solidify at about 1450°C [10]. Here M is a substitute for the metallic (Nb, Cr, and Fe). In other words, the complex carbide core was encircled by NbC carbide. All these carbide particles are surrounded by the low-carbon and high carbon martensite to form interlaced microstructure.

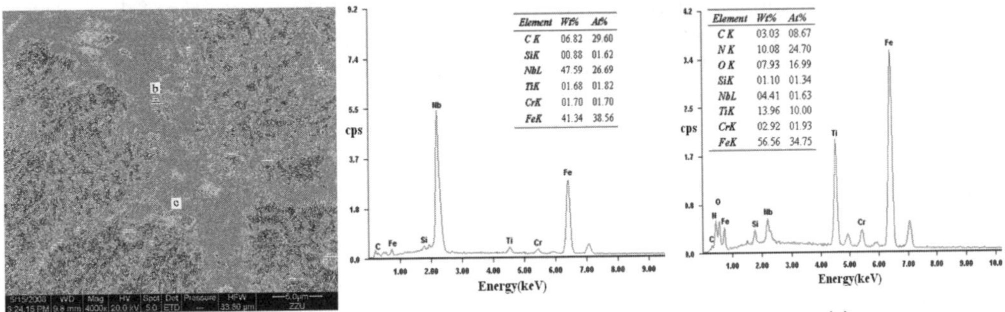

(a) (b) (c)

Fig. 3: Microstructures and typical composition of carbides, based on standardless energy dispersive spectroscopy analysis: (a) SEM image of the hardfacing layer of sample B, (b) the chemical composition of area b, the white particles, and (c) the chemical composition of area c, the black core of white particles.

In the hardfacing layer, the matrix was strengthened by Cr, Mo, Mn, Si and C element via solid solution. As a result, the microstructure of hard martensite matrix has high hardness and good wear resistance. Therefore, in Table 3, the macrohardness of B alloy is 59.7HRC which is a little bit higher than that of A alloy. In view of the foregoing, the developed hardfacing deposits of the Fe–Cr–C–Nb/Ti system can be said to have improved wear resistance when the number of NbC increase. It can promote the nucleation rate of austenite to prevent coarsening of the matrix metal [11]. The excellent crack resistance in the B alloy is attributed to the role of low-carbon martensite and the small metal cells in restraining the crack propagation. In the progress, macroscopic examination of the hardfacing alloy surface shows that it does not exhibit cracks in the area 400mm×200mm with macroscopically observable. This is one of the most advantages of Fe-Cr-C-Nb/Ti type hardfacing alloys that cracking resistance was compared with conventional Fe-Cr-C alloys. Furthermore, Nb and Ti are strong carbide forming elements. With an increase in the amount of Nb and Ti, the metallurgical reaction formation of NbC and TiC particles increases in the hardfacing layers, as a result the hardness increases [12]. After the primary complex particle solidification, there was not enough carbon left for forming lots of high carbon martensite. Consequently, the low carbon martensite was formed. This soft phase enables the matrix to suppress the cracking tendency.

For comparison, the wear loss experiments and wearability of two different layers were evaluated, and the results are shown in Table 3. It can be easily observed that the hardfacing layer shows excellent wear resistance compared with to Fe–Cr–C alloy.

From microstructural considerations, the hard high-carbon martensite and soft low-carbon martensite were formed together in the hardfacing layers. Furthermore, little pliable retained austenite and fine hard particles were existed in the layers. The best abrasion resistance was obtained in the complex carbides deposit, in which the elevated volume fraction of coarse M_7C_3 carbides provided a barrier against indentation, grooving and cutting [9]. The carbides particles which are exposed on the surface can play a role of blockage to hinder the abrasion from happening.

In Table 3, it can be seen that the wear resistance of B alloy is much greater than that of A alloy. This beneficial effect is determined by the microstructures of hardfacing alloy. Firstly, the complex carbides particles have high bonding strength with the matrix, and it will be defined as a skeleton. They can form a good carrier capacity system together, which distinctively reduces the load intensity carried by the particle, and effectively restrain the forming and development of microcutting, and improve the wear resistance and anti-friction properties of the hardfacing layer. Therefore the alloyed layer including hard particles has excellent anti-wear property [13]. Secondly, the wear resistance of B alloy is quite improved, as the selective wear of the low carbon martensite. Matrix is considerably affected by the microstructure of the surface layer. The soft low matrix can hold carbides effectively and inhibit the fall-off carbides effectively, thereby leading to homogeneous wear of the matrix and carbides. Another important factor in abrasion resistance is that the hard high carbon martensite provides the necessary support for these carbides. The dispersedly distributed carbides and martensite are the microbase of the surfacing layer with satisfactory abrasive resistance and well cracking resistance [14].

Conclusions

The results of this investigation can be summarized as follows:

(1) A new kind of hardfacing electrodes was developed by adding Nb, Ti as a carbide forming element and Cr, Mo as a matrix hardening agent and the relative wearability is 1.7 times of commercial Fe-Cr-C alloy. In addition, the crack resistance of the Fe–Cr–C–Nb/Ti alloy is much greater than that of commercial Fe-Cr-C alloy.

(2) The present study shows that complex carbides in the martensite matrix provide a composite effect on the hardness and toughness of the hardfacing alloy in the course of wear. As the martensite matrix is mainly to provide mechanical support during the abrasion process, the relative wearability is very high, especially under low stress abrasion.

(3) Microstructural analyses indicate that spherulitic carbide particles, which presented a TiC core encircled by a NbC, were homogeneously distributed in the martensite matrix.

Acknowledgement

This research was supported by the support fund for postgraduate education of Zhengzhou University.

References

[1] P. Crook: *Friction and wear of hardfacing alloys*. ASM Handbook Vol. 18: Friction, Lubrication, and Wear Technology (1992), pp. 758-765.

[2] Kim chang Kyu, Euh Kwangiun, Lee Sunghak, Jung jae-Young, Ahn Sangho: *Correlation of Microstructure with the Wear Resistance and Fracture Toughness of Hardfacing Alloys Reinforced with Complex Carbides*. Metallurgical and Materials Transactions A: Physical Metallurgy and Materials Science, 131, 12(2000), pp. 3041-3052.

[3] K.G. Budinski: *Surface Engineering for Wear Resistance*. Prentice-Hall, Englewood Cliffs, NJ, (1988), pp. 242-287.

[4] M. H. Staia, A. Fragiel, M. Cruz, et al: *Characterization and wear behavior of pulsed electrode surfacing coatings*. Wear, 251, 10(2001), pp. 1051-1060.

[5] B. Djerdjare, S. Lebaili, S. Lay: *Microstructural study of (Co–Ni–Fe) based alloys*. Materials Science and Engineering: A Vol. 475 (2008), pp. 336-342.

[6] M.X. Yaoa, J.B.C. Wub, Y. Xiec: *Wear, corrosion and cracking resistance of some W- or Mo-containing Stellite hardfacing alloys*. Materials Science and Engineering: A Vol. 407 (2005), pp. 234-244.

[7] Dwivedi, D.K.: *Abrasive Wear behavious of iron based hard surfacing alloy coatings developed by Welding*. Surface Engineering, 20, 4(2004), pp. 87-92.

[8] Yang Shang-lei, Lv Xue-qin, Zou Zeng-da, Lou Song-nian: *Investigation of surfacing electrode with high hardness based on lath martensite*. Materials Science and Engineering: A, Vol. 438–440 (2006), pp. 281-284.

[9] M.F. Buchely, J.C. Gutierrez, L.M. Le on, A. Toro: *The effect of microstructure on abrasive wear of hardfacing alloys*. Wear Vol. 259 (2005), pp. 52-61.

[10] V.E. Buchanan, P.H. Shipway, D.G. McCartney: *Microstructure and abrasive wear behaviour of shielded metal arc welding hardfacings used in the sugarcane industry*. Wear Vol. 263 (2007), pp. 99-110.

[11] H. Fujimura, A. Notomi, Y. Kamito, M. Kinoshita: *Development of hard overlay welding for high wear resistance*, IIW Document No.XII-1071-88.

[12] X.H.Wang, Z.D.Zou, S.Y.Qu, S.L.Song: *Microstructure and properties of the TiC/Fe-based alloy hardfacing layers*. JOURNAL OF MATERIALS SCIENCE Vol. 40 (2005), pp. 3629-3633.

[13] A.K. Jha, B.K. Prasad, O.P. Modi, S. Das and A.H. Yegneswaran: *Correlating microstructural features and mechanical properties with abrasion resistance of a high strength low alloy steel*. Wear Vol. 254 (2003), pp. 120-128.

[14] Yang Shang-lei, Lv Xue-qin, Zou Zeng-da, Lou Song-nian: *Investigation of surfacing electrode with high hardness based on lath martensite*. Materials Science and Engineering: A Vol. 438–440 (2006), pp. 281-284.

Advanced Materials Research Vol. 279 (2011) pp 132-135
© (2011) Trans Tech Publications, Switzerland
doi:10.4028/www.scientific.net/AMR.279.132

Synthesis and Crystal Structure of a Novel Copper Complex $Cu(phen)(C_8H_7O_2)_2$

Ping Xia

Department of Chemistry, Huzhou Teachers College, Huzhou, 313000, P. R. China

xp@hutc.zj.cn

Keywords: copper, 2-mehyl benzoate complex, aryl π-π packing interaction

Abstract. Reactions of a freshly prepared $Cu(OH)_2{\cdot}yH_2O$ precipitate, 2-methyl benzoic acid with 1,10'-phenanthroline in C_2H_5OH/H_2O afforded $Cu(phen)(C_8H_7O_2)_2$. The title compound was structurally characterized by X-ray diffraction methods. The unit of complex molecule was proposed 4+2 coordination sphere. In the crystal, molecules are connected by aryl π-π packing forming chains along [010] direction, the supramolecular double chains are further constructed two-dimension layers which parallel to (011) by π-π packing interactions.

Introduction

In the past decade, a variety of supramolecular architectures based on non–covalent intermolecular interactions such as hydrogen bonding, van der Walls forces and π-π interactions have been achieved by using transition metal centers and organic ligands due to their possible intriguing structural topologies and potential applications in optics, catalysis, ion exchange, gas storage, and the molecular–based magnetic materials[1,2]. Simultaneous coordination of rigid benzoate ligands and the heteroaromatic N-donor ligands as sources for π-π packing interactions resulted in a variety of structural motifs. For example, self-assembly of Cu^{2+} cations with phenanthroline and benzoic acid led to mononuclear complex molecules such as $Cu(phen)(H_2O)(C_7H_5O_2)_2$[3]. 2-Methyl benzoato ligands have been actively utilized as construction units to obtain many supramolecular complexs[4,5]. To the best of our knowledge, no 2-methyl benzoate copper complex constructed only by aryl π-π packing interactions. In the present work, we are interested in self–assemblies of Cu^{2+} ions and 1,10'–phenanthroline with 2-methyl benzoic acid, further report the synthesis and structure of $Cu(phen)(C_8H_7O_2)_2$ in detail.

Experimental Section

Synthesis. Dropwise addition of 2.0 ml of 1.0 M aqueous NaOH to a stirred aqueous solation of $CuCl_2{\cdot}2H_2O$ (0.1705 g, 1.0mmol) in 3 ml H_2O produced a blue precipitate, $Cu(OH)_2{\cdot}yH_2O$, which was centrifuged and washed with water until no chroride anions were detected in the supernatant. The blue precipitate was added to a mixed solution of 2-methyl benzoic acid (0.2725 g, 2.0 mmol) and 1,10'–phenanthroline (0.1982 g, 1.0 mmol) in 30 ml $C_2H_5OH–H_2O$(1:1,v/v). The solution was stirred and filtered. The blue solution (pH = 4.28) was allowed to stand at room temperature. Slow evaporation during several days afforded blue block crystals.Yield:42% based on initial $CuCl_2{\cdot}2H_2O$.
Characterization. Suitable single crystals were selected under a polarizing microscope and fixed with epoxy cement on respective fine glass fibers, which were then mounted on a Rigaku R-Axis Rapid IP X-ray diffractometer with graphite-monochromated Mo $K\alpha$ radiation (λ = 0.71073 Å) for cell determination and subsequent data collection. The reflection intensities in the θ range 3.01–25.00° were collected at 293 K using the w scan technique. The employed single crystals exhibit no detectable decay during the data collection. The data were corrected for Lp and absorption effects. The SHELXS–97 and SHELXL–97 programs were used for structure solution and refinement [6,7]. The structures were solved by using direct methods. Subsequent difference Fourier syntheses enabled all non-hydrogen atoms to be located. After several cycles of refinement, all hydrogen atoms

associated with carbon atoms were geometrically generated, and the rest of the hydrogen atoms were located from the successive difference Fourier syntheses. Finally, all non-hydrogen atoms were refined with anisotropic displacement parameters by the full-matrix least-squares technique and hydrogen atoms with isotropic displacement parameters set to 1.2 times of the values for the associated heavier atoms. Detailed information about the crystal data and structure determination is summarized in Table 1.

Table 1: Crystallographic and experimental data for $Cu(phen)(C_8H_7O_2)_2$

Compound	$Cu(phen)(C_8H_7O_2)_2$	Volume [Å^3]	2283.7(8)
Empirical formula	$C_{28}H_{22}CuN_2O_4$	Temperature[K]	293(2)
Formula weight	514.02	D_{Calc} [g·cm^{-3}]	1.495
Crystal system	monoclinic	$F(000)$	1060
Space group	$P2_1/c$	μ[mm^{-1}]	0.996
Description	Blue block	θ Range [°]	3.01-25.00
Crystal size [mm]	0.10 × 0.10 × 0.10	Reflections collected	17441
a (Å)	16.245(3)	Unique	0.0174
b (Å)	10.136(2)	Data,restraints,	4021, 0, 319
c (Å)	14.048(3)	Goodness of fit on F^2	1.130
α(°)	90	R_1,wR_2 [$I \geq 2\sigma(I)$]	0.0502, 0.1022
β (°)	99.15(3)	R_1,wR_2 (all data)	0.1019, 0.1680
γ (°)	90	A, B values in	0.0447, 4.77
Z	4	$\delta\rho_{max}$, $\delta\rho_{min}$ [e·Å^{-3}]	0.737, -1.038

$R_1 = \sum(|F_o| - |F_c|)/\sum|F_o|$, $wR_2 = [\sum w(F_o^2 - F_c^2)^2/\sum w(F_o^2)^2]^{1/2}$, and $w = [\sigma^2(F_o^2) + (0.0447P)^2 + 4.77P]^{-1}$ where $P = (F_o^2 + 2F_c^2)/3$

Table 2: Selected bond lengths (Å) and angles (°) for $Cu(phen)(C_8H_7O_2)_2$

Cu–N1	2.025(4)	Cu–O1	1.927(4)	Cu–O2	2.608(4)
Cu–N2	2.011(4)	Cu–O3	1.919(4)	Cu–O4	2.666(4)
O1–Cu–O3	93.4(2)	N2–Cu–O3	170.7(2)	O1–C13–O2	122.1(5)
N1–Cu–N2	81.6(2)	N1–Cu–O3	91.9(2)	O3–C21–O4	122.7(5)
N1–Cu–O1	174.0(2)	N2–Cu–O1	93.5(2)		

Results and Discussions

The stereochemistry of the monomeric complex $Cu(phen)(C_8H_7O_2)_2$ could be described as pseudo-octahedral with the copper atom on the inversion centre. Equatorial plane is constituted of two nitrogen atoms from the 1,10'–phenanthroline ligands (Cu–N1 2.025(4) Å, Cu–N2 2.011(4)Å) and two oxygen atoms from the two 2-methylbenzoate ligands(Cu–O1 1.927(4) Å, Cu–O3 1.919(4)Å), as shown in Fig. 1. The distances of the remaining carboxylate oxygens (Cu–O2 2.608(4)Å, Cu–O4 2.666(4)Å) are substantially longer than Cu–O1 distance, but shorter than the sum of the van der Waals radii, which indicates a weak interaction between the two atoms. Therefore, 4+2 coordination motif was proposed[8]. The molecules are assembled into one-dimension chains extending along the [010] direction through edge-to-face aryl π-π packing interactions, and the centroid distance of 3.64 Å between edge of phen and rings of benzyl was proved to be aryl π-π packing interactions, as shown in Fig. 2. The supermolecular chains are further constructed two-dimension layers which parallel to (011) by offset face-to-face aryl π-π packing interactions. The centroid distance of 3.53 Å between parallel rings of phen was proved to be aryl π-π packing interactions. The resulting layers are constructed framework along the [100] direction by van der Waals interactions, as shown in Fig. 3. Selected interatomic distances and bond angles are given in Table 2.

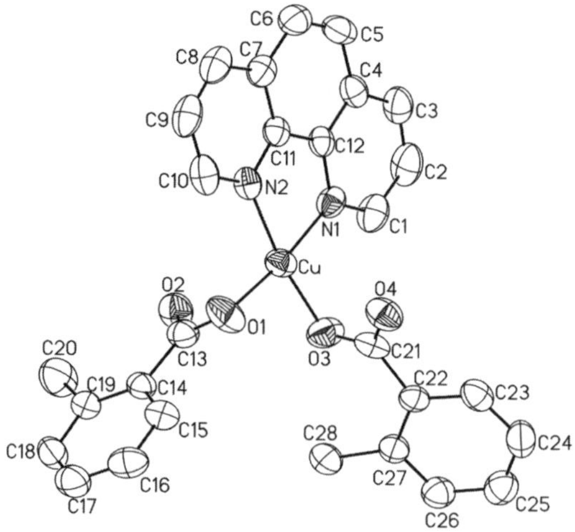

Fig. 1: ORTEP view of Cu(phen)$(C_8H_7O_2)_2$(45% probability level)

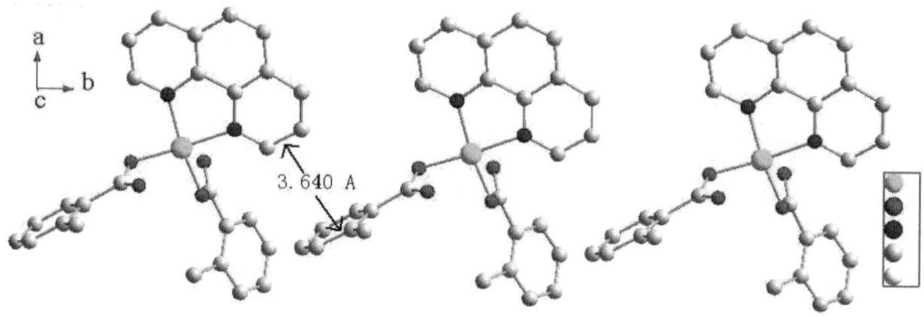

Fig. 2: 1D chain structure linked along [010] direction by π-π packing interactions in Cu(phen)$(C_8H_7O_2)_2$

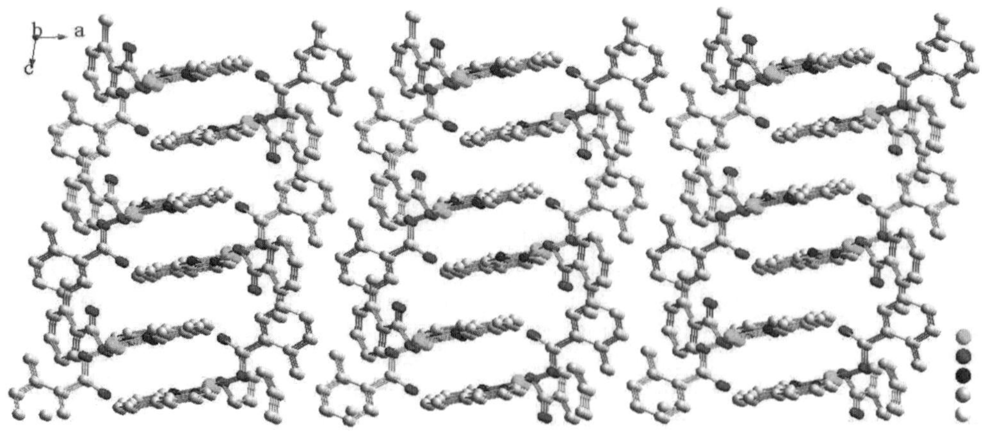

Fig. 3: 3D framework structure of Cu(phen)$(C_8H_7O_2)_2$

Conclusions

The asymmetric unit of the novel compound contains a mononuclear Cu(phen)$(C_8H_7O_2)_2$ complex molecule. The Cu^{2+} cation is in a pseudo-octahedral coordination sphere defined by two nitrogen atoms of 1,10'-phenanthroline ligand and four oxygen atoms of the two 2-methyl benzoate anions. In the crystal, the molecules are assembled into one-dimension chains extending along the [010] direction through edge-to-face aryl π-π packing interactions, the supermolecular double chains are further constructed two-dimension layers which parallel to (011) by aryl π-π packing interactions, and the resulting layers are constructed framework along the [100] direction by van der Waals interactions.

References

[1] A. de Bettencourt-Dias: *Isophthalato-based 2D coordination polymers of Eu(III), Gd(III), and Tb(III): enhancement of the terbium-centered luminescence through thiophene derivatization.* Inorganic Chemistry Vol. 44 (2005), pp.2734-2741.

[2] F. F. Chen, Z. Q. Bian, Z. W. Liu, D. B. Nie, C. H. Huang: *Highly efficient sensitized red emission from Europium (III) in Ir-Eu bimetallic complexes by 3MLCT energy transfer*. Inorganic Chemistry Vol. 47 (2008), pp.2507-2513.

[3] M. Devereux, D. O'Shea, A. Kellett, M. McCann, C. Deegan, K. Kedziora: *Synthesis, X-ray crystal structures and biomimetic and anticancer activities of novel copper(II)benzoate complexes incorporating 2-(40-thiazolyl)benzimidazole (thiabendazole), 2-(2-pyridyl)benzimidazole and 1,10-phenanthroline as chelating nitrogen donor ligands*. Journal Inorganic Biochemistry Vol. 101 (2007), pp. 881-892.

[4] X. M. He, C. H. Li, Y. Q. Yang, W. Li: *Hydrothermal synthesis,crystal structure and electrochemical properties of the complex Cu(o-methylbenzioc acid)2(2,2'-bipy).(H_2O)*. Chinese Journal Structure Chemistry Vol. 26(2007), pp.206-210.

[5] H. Busskamp, G. B. Deacon, M. Hilder, P. C. Junkr, D. R. Turner: *Structural variations in rare earth benzoate complexes.* CrystEngComm Vol. 9 (2007), pp.394-411.

[6] G. M. Sheldrick: *SHELXS-97 , Program for X-ray crystal structure solution.* Gottingen University, Germany 2008.

[7] G. M. Sheldrick: *SHELXS-97, Program for X-ray crystal structure refinement.* Gottingen University, Germany 2008.

[8] B. Kozlevcar, N. Lah, D. Zlindra, I. Leban, P. Segeden: *Copper(II) benzotates and acetates with 2-aminopyridine.* Acta Chemisty Slovenia Vol.48 (2001), pp.363-374.

Advanced Materials Research Vol. 279 (2011) pp 136-141
© (2011) Trans Tech Publications, Switzerland
doi:10.4028/www.scientific.net/AMR.279.136

A Facile Route to Prepare Hydrophilic Copper Nanoparticles and Their Thermal Conductivities

Dan Li[1,a]

[1]Department of Chemistry and Chemical Engineering, Weifang University, Weifang 261061, Shandong Province, China

[a]danli830109@163.com

Keywords: copper, hydrophilic, surface coating, thermal conductivity

Abstract. Well-dispersed and hydrophilic copper nanoparticles have been synthesized from an organometallic precursor in an organic (or oil) phase. There are two layers coated on the surface of the copper nanoparticles. The coating layer can not be dissolved or rinsed off by organic media. The changing of viscosity and enhancement of the thermal conductivity of copper nanofluids is also presented.

Introduction

Heat transfer fluids are presently employed in the transportation industry, heating and cooling systems in buildings, and industrial processing of goods including textiles, paper products, food, chemicals, and many others. Nanofluid is a novel heat transfer fluid prepared by suspending solid nanoparticles in a traditional fluid [1,2]. As compared to the traditional fluid, the nanofluid exhibits an enhanced thermal conductivity. However, recent efforts have mainly been focused on the preparation and evaluation of water or EG-based nanofluids and very few reports of the synthesis of oil-based nanofluids have been found [3,4].

Copper is one of the most widely used materials in the world. It has a great significance in all industries, particularly in the electrical sector due to low cost. It is well known that metallic copper have higher thermal conductivities than traditional fluids (such as water, ethylene glycol, and oil). Therefore, fluids containing suspended copper particles can be reasonably expected to have higher thermal conductivities than pure fluids. There is a strong need in many industrial fields to develop heat transfer fluids with significantly high thermal conductivity. Nanofluids consisting of metallic nanoparticles have attracted research attention. Copper nanoparticles have been synthesized and characterized by different methods. Stability and reactivity are the two important factors that impede the use and development of the metal cluster in a new generation of nanoelectronic device [5,6].

In this work, stable oil-based nanofluids with copper nanoparticles were prepared. Synthesis of oleic acid coated copper nanoparticles has been carried out by use of ascorbic acid in organic medium. Nano-copper can be effectively coated with oleic acid. The products obtained from this method can be hydrophilic, which can be used either with oil phase or in aqueous phase for further applications. The viscosity and thermal conductivity for the nanofluids are investigated.

Experimental procedures

Materials. All agents used in this work were analytical grade, including $CuSO_4\cdot 5H_2O$; KOH; oleic acid; ascorbic acid; ethanol.

Synthesis. In a typical experiment, cupric ions are converted to copper oleate through neutralisation of oleic acid with the precipitated copper hydroxide. The copper oleates formed in the organic phase (n-hexane or toluene) were reduced by an excess of ascorbic acid solution to achieve complete reduction. The reaction solution turned to brown copper colloids. Reaction solution was kept under agitation at a moderate speed to maintain the uniformity of the system and to keep the precipitated particles dispersed until the reduction was completed. After 30 min of ageing, the precipitation was separated from the suspension by centrifugation, and subsequently was washed with alcohol for three times. The ultimate powders were then dried in the vacuum stove at 45 °C for 4 h.

Characterization. The phase purity of the products was characterized by X-ray powder diffraction (XRD) using a Thermo X-ray diffractometer (Bruker, Germany) with monochromatized Cu Kα radiation (λ=1.5405 Å). A Nexus 470 Fourier transform infrared spectrometer (NICOLET, USA) was used to observe the changes of organic functional groups. Transmission electron microscope (TEM) images were taken with JEM-200CX (JEOL, Japan) instrument using an operating voltage of 160 kV and the electron diffraction pattern was also recorded for the selected area. A drop of the prepared nanofluid was allowed to dry on a copper grid coated with carbon film for the TEM examination.

Thermal properties. A capillary viscometer was used to determine the viscosities of the copper nanofluids. The capillary viscometer was submerged in a glass water tank. Water from a constant-temperature bath entered the bottom of the tank by a pump and returned to the bath through an over flow pipe. The water tank was maintained at a prescribed constant temperature by the water circulation. The vertical angle of the viscometer was accurately controlled with a special tripod. The transient hot-wire method is used in this study to measure fluid thermal conductivity.

Results and discussion

Characterization of Copper Nanoparticles. Oleic acid-coated copper nanoparticles were prepared with a mole ratio of oleic acid/cupric ions = 3/1. Reaction temperature range is from 30 to 70 °C. It is difficult to occur for the reaction below 45 °C. Heating (~50 °C) was continued for about 2 more hours with continuous stirring. When the reaction temperature was up to 60 °C, the copper nanoparticles were reduced within 0.5 h completely. The XRD pattern of a typical sample is shown in Fig. 1. The diffraction peaks can be indexed to those of pure facecentered cubic (FCC) Cu, corresponding to the (111), (200), and (220) planes, respectively. Fig. 2 presents the FTIR spectra of copper oleate and oleic acid coated copper nanoparticles prepared. The characteristic IR bands for metal oleate were in the range of 1650-1510 cm^{-1} for the asymmetrical vibrations and 1400-1280 cm^{-1} for the symmetrical vibrations. The FTIR spectrum of copper nanoparticles prepared showed vibrations at 1650 and 1465 cm^{-1}, but no band at 1710 cm^{-1}. The FTIR spectrum collected from the copper oleate contains a strong band at 1710 cm^{-1}. This means that unreacted oleic acid has been removed.

To investigate the influence of reaction temperature on the formation of copper nanoparticles, different reaction temperatures, ranging from 30 °C to 70 °C were applied. Fig. 3 depicts TEM images of copper nanoparticles prepared at different temperature and the corresponding selected area electron diffraction (SAED) pattern of the as-prepared product, respectively. Copper nanoparticles prepared in organic phase disperse very well in the ethanol water and apor solvent, and the sizes are less than 50 nm, suggesting that the method is effective. The as-prepared copper nanoparticles are of uniform size and the obtained nanofluids have good stabilization.

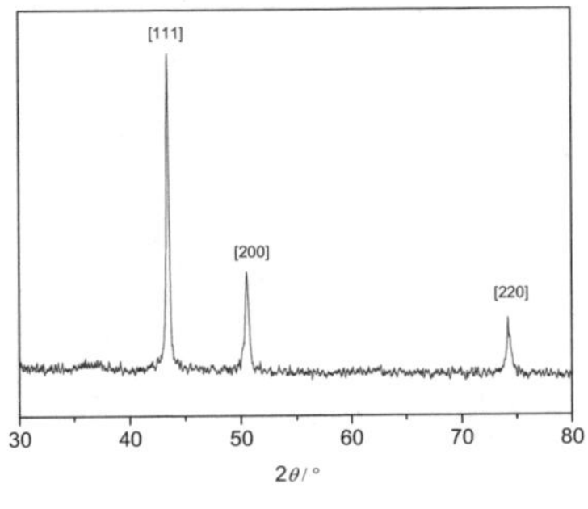

Fig. 1: X-ray diffraction pattern of the as-prepared typical sample.

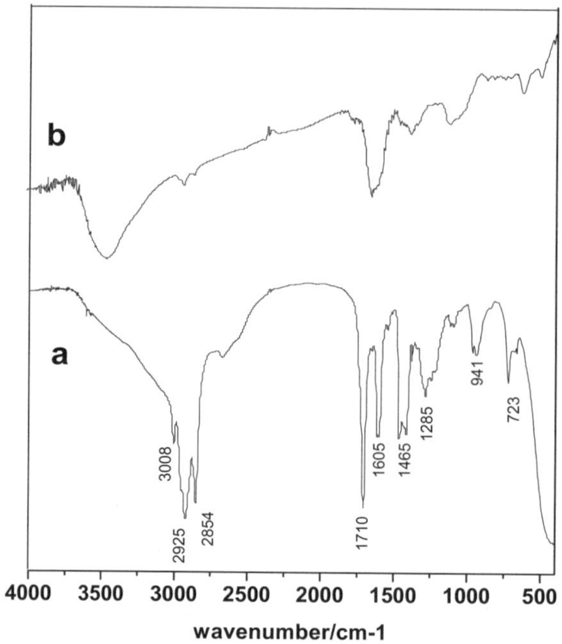

Fig. 2: The FTIR spectra of a) Copper oleate; b) Oleic acid coated copper nanoparticles

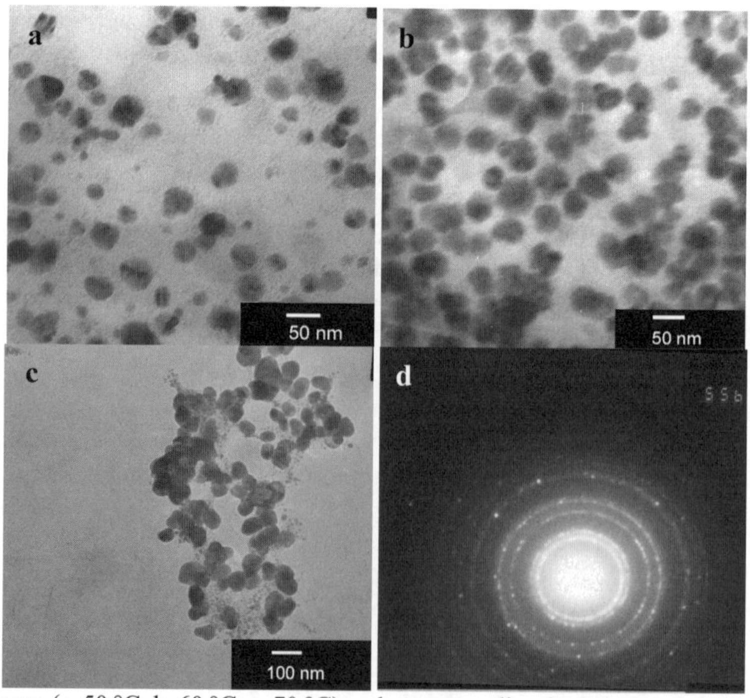

Fig. 3: TEM image (a, 50 °C; b, 60 °C; c, 70 °C) and corresponding SAED pattern of Cu nanoparticles prepared at 60 °C (d) of a Cu nanofluid prepared under a typical procedure.

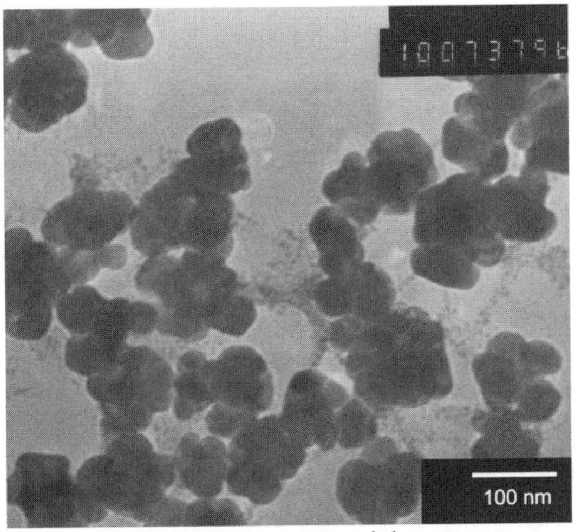

Fig. 4: TEM image of copper nanoparticles were prepared with a mole ratio of oleic acid/cupric ions = 2/1

Copper oleate was prepared through neutralisation of oleic acid with the precipitated copper hydroxide. The mole ratio of oleic acid to copper is more than 2:1. When the mole ratio of oleic acid to copper is just 2:1, the resultant particles are also hydrophobic, but the copper nanoparticles aggregated as the TEM image shown in Fig. 4. Oleic acid acts as a surfactant to coat the new generated particles surface. During the growth process, oleic acid prevented the copper particles from aggregation and made the resultant copper particles. In addition, preventing the copper particles from oxidation is another role in this process. When the mole ratio of oleic acid to copper is 2:1, the oleic acid is just used to produce copper oleates and no more oleic acid is used for preventing the copper particles from aggregation. As shown in Fig. 4, aggregation appears between the copper nanoparticles.

Oleic acid plays an important role throughout this process and the oleic acid seems to form a reversed-micelle in reaction system. In this process, oleic acid acts as both a particle protector coordinating their carboxyl end groups on the newly generated copper particles surface, and the hydrophobic carbon tails of the oleic acids are pointed outwards from the surface of particles [7-10]. Therefore the oleic acid-coated copper nanoparticles should be hydrophobic, but the resulted copper nanoparticles are hydrophilic. So the hydrophilic groups of the oleic acid are pointed outwards. The coating layer of the copper nanoparticles is not only one. Fig. 5 schematically illustrates the surface structure of an oleic acid-coated copper nanoparticle. Interaction between the oleic acid and copper nanoparticles may play an important role, because washing with water and alcohol several times did not remove the oleic acid from the surface of the copper particles. Even heating to boiling, the oleic acid layers on the surface of the copper nanoparticles were still not removed. In the other hand, this coating layer also plays an important role for preventing the newly generated copper particles from oxidation. After ultrasonic washing, it is found that the nanopaticles turned from hydrophilic to hydrophobic. This means that the layer of the hydrophilic tails of the oleic acids pointed outwards (the outside layer) was rinsed off in this process. Then the hydrophobic nanoparticles in the system were kept ultrasonic washing for a while. It is found that the nanoparticles are still hydrophobic, which means that the inside layer was not rinsed off. The above experiments indicated the surface of the copper nanoparticles coated two layers of oleic acids, as shown in Fig. 5.

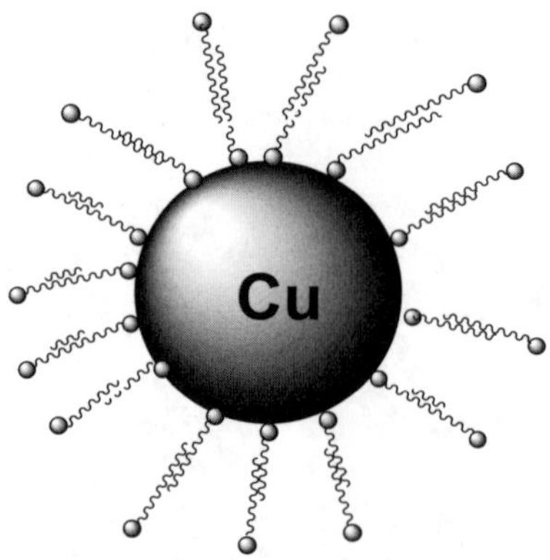

Hydrophilic group hydrophobic group

Fig. 5: Schematic illustration of an oleic acid-coated copper nanoparticle

Viscosity and thermal conductivity. The viscosity of nanofluids was investigated by measuring the viscosity of nanofluids for different temperatures ranging from 25~60 °C. Fig. 6 illustrates the viscosity against temperature for the prepared nanofluids. Thermal conductivity of the prepared nanofluids as a function of mass fraction of nanoparticles for different temperatures ranging from 25~40 °C is presented in Fig. 7. It can be seen that the thermal conductivity of copper nanofluid increases with increasing the mass fraction of nanoparticle for different based nanofluids. The relationship between the thermal conductivity enhancement and the mass fraction is nonlinear.

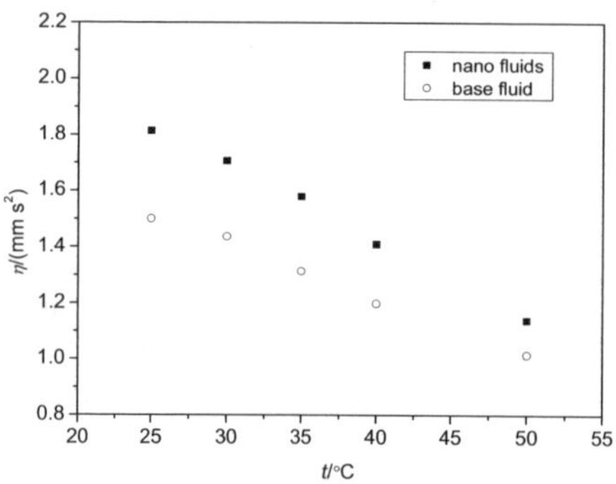

Fig. 6: The variety of the viscosity of nanofluids with temperature change

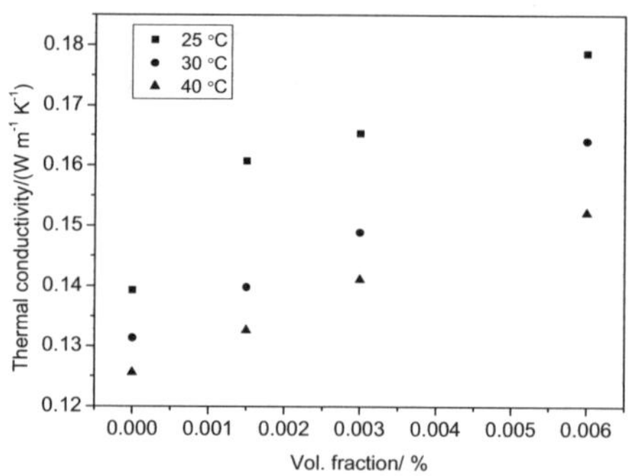

Fig. 7: Thermal conductivity of prepared Cu Nanofluids

Summary

This study illustrates simple and convenient methods for the synthesis of copper nanoparticles through the reduction of an organometallic precursor in an organic phase. The precursor is completely transformed to copper nanoparticle coated by oleic acid. This method for the preparation of hydrophilic nanoparticle avoids production of the inorganic ions that are usually associated with aqueous methods. The oleic acid prevents the growth and aggregation of copper nanoparticles in the preparation, resulting in a stable copper nanofluid. The current copper nanofluid has a higher thermal conductivity. This showed that the product obtained from this method can be hydrophilic and can be used either with aqueous medium or in polar organic medium for further applications.

References

[1] Y. Xuan, Q. Li: *Heat transfer enhancement of nanofluids*. International Journal of Heat and Fluid Flow, 21, 1(2000), pp. 58-64.

[2] M. Liu, M. C. Lin, C.Y. Tsai, C. Wang: *Enhancement of thermal conductivity with Cu for nanofluidsn using chemical reduction method.* International Journal of Heat and Mass Transfer, 49, 17(2006), pp. 3028-3033.

[3] C. Choi, H. S. Yoo, J. M. Oh: *Preparation and heat transfer properties of nanoparticle-in-transformer oil dispersions as advanced energy-efficient coolants.* Current Applied Physics, 8, 6(2008), pp. 710-712.

[4] P .K. Khanna, S. Gaikwad, P. V. Adhyapak, N. Singh, R. Marimuthu: *Synthesis and characterization of copper nanoparticles*. Materials Letters, 61, 25(2007), pp. 4711-4714.

[5] P. Kanninen, C. Johans, J. Merta, K. Kontturi: *Influence of ligand structure on the stability and oxidation of copper nanoparticles*. Journal of Colloid and Interface Science, 318, 1(2008), pp. 88-95.

[6] S. A. AL-Thabaiti, F.M. Al-Nowaiser, A.Y. Obaid, A.O. Al-Youbi, Z. Khan: *Formation and characterization of surfactant stabilized silver nanoparticles: A kinetic study*. Colloids and Surfaces B: Biointerfaces, 67, 2(2008), pp. 230-237.

[7] P. K. Khannaa, T. S. Kale, M. Shaikh, N. K. Raob, C.V.V. Satyanarayana: *Synthesis of oleic acid capped copper nano-particles via reduction of copper salt by SFS.* Materials Chemistry and Physics, 110, 1(2008), pp. 21-25.

[8] J. Yang, Y. Zhou: *A new method for preparing hydrophobic nano-copper powders*. J Mater Sci, 42, 18(2007), pp. 7638-7642.

[9] S. Bucak, A. Pugh-Jones, C. Lewis, D.C. Steytler: *Metal nanoparticle formation in oil media using di(2-ethylhexyl) phosphoric acid (HDEHP)*. Journal of Colloid and Interface Science, 320, 1(2008), pp. 163-167.

[10] C. Wu, B. P. Mosher, T. Zeng: *One-step green route to narrowly dispersed copper nanocrystals.* Journal of Nanoparticle Research, 8, 6(2006), pp. 965-969.

[11] H. Zhu, Y. Lin, Y. Yin: *A novel one-step chemical method for preparation of copper nanofluids*. Journal of Colloid and Interface Science, 277, 1(2004), pp. 100-103.

Advanced Materials Research Vol. 279 (2011) pp 142-146
© (2011) Trans Tech Publications, Switzerland
doi:10.4028/www.scientific.net/AMR.279.142

New technology research of nondestructive testing for aero aircraft composite materials

Ning Shan[1, a], Guo Heng[2,b]

[1]Department of Equipment and Transportation, Engineering College of Armed Police Force, ShannXi, Xi'an, China 710086

[2] School of Electric Power, North China University of Water Conservancy And Electric Power, Henan, Zheng Zhou, China 450011

[a]ssnn3193@163.com, [b]ning25531@163.com

Keywords: Composite materials, aero aircraft, nondestructive testing, continuous distributed sensing technology

Abstract. Composite materials have many advantages such as big specific strength, high specific stiffness, good disrepair safety characteristic, large area molding conveniently and easy to forming complex shape. It becomes important structural materials of aero aircraft equipment necessarily and is used widely in aviation industry. Composite materials components of aero aircraft with complex and delicate structure chronically work in harsh environment. Early testing of its key component's minute defects has been an urgent and necessary task. In this paper, its damage model and characteristic are analyzed and studied. Considering the limitation of composite materials structure's detection methods at present, a number of existing problems are presented and development directions are pointed out. Laser ultrasound detection technique, continuous distribution sensing technique and optical fiber sensing technique are studied. The method of damage detection of aero aircraft's composite materials components is put forward based on the organic combination of the three techniques. The results show that the system can be used to detect multi-ultrasound signals of large composite materials structure. Its structure is simple. It has small bulk and low cost. It is characterized by easy realization.

Introduction

With the increasing advance of requirements of materials quality by the rapid development of Aero-technology, the Single-material materials do not meet the requirements of high-performance and high-index for Aerospace systems. The composite materials are really featured by high specific strength, high strength-to-weight ratio, corrosion-resistant, fatigue resistibility, good security of damage, easy design, molding of large areas easily, shaping into complex forms easily, etc. The composite materials can meet the high requirements of safety, economy and comfort for aeronautic vehicles, and the materials have been necessary and important structural materials of aviation engineering[1-6].

Study on the ultrasonic nondestructive detection method of composite materials

The composite materials of large structure for aeronautic vehicles are used in long-term high-temperature, high pressure, and high revolving environment, and the materials are very sensitive to fatigue crack, corrosion damage and impact flaw. It's easy to lead to the failure of key structures, the catastrophic accident and a heavy loss. Also, the high cost of maintenance and safeguard increase the work cost of composite materials sharply. For example, composite materials dosage of a certain type aircraft is up to 35%. Just off-line testing of composite skins for aircraft fuselage needs to take scores of man-hours, and a large amount of money is spent. If all the composite materials structures for aircraft are tested, the more times and cost will be taken. So, the study on new theory and technology of health test for large composite materials structures, which is used in the rapid and real-time test of engineering structures damages and the reasonable assessment of damages, has important theoretical meaning and broad application prospect for ensuring the running safety of engineering structures, forecasting catastrophic accident, setting a reasonable damage tolerance, decreasing the cost of safeguard, extending work life of engineering structures, and so on. Now the methods of nondestructive testing for composite materials include noninstrumental method, liquid penetrant method, ray test method, thermal wave imaging method, acoustic emission method, eddy current method, microwave test method, and ultrasonic test method. The ultrasonic test method can test many hazardous defects of structures, such as loose, hierarchie, inclusion, pore, crack, debonding, and so on, and it has high sensibility and broad using scope. The ultrasonic method is used in arbitrary materials and outer circumstances, and then it can be used in testing the mechanical characterization of new composite materials. So, the ultrasonic test method is increasingly used in the nondestructive testing of composite materials.

If there are defects inside the materials, the inside mediums of materials are discontinuous. The ultrasonic wave encounters defects in the progress of propagation, and it's equal with inserting into another medium from one medium. Now the refraction and reflection are caused at the two interfaces, under perpendicular incident condition, the sound pressure reflectance rp and transmittance tp are Eq. 1 and Eq. 2 respectively.

$$r_p = \frac{P_r}{P_c} = \frac{Z_2 - Z_1}{Z_2 + Z_1} \tag{1}$$

and

$$t_p = \frac{P_t}{P_c} = \frac{2Z_2}{Z_2 + Z_1} \tag{2}$$

Here, P_c, P_r and P_t are the sound pressure of incident wave, reflected wave and transmitted wave respectively; Z_1 and Z_2 are the acoustic impedance of the two interfaces.

Limitation of ultrasonic nondestructive detection

In the initial development period of composite materials, the structures are smaller in size and amount. The manual ultrasonic testing and single automatic ultrasonic testing are mainly used, it takes more times and energies, and then it has low efficiency and poor reliability. With the rapid development of composite materials technology, the structures in large size are widely used. To meet the high requirements of test accuracy and efficiency, ultrasonic phased array, air-coupled ultrasonic and water coupled ultrasonic have been arisen. Ultrasonic phased array adopts multi-chip probe to ultrasonic excitation, and it can test large area single detection. The testing efficiency is improved. But with the increasing of testing area, the size of probe increases. It is difficult to process, and the cost increases greatly, too. The testing accuracy is decreased. Air-coupled ultrasonic adopts air as_the coupling medium, it has large attenuation of ultrasonic signals and low accuracy, and it can't achieve real-time online detection. By water coupled ultrasonic, because the sample is placed in a special tank and sprinkler system provides coupled water column, the size of tested specimen is restricted. It can't achieve in-situ detection, and water stain is removed difficultlly. What's more, the above traditional ultrasonic testing all adopt piezoelectric transducer to generate ultrasonic, and the same probe usually only produce a waveform. It has narrow band, usually tens of megahertz, and the testing accuracy is lower, mostly millimeter level[7,8].

At home and abroad, the ultrasonic nondestructive testing technology is as pilot project of the study and application of composite materials, a high attention is regarded and some results are obtained. But the ultrasonic testing technology of composite materials develops incompletely, the system is incomplete, the advanced equipments are less and impractical, and the gap is widening between the equipments and the rapid development of composite materials technology. Now, the major problems that affect the development of ultrasonic testing technology for composite materials, are obtaining of ultrasonic signal by piezoelectric ultrasonic sensors and optical ultrasound measuring device[9-11]. However, a lot of engineering practice has proved that environmental factors such as humidity and electromagnetic fields can badly affect the detection accuracy and long-term stability of commonly used piezoelectric ultrasonic sensors. The frequency of piezoelectric ultrasonic sensors is low and the sensing area is small, which is going against to early detection of small defects to large-scale structures. The clearing of coupling agent is difficult while using, so the on-line health testing isn't make to most engineering structures. The traditional optical ultrasound measuring devices are bulky. It has the disadvantages such as the stallation inconvenient, movement difficulty, equipment complexity, operation difficulty and anti-interference ability poor. The traditional optical ultrasound measuring devices can not adapt to specificity of large composite structures and high detection accuracy.

Design of continuous distributed optical fiber ultrasound sensing system

To meet the rapid development of composite materials technology and solve the rapid testing of large composite structures, the fiber f-p sensor technology is combined with the distributed continuous sensing technology in this paper. The multiple ultrasonic information of testing damage of large composite structures is established, and the continuously distributed optical fiber ultrasonic sensor system can localize the damage sources effectively and precisely. As shown in Fig. 1, the multi-sensor array is composed of a number of optical fiber sensing probe, and one amplified spontaneous emission (ASE) broadband light source provides the light source needed by the multi-sensor array. The light from light source transmits to the fiber-coupled array by the cluster fiber, and then into the probe array. The reflected light array carrying multiple ultrasonic information transmits to the wavelength division multiplexing(WDM) array by the fiber coupler again. The needed orthogonal optical signals are filtered out according to the different structural parameters of each probe. The electrical signal is converted by the photo-detector array. The signals are input to computer after acquiring and processing. The ultrasonic signal testing by a single optical fiber F-P probe is shown in Fig. 2.

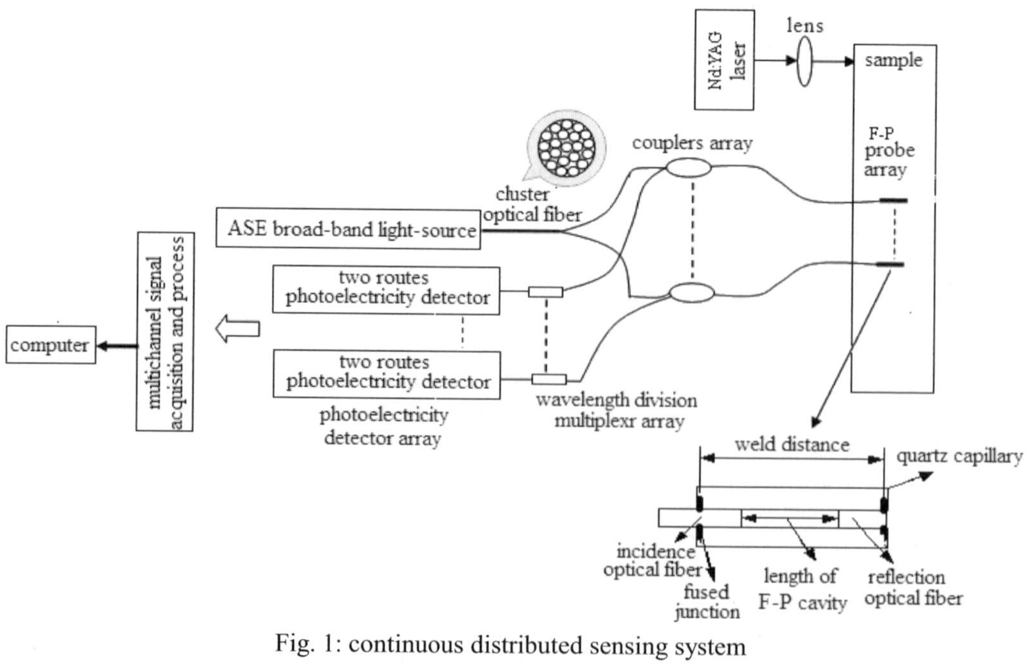

Fig. 1: continuous distributed sensing system

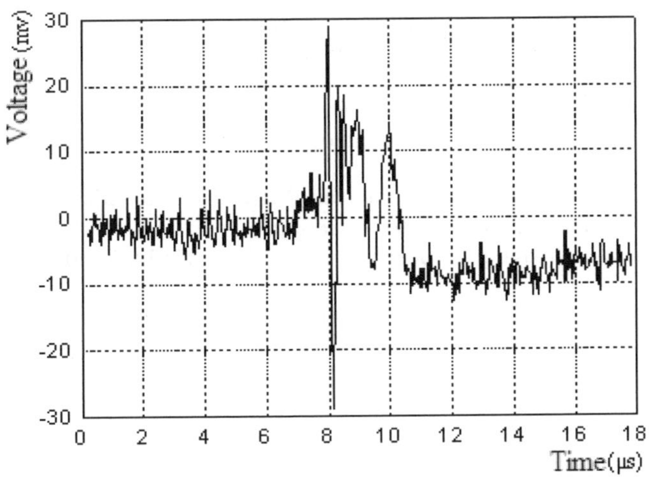

Fig. 2: ultrasound detected by single optical fiber F-P sensor

Conclusion

In this paper, the composite materials's damage model and characteristic are analysed and studied. According to the limitation of composite materials structure's detection methods at present, a number of existing problems are presented and development directions are pointed out. Laser ultrasound detection technique, continuous distribution sensing technique and optical fiber sensing technique are studied. The method of damage detection of aero aircraft's composite materials components is put forward based on three techniques organic combination. The results show that the system can be used to detect multi-ultrasound signals of large composite materials structure. Its structure is simple. It has small bulk and low cost. It is characterized by easy realization.

Acknowledgments

This work is supported by China Postdoctoral Science Foundation Grant # 20100481509.

References

[1] Du Shanyi: *Advanced composite materials and aerospace engineering*. Acta Materiae Compositae Sinic, 24, 5(2007), pp.1-12.

[2] Major Erich S. Johnson: *RAH-66 Comanche-Eyes and Ears for the 21st Century*. Field Artillery, Vol.5-6(1996), pp.22-25.

[3] Schwarz K., *Sukhoi S-37 Berkut Unveiled*. Flug Revue, Vol.11(1999), pp.58.

[4] Du Shanyi, Guan Zhidong: *Strategic considerations for development of advanced composite technology for large commercial aircraft in China*. Acta Materiae Compositae Sinic,25, 1(2008), pp.1-10.

[5] Yang nai bin: *Composite Structures for New Generation Large Commercial Jet*. Acta Aeronautica Et Astronautica Sinica, 25, 3(2008), pp.596-604.

[6] Zhang Jifeng, Zhang Boming, Du Shanyi: *Improvement an parameter design of advanced composite grid plate structure*. Acta Materiae Compositae Sinica, 23, 3(2006), pp.153-157.

[7] Zhou Zhenggan, Xiang Shan, Wei Dong: *Ultrasonic Testing Technologies for Composites*. Aeronautical Manufacturing Technology, Vol84(2009), pp.70-73.

[8] Thomas S., Bongiovann C., Nutt S.R.: *In situ estimation of through-thickness resin flow using ultrasound*. Composites Science and Technology, Vol15-16(2008), pp.3093-3098.

[9] Rousseau Guy, Blouin Alain, Monchalin Jean-Pierre: *Ultrasound-modulated optical imaging using a confocal Fabry-Perot interferometer and a powerful long pulse laser*. Progress in Biomedical Optics and Imaging - Proceedings of SPIE, Vol7177(2009), pp.71771E.

[10] Thursby G., McKee C., Culshaw B.: *A comparison of three optical systems for the detection of broadband ultrasound*. Proceedings of SPIE - The International Society for Optical Engineering, Vol7648(2010), pp.76480T.

[11] Sorazu B., Thursby G., Culshaw B.: *Optical generation and detection of ultrasound*. Strain, Vol3(2003), pp.111-114.

Advanced Materials Research Vol. 279 (2011) pp 147-150
© (2011) Trans Tech Publications, Switzerland
doi:10.4028/www.scientific.net/AMR.279.147

Study on the Adsorption Capability of Nano-barium-strontium Titanate Immobilized on Aluminum Oxide for Zinc Ion

Dong Zhang[1, a], Yanli Zhang[2, b]

[1]School of Environmental and Chemical Engineering Shenyang Ligong University, Shenyang 110159, China

[2]School of Foreign Languages, Shenyang Ligong University, Shenyang 110159, China

[a]sylgdxdong@sina.com, [b]ellezhang3218@sina.com

Keywords: Nano-barium-stroutium titanate, Aluminum oxide, Adsorption, Preconcentration, Zinc, Water

Abstract. Zinc ion was adsorbed and enriched by nano-barium-strontium titanate immobilized on aluminum oxide(ABST). The adsorption and elution conditions were investigated. The results showed that ABST had strong adsorption ability for zinc ion in water in the medium at pH more than 4.0. The maximum adsorption capability was up to 16.38 mg/g. The adsorbed zinc on ABST could be eluted and reclaimed with 0.5 mol/L of HNO_3. Zinc was determined by flame atomic absorption spectrophotometry (FAAS). The detection limit of this method for zinc ion was 2.5 μg/L. The method had been applied to the determination of trace zinc in surface water and tap water samples with recovery of 92.5%-98.7%. The results were found to be in agreement with the method of dithizone spectrophotometry.

Introduction

As we all know, zinc is an indispensable beneficial element to human life. However, zinc can be accumulated in the tissues of human, animals and plants, and it has certain toxicity at high concentration. At present, the methods for determining zinc ion mainly include spectrophotometry and atomic absorption spectrometry. For low concentration and high saline samples, enrichment and separation are also needed [1]. Barium-strontium titanate is a kind of important electronic ceramic materials with excellent optical, electrical and magnetic properties, and it has been widely applied to minitype capacitor, ferroelectric memory and electro-optical device, etc. Using $BaCl_2$, $SrCl_2$ and $TiCl_4$ as the sources of Ba, Sr and Ti, respectively, nanosized barium-strontium titanate powder by oxalate chemical coprecipitation was prepared and successfully used in adsorption and enrichment determination of heavy metal ions in water including lead, cadmium and chromium, etc[2-4]. However, the nano-barium strontium titanate particles are so small that, when used in adsorption of metal ions, it was easy to coacervate and lose activity, and was difficult to recover. Fixing nanoparticle onto different substrates, such as silica gel G, corrects these problems with satisfactory results [5,6]. In our previous work, the immobilized nano- barium strontium titanate onto aluminum oxide has been studied. But nano-barium-strontium titanate immobilized on aluminum oxide (ABST) was applied to the adsorption and enrichment determination of zinc ion in water has not been reported. In this work, ABST was used as a solid-phase extraction agent to concentration and separation of zinc in surface water and tap water and determined by flame atomic absorption spectrometry.

Experimental

Main equipments and reagents. WYX-9003A type flame atomic absorption spectrophotometry (Shenyang Analytical Instrument Factory); A PHS-3C acidometer (Shanghai REX Instrument Factory, Shanghai, China) supplied with a combined electrode. A constant temperature bath shaker (Jintan Experiment Instrument Factory, Jiangsu China).

Nano-barium-strontium titanate immobilized on aluminum oxide(ABST) was synthesized and characterized according to reference [7].

Zinc standard stock solution: 1 g/L, which was prepared using high purity zinc particle by conventional method and diluted to required concentration when used.

Static adsorption experimental method. Certain Zn^{2+} solution was added into a 10 mL centrifuge tube with plug and scale, and then diluted to scale with water at pH 6. After 0.5 g of ABST was added, the centrifuge tube was pluged, and then shaken for adsorbing at constant temperature for 10 min. The suspension was centrifugalized, the content of Zn^{2+} in supernate was determined using flame atomic absorption spectrophotometry and the adsorption capacity was calculated.

Elution. After adsorption, the supernate was abandoned and the precipitation was washed twice by water. After 5 mL of 0.5 mol/L nitric acid was added, the solution was shaken and eluted for 5min. Then the suspension was centrifugalized and the content of Zn^{2+} in supernate was determined by flame atomic absorption spectrophotometry. The recoveries were calculated.

Results and discussions

Effect of pH on adsorption. The pH affects the distribution of active sites on a titanate surface. At high pH, the OH^- on the surface provides the higher potential of binding cations. The decrease of pH leads to lower OH^- concentration on the surface, so the adsorption of cations onto calcium titanate decreases accordingly. 0.2 mg of Zn^{2+} was added to a centrifuge tube and was diluted to 10mL with water at different pH values (prepared by mixing 0.1 moL/L of hydrochloride with 0.1 moL/L of ammonia determined on acidimeter). The adsorption capacities were determined according to the experimental method and the recoveries were calculated, as shown in Fig. 1.

The results showed that the pH value of solution had great influence on adsorption ratios. When pH value was higher than 4, the adsorption ratio was up to the maximum. pH 6 was used in this experiment.

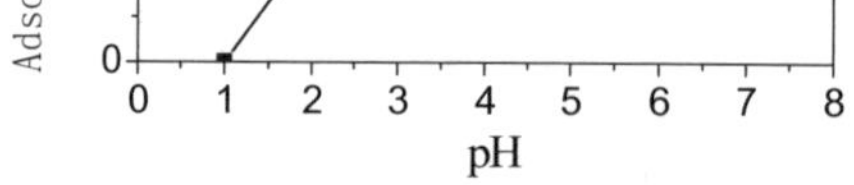

Fig. 1: Effect of pH on the adsorption percentage

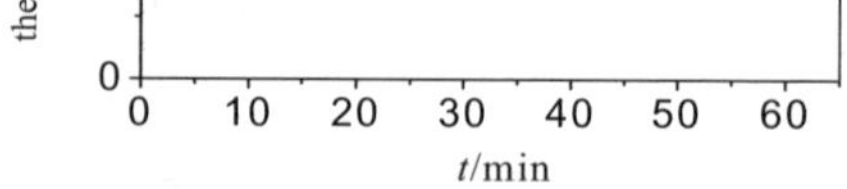

Fig. 2: Effect of contact time on zinc ion adsorption

Effect of shaking absorption time on adsorption capacity. In the dissolution media at pH=6, the adsorption capacities were determined at different shaking absorption time, respectively. The results were showed in Fig. 2.

As seen in Fig.2, shaking absorption time had a great influence on the adsorption capacity. With the increase of shaking absorption time, the adsorption capacity was increased. The maximum adsorption capacity was observed after 10 min, beyond which there was almost no further an increase in the adsorption. In this paper, the shaking time was 10 min.

Elution experiment. After adsorption under optimum adsorption conditions, Zn^{2+} was eluted by 5 mL of 0.05-2.5 mol/L nitric acid solution. The recoveries were calculated. The results showed that when Zn^{2+} was eluted by 5 mL of 0.5 mol/L nitric acid solution, with shaking eluting time of 5 min, the recoveries were up to 95%.

0.5 mol/L nitric acid was used as eluent in the experiments. After elution, ABST was used repeatedly for three times and the recoveries had no significant decrease.

Static adsorption capacity. According to Fig. 2, under the optimized adsorption conditions, the static adsorption capacity of ABST to zinc ion was 16.38 mg/g. ABST had a great adsorption capacity to zinc ion and it was an ideal material for adsorption and enrichment.

Enrichment multiple, detection limit and precision. 50 mL colorimetric cylinder was used instead of centrifuge tube and filled with 50 mL of water at pH=6. After 0.5 g of ABST was added, the adsorption was determined according to the experimental method. Then 5 mL of eluent was used for elution. The recoveries were all higher than 95% after enriched by ABST. The detection limit of this method (3σ20) was 2.5 μg/L. For Zn^{2+} solution with initial concentration of 0.01 mg/L, independent adsorption experiments were conducted for 6 times and the calculated RSD was 2.8%.

Effect of concurrent ions. Different amounts of each ion were added to 50mL colorimetric cylinders containing 5 μg Zn^{2+}, respectively. After the pH was adjusted, the solution was diluted to 50 mL with water at pH 6, and then 0.5g of ABST was added. According to the experimental method, the recoveries of Zn^{2+} were determined. The errors were controlled in ±5%. The allowable amounts of concurrent ions were (in mg): Na^+, K^+, NH_4^+, NO_3^-(10); Cl^-, Cd^{2+}, Pb^{2+}(5); Mg^{2+}, Fe^{3+}, Ni^{2+}, Cr(III), Ca^{2+}, Al^{3+}(2); Cu^{2+}(1).

Sample analyses. Water sample was pretreated by conventional method [1]. Proper amount of the sample was decanted into a 50mL colorimetric cylinder, adjusting the pH value. The solution was diluted to scale by water at pH 6 and then 0.5 g of adsorbent was added. The adsorption and enrichment was determined by experimental method. The recovery of standard addition test was carried out and compared with the results determined by dithizone spectrophotometry [1] (showed in Table 1). The results showed that the recoveries of standard addition were between 92.5% and 98.7%, and the results were in good agreement with those obtained by dithizone spectrophotometry.

Table 1: Determination results of Zn^{2+} in water samples mg/L

Water Sample	Found	Added	Recovered	Recovery [%]	Found by dithizone spectrophotometry
		0.02	0.0981	92.5	
River water	0.0796	0.05	0.1278	96.4	0.0764
		0.10	0.1783	98.7	
		0.02	0.0433	97.5	
Tap water	0.0238	0.05	0.0722	96.8	0.0240
		0.10	0.1201	96.3	

Conclusions

According to the research results in this paper, it could be concluded that nano-barium-strontium titanate immobilized on aluminum oxide (ABST) had strong adsorption ability for zinc ion in water in the medium at pH more than 4.0. The maximum adsorption capability was up to 16.38 mg/g. The adsorbed zinc on ABST could be eluted and reclaimed with 0.5 mol/L of HNO_3. Zinc was determined by flame atomic absorption spectrophotometry (FAAS). The detection limit of this method for zinc ion was 2.5 μg/L. The method had been applied to the determination of trace zinc in surface water and tap water samples with satisfied result.

References

[1] State Environmental Protection Administration of China: *Analysis Method of Water and Waste Water*. China Environmental Science Press, fourth ed., Beijing, (2002).

[2] H. D. Su, D. Zhang: *Preparation of nanometer-sized BST powder and its adsorption behavior for cadmium ion*. J. Chem. Ind. Eng., 57, 12(2006), pp. 2892-2896.

[3] D. Zhang, H. D. Su, H. Gao: *Preparation of barium-stroutium titanate nano-powder and its adsorption properties for chromium(VI) and chromium(III)*. Metal. Anal., 27, 4(2007), pp.7-10.

[4] D. Zhang, H. D. Su, H. Gao: *Study on Adsorption Behavior of Nanosized Barium-Strontium Titanate Powder for Lead Ion in Water Using FAAS*. Spectrosc. Spect. Anal. 28, 1(2008), pp.218-221.

[5] D. Zhang, W.J.Zhang, X. Guan, H.Gao, H.B.He: *Adsorption Behavior of Immobilized Nanometer Barium-Strontium Titanate for Cadmium Ion in Water*. Spectrosc. Spect. Anal. 29, 3(2009), pp.824-828.

[6] D. Zhang, Z. G. Yuan, X. Guan, W.J.Zhang: *Preparation of Strontium-Barium Titanate Based on Glass Fiber Filter and Its Adsorption Behavior for Cadmium Ion*. Rare Metal Materails Eng., 38, 12(2009), pp.2207-2211.

[7] D. Zhang, C.L. Zhang: *Preparation and characterization of nanosized barium-strontium titanate immobilized on aluminum oxide and its application in adsorption*. Applied Mechanics and Materials Vol. 26-28 (2010), pp. 682-685.

Advanced Materials Research Vol. 279 (2011) pp 151-156
doi:10.4028/www.scientific.net/AMR.279.151

Fabrication and Characteristics of elliptical-holes and near elliptical core hexangular lattice photonic crystal fibers based on polymer

Fei Liu[a], Hongyan Gao[b], Qiang Xu[c] and Yani Zhang[d]

Department of Physics and information Technology, Baoji College of Arts & Science, Baoji, China 721007

[a]bwllf@163.com,[b]bwlghy@163.com, [c]xuqiangwlx@163.com, [d]zhangyn@opt.ac.cn

Keywords: guided-wave optics, optical materials, polarization properties, photonic crystal fiber

Abstract. The polarization properties of a new structure of high birefringence polarization maintaining photonic crystal fibers were explored, which were consisted of elliptical-holes and near elliptical core triangle lattice of hexangular structure based on polymer. The optical birefringence and the first order polarization mode dispersion were studied based on a full-vector localized basis function method. The numerical results show that the polarization properties are strongly dependent on structural parameters of the PCF. Furthermore, the proposed fiber was fabricated based on MMA monomer polymerization method in situ. So the proposed PCFs may be useful in the field of high birefringence polarization-maintaining optical communication.

Introduction

Photonic Crystal Fiber (PCF)[1] consisting of a central defect region surrounded by multiple air holes running along its length has been the subject of much interest in recent years because of their unique properties that are not realized in conventional optical fibers. According to the recent literatures, high birefringence in PCFs can be produced by combining the asymmetric core and the large core-cladding index contrast[2], by introducing two-size air holes around the fiber core[3-5], and by selectively filling the air holes with polymer to obtain an asymmetric structure[6-8]. The principle of the above methods is to produce birefringence, which is realized by breaking the 60 symmetry of the triangular lattice PCFs. Comparing with those of the conventional polarization maintaining fibres (PMFs), such as PANDA and bow-tie fibres, for which modal birefringence is about 5×10^{-4}, the introduced birefringence in PCFs is one order of magnitude larger than that of conventional PMFs.

Because of the difficulty to fabricate the shape of elliptical holes, in this paper, we proposed a new structure high birefringence polarization maintaining PCF with near elliptical core design by three adjacent central holes missing and triangular lattice squeezed hexangular structure air holes cladding. The polarization properties of the proposed PCF, ncluding to the modal birefringence and the first order polarization mode dispersion, were explored by using a full-vector localized basis function method[9]. The result shows that the polarization properties are strongly dependent on structural parameters of the PCF. Furthermore, the proposed fiber was fabricated by using the polymerization method in-situ and the sleeving draw-stretching based on methyl methacrylate (MMA) monomer[10]. The optical properties of fabricated fiber were measured and modal birefringence was well consistent with numerical calculation.

Design principle and theoretical model

The cross-section of near elliptical core squeezed hexangular structure PCF is indicated in Fig. 1. It is the polymer PCF that has an array of air holes running along its length. The elliptical air holes can be specified by transforming the circular air holes with a specified stress factor: *b/d=d/a* in Fig. 2. The structure parameters are characterized by holes diameter *d* and holes pitch Λ respectively. The refractive index of the polymer PMMA region is 1.49. To simulate the propagation properties, especially modal birefringence and polarization mode dispersion, the scalar approximation becomes inadequate. Because the distributions of electro-magnetic fields are intensively localized, a localized basis function was introduced to get the solution of Maxwell's equations. And then Maxwell's equations are reformulated for a medium, which is translational invariant along the Z-axis as an eigen-value problem for the propagation constant, i. e.

$$(\nabla^2+k^2\varepsilon)\ h+(\nabla \ln\varepsilon)\times(\nabla\times h)=\beta^2 h \tag{1}$$

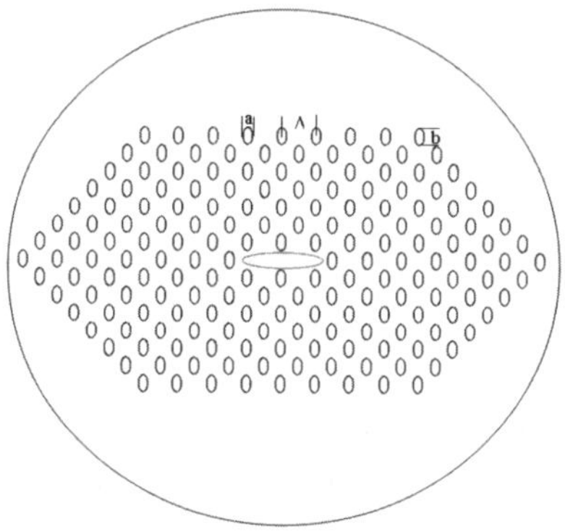

Fig. 1: Cross section of elliptical-holes and near elliptical core hexangular lattice PCF

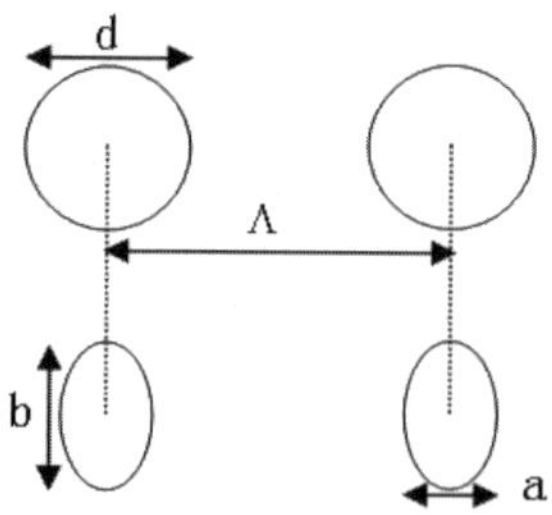

Fig. 2: Illustration of the structure parameters

For the optical fibers, the distributions of electro-magnetic fields could be simplified to two dimensions plane-wave problems, so Eq. 1 could be transformed into the following Eq. 2.

$$(\nabla_t^2+k^2\varepsilon)\ h_t+(\nabla_t \ln\varepsilon)\times(\nabla_t\times h_t)=\beta^2 h_t \tag{2}$$

Among Eq. 2

$$h_t = h_x x + h_y y \tag{3}$$

$$\nabla_t = \frac{\partial}{\partial x}x+\frac{\partial}{\partial y}y \tag{4}$$

k is free-space wave-number, ε is the dielectric constant of materials, hx and hy are the magnetic field in the direction x and y. For the polarization optical fibers, the two linear polarization modes (hx and hy) are mutually orthogonal, and their propagation constants are βx and βy. Eq. 2 could be decomposed the following Eq. 5 and Eq. 6.

$$(\nabla_t^2+k^2\varepsilon)\ h_x+(\nabla_t \ln\varepsilon)\times(\nabla_t\times h_x)=\beta_x^2 h_x \quad (5)$$

$$(\nabla_t^2+k^2\varepsilon)\ h_y+(\nabla_t \ln\varepsilon)\times(\nabla_t\times h_y)=\beta_y^2 h_y \quad (6)$$

In order to get the solutions of Eq. 5 and Eq. 6 for the photonic crystal polarization optical fibers, we must get such a localized basis function that is continuous in the whole x, y plane, and whose differential coefficient is also continuous. The Hermite-Gaussian functions meets the conditions, and it is in the following Eq. 7.

$$H_n(x)=(-1)^n\exp(x^2)\frac{d^n(\exp(-x))}{dx^n} \quad (7)$$

Here n is integer 0, 1, 2 , $Hn(x)$ is continuous in the range of x from $-\infty$ to $+\infty$ and could differentiated, its differential coefficient is also continuous. Since the two linear polarization of hx and hy are orthogonal, the localized basis function must be orthogonalized, and the orthogonalizing factor $O(x)=exp(-x^2)$ is founded.

$$\int_{-\infty}^{+\infty} O(x)H_p(x)H_q(x)dx=0 \qquad (p=q) \quad (8)$$

$$\int_{-\infty}^{+\infty} O(x)H_p(x)H_q(x)dx=0 \qquad (p\neq q) \quad (9)$$

The space distance between two holes center is Λ, which is the period constant of the lattice of photonic crystal fiber. The relationship between the orthogonal localized function and Λ for photonic crystal polarization optical fibers is as follows Eq. (10).

$$\phi_{mn}=\exp[-(x^2+y^2)/(2\Lambda^2)]H_m(x/\Lambda)H_n(y/\Lambda) \quad (10)$$

The functions Φ_{mn} are mutually orthogonal and form a complete system in the whole x,y plane. We could make the certain region of the fiber core vicinities as a super-cell, and then we can supposed that the whole fiber are composed of many super-cells of this kind, the kind of super-cells consist of two dimensional periodical structure. So the Eq. 5 and Eq. 6 turn into the propagating constants β_x and β_y algebraic eigen-value problems of Eq. 11.

$$\sum_{kl} L_{k,l}^{m,n} h_x^{k,l}=\beta_x^2 h_x^{mn}, \qquad \sum_{kl} L_{k,l}^{m,n} h_y^{k,l}=\beta_y^2 h_y^{mn} \quad (11)$$

Among them, $L_{k,l}^{m,n}$ are the matrix coefficients of the operator on the left sides of the Eq. 5 and Eq. 6. From the Eq. 1 to Eq. 11, the solutions of propagation constants βx and βy could be got. The modal birefringence of photonic crystal polarization maintaining optical fiber could be calculated by Eq. (12)

$$B=(\beta_x-\beta_y)/k \quad (12)$$

For pulsed propagation in birefringent systems, the walk-off parameter is defined as Eq. 13

$$\Delta n_g=n_g^y-n_g^x \quad (13)$$

It determines the rate at which orthonormal polarized pulses separate on propagation. In fact,

Δn_g is the first-order polarized mode dispersion (PMD) and may be expressed as (14)

$$\Delta n_g = \Delta n - \lambda \frac{d\Delta n}{d\lambda} \tag{14}$$

Results and discussions

Generally the rings of air holes around the core region in PCF should be greater than five rings so that the confinement loss[11] could be as low as possible and guided mode can propagate stably in the fiber. Here, five rings of air hole in the cladding are arranged in the near elliptical core squeezed hexangular structure PCF with $\Lambda = 2\mu m$ and $d/\Lambda = 0.4$. Modal birefringence and walk-off parameter can be calculated in the wavelength range from 0.3 to 8 μm by aforementioned modeling algorithm. Fig. 3 shows variation of modal birefringence and walk-off parameter with wavelength. There is a maximum in the curve of modal birefringence. The decreasing tendency in modal birefringence can be explained as follows. An interface with the highest modal birefringence locates between core region and the cladding in near elliptical core squeezed hexangular structure PCF. If modal field distribution is close to the interface, the higher modal birefringence can be obtained. At the short wavelength, modal field can be confined well in the core region that is far away from the interface, so modal birefringence has a tendency to zero. Because more and more modal field permeated through the cladding, the modal birefringence reduced. Point P is the cross point of solid line Δn and dashed line Δn_g and its corresponding wavelength is the wavelength of the highest birefringence point. This can be explained by Eq. 14. According to the maximum condition i.e. $d\Delta n/d\lambda=0$, the relation $\Delta n_g=\Delta n$ can be deduced while modal birefringence reach to the maximum point P at the wavelength λp.

To study a general birefringence behavior of near elliptical core squeezed hexangular structure PCF, Fig. 4 depicts the influence of air-hole size d, hole pitch Λ, and the operating wavelength λ on modal birefringence. Clearly, birefringence has a strong dependence on air-hole size. Notably, for $d/\Lambda = 0.3$ or less, the fiber is endlessly single mode. In the single-mode domain, for a PCF with $d/\Lambda=0.7$, the birefringence is seen to be maximum, i.e., $B=1.0\times10^{-2}$ at a normalized frequency of $\Lambda\lambda. = 0.65$. However, as d/Λ decreases, the optical field flare more into the air holes, and at the birefringence peaks, the field in air holes is strongest. Thus, the model confinement is less at these maxima.

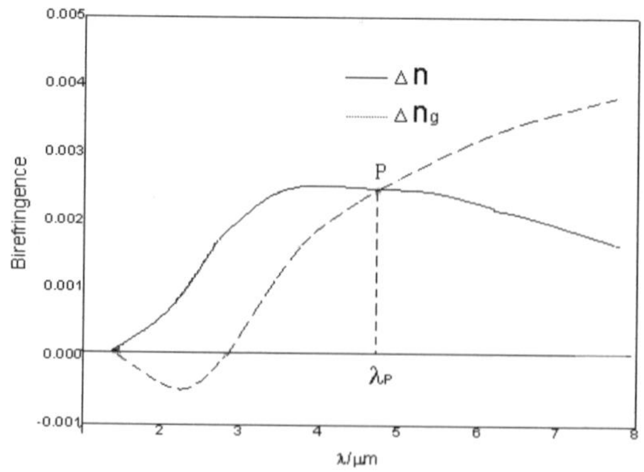

Fig. 3: Δn and Δn_g as a function of wavelength for the structure PCF in $\Lambda=2\mu m$ and $d/\Lambda=0.4$

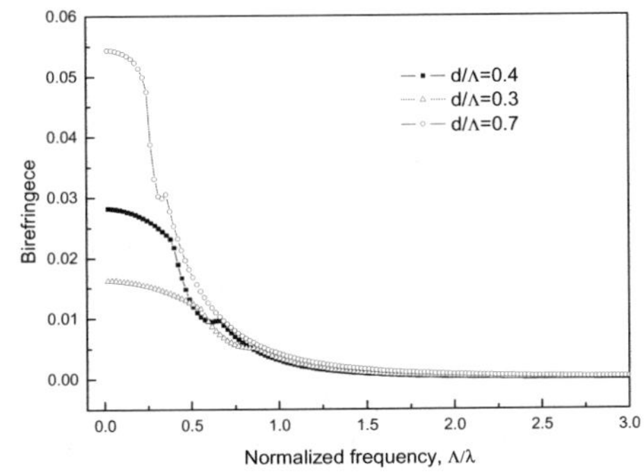

Fig. 4: Dependent of modal birefringence on hole size

Table 1: The relation of the leakage loss versus the air hole rings

The rings of air holes	Confinement loss [dB/m]
5	20.5
7	2.23
9	0.36
10	0.15

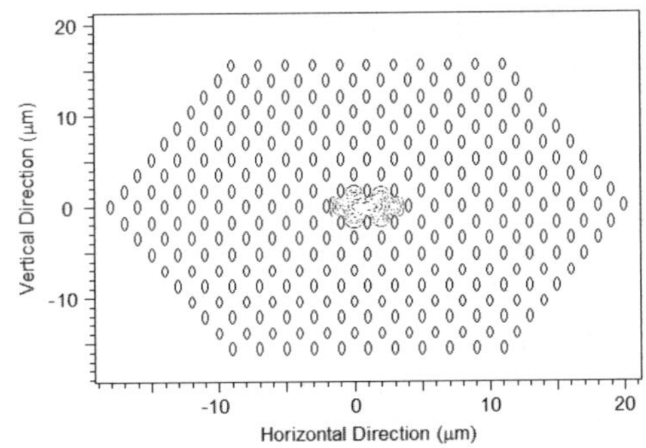

Fig. 5: Field profile of the structured PCFs at wavelength 650nm

Although it can be seen that the birefringence can be increase to the order of 10^{-2}, the polarization dependent leakage loss still needs to be evaluated before one can conclude the fiber structures to be useful. The leakage loss can be determined according to the following formulation:

$$\text{Leakage loss} = \frac{2\cdot 10^7}{\ln(10)}\cdot\frac{2\pi}{\lambda}\cdot \mathrm{Im}[n_{eff}](dB/m) \tag{15}$$

The cladding of PCFs contains a limited number of air holes, hence the guided wave mode in the core of the fibers is actually leaking in the transmission process, which is polarization dependent leakage loss. The full-vector FEM with hybrid edge/nodal element as well as the anisotropic perfectly-Matched-Layers (PMLs) boundary condition are applied to study the proposed structure, and the effective index n_{eff} of the fundamental mode can be obtained form the corresponding complex eigenvalue. Then according to the Eq.15, the leakage loss is counted at the work wavelength of 0.65μm. The leakage loss changes with the air hole rings, as shown in Table 1.

Obviously, with the increase of the number of air-hole rings, the leakage loss reduced significantly. When the air-hole rings added up to 10 rings, the leakage loss could be reduced to 0.15 dB/m. Fig. 5 shows the mode Field profile of rectangular-lattice PCFs at wavelength 650nm. The mode is very good constraint in the core area.

Conclusion

In conclusion, the polarization properties of near elliptical core squeezed hexagonal structure PM-PCF were explored by a full vector model localized basis function method, and the optimum modal birefringence dependence on holes pitch was found with $d/\Lambda = 0.4$ and $d=2\mu m$. The measurement birefringence of sample fiber becomes increase with wavelength increase, and this result is good agreement with numerical analyses. The research result indicates that it is possible to develop high birefringence PM-PCF based on polymer at visible wavelength.

Acknowledgements

Project supported by National Science Foundation for Post-doctoral Scientists of China (Grant No. 20080431258), the Science and Technology Project of Shaanxi Province, China (Grant No. 2010K01-078) and the Major scientific research project of Baoji College of Arts & Science (Grant Nos. ZK1016, YK1025, ZK1015)

References

[1] T. A. Birks, J. C. Knight, B. J. Mangan, and P. S. J.P. St. J: *Russell, Photonic crystal fibers: an endless variety*. IEICE Trans. Electron, E84-C(2001), pp. 85-592.

[2] I. K. Hwang, Y. H. Lee, K. Oh and D. N. Payne: *High birefringence in elliptical hollow optical fiber*. Opt. Express, Vol.12 (2004), pp.1916-1921.

[3] T. P. Hansen, J. Broeng, S.E.B. Libori, E. Knudsen: *Highly birefringent index-guiding photonic crystal fibers*. IEEE Photon. Technol. Lett. Vol.13 (2001), pp.588-592.

[4] C. Kerbage, P. Steinvurzel, P. Reyes: *Highly tunable birefringence microstructured optical fiber*. Opt. Lett, Vol.27 (2002), pp.842-848.

[5] M. Koshiba, Y. Tsuji: *Curvilinear hybrid edge/nodal elements with triangular shape for guided-wave problems*. J. Lightwave Technol, Vol.18 (2000), pp.737-741.

[6] ZHANG Ya-ni. *Polarization Properties of Rectangular Lattice Microstructured Polymer optical Fiber*. Acta Photonic Sinica, 38, 1(2009), pp.34-38.

[7] ZHANG Ya-ni, WANG Li-li, WANG Xue-zhong: *Chemistry Fabrication of polymer Photonic Crystal Fibers with High Polarization Maintaining Properties*. Acta Photonic Sinica, 35, 9(2006), pp.1349-1353.

[8] ZHANG Ya-ni, MIAO Run-cai, Ren Li-yong: *High birefringence polymer photonic crystal fiber with elliptical core non-hexangular symmetry*. Acta Photonic Sinica, 36, 6(2007), pp.749-754.

[9] N. Guan, S. Habu, K. Takenaga: *Boundary element method for analysis of holey optical fiber*. Lightwave Technol, Vol.21 (2003), pp.1787-1792

[10] Y. Zhang, K. Li, L. Wang: *Casting preforms for microstructured polymer optical fibre fabrication*. Opt. Express, Vol.14 (2006), pp.5541-5547.

[11] W. Zhi, R. Guobin, L. Shuqin: *Supercell lattice method for photonic crystal fibers*. Opt.Express, Vol.11 (2003), pp.980-991.

Advanced Materials Research Vol. 279 (2011) pp 157-160
© (2011) Trans Tech Publications, Switzerland
doi:10.4028/www.scientific.net/AMR.279.157

Alignment of graphite nanosheet induced by electic field

Haiquan Wang[1, a], Lintong Wang[1, b] and Hongyan Zhang[1, c]

[1]Chemistry and Chemical Engineering College, Weifang University, Weifang, Shandong Province, China

[a]haiquan202@hotmail.com, [b]wanglt6408@sina.com, [c]island_maritime@hotmail.com

Keywords: conducting composite, graphite nanosheets, orientation, percolation threshold

Abstract. Unsaturated polyester resin/Graphite nanosheet composites were prepared using in situ polymerization under AC electric field, which was applied during composite curing to induce orientation of graphite nanosheets. Scanning electron microscopy (SEM) has been applied to demonstrate the orientation of graphite nanosheets. As a result, the percolation threshold was decreased to 0.25 wt% due to the orientation of the graphite nanosheets.

Introduction

Graphite nanosheets(GN), nanoscale conductive filler has attracted significant attention as the low cost alternative to metal- and carbon-based conventional electrically conducting reinforcements for conducting polymer composites, capable of dissipating electrostatic charges and shielding devices from electromagnetic radiation. Previous experimental studies [1-5] suggested that as the concentration (volume fraction of conducting phase) of the conductive component is increased above a threshold value (the percolation threshold) [6-7], the composites underwent an insulator-to-conductor transition and a typical percolation behavior was observed. This behavior is attributed to the formation of multiple, continuous electron paths, or conducting network, in the polymer composites with the filler content at or above the percolation threshold.

The percolation threshold is greatly affected by the properties of the fillers and the polymer matrices, processing methods, temperature, and other related factors [8]. The percolation threshold of GN reinforced polymer nanocomposites is much lower than the conventional fillers, such as graphite powder and carbon black, due to the extremely large surface area and high aspect ratio of GN. Recently, we have found that GN tends to align in polymer matrix induced by shear-force, resulting in the polymer-based nanocomposites with unique piezoresistivity [9]. An electric field is an efficient and direct means to orient and align the filler in the composites.

Many references [10-13] have reported the alignment of layered silicates, carbon nanotubes, carbon black and carbon fibers in the polymers induced by the electric field, while the orientation and alignment of this new GN conducting filler seems to be a virgin field. Herein, we describe a strategy to orient the GN in one direction in unsaturated polyester resin by an AC electric field. Random dispersion of GN in unsaturated polyester prepolymer solution was subjected to an electric field, followed by the cross-linking of the system.

Experimental

Material. The conductive filler used here consists of graphite nanosheets prepared by the ultrasonic techniques with an average thickness of 50 nm and an average diameter of 12μm. Unsaturated polyester resin was supplied by Quanzhou Unsaturated Polyester Resins Factory (Quanzhou, China). To solidify the fluid, Methyl ethyl ketone peroxide (MEKP) and Cobalt (II) naphthenate were used as the hardener and accelerator, respectively.

Procedures. Preweighted unsaturated polyester resin and graphite nanosheets were mixed together and sonicated for 20 minutes to randomly disperse the inclusions. Afterwards the hardener2wt% and accelerator1wt% were added and dispersed by conventional stirring. The suspension was then cast into the setup and an AC electric field was applied till the matrix was completely crosslinked. At the initial stage of in situ polymerization, the viscosity of matrix was low. After about 13 minutes, the viscosity of matrix sharply increased, and the crosslink of matrix finished rapidly.

Measurement. Electronic microscopy study was carried out with LEO-1530 field-emitting scanning electronic microscopy (Germany) at the operating voltage of 20KV to observe the orientation of the graphite nanosheets. In order to prevent charging of the specimens, a thin coating of gold was deposited on the surfaces of the specimens.

The resistivity of unsaturated polyester resin/GN composites was measured using ZC36 high-resistance electrometer (Shanghai No.6 Electrometer Factory) and UT 70A digital multi-Purpose Meter. To ensure good electrical contact, the surfaces of the specimens were polished with 1500 grit sandpaper and conductive paint was coated between the electrodes and the specimens.

Results and Discussion

Structure of the composite.

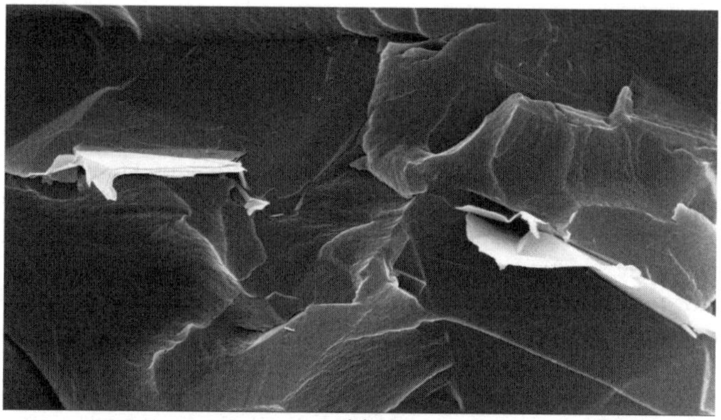

(a)

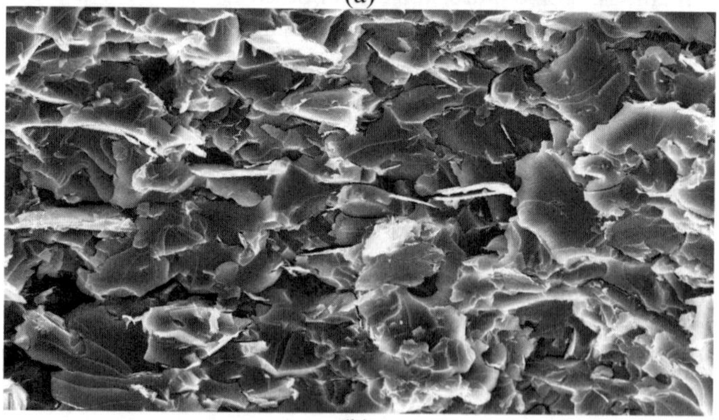

(b)

Fig. 1: Scanning electron microscopy image of a fracture surface from a cured unsaturated polyester resin/ GN composite containing 0.3wt% GN

SEM was employed to examine the morphology of the unsaturated resin/GN composites. Fig. 1(a) and (b) show the SEM images of a composite fracture surface perpendicular to electric field with a filler content of 0.3wt%. It is seen that graphite nanosheets embedded in the matrix were oriented with their flakes along the electric field induced by the electric field.

Electrical conducting properties.

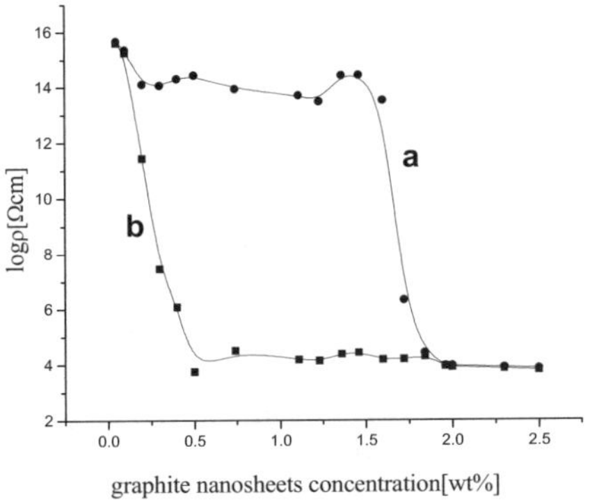

Fig. 2: The percolation threshold of the unsaturated polyester resin /GN composites without (a) application of electric field and (b) with application of electric field

The electrical resistances of unsaturated polyester resin/GN composites with respect to the filler content are displayed in Fig. 2. For both of the two composites, an abrupt conductivity transition occurs at a critical filler content, which can be called as the percolation threshold. The composite containing oriented GN shows a percolation threshold about 0.25wt%; whereas the composite containing randomly dispersed GN posses a much higher threshold about 1.65wt%. This decrease in the percolation threshold can be attributed to the redistribution of GN in the composites. Under an AC electric filed, GN was oriented along the electric filed direction from a random state, which increases the effective aspect ratio of GN inclusions. This increased effective aspect ratio decreases the percolation threshold. This decrease in the percolation threshold also can be explained by the more efficient percolation path/or the decrease of disorder due to the orientation of the GN. When the graphite nanosheets were oriented along the electric filed in the polymer matrix, the attractive interaction (dielectrophoresis) makes the GN form conductive connections, and then more sheet-sheet contacts can be formed at a given loading level, resulting in a higher-conductivity polymer composite with a lower percolation threshold.

In situ current measurements.

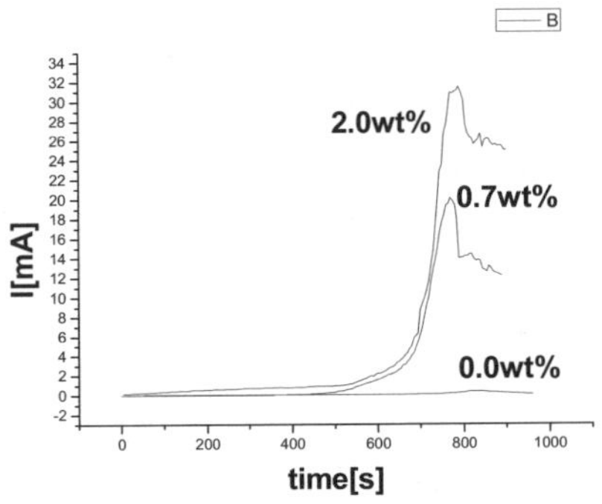

Fig. 3: Electric current measurement of nanosheets graphite/UR composites specimen containing 2.0wt%, 0.7wt% and none graphite nanosheets

To make a further study of the orientation induced by the electric field, the current of the specimen was measured in situ during the curing. Fig. 3 shows the resulting current as a function of time for samples containing 0.0, 0.7 and 2.0 wt% of GN under 25V/mm electric field, respectively. For its lower conductivity of the unsaturated polyester resin, the pure resin system showed a low current under the applied electric field in the whole process. In the case of 0.7wt% and 2.0wt%, the electric field led to an increase in the current, which can be attributed to the GN network formation due to the orientation of GN. At higher loading of GN, the probability of nanosheets contact is increased, resulting in a greater number of more conductive pathways and a higher maximum current. At the end of the curing, there was a decrease in the current, which was attributed to the cross-linking of the matrix, i.e., the decrease of the current was attributed to the distortion of the GN network as a result of the shrinkage due to the final cross-linking of the matrix.

Summary

Unsaturated polyester resin/GN composites with oriented GN were prepared using in situ polymerization under the AC electric field. The in situ current measurement showed that the electric current came to a climax when the GN was oriented completely. The percolation threshold of the oriented composite was decreased to 0.25wt% from 1.65wt% of the randomly dispersed composite, which was attributing to the better conducting network due to the orientation of GN.

References

[1] WG Weng, GH Chen, DJ Wu, and WL Yan: *HDPE/expanded graphite electrically conducting composite*. Composit Interfaces, 11, 2(2004), pp. 131-431.

[2] H Fukushima, LT Drzal: *Graphite nanocomposites: structural and electrical properties*. Proc. 14th Int. Conf. Composit. Mater, California, USA, (2003).

[3] GH Chen, DJ Wu, WG Weng, and WL Yan: *Preparation of polymer/graphite conducting nanocomposite by intercalation polymerization*. J. Appl. Polym. Sci., 82, 10(2001), pp. 2506-2513.

[4] W Zheng, SC Wong: *Electrical conductivity and dielectric properties of PMMA/expanded graphite composites*. Compos. Sci. Technol., 63, 2(2003), pp. 225-235.

[5] XM Chen, JW Shen, WY Huang: *Novel electrically conductive polypropyrene/graphite nanocomposites*. J. Mater. Sci. Let., 21 (2002), pp. 213-214.

[6] S Kirkpatrick: *Electrical conduction in a nonconjugated polymer doped with* $SnCl_4$ *and* $SbCl_5$. Rev Mod Phys, 45 (1973), pp. 574-588.

[7] T Prasse, L Flandin, K Schulte, and W Bauhofer: *In situ observation of electric field induced agglomeration of carbon black in epoxy resin*. Appl Phys Lett, 72 (1998), pp. 2903-2905.

[8] F Carmona: *Conducting filled polymers*. Physica A, 157 (1989), pp. 461-469.

[9] JR Lu, WG Weng, XF Chen, DJ Wu, CL Wu, and GH Chen: *Piezoresistivity material from directed shear-induced assembly of graphite nanosheets in Polyethylene*. Adv Funct Mater, 15, 8(2005), pp. 1358-1363.

[10] MK Schwartz, W Bauhofer, K Schulte: *Alternating electric field induced agglomeration of carbon black filled resins*. Polymer, 43 (2002), pp. 3079-3082.

[11] XY Tai, GZ Wu, Y Tominaga, S Asai, and M Sumita: *An Approach to One-Dimensional Conductive Polymer Composites*. J Polym Sci Pol Phys, 43 (2005), pp. 184-189.

[12] T Prasse, JY Cavaille, W Bauhofer: *Electric anisotropy of carbon nanofibre/epoxy resin composites due to electric field induced alignment*. Compos Sci Technol, 63 (2003), pp. 1835-1841.

[13] T Takahashi, T Murayama, A Higuchi, H Awano, and K Yonetake: *Aligning vapor-grown carbon fibers in polydimethylsiloxane using dc electric or magnetic field*. Carbon, 44 (2006), pp. 1180-1188.

Advanced Materials Research Vol. 279 (2011) pp 161-165
© (2011) Trans Tech Publications, Switzerland
doi:10.4028/www.scientific.net/AMR.279.161

Photoluminescence of ZnO Whiskers with various excitation intensity at room temperature

GUO Xing-Yuan[1, a], ZHOU Jing[1, b], DING Zhan-Hui[1, c], XUE Yan-Feng[1, d], YAO Bin[1, e]

[1]College of Physics, Jilin University, Chang chun 130012, China

[a]guoxy@jlu.edu.cn, [b]397007350@qq.com, [c]dingzh@jlu.edu.cn, [d]xueyf@jlu.edu.cn, [e]binyao@jlu.edu.cn

Keywords: ZnO, whiskers, X-ray diffraction, Photoluminescence, UV

Abstract. Hexagonal ZnO whiskers were prepared by using an optical floating zone furnace. Most of the as-grown whiskers have diameters of about 20-30 μm and lengths of about 2~3 mm. The characteristic peaks of X-ray diffraction (XRD) verify that the as-grown ZnO whiskers have excellent hexagonal wurtzite phase and (001) orientation. The room-temperature photoluminescence spectra show a near-UV emission band located at 380 nm and a green light emission band centered at 515 nm. The threshold of near-UV emission and lasing were observed.

Introduction

Zinc oxide is a direct band gap semiconductor with optical properties similar to GaN. The strain-free ZnO has a wide band gap of 3.37 eV, a relatively large exciton binding energy of 60 meV and a lower lasing threshold at room temperature [1-3]. These properties demonstrate that it can be used to produce light emitting devices in the ultra-violet region [4,5]. Recently, micro/nanostructure with hexagonal cross section have drawn considerable interest for optical resonator applications due to their uniform geometry. However, the synthesis of high quality hexagonal structure with an isolated, undamaged, single-structure remains very difficult. Therefore, it is of paramount importance to understand the optical properties of ZnO. For this reason, we prepared some ZnO whiskers by using optical floating zone furnace [6].

Experimental Procedure

High purity ZnO powder was used as starting material. The powder was sealed in a rubber tube and molded into a rod. Then under a hydrostatic pressure of 60 Mpa, the molded specimen was slowly introduced into a vertical tube furnace for sintering in a programmed manner so that the rod passed the hottest zone, which was kept at 1200 ^{0}C, for 2 h. After sintering, the sintered rod were fixed to shaft, and the growth apparatus was infrared convergence type with four ellipsoidal mirrors. The oxygen partial pressure of growth atmosphere was applied. The morphologies and microstructures of the microrods were characterized by scanning electron microscopy (SEM) and polarization microscopy (PM). X-ray diffraction (XRD) analyses were performed on Rigaku D/Max-rA diffractometer with Cu Kα radiation (λ=1.5418 Å). The photoluminescence (PL) spectra at room temperature (RT) were investigated by using JY-U-1000 spectrometer with a LN2 cooled CCD camera. The microrods are optically pumped by the fourth harmonics (λ=266 nm) of a Nd:YAG laser (10 Hz repetition rate, 7 ns pulse width). The pump beam is focused to a ~20 um spot on the sample.

Results and Discussion

Fig. 1 shows the morphology of as-grown ZnO whiskers, which like a fruit of dandelion. Fig. 2 shows the microscopy image of one whisker. From Fig. 2 we can see that, the ZnO whiskers have an excellent hexagonal structure with high aspect ratio. We estimate that, most of the ZnO whiskers have diameter of about 20-30 μm and length of about 2~3 mm. The crystallinity and orientation of the ZnO whiskers were investigated by X-Ray diffraction as show in Fig. 3. As it is seen from Fig. 3, all diffraction peaks match the hexagonal structure of wurtzite ZnO exactly. In addition, no impurity related peaks were observed in Fig. 3, and this fact shows clearly that the as-grown ZnO whiskers have high crystal quality.

Fig. 1: Photograph of ZnO whiskers

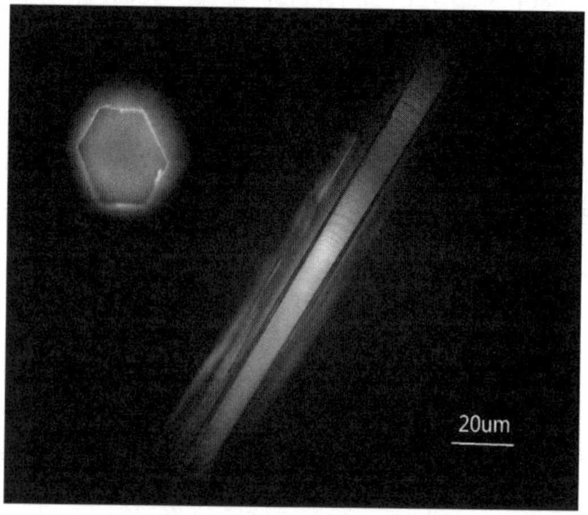

Fig. 2: Polarizing microscopy of ZnO whisker

Fig. 4 shows the room-temperature PL spectra of the ZnO whiskers under various excitation intensities. From Fig. 4 we see that the visible green fluorescence spectra (central wavelength 515 nm) appeared clearly at low excitation intensities (<2 kw/cm^2). When the excitation intensity increasing up to 2 kw/cm^2, an additional emission peak located in the near-UV range at 380 nm and above-mentioned peak of 515 nm both were observed simultaneously. As the excitation intensity is increased from 1 kw/cm^2 to 100 kw/cm^2, the emission peak at 380 nm appears clearly. With the further increasing excitation intensity, the emission intensity of the UV peak increases rapidly, and the line width narrowing.

It is well known that, the green emission peak is commonly referred to as a deep-level or trap-state emission, and it has been attributed to the singly oxygen vacancy and the radioactive recombination of a photon-generated hole with a electron occupying of oxygen vacancy [7-9].

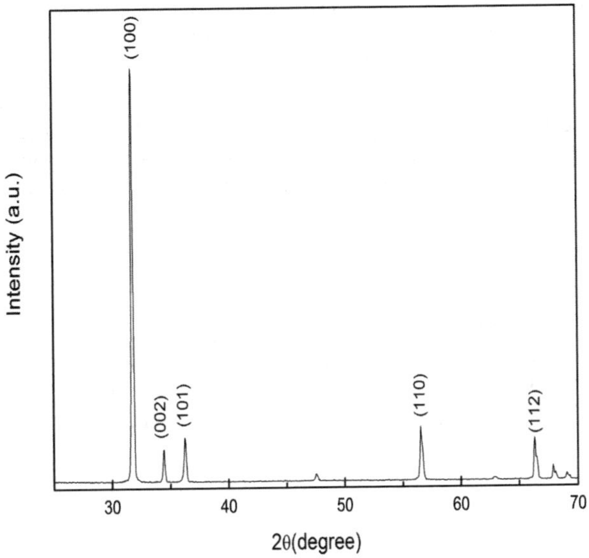

Fig. 3: XRD pattern of ZnO whiskers

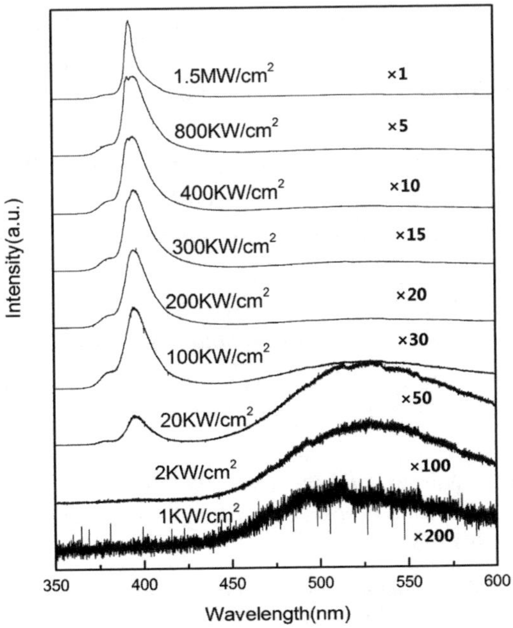

Fig. 4: PL spectra of ZnO whiskers under various excitation intensities

For the emission peak at 380 nm, the mechanism of emission at room temperature were studied by several groups in detail [10-12]. In our work, at excitation intensities below 100 kw/cm^2, the emission peak at 380 nm agrees with the band gap of bulk ZnO [13-15], which comes from the recombination of free excitons [16], though the peak position is shifted to energies lower than the expected room-temperature value as a result of heating by the excitation laser. At this stage, the emission intensities of FX appear a tendency of linear growth that is expected of spontaneous emission (P) processes. At excitation intensities above 100 kw/cm^2 the P band splits into two peaks, near high energy band is called P_2 band, and P_∞ band located in the side of low energy band [17]. With further increasing excitation intensities will lead to overlap of exciton densities, and forming electron-hole plasma, the excitons lose their original characteristics [18]. As the excitation intensities above 1.5 MW/cm^2, the PL peak at 380 nm superlinearly increases. Considering that the PL peaks appear above a certain threshold power, it is concluded that the stimulated emission and lasing occurs in the ZnO whiskers.

Conclusions

In summary, using FZ method, it does not require any catalyst. Therefore, the as-grown ZnO whiskers have very high purity and exact hexagonal shape. These advantages make it possible to grow ZnO whiskers with good optical quality for many fundamental measurements on the optical and electrical area. The PL spectra under various excitation intensities show the ZnO whiskers by FZ method to have lasing characteristic.

Acknowledge

This paper was supported by the Fundamental Research Funds for the Central Universities (No.200903324) and the Natural Science Foundation of Jilin Province, China (Grant No.20101555).

References

[1] Y.F.Chen, D.M.Bagnall, H.J.Koh, K.T.Park, K.J.Hiraga, Z.Q.Zhu and T.Yao: *Plasma assisted molecular beam epitaxy of ZnO on c -plane sapphire: Growth and characterization.* J. Appl. Phys. Vol. 84 (1998), p.3912.

[2] W.Y.Liang and A.D.Yoffe: *Transmission Spectra of ZnO Single Crystals.* Phys. Rev. Lett. Vol. 20 (1968), p.59.

[3] D.C.Reynolds, D.C.Look, B.Jogai, C.W.Litton, G.Cantwell and W. C.Harsch: *Valence-band ordering in ZnO.* Phys. Rev. B Vol. 60 (1999), p.2340.

[4] D.C.Look: *Recent advances in ZnO materials and devices.* Mater. Sci. and Engin. B Vol. 80 (2001), p.383.

[5] M.H.Huang, S.Mao, H.Feick, H.Yan, Y.Wu, H.Kind,E.Weber, R.Russo and P.Yang: *Room-Temperature Ultraviolet Nanowire Nanolasers.* Science Vol. 292 (2001), p. 1897.

[6] X.Y.Guo, Y.N.Yu, D.P.Xu, Z.H.Ding, W.H.Su: *Growth of ZnO (Wurtzite) Whiskers Using the Floating Zone Method at High Oxygen Pressure.* Chem. J. Chinese University Vol. 27 (2006), p.1811.

[7] K.Vanheusden, W.L.Warren, C.H.Seager, D.R.Tallant, J.A.Voigt and B.E.Gnade: *Correlation between photoluminescence and oxygen vacancies in ZnO phosphors.* J. Appl. phys. Vol. 79 (1996), p.7983.

[8] Y.Li, G.S.Cheng and L.D.Zhang: *Fabrication of highly ordered ZnO nanowire arrays in anodic alumina membranes.* J. Mater. Res. Vol. 15 (2000), p.2305.

[9] M.H.Huang, Y.Wu, H.peick, N.Tran, E.Weber and P.Yang: *Catalytic Growth of Zinc Oxide Nanowires by Vapor Transport.* Adv. Mater. Vol. 12 (2001), p. 113.

[10] D. M. Bagnall, Y. F. Chen, Z. Zhu, T. Yao, S. Koyama, M. Y. Shen and T. Goto: *Optically pumped lasing of ZnO at room temperature.* Appl. Phys. Lett. Vol. 70 (1997), p.2230.

[11] K.Bando, T.Sawabe, K.Asaka and Y.Masumoto: *Room-temperature excitonic lasing from ZnO single nanobelts.* J. Lumin. Vol. 108 (2004), p.385.

[12] Z.K.Tang, M.Kavasaki, A.Ohtomo, H.Koinuma and Y.Segawa: *Self-assembled ZnO nano-crystals and exciton lasing at room temperature.* J.Crys. Growth Vol. 287 (2006), p.169.

[13] U.Koch, A.Fojtik, H.Weller and A.Henglein: *Photochemistry of semiconductor colloids. Preparation of extremely small ZnO particles, fluorescence phenomena and size quantization effects.* Chem. Phys. Lett. Vol. 122 (1985), p.507.

[14] S.Monticone, R.Tufen and A.V.Kanaev: *Complex Nature of the UV and Visible Fluorescence of Colloidal ZnO Nanoparticles.* J. Phys. Chem. B Vol. 102 (1998), p.2854.

[15] E.M.Wong and P.C.Searson: *ZnO quantum particle thin films fabricated by electrophoretic deposition.* Appl. Phys. Lett. Vol. 74 (1999),p.2939.

[16] B.D.Yao, H.Z.Shi, H.J.Bi and L.D.Zhang: *Optical properties of ZnO loaded in mesoporous silica.* J. Phys. Condens. Matter. Vol. 12 (2000), p.6265.

[17] C. Klingshirn: *Properties of the electron-hole plasma in II–VI semiconductors.* J. Cryst.Growth Vol.117 (1992), p.753.

[18] G.Xiong, J.Wilkinson, K.B.Ucer and R.T.Williams: *Time-of-flight study of bound exciton polariton dispersive propagation in ZnO.* J. Phys. Condens. Matter. Vol. 17 (2005), p.7287.

Advanced Materials Research Vol. 279 (2011) pp 166-169
© (2011) Trans Tech Publications, Switzerland
doi:10.4028/www.scientific.net/AMR.279.166

Preconcentration and Separation Capabilities of Nano-barium-strontium Titanate Powder Coated by Dithizone for Trace Cobalt in Water

Dong Zhang

School of Environmental and Chemical Engineering Shenyang Ligong University, Shenyang 110159, China

sylgdxdong@sina.com

Keywords: Nano-barium-strontium titanate, Dithizone, Cobalt, Preconcentration, Separation

Abstract. A new method for the preconcentration and separation of cobalt ion in water was described. The methodology combines determination using a flame atomic absorption spectrometry (FAAS) with adsorption and desorption of cobalt on the nano-barium-strontium titanate powder coated by dithizone (BST-dithizone). The operation conditions for the pre-concentration and separation of the cobalt, including pH, contact time, eluent condition and co-existing ions have been investigated. The result showed that the cobalt ion could be adsorbed by BST-dithizone. The adsorption quantity was affected by the pH value of medium and shaking time. In the medium of pH 6.0, the adsorption capacity of BST-dithizone to cobalt ion was 8.92 $mg \cdot g^{-1}$ when the shaking time was more than10 min. The cobalt adsorbed on the adsorbent could be completely eluated and recovered using 10 $g \cdot L^{-1}$ ethylene diamine tetraacetic acid (EDTA) solution. The enrichment factor was 100. The detection limit of cobalt was 0.53 $ng \cdot L^{-1}$. The purposed method has been applied to the pre-concentration and separation of cobalt in the water sample with satisfied results.

Introduction

Cobalt is present in electronics, metallurgical, mining, electroplating and paint industries [1]. Cobalt as a pollutant has many toxic effects such as imparting neurotoxicological disorders, genotoxicity, cardiomyopathy, carcinogenicity and bronchial asthma. At the same time, cobalt is an essential element for the functioning of many vital processes. It is extremely important in the processes of blood formation, stimulation of hemoglobin synthesis, and the functioning of vitamins, enzymes and hormones. This metal has a very positive influence on the metabolism of vitamins, such as ascorbic acid and vitamin B12. Cobalt is an essential element in the human body as a component of vitamin B12 and it is clear that the monitoring of body fluids for cobalt is important for the control of nutritional deficiencies and, perhaps, the prevention of its toxic effect in cases of occupational exposure. Finally, it is also necessary for the synthesis of a number of hormones, neurotransmitters, and other compounds, such as bile acids and DNA[2]. Therefore it is necessary to determine cobalt in food and water. The determination of cobalt is usually carried out by flame atomic absorption spectrometry [3]. But it is difficult to direct determination the trace amounts of cobalt in materials, food and water samples. So, the development of new methods for selective separation, concentration and determination of cobalt in low levels was still a challenging problem. At present, commonly used methods include liquid–liquid extraction [4] and solid phase extraction (SPE) [5]. Solid-phase extraction techniques offer rapid phase separation, high enrichment factor and lower organic solvent consumption as well as the ability to combine with different detection techniques in on-line mode. As a new adsorbent agent, the nano-barium-strontium titanate powder coated by dithizone (BST-dithizone) has been used in the adsorption of lead and cadmium with satisfied results [6]. But the BST-dithizone use in solid-phase extraction of cobalt in the water samples has not been studied. In this work, BST-dithizone was used as a solid-phase extraction agent to preconcentration and separation of cobalt in the water.

Experimental

Apparatus and reagents. Nano-barium-strontium titanate powder coated by dithizone (BST-dithizone) was prepared according to our previous work [6]. A stock standard solution of cobalt chloride was prepared by dissolving an accurate weight in water and completed to 100 mL in a 100 mL measuring flask which had been standardized gravimetrically. Working standard solutions were obtained by appropriate dilution of the stock standard solution.

A flame atomic absorption spectrometry (FAAS), WYX-9003A-model (Shenyang Yi Tong Analytical Instrument Co., Ltd.), a cobalt hollow cathode lamp was used as a light source. The wavelength for cobalt was 240.7 nm with a spectral band-pass of 0.2 nm.

The pH values were measured with a PHS-3C acidometer (Shanghai REX Instrument Factory, Shanghai, China) supplied with a combined electrode.

A constant temperature bath shaker (Jintan Experiment Instrument Factory, Jiangsu China) was used for adsorption and desorption experiments.

All chemicals used were of analytical grade unless otherwise stated. All of the solutions were prepared with distilled water

Adsorption experiments. A 50 mL calibrated colorimetric cylinder with plug was prepared and cobalt solution was added into it. The pH was adjusted to 6.0 with hydrochloric acid solution or ammonia solution. After dilution to the mark with water, 0.5g of BST-dithizone was added in it. After shaking in the bath shaker for 10 min, the solution was separated from the adsorbent. Then the concentrations of cobalt were determined by the falme atomic absorption spectrometer and absorption capacity was calculated.

After adsorption, the cobalt loaded adsorption agent was washed with water to remove any unadsorbed ion. The adsorbed cobalt was eluted with 5 ml of 10 $g \cdot L^{-1}$ ethylene diamine tetraacetic acid(EDTA) solution. The concentration of the cobalt ion in the eluent was determined by flame atomic absorption spectrometry, and the recoveries were calculated from concentrations of eluent.

Results and discussion

Effect of pH on the adsorption. The pH values of the medium governing the complexation of dithizone with metal ion. The pH value has been the most important variable governing cobalt ion adsorption onto BST-dithizone. The effect of pH values on the adsorption of cobalt ion onto BST-dithizone are shown in Fig. 1. It could be seen that the adsorption capacity increased with increasing pH value of medium and reached a plateau value at pH ranges of 4.0-8.0. So, the pH 6.0 was selected as the optimal condition.

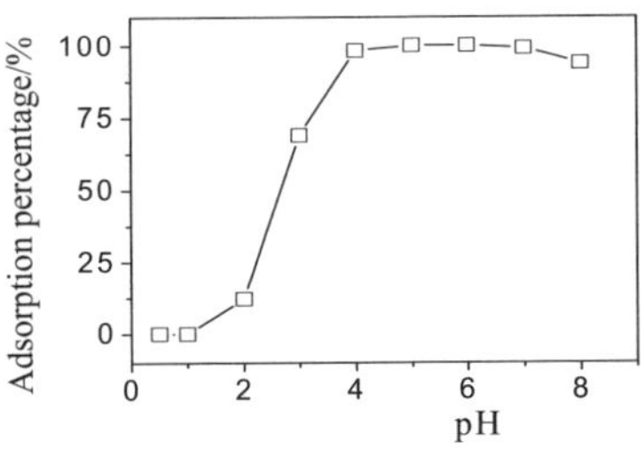

Fig. 1: Effect of pH on the adsorption

Effect of time on the adsorption. Fig. 2 shows the influence of shaking time on the adsorption. It could be seen that the amount of adsorption increased with increasing the shaking time. The maximum adsorption capacity was observed after 10 min. So, in this work, the shaking contact time chose 10 min.

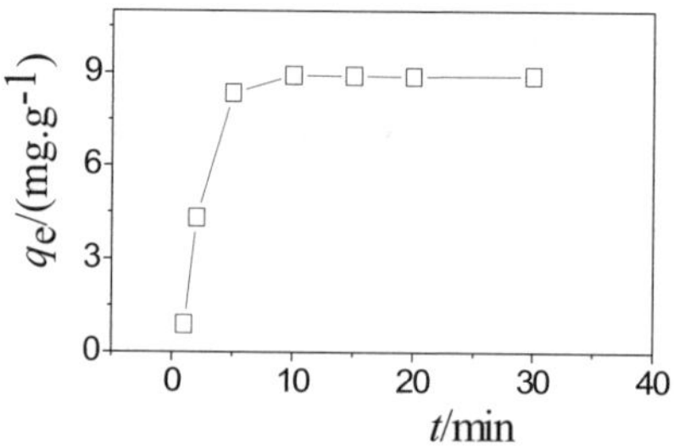

Fig. 2: Effect of contact time on adsorption

Adsorption capacity. According to Fig. 2, under the optimized adsorption conditions, the adsorption capacity of BST-dithizone to cobalt was 8.92 mg·g^{-1}.

The condition of elution. For choosing the proper eluent for desorbing the cobalt from dithizone-BST, after the cobalt was adsorbed under the optimized adsorption conditions, 5 mL of selected eluent solution, such as hydrochloric acid solution, nitric acid solution, thiourea and ethylene diamine tetraacetic acid(EDTA) solution at different concentration were used for elution by shaking 5 minute, respectively. The recovery results are showed in Table 1.

Table 1: Elution conditions

Desorption agents	Concentration	Recovery/%
HNO_3	1 moL·L^{-1}	38.0
HNO_3	2 moL·L^{-1}	65.8
HNO_3	3 moL·L^{-1}	69.0
HCl	1 moL·L^{-1}	82.1
HCl	2 moL·L^{-1}	90.4
HCl	3 moL·L^{-1}	91.7
thiourea	5g·L^{-1}	92.3
thiourea	10 g·L^{-1}	93.9
EDTA	5g·L^{-1}	95.2
EDTA	10 g·L^{-1}	99.7

The recovery results showed that recovery of using 10 g·L^{-1} ethylene diamine tetraacetic acid(EDTA) solution were up to 95%. So, 5mL 10 g·L^{-1} ethylene diamine tetraacetic acid(EDTA) solution was chosen.

After elution, the BST-dithizone was dried under 60°C and reused 10 times. The performance of the adsorbent did not decreased and these showed that BST-dithizone was stable.

Enrichment factor and detection limit. For study the possibility of enriching low concentrations of cobalt from large volumes water, 1.0 μg of cobalt was dissolved in various volumes (50 mL, 100 mL, 200 mL, 300mL, 500 mL) of water, respectively. The amounts recovered are listed in Table 2. The solution volume was 500 mL, the concentration of cobalt was 2.0 μg·L^{-1}, and the recovery was above 95%. The enrichment factor was 100. The detection limit of cobalt was 0.53 ng·L^{-1}.

Table 2: Preconcentration and recovery of cobalt

Volume of the solution/[mL]	The quantity in eluent/[μg]	Recovery/[%]	Enrichment factor
50	0.978	97.8	10
100	0.990	99.0	20
200	0.975	97.5	40
300	0.962	96.2	60
500	0.958	95.8	100

Effects of coexisting ions. The effects of coexisting ions on the recovery of cobalt were studied. The potential interfering ions were added to solution of 0.2 mg·L^{-1} cobalt. The experimental results showed that recoveries of the cobalt remained above 95% -105% even in the presence of ions in the following concentrations: more than 1000 mg·L^{-1} for Na^+, K^+, Cl^- and NO_3^-; 500 mg·L^{-1} for Ca^{2+}, NH_4^+ and Mg^{2+}; 200 mg·L^{-1} for Cr(III) , Al^{3+}, Ag^+, Bi(III), $Ni^{2+}$and Cu^{2+}; 100 mg·L^{-1} for Cr(VI) and 50 mg·L^{-1} for Cd^{2+} , Pb^{2+}, Fe^{3+} and Zn^{2+}. It will be seen that these ions produce no interference in the extraction of cobalt.

Analysis application. The BST-dithizone was applied to the pre-concentration and separation of cobalt in the water. The river water was come from the Hunhe river(Shenyang, China), the lake water was come from the Nanhu lake (Shenyang, China) and the tap water was collected from our lab. According to normal method, the samples were pretreatment, and the cobalt level found in the samples. The results obtained are lined in Table 3.

Table 3: Determination of cobalt in water samples (n=6)

Samples	Base value [μg·L^{-1}]	RSD [%]	added [μg·L^{-1}]	found [μg·L^{-1}]	Recovery [%]
River water	68.37	2.47	50	120.68	104.62
Lake water	72.69	3.16	50	121.44	97.50
Tap water 1	29.51	2.98	50	77.09	95.16
Tap water 2	24.28	3.14	50	73.26	97.96

Conclusions

The nano-barium-strontium titanate powder coated by dithizone (BST-dithizone) has been successfully to use for the preconcentration and separation of cobalt, and then its determination by FAAS. In the medium of pH 6.0, the adsorption time was 10 min, and the cobalt could be adsorbed on the BST-dithizone. The cobalt adsorbed on the BST-dithizone could be completely eluated by using 5 mL ethylene diamine tetraacetic acid(EDTA) solution. The proposed method has been applied to the pre-concentration and separation of cobalt in water samples with satisfied results.

References

[1] B.J. Alloway, D.C.Ayres: *Chemical Principles of Environmental Pollution*. Chapman & Hall, Oxford, (1993).

[2] Jeredic, Vucetic: *Mikroelementiu bioloskom materijalu*, Privredni pregled. Serbian, Beograd, (1982).

[3] State Environmental Protection Administration of China: *Analysis Method of Water and Waste Water*. China Environmental Science Press, fourth ed., Beijing, (2002).

[4] S.G. Sarkar, P.M. Dhadke: *Solvent extraction separation of antimony(III) and bismuth(III) with bis(2,4,4-trimethylpentyl) monothiophosphinic acid (Cyanex 302)*. Sep. Purif. Technol. 15, 2 (1999), pp. 131-138.

[5] N. Tokman, S. Akman, M. Ozcan: *Solid-phase extraction of bismuth, lead and nickel from seawater using silica gel modified with 3-aminopropyltriethoxysilane filled in a syringe prior to their determination by graphite furnace atomic absorption spectrometry*. Talanta, 59, 1 (2003), pp. 201-205.

[6] D. Zhang, H. D. Su, H. Gao: *Study on the Adsorption Capability of Barium-Strontium Titanate Powder Coated with Dithizone for Cadmium Ion in Water*. Spectrosc. Spect. Anal. 28, 3(2008), pp.693-696.

Advanced Materials Research Vol. 279 (2011) pp 170-173

doi:10.4028/www.scientific.net/AMR.279.170

Investigation of photo-induced C-H bond activating by $CpRh(CO)_2$ complex using density functional theory

Qinglong Liu[1,a], Shuwei Wang[1,b], Li Lv[2,c] and Xiaojing Wang[1,d,*]

[1]College of Chemistry and Chemical Engineering, Inner Mongolia University, Huhehaote city 010021, P.R. China

[2]Chemical Engineering College of Inner Mongolia University of Technology, Huhehaote city 010051, P. R. China

[a]lql55745@yahoo.com.cn, [b]wsw0719@163.com, [c]lvli1997@126.com
[d]wang_xiao_jing@hotmail.com

Keywords: Photochemical behaviors, $CpRh(CO)_2$, C-H bond activating, excited state

Abstract. The charge transfer and structural distortions that occurred in the complex $CpRh(CO)_2$ upon excitation with an light irradiation were studied by density functional theory (DFT). The calculations showed that the electrons transferred from Cp to CO ligands with the transition of $CpRh(CO)_2$ from ground state to the first excited state. Accompanying with this transfer process, $CpM(CO)_2$ became distorted and the linear bond of M-CO became bent upon excitation. The second excitation is the strongest excitation which is identified to be metal to ligand CO charge transfer (MLCT) excitations. We also found the lowest excited state has little effect for the M-CO bond photoactivation while the photodissociation of CO from $CpM(CO)_2$ can be achieved in the second excited state.

Introduction

C–H bonds are among the strongest and most inert of single bonds. Consequently, it is a major challenge to chemists to activate C–H bonds and to convert them to organic functional groups that are more useful. Only a few transition metals have been observed to insert into C-H bonds in saturated hydrocarbons via an oxidative addition mechanism. Various experiments suggest a photo-induced excitation process as the initial step of activation C-H bond in the compounds of saturated hydrocarbons [1-2]. The d^8 transition-metal dicarbonyls series compounds such as $CpM(CO)_2$ and $Cp^*M(CO)_2$, ($Cp=C_5H_5$, $Cp^*=C_5Me_5$, M=Rh and Ir) have been found to activate C-H bonds of alkanes in an photo-induced process [3]. Spectroscopic data and quantum efficiency values for photosubstitution and intermolecular C–H/Si–H activation reactions involving d^8 transition-metal dicarbonyls series compounds in various room-temperature solutions were reported [4].The intermediates generated by photolysis of trans-$RhCl(CO)(PR_3)_2$ appears to participate directly in C-H activation [5]. Recent researches found that the quantum efficiencies are depended on the nature of ligands and metal [6]. In order to understand the structures, spectra, and the details of the photoreaction mechanisms, in this work we investigated the structure change during the excitation from the ground state to the excited state. Furthermore we discussed the charge transfer direction and degree during the excitation by comparing the Mulliken population both in ground state and excited state.

Computational details

All calculations in this investigation have been performed by using Amsterdam density functional program package (ADF2000). The PerdewandWang's 1991 exchange and correlation functional (PW91XC) was taken at the generalized gradient approximation (GGA), and the basis sets of triple-ζ STO with polarization function were applied to the valence electrons of all atoms. The inner electrons within the atomic shells (i.e. 1s for C and O, 3d for Rh) were handled as a frozen-core approximation.

The relativistic terms were calibrated by a combined scalar relativistic zero order regular approximation (ZORA). The geometrical and electronic properties of excited states were obtained using orbital occupations method by controlling the assignment of spin-up and spin-down electrons to the corresponding molecular orbitals. The pictures of the molecular orbitals and optimized structures were illustrated using a MOLEKEL4.1 molecular visualization program.

Results and discussion

The structures of $CpRh(CO)_2$ in ground state were optimized by minimizing the total energy using ADF2000 and the optimized stereoviews are depicted in Fig. 1. Several selected bond lengths and bond angles are listed in Table 1. The angles between two Rh-C-O are 179.2◦ and 178.4◦, and the two Rh-CO are 1.858 and 1.859, respectively. It is clear that $CpRh(CO)_2$ shows symmetric structure and Rh-CO is linear. The optimization of $CpRh(CO)_2$ in the first triplet excited state and the second excited state were performed with a model by promoting one electron from highest occupied molecular orbital (HOMO) to the corresponding lowest unoccupied molecular orbital with the opposite spin using ADF2000 program. The optimized structures showed in Fig. 1. It can be found a significant rotation of the CO ligands in the lowest energy excited states. The bonds of M-CO are bent and the η^5-Cp coordination is changed to be η^3-Cp coordination during the excitation from the ground state to the first excited state. However, in the second excited state, the M-CO is still linear but the coordination is changed to be η^2-Cp. The bonds of the two M-CO are 1.957Å showing a longer when the electron transfer from the ground state to the second excited state. The Mulliken charges of $CpRh(CO)_2$ in the stable ground states as well as in its two lowest excited states are calculated (Table 2). It is observed that the distribution of electronic density is different in the ground and excited states. The difference of the charge between ground state and the first excited state shows that during the excitation from ground state to excited state, the charge transfer are assigned to be the excitation from metal to Cp ligands, that is, the MLCT excitation. However, the electrons transfer from the Cp ligand to metal and to CO ligand during the excitation from the ground state to the second excited state, that is, the LLCT excitation occurs.

Table 1: Several selected bond lengths and bond angles of $CpRh(CO)_2$ and the excited states

	$CpRh(CO)_2$	Excited State(1)	Excited State(2)
Rh-C	2.304	2.434	2.291
Rh-center	1.957	2.110	2.110
Rh-CO	1.858	1.907	1.957
Rh-CO	1.859	1.909	1.992
C-O	1.162	1.163	1.173
C-O	1.162	1.163	1.172
∠Rh-C-O	179.2	166.6	176.7
∠Rh-C-O	178.4	164.9	177.6
∠Cp-Rh-CO	135.7	125.1	127.7
∠Cp-Rh-CO	134.2	126.6	129.0

Table 2: Tthe Mulliken population of $CpRh(CO)_2$ and the excited states

	$CpRh(CO)_2$	Excited State(1)	Excited State(2)
Rh	0.5511	0.7150	0.5334
CP	-0.1052	-0.209	0.0042
CO1	-0.2214	-0.1933	-0.2686
CO2	-0.2244	-0.1945	-0.2695

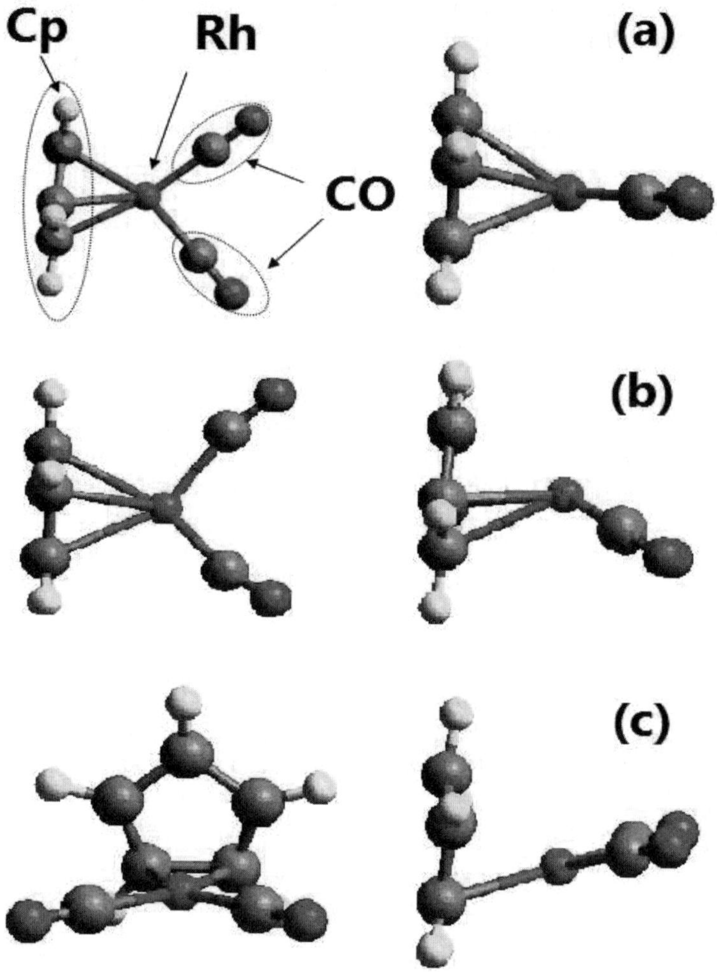

Fig. 1: The optimization structure view from the different direction of (a) $CpRh(CO)_2$ (b) the first excited state of $CpRh(CO)_2$ (c) the second excited state of $CpRh(CO)_2$

To gain insight into the excited state processes involved, the orbital compositions of $CpRh(CO)_2$ in ground state and in the lowest excited states are analyzed based on the Mulliken population. From Fig. 2, it can be seen the electrons transfer from the bonding orbital which formed by Rh d orbital and the CO π orbital with a slight mixing of Cp ligands to the bonding orbital formed by Rh and the CO π orbital. Thus the electrons excitation will not result in the dissociation of CO from the metal. However, it is obviously that the electrons transfer from the Rh-CO bonding orbital to the antibonding orbital during the excitation from the ground state to the second excited state, indicating a weaken of Rh-CO bonding, which may resulting in the dissociation of CO from the metal.

Conclusions

The molecular structures and Mulliken population of $CpRh(CO)_2$ in the ground state and the lowest excited states were investigated by density functional theory in this paper. Large geometry differences have been found with a linear Rh-C-O bond in the ground state and a bent Rh-C-O bond in the first excited state. The electrons transfer from the metal to the Cp ligand during the excitation from ground state to the first excited state. And the LLCT occurs when electrons excited to the second excited state. The lowest excited state has little effect for the Rh-CO bond photoactivation. The photodissociation of CO from $CpRh(CO)_2$ can be achieved in the excited state of higher energy level. Thus the photochemistry has been understood by invoking different pathways following excitation at short and long wavelengths regions.

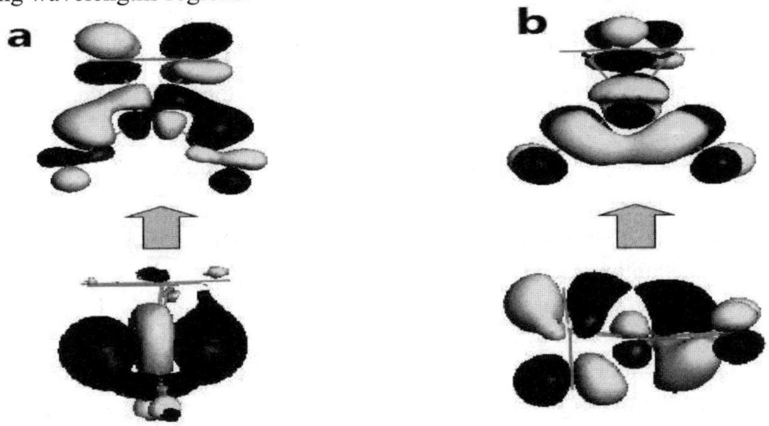

Fig. 2: Tthe electron excitation process of (a) from the ground state to the first excited state (b) from the ground state to the second state

Acknowledgment

This work is supported by the National Natural Science Foundation of China (Nos. 20863004 and 20967004), the Chunhui Program of Ministry of Education (Z2004-2-15028 and Z2007-1-01034), the inner Mongolia National Science Foundation (2010Zd08), the Science Research Foundation of Inner Mongolia Education Bureau (NJ05027), the scientific research foundation for the returned overseas Chinese scholars of Ministry of Education (2006-331),the Project of Scientific and Technological Innovation Team of Inner Mongolia University (12110614).

References

[1] W. Jones: *Isotope Effects in C−H Bond Activation Reactions by Transition Metals*. Acc. Chem. Res. 36, 2(2003), pp. 140-146.

[2] A. J. Lees: *Quantitative photochemistry of organometallic complexes: insight to their photophysical and photoreactivity mechanisms*. Coord. Chem. Rev. 211, 1(2001), pp. 255-278.

[3] A. J. Esswein, D. G. Nocera: *Hydrogen Production by Molecular Photocatalysis*. Chem. Rev. 38, 50(2007), pp. 4022-4047.

[4] N. Dunwoody, S. Sun, A. J. Lees: *Quantitative Photochemistry and Mechanisms for a Series of Rhodium Dicarbonyl Derivatives*. Inorg. Chem. 39, 20(2000), pp. 4442-4451.

[5] J. S. Bridgewater, T. L. Netzel, J. R. Schoonover, S. M. Massick, and P. C. Ford: *Time-Resolved Optical and Infrared Spectral Studies of Intermediates Generated by Photolysis of trans-RhCl(CO)(PR$_3$)$_2$. Roles Played in the Photocatalytic Activation of Hydrocarbons*. Inorg. Chem. 40, 7(2001), pp. 1466-1476.

[6] A. Vlcek, S. Zalis: *Modeling of charge-transfer transitions and excited states in d^6 transition metal complexes by DFT techniques*. Coord. Chem. Rev. 251, 3-4(2007), pp. 258-287.

Advanced Materials Research Vol. 279 (2011) pp 174-178
doi:10.4028/www.scientific.net/AMR.279.174

Synthesis and Crystal Structure of $[Zn(C_{14}H_{12}N_2)(HCO_2)_2]\cdot 2H_2O$

Shengliang Ni[a], Yue Meng[b] and Peisong Tang[c]

Department of Chemistry, Huzhou Teachers College, Huzhou, 313000, P. R. China

[a]nisl@hutc.zj.cn, [b]mengyue@hutc.zj.cn, [c]tangps@hutc.zj.cn

Keywords: zinc, 2,9'–dimethyl–1,10'–phenanthroline complex, formato ligand

Abstract. Reactions of a freshly prepared $Zn(OH)_{2-2x}(CO_3)_x\cdot yH_2O$ precipitate, formic acid with 2,9'-dimethyl-1,10'-phenanthroline in CH_3OH/H_2O afforded $[Zn(C_{14}H_{12}N_2)(HCO_2)_2]\cdot 2H_2O$. The title compound was structurally characterized by X-ray diffraction methods. It consists of complex molecules $[Zn(C_{14}H_{12}N_2)(HCO_2)_2]$ in which Zn atoms are hexa-coordinated by two N atoms of one phenanthroline ligand and four O atoms of two bidentate formate groups. In the crystal, molecules are connected by O–H···O hydrogen bonds forming layers parallel to (010), and the resulting layers are further linked 3D framework along [100] by π-π packing interactions.

Introduction

In the past decade, a variety of supramolecular architectures based on non–covalent intermolecular interactions such as hydrogen bonding, van der Walls forces and π-π interactions have been achieved by using transition metal centers and organic ligands due to their possible intriguing structural topologies and potential applications in optics, catalysis, ion exchange, gas storage, and the molecular–based magnetic materials[1,2]. Simultaneous coordination of flexible α, ω-dicarboxylate spacer ligands and the heteroaromatic N-donor ligands as sources for π-π packing interactions resulted in a variety of structural motifs. For example, self-assembly of Zn^{2+} cations with phenanthroline and α, ω–dicarboxylic acids led to mononuclear complex molecules such as {[Zn(phen)(HL)$_2$]} with H_2L = pimelic, suberic acids[3]. Formato ligands have been actively utilized as construction units to obtain many supramolecular complexs[4,5]. In the present work, we are interested in self–assemblies of Zn^{2+} ions and 2,9'–dimethyl–1,10'–phenanthroline with formic acid, further report the synthesis and structure of $[Zn(C_{14}H_{12}N_2)(HCO_2)_2]\cdot 2H_2O$ in detail.

Experimental Section

Synthesis. Dropwise addition of 2.0 ml of 1.0 M aqueous Na_2CO_3 to a stirred aqueous solation of $ZnCl_2$ (0.1363 g, 1.0 mmol) in 5.0 ml H_2O produced a white precipitate, $Zn(OH)_{2-2x}(CO_3)_x\cdot yH_2O$, which was separated by centrifugation and washed several times with doubly distilled water until no Cl^- ions were detected in the supernatant. The precipitate was added to a stirred aqueous methanolic solution of 2,9'–dimethyl–1,10'–phenanthroline in 30 ml CH_3OH/H_2O(1:1 v/v). Subsequently, 1.77 ml of 1.0 M aqueous solution of formalic acid was dropwise added to the above mixture and stirred continuously until dissolved of the white precipitate. The colourless solution (pH = 4.52) was allowed to stand at room temperature. Slow evaporation over several days afforded well-shaped colorless block crystals. Yield: 47% based on initial $ZnCl_2$.

X-ray structural analysis. Suitable single crystals were selected under a polarizing microscope and fixed with epoxy cement on respective fine glass fibers, which were then mounted on a Rigaku R-Axis Rapid IP X-ray diffractometer with graphite-monochromated Mo $K\alpha$ radiation (λ = 0.71073

Å) for cell determination and subsequent data collection. The reflection intensities in the θ range 3.00–27.48° were collected at 293 K using the ω scan technique. The employed single crystals exhibit no detectable decay during the data collection. The data were corrected for *Lp* and absorption effects. The SHELXS–97 and SHELXL–97 programs were used for structure solution and refinement[6,7]. The structures were solved by using direct methods. Subsequent difference Fourier syntheses enabled all non-hydrogen atoms to be located. After several cycles of refinement, all hydrogen atoms associated with carbon atoms were geometrically generated, and the rest of the hydrogen atoms were located from the successive difference Fourier syntheses. Finally, all non-hydrogen atoms were refined with anisotropic displacement parameters by the full-matrix least-squares technique and hydrogen atoms with isotropic displacement parameters set to 1.2 times of the values for the associated heavier atoms. Detailed information about the crystal data and structure determination is summarized in Table 1.

Tabel 1: Crystallographic and experimental data for$[Zn(C_{14}H_{12}N_2)(HCO_2)_2]\cdot 2H_2O$

Compound	$[Zn(C_{14}H_{12}N_2)(HCO_2)_2]\cdot 2H_2O$	Volume [Å^3]	1700.7(6)
Empirical formula	$C_{16}H_{18}ZnN_2O_6$	Temperature[K]	293(2)
Formula weight	399.69	D_{Calc} [g·cm^{-3}]	1.561
Crystal system	monoclinic	$F(000)$	824
Space group	$P2_1/c$	μ[mm^{-1}]	1.479
Description	Colourless block	θ Range [°]	3.0-27.48
Crystal size [mm]	0.12 × 0.10 × 0.08	Reflections collected	3978
a [Å]	7.505(2)	Unique	0.0451
b [Å]	17.820(4)	Data,restraints,	2987, 1526, 232
c [Å]	13.023(3)	Goodness of fit on F^2	1.029
α[°]	90.00	R_1,wR_2 ($I \geq 2\sigma(I)$)	0.0501, 0.1026
β[°]	102.45(3)	R_1,wR_2 (all data)	0.0821, 0.1695
γ[°]	90.00	A, B values in	0.0505, 0.0000
Z	4	$\delta\rho_{max}$, $\delta\rho_{min}$ [e·Å^{-3}]	0.382, -0.304

$R_1 = \sum(|F_o| - |F_c|)/\sum|F_o|$, $wR_2 = [\sum w(F_o^2 - F_c^2)^2/\sum w(F_o^2)^2]^{1/2}$, and $w = [^2(F_o^2) + (0.0505P)^2 + 0.0000P]^{-1}$ where $P = (F_o^2 + 2F_c^2)/3$

Results and Discussions

The asymmetric unit of the title compound consists of one Zn^{2+} ion, one 2,9'-dimethyl–1,10'–phenanthroline molecule, two O,O'-chelated formate anions, and two lattice H_2O molecules, as shown in Fig. 1. Each Zn^{2+} atoms is coordinated two N atoms from one 2,9'–dimethyl–1,10'–phenanthroline and four O atoms from two formate anions to complete a distorted octahedral ZnN_2O_4 chromophore. The Zn-N/O distance of 2.055(2)-2.459(2) Å fall in the normal region[8,9], both the *cisoid* and *transoid* bond angles in the range 52.4(2)–122.7(2) Å and 114.1(2)–150.5(2) Å, respectively, indicating that the octahedral ZnN_2O_4 geometry is a highly distorted one. For two formate anions, the angle (O1–C15–O2, 121.6(8)°) is smaller than the angle(O3–C16–O4,122.5(9)°). The 2,9'–dimethyl–1,10'–phenanthroline ligand is almost coplanar with the mean square deviations 0.0245 Å. Two solvent water molecules O5 and O6 are not coordinated to Zn and the distance between zinc and water oxygen atom of 8.338 Å and 4.476 Å, respectively. However, O6 is linked to the complex molecule in the asymmetric unit by O6-H61-O3 hydrogen bond. Selected interatomic distances and bond angles are given in Table 2.

Tabel 2: Selected bond lengths (Å) and angles (°) for $[Zn(C_{14}H_{12}N_2)(HCO_2)_2]\cdot 2H_2O$

Zn–N1	2.071(4)	Zn–O1	2.150(6)	Zn–O3	2.055(5)
Zn–N2	2.073(4)	Zn–O2	2.086(2)	Zn–O4	2.440(5)
N1–Zn–O1	122.6(2)	N1–Zn–O2	108.8(8)	N1–Zn–O3	94.2(2)
N1–Zn–O4	139.5(2)	N2–Zn–N1	81.5(2)	N2–Zn–O1	96.3(2)
N2–Zn–O2	148.0(8)	N2–Zn–O3	114.2(2)	O1–Zn–O2	52.3(2)
O1–Zn–O3	135.8(2)	O1–Zn–O4	97.8(1)	O2–Zn–O4	93.0(2)

Hydrogen bonding contacts

D–H···A	d(D–H)	d(H···A)	∠DHA	d (D···A)
O5-H51···O1$^{\#1}$	0.85	1.84	174	2.688
O5-Hh51···O4$^{\#}$	0.84	1.98	157	2.774
O6-H61···O3$^{\#3}$	0.85	1.96	176	2.809
O6-H62···O5	0.85	2.33	152	3.098

Symmetry transformations used to generate equivalent atoms: #1 = x+1, y, -z ; #2 = -x+1, -y, -z+1; #3 = -x, -y, -z+1.

The molecules are connected by (O5–H51···O4$^{\#1}$, O5–H52···O1$^{\#2}$ O6–H62···O5$^{\#3}$) hydrogen bonds forming layers parallel to (010), as shown in Fig. 2. The resulting layers are further linked 3d framework along [100] by π···π packing interactions with centriod-centroid distances of 3.47(2) Å with 2,9'–dimethyl–1,10'–phenanthroline aromatic cycle orientated inward.

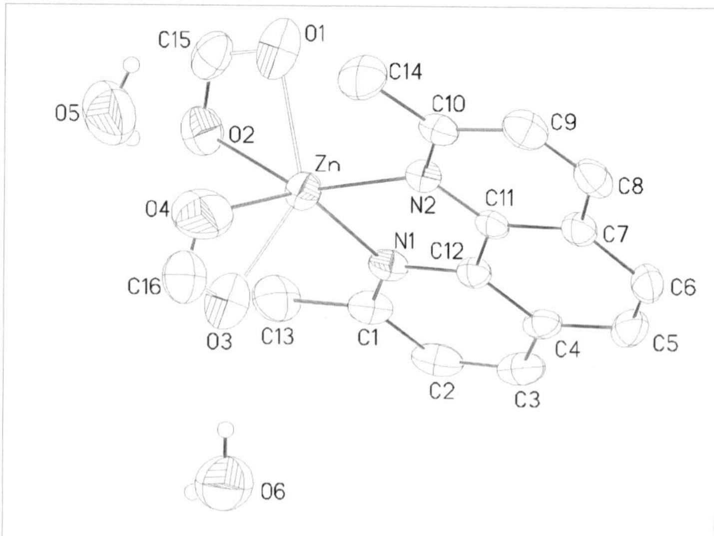

Fig. 1: ORTEP view of the title compound(45% probability level)

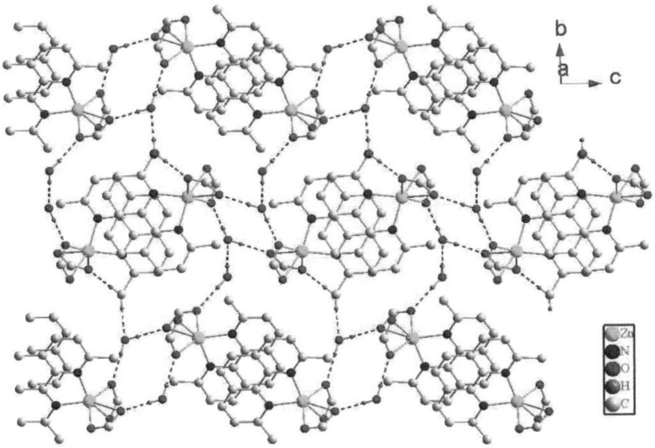

Fig. 2: 2D layer structure of the title compound, the hydrogen bonds are indicated by dashed lines.

Conclusions

The asymmetric unit of the novel compound contains a mononuclear $Zn(C_{14}H_{12}N_2)(HCO_2)_2$ complex molecule hydrogen bonded to two lattice water molecules. The Zn^{2+}cation is in a distorted octahedral coordination sphere defined by two N atoms of 2,9'-dimethyl-1,10'-phenanthroline ligand and four O atoms of the two formate anions. In the crystal, molecules are connected by O–H···O hydrogen bonds forming layers parallel to (010). The resulting layers are further linked 3d framework along [100] by π-π packing interactions with centriod-centroid distances of 3.47(2) Å with 2,9'-dimethyl-1,10'-phenanthroline aromatic cycle orientated inward.

Acknowledgements

This work was financially supported by the Foundation of Zhejiang Education Department of (ZC200805662) and the Zhejiang Province Natural Science Foundation (Y4100471).

References

[1] F. F. Chen, Z. Q. Bian, Z. W. Liu, et al.: *Highly Efficient Sensitized Red Emission from Europium (III) in Ir−Eu Bimetallic Complexes by 3MLCT Energy Transfer.* Inorg. Chem., 47, 7(2008), pp. 2507-2513.

[2] A. de Bettencourt-Dias: *Isophthalato-Based 2D Coordination Polymers of Eu(III), Gd(III), and Tb(III): Enhancement of the Terbium-Centered Luminescence through Thiophene Derivatization.* Inorg. Chem., 44, 8(2005), pp. 2734-2741.

[3] D. Y. Wei, Y. Q. Zheng, J. L. Lin: *Self-assembly of zinc ions with suberic acid and phenanthroline - crystal Structures of five new zinc phenanthroline suberato complexes.* Zeitschrift für anorganische und allgemeine Chemie, 628, 9-10(2002), pp. 2005-2012.

[4] Q. Sun, Q. Yue, J. Y. Zhang, et al.: *Hydrothermal synthesis, structures, and properties of Manganese(II) and Zinc(II) coordination polymers with different phenylenediacrylates.* Crystal Growth and Design, 9, 5(2009), pp. 2310-2317.

[5] J. Boonmak, S. Youngme, T. Chotkhun, C. Engkagul, N. Chaichit, G. Albada, J. Reedijk: *Polynuclear copper(II) carboxylates with 2,2'-bipyridine or 1,10-phenanthroline: Synthesis, characterization, X-ray structures and magnetism.* Inorganic Chemistry Communications, 11, 10(2008), pp. 1231-1235.

[6] G. M. Sheldrick: *SHELXS-97, Program for X-ray crystal structure solution.* Gottingen University, Germany, (1997).

[7]G. M. Sheldrick: *SHELXS-97, Program for X-ray crystal structure refinement.* Gottingen University, Germany, (1997).

[8]Y. Q. Zheng, and J. L. Lin: *Two heteroleptic Zinc(II) complexes: $[Zn(phen)(C_9H_{15}O_6)_2]$ and $[Zn_2(phen)_2(H_2O)_2(C_{10}H_{16}O_4)_2]\cdot 3H_2O$.* Zeitschrift für anorganische und allgemeine Chemie, 629, 6(2003), pp. 1007-1010.

[9]X. C. Fu, X. Y. Wang, M. T. Li, C. G.. Wang: *Dibenzoato(1,10-phenanthroline)zinc(II).* Acta Crystallographica Section E, Vol. 62, Part 4(2006), pp. m773-m775.

II. Solid Mechanics

Advanced Materials Research Vol. 279 (2011) pp 181-185
© (2011) Trans Tech Publications, Switzerland
doi:10.4028/www.scientific.net/AMR.279.181

Numerical Simulation of Thermoplastic Composite Material by ANSYS

Guohua Zhao[1, a], Qinglian Shu[1,b] and Bosheng Huang[1,c]

[1]School of Civil Engineering and Architecture, University of Jinan, Jinan 250022, P.R. China

[a]cea_zhaogh@ujn.edu.cn, [b]cea_shuql@ujn.edu.cn, [c]cea_huangbs@ujn.edu.cn

Keywords: thermoplastic composite material, AS4/PEEK, ANSYS, material model

Abstract. This paper proposes a material model of AS4/PEEK, a typical thermoplastic composite material, for the general purpose finite element code—ANSYS, which can be used to predict the mechanical behavior of AS4/PEEK composite structures. The computational result using this model has a good agreement with the test result. This investigation can lay the foundation for the numerical simulation of thermoplastic composite structures.

Introduction

PEEK (poly-ether-ether-keton) is one of the most popular thermoplastic matrixes, which was developed by Imperial Chemical Industries (Great Britain) in 1980's. PEEK matrix has many attractive features, such as high temperature resistance, chemical resistance, high strength, high toughness, excellent machinability, which make it a priority in the development of advanced composite. AS4/PEEK is PEEK matrix composite reinforced by graphite fibers (AS4). It can normally work under temperatures up to 250^0C and is extensively applied in aerospace and military fields. The wide use of AS4/PEEK has given rise to the research of its mechanical properties and constitutive equations [1-7].

Finite element method (FEM) is a promising numerical simulation technique. When applied to engineering field, it can greatly decrease the cost of design and test of products. Especially in recent 20-30 years, some general purpose finite element codes, such as ANSYS, ABAQUS, NASTRAN, were developed, which largely lowered the difficulties of FEM application. Even so, due to the complexity of its constitutive equations, it is difficult to build a proper material model of AS4/PEEK for FEM calculation.

In this paper, the authors try to establish a material model of AS4/PEEK for the general purpose finite element code—ANSYS, which can lay the foundation for the numerical simulation of thermoplastic composite structures.

The Material Model of AS4/PEEK

PEEK matrix is a typical plastic material and its stress/strain relationship is nonlinear. So the mechanical behavior of AS4/PEEK is quite different from that of thermoset composite. For instance, plastic deformation occurs at the beginning of loading. Therefore the plastic deformation must be considered when using FEM to predict the behavior of AS4/PEEK laminates.

Plastic and viscoplastic constitutive equations of AS4/PEEK in quasi-static condition and at different strain rates were developed through tests by C.T. Sun and his co-workers [1-5]. In this paper, the authors try to apply these constitutive equations to general purpose finite element code. The simulation were performed by ANSYS (ver.12.0), a famous general purpose finite element package.

AS4/PEEK is an orthotropic elastoplastic material. To conduct the calculation of a elastoplastic material in ANSYS, the following three ingredients must be considered: yield criterion, flow rule and hardening rule [8].

Yield criterion. As a typical thermoplastic material, AS4/PEEK has the following special characteristics.

1. When loading in fiber direction, the deformation of the material is linear, and almost no plastic deformation exists before failure.

2. When loading perpendicular to the fiber direction, the deformation is nonlinear, and plastic deformation occurs almost at the beginning of loading.

Based on test results, incorporated the two characteristics, C. A. Weeks and C. T. Sun developed a yield criterion for AS4/PEEK laminates [5]:

$$\sigma_e = \sqrt{\frac{3}{2}\left[(\sigma_2 - \sigma_3)^2 + 4\tau_{23}^2 + 2a_{66}\left(\tau_{13}^2 + \tau_{12}^2\right)\right]} = \sigma_0 \tag{1}$$

Where:

σ_e—effective stress, it is the combination of the stress components of one point;

1, 2, 3—the three directions of material main axes;

a_{66}—a parameter which indicates the anisotropy in plasticity. Referred to test results, $a_{66} = 0$ is suitable for most conditions;

σ_0—reference yield stress.

But the yield criterion is not existent in ANSYS material library. Instead, the Hill criterion was adopted as the yield criterion for orthotropic elastoplastic material:

$$\sigma_e = \sqrt{H(\sigma_1 - \sigma_2)^2 + F(\sigma_2 - \sigma_3)^2 + G(\sigma_3 - \sigma_1)^2 + 2N\tau_{12}^2 + 2L\tau_{23}^2 + 2M\tau_{31}^2} = \sigma_0 \tag{2}$$

Where:

H, *F*, *G*, *N*, *L*, *M*—constants related to the specific material.

Compared Eq. 1 to Eq. 2, it can be seen that Eq. 1 is a special form of Eq. 2. So Hill yield criterion can be adopted approximately as the yield criterion of AS4/PEEK in ANSYS nonlinear computation.

Flow rule. Flow rule describes the direction of the plastic strain when yielding, that is, how the plastic strain components develops along with yielding. Specifically, the flow rule can be described as the following flow equation:

$$\left\{d\varepsilon^{pl}\right\} = d\lambda\left\{\frac{\partial Q}{\partial \sigma}\right\} \tag{3}$$

Where:

$\{d\varepsilon^{pl}\}$ —plastic strain increment, which can be denoted as the six components $\{d\varepsilon_1{}^{pl}, d\varepsilon_2{}^{pl}, d\varepsilon_3{}^{pl}, d\gamma_{12}{}^{pl}, d\gamma_{23}{}^{pl}, d\gamma_{13}{}^{pl}\}$— the superscript '*pl*' stands for 'plastic strain';

$d\lambda$— plastic multiplier, which determines the magnitude of plastic straining;

Q— plastic potential, which is function of stress.

If Q is the yield function, as is normally assumed, the flow rule is termed associative and the plastic strain occurs in a direction normal to the yield surface. If Q is other kind of function instead of yield function, the flow rule is termed non-associative. Referring to literature [5], the associative flow rule is adopted. So the expression of plastic potential Q can be deduced from Eq. 1.

Hardening rule. The hardening rule describes the changing of yielding surface with progressive yielding, so that the conditions (i.e. stress state) for subsequent yielding can be established. In literature [5], the power function hardening rule based on isotropic hardening was developed. But this hardening rule is not included in ANSYS. So we had to give up it and adopt the Voce hardening rule in ANSYS, which is a isotropic hardening rule. The hardening process is termed nonlinear isotropic hardening or NISO in ANSYS.

The hardening rule takes the form:

$$\sigma_e = C_1 + C_2\varepsilon_e{}^{pl} + C_3[1-\exp(-C_4\varepsilon_e{}^{pl})] \tag{4}$$

Where:
C_1, C_2, C_3, C_4—material constants;
ε_e^{pl} —effective strain, defined as:

$$\varepsilon_e^{pl} = \sqrt{\frac{1}{6}\left[\left(\varepsilon_2^{pl} - \varepsilon_3^{pl}\right)^2 + \frac{2}{a_{66}}\left(\gamma_{12}^{pl}\right)^2 + \frac{2}{a_{66}}\left(\gamma_{13}^{pl}\right)^2 + \left(\gamma_{23}^{pl}\right)^2\right]} \tag{5}$$

By fitting the experiment data in reference [5], the relationship of effective stress and effective plastic strain can be derived:

$$\sigma_e = 8.826 + 3469\varepsilon_e^{pl} + 116.4\left[1-\exp\left(-1291\varepsilon_e^{pl}\right)\right] \text{ (MPa)} \tag{6}$$

The experiment curve and fitting curve generated according to equation (6) are plotted in Fig. 1. It can be seen that two curves is very close.

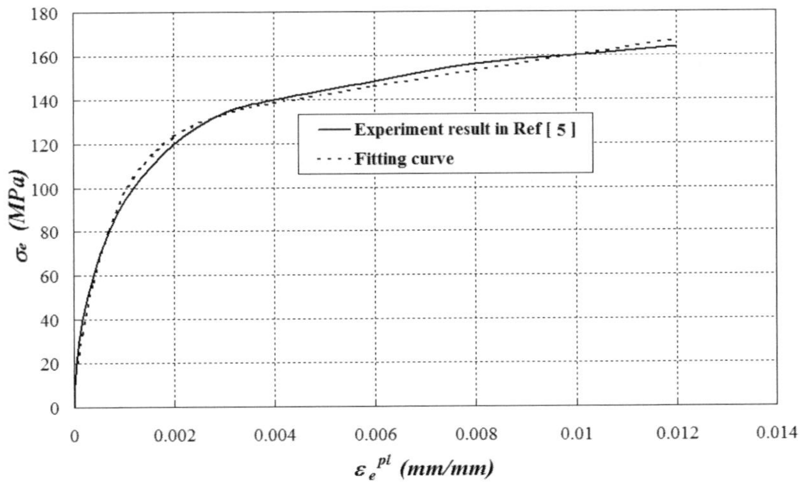

Fig. 1: Comparison of experimental curve and fitting curve

It should be noticed that the experiment curve was obtained at the strain rate of 1×10^{-5}/s (quasi-static). In this paper, only the static loading is considered, so the experiment result at this strain rate is adopted.

The verification of the material model

To verify the used material model, stress-strain relationships of some AS4/PEEK laminas will be plotted using the FEM prediction results and compared with the curves obtained by experiment results.

Computational model. Four AS4/PEEK orthotropic laminas were employed to calculate stress-strain relationships. These models were under uniaxial tension, and the lay-up angles of fibers in laminas were 14^0, 30^0, 45^0 and 90^0, respectively. All laminas have the same dimension, which is shown in Fig. 2.

The boundary conditions are as follows:

1. The displacement in x direction at plane x=0 is set to zero, that is, $u_x(0, y, z) = 0$;
2. At the plane x=0, the various displacements are applied to simulate various strains for the lamina;
3. To prevent rigid body movement in y and z directions, let $u_y(0,0,0) = u_z(0,0,0) = 0$.

The elastic properties of orthotropic lamina was excerpted from the reference [5] and shown in Table 1.

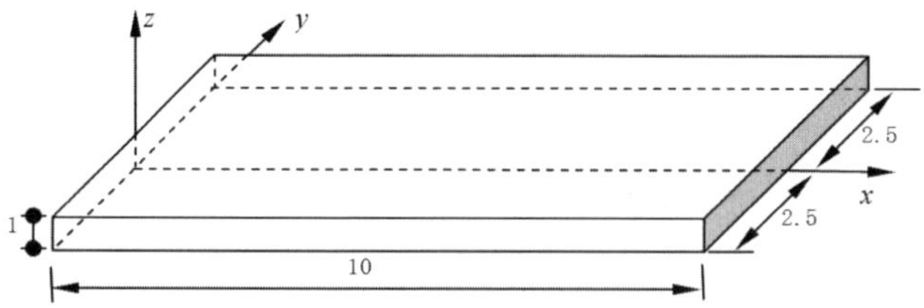

Fig. 2: Computational model and adopted Cartesian coordinate (Units:mm)

Table 1: Elastic properties of AS4/PEEK lamina

E_1 [GPa]	E_2=E_3 [GPa]	G_{12}=G_{13} [GPa]	G_{23} [GPa]	μ_{12}=μ_{13}	μ_{23}
127.6	10.3	6.0	3.45	0.32	0.49

Comparison of computational and experimental results. In ANSYS, NLISO material model can only be supported by 8-node three-dimension solid brick element SOLIDE185 and 20-node three-dimension solid brick element SOLIDE186. To obtain more accurate result, SOLIDE186 was employed. The computational results and experimental result curves were plotted in Fig. 3. It can be seen that the computational result and experimental result fitted well.

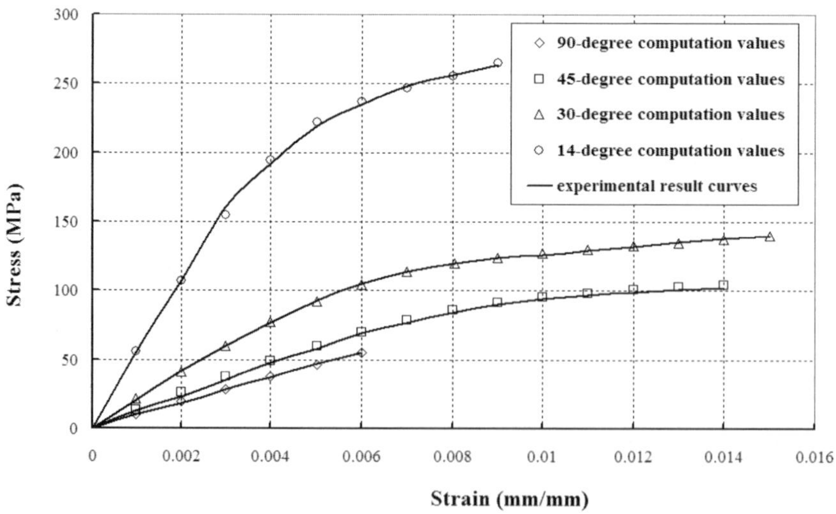

Fig. 3: Comparison of computational results and experimental curves

Conclusion

The above verification shows the material model we adopted is feasible and can be used in ANSYS to predict the mechanical behaviors of AS4/PEEK structures.

Actually, the authors have applied the above material model to the calculation of interlaminar stresses in AS4/PEEK structures and the results were in good accord with experimental results [9,10].

Acknowledgement

The authors were thankful for the funding of Natural Foundation of University of Jinan (Granting Number: XBS1013).

References

[1] T.S. Gates and C.T. Sun: *Elastic/viscoplastic constitutive model for fiber reinforced thermoplastic composites*, AIAA Journal, Vol. 29 (1991), pp. 457-463.

[2] C.T. Sun and J.L. Chen: *A simple flow rule for characterizing nonlinear behavior of fiber composites*, Journal of Composite Material, Vol. 23 (1989), pp.1009-1020.

[3] K. J. Yoon and C.T. Sun: *Characterization of elastic-viscoplastic properties of an AS4/PEEK thermoplastic composite*, Journal of Composite Material, Vol. 25 (1991), pp.1277-1296.

[4] J.L. Chen and C.T. Sun: *A plastic potential function suitable for anisotropic fiber composites*, Journal of Composite Material, Vol. 27 (1993), pp.1379-1390.

[5] C.A. Weeks and C.T. Sun: *Modeling non-Linear rate-dependent bBehavior in fiber-reinforced composites,* Composites Science and Technology, Vol. 58 (1998), pp.603-611.

[6] T.J. Vogler and S. Kyriakides: *Inelastic behavior of an AS4/PEEK composite under combined transverse compression and shear, part I: experiments*, International Journal of Plasticity, Vol. 15 (1999), pp.783-806.

[7] N. Papadakis, N. Reynolds, M.W. Pharaoh, P.K.C. Wood, etc: *Strain rate effects on the shear mechanical properties of highly oriented thermoplastic composite material using displacement measurement methodology, Part A: elasticity and shear strength*, Composites Science and Technology 64 (2004), pp.729-738.

[8] ANSYS Corporation: *ANSYS online help*, 2010.

[9] G.H. Zhao, J.W. Tong, M. Shen, etc: *Numerical analysis of interlaminar stresses of angle-ply AS4/PEEK laminate with a central hole*, Journal of Thermoplastic Composite Materials, Vol. 22 (2009), pp.383-406.

[10] G.H. Zhao, J.W. Tong and M. Shen: *Numerical analysis and experimental validation of interlaminar stresses of quasi-isotropic APC-2/AS-4 laminate with a central hole loaded in tension*, Journal of Thermoplastic Composite Materials, Vol. 23 (2010), pp.413-433.

Advanced Materials Research Vol. 279 (2011) pp 186-193
© (2011) Trans Tech Publications, Switzerland
doi:10.4028/www.scientific.net/AMR.279.186

Shape Control for Composite Structures of Photostrictive Actuators using Topology Optimization Method

Yu Wang [a] and Zhen Luo [b,*]

Elec, Mech and Mechatronic Systems, University of Technology, Sydney, NSW 2007, Australia

[a] wangyu871114@gmail.com, [b] zhen.luo@unsw.edu.au

Keywords: Topology optimization, Shape control, Composite structures, Photostrictive actuators

Abstract. This paper aims to present an alternative design method for remote wireless shape control of laminated composite structures using topology optimization. The photostrictive material is introduced to implement the active control of the structure by making use of its photostriction mechanism, which is actually the superposition of photovoltaic effect and converse piezoelectric effect when exposed to the illumination of near ultraviolet light. The finite element formulation including multiphysics effects of photovoltaic, pyroelectric and thermal expansion is developed to model composite structures of ferroelectric materials, based on the Mindlin plate theory of first-order shear deformation. The topology optimization method is used to seek the optimal topologies for material layouts of both the smart actuation and elastic host layers. A typical numerical example is used to demonstrate the feasibility of this method in shape control of composite structures.

Introduction

Photostrictive materials [1,2] can be regarded as a new family of smart materials able to generate strains under irradiation of high energy lights with specific wavelengths, which has shown notable potential in a wide range of wireless and remote control applications of future generation devices [3,4]. The photostriction deformation of photostrictive materials involves a multiphysics process to generate light-induced strain because of the superposition of photovoltaic effect and converse piezoelectric effect. When exposed to illumination of ultraviolet light irradiation, a polarized photostrictive material can produce high photovoltage exceeding band-gap energy and then result in mechanical strain.

Actuators with photostrictive materials [3,4] show many advantages such as remote control, noncontact actuation, immunity to electromagnetic disturbances, free of complex electric wires and circuits, lightweight and miniaturization, compared to conventional piezoelectric actuators. The lead lanthanum zirconate titanate (PLZT) ceramic [5,6] is an attractive photostrictive material due to its excellent photovoltaic properties such as high piezoelectric coefficient, large photostriction and easiness of fabrication. PLZT ceramics can be regarded as a specific family of piezoelectric (PZT) materials with the photovoltaic effect. Shape control of smart composite structures [7,8] is a very interesting research area. However, it can be found that most previous studies focused on the shape control is in the context of piezoelectric materials [9-10], which are characterized with hard-wired systems that might contaminate the control signals due to the electric noise caused by the metallic signal wires. Hence, it can be seen that this research to apply photostricitve actuators to structural shape control is relatively rare [11,12].

One main objective for design optimization of PLZT shape control is to optimize some control parameters and configurations of PLZT actuators for morphing shapes so that the actuated shape can match the desired one as closely as possible [9,11]. Recently, the topology optimization method [13] has been applied to structural shape control [12,14] as a powerful alternative. The aim of topology optimization is to seek the best distribution of a given amount of material via an iterative numerical process to extremize structural performance under specific constraints. Amongst many topology optimization methods, SIMP [15,16] has obtained great popularity due to its conceptual simplicity and practical easiness. This paper applies the topology optimization to shape control of composite structures with 0-1 polarization subjected to multiphysics fields, based on the laminated plate theory of first-order shear deformation (FSDT) [17].

Constitutive equations with 0-1 polarization

When a uniformly distributed light is applied to the surface of a PLZT material, the material wafer will be polarized in 0-1 direction and an electric field E_1 opposite to the polarization direction is generated due to the photovoltaic effect. This study is mainly concerned with the steady state. The electric field produces a photovoltaic voltage between the two metalized electrodes which results in mechanical strain ε_{11} because of the converse piezoelectric effect.

According to the principle of virtual work of a deformed medium, the weak formulation of the equilibrium equation can be given as:

$$\int_{\Omega}\sigma_{ij}\delta\varepsilon_{ij}d\Omega-\int_{\Omega}p_i\delta u_i d\Omega+\int_{S}\tau_i\delta u_i dS=0 \tag{1}$$

where σ_{ij} represents the symmetric Cauchy stress tensor. p_i and τ_i represent mechanical body force and mechanical traction components, and $\delta\varepsilon_{ij}$ and δu_{ij} are admissible virtual strains and displacements, respectively. For the linear elastic materials in this study, the stress σ_{ij} is related to (1) Lagrangian symmetric tensor ε_{kl} caused by converse piezoelectric effect of the photostrictive material, (2) electric field vector E_k induced by the light illumination, and (3) the thermal strain $\varepsilon^{\mathrm{T}}{}_{kl}$ created by the temperature variation due to light illumination. The constitutive equation to be discussed is based on the assumption that the total strain in the actuator is the sum of strains caused by light. The time dependence and voltage leakage influence of the light $I(t)$ are not included. Hence the constitutive equation under quasi-static models can be written in tensor form as:

$$\sigma_{ij}=\mathrm{Q}_{ijkl}(\varepsilon_{kl}(u)-\varepsilon^{\mathrm{T}}{}_{kl}(T,\Delta T))-e_{kij}E_k(I(t)) \tag{2}$$

where E_k can be further expressed via the electric potential $\nabla\phi_k$ as:

$$E_k(I(t))=-\nabla\phi_k(I(t)) \tag{3}$$

For a PLZT wafer, the initial photocurrent density decreases with light penetration depth [1,4,6].

$$I(\zeta_d)=I_o e^{-\alpha\zeta_d} \tag{4}$$

where I_o is the initial photocurrent density. α and ζ_d are light absorption coefficient and the light penetration depth that is measured from the light irradiation surface.

In the above equations, Q_{ijkl} and e_{kij} represent the elastic and coupling piezoelectric constants. T and ΔT are the body temperature and the temperature change, respectively. The induced electric field E_k can be calculated according to the illumination intensity of the light intensity I, which includes both the photovoltaic electric field and pyroelectric electric field caused by converse piezoelectric effect. For 0-1 polarization, the induced field strength E_1 is uniform through thickness because of the same electrical potential happened between two electrodes on left and right sides (Fig. 1). Hence, the strain is uniformly distributed through the thickness of the PLZT wafer.

As a result, the total strain $\bar{\varepsilon}_{ij}$ includes three components which are generated by photovoltaic, pyroelectricity and thermal expansion, respectively.

$$\bar{\varepsilon}_{ii}=\varepsilon_{ii}^{(1)}+\varepsilon^{T(2)}_{ii}(T)+\varepsilon^{T(3)}_{ii}(\Delta T),\ (i=1,2,3) \tag{5}$$

where $\varepsilon_{ii}^{(1)}$ is the strain caused by converse piezoelectric effect of photostrictive materials, $\varepsilon^{T(2)}_{ii}$ is the strain related to pyroelectricity, and $\varepsilon^{T(3)}_{ii}$ denotes the thermal strain caused by body temperature change. It is noted that $\varepsilon^{T(2)}_{ii}$ and $\varepsilon^{T(3)}_{ii}$ in PZT materials are uniformly distributed via the thickness direction while $\varepsilon_{ii}^{(1)}$ is not because it depends on the direction of the polarization. They are given as follows [1-2,4], respectively:

$$\varepsilon_{ii}^{(1)}(I_o) = d_{1i}E_1 = d_{1i}\left(\frac{\xi_1 I_0(1-e^{-\alpha H})}{w_\alpha\left[\gamma_d H + \gamma_{ph} I_0(1-e^{-\alpha H})/\alpha\right]}\right) \ (i=1) \tag{6}$$

$$\varepsilon^{T(2)}_{ii}(T) = d_{ij}\left(\frac{\Theta_n}{\lambda_1^{(1)}}T\right) \ (i=j=1) \text{ where } T \geq T_{trig} \tag{7}$$

$$\varepsilon^{T(3)}_{ii}(\Delta T) = (\alpha_T)_{ij}\Delta T, \ (i=j=1) \tag{8}$$

where Θ_n and $\lambda_1^{(1)}$ are pyroelectric constant and the 1st order permittivity component in 0-1 polarization. T_{trig} and α_T are the trigger temperature and the thermal expansion coefficient. E_1 is electrical field strength, and d_{11} is piezoelectric strain constant with 0-1 polarization.

Finite element formulation

With the finite element method based on FSDT, the equilibrium equations are written as

$$\mathbf{KU} = \mathbf{F} + \mathbf{f} \tag{9}$$

where $\mathbf{K}$ is the element stiffness matrix and $\mathbf{U}$ is the vector of nodal displacements. $\mathbf{F}$ and $\mathbf{f}$ are the vectors of applied nodal forces and nodal vector generated by light. For photostrictive actuators applied to composite plates, the light is supposed to act only on the photostricitve material layers. In this study, eight-node isoparametric plate elements are used to discretize the design domain, 2D quadratic shape functions are applied to interpolate the generalized displacements [17]. Then, the light induced load can be given as follows:

$$\mathbf{f}^{\mathrm{e}} = (\mathbf{f}^{\mathrm{e}}_{\mathrm{opt}})_1 + (\mathbf{f}^{\mathrm{e}}_{\mathrm{opt}})_2 + (\mathbf{f}^{\mathrm{e}}_{\mathrm{thermo}})_1 + (\mathbf{f}^{\mathrm{e}}_{\mathrm{thermo}})_2 \tag{10}$$

where each components are calculated based on element areas S^e as:

$$(\mathbf{f}^{\mathrm{e}}_{\mathrm{opt}})_1 = \sum_{k=1,NL} ({}^{(k+1)}z - {}^{(k)}z)\int_{S^e} ({}^{(k)}\mathbf{A}^{\mathrm{T}})^e ({}^{(k)}\overline{\mathbf{e}})^e ({}^{(k)}\mathbf{E})^e dS \tag{11}$$

$$(\mathbf{f}^{\mathrm{e}}_{\mathrm{opt}})_2 = \sum_{k=1,NL} \frac{1}{2}({}^{(k+1)}z^2 - {}^{(k)}z^2)\int_{S^e} ({}^{(k)}\mathbf{C}^{\mathrm{T}})^e ({}^{(k)}\overline{\mathbf{e}})^e ({}^{(k)}\mathbf{E})^e dS \tag{12}$$

$$(\mathbf{f}^{\mathrm{e}}_{\mathrm{thermo}})_1 = \sum_{k=1,NL} ({}^{(k+1)}z - {}^{(k)}z)\int_{S^e} ({}^{(k)}\mathbf{A}^{\mathrm{T}})^e ({}^{(k)}\overline{\mathbf{Q}})^e ({}^{(k)}\boldsymbol{\varepsilon}(T,\Delta T))^e dS \tag{13}$$

$$(\mathbf{f}^{\mathrm{e}}_{\mathrm{thermo}})_2 = \sum_{k=1,NL} \frac{1}{2}({}^{(k+1)}z^2 - {}^{(k)}z^2)\int_{S^e} ({}^{(k)}\mathbf{C}^{\mathrm{T}})^e ({}^{(k)}\overline{\mathbf{Q}})^e ({}^{(k)}\boldsymbol{\varepsilon}(T,\Delta T))^e dS \tag{14}$$

where $\boldsymbol{\varepsilon}(T,\Delta T) = \boldsymbol{\varepsilon}^{(2)}(T) + \boldsymbol{\varepsilon}^{(3)}(\Delta T)$ is the strain related to thermal expansion coefficient $\boldsymbol{\alpha}$, temperature and temperature change [11]. $\mathbf{E}$ is in 0-1 direction as $E_1(I_o,T,\Delta T)$. The elastic matrix $\mathbf{Q}$, thermal expansion coefficient $\boldsymbol{\alpha}$ and piezoelectric stress constant $\mathbf{e}$ need to be transformed to the laminate coordinates [17]. The element stiffness matrix can be written as

$$\mathbf{K}^e = (\mathbf{K}_A)^e + (\mathbf{K}_{AC})^e + [(\mathbf{K}_{AC})^e]^{\mathrm{T}} + (\mathbf{K}_C)^e \tag{15}$$

where $$(\mathbf{K}_A)^e = \sum_{k=1}^{NL} ({}^{(k+1)}z - {}^{(k)}z) \int_{S^e} ({}^{(k)}\mathbf{A}^{\mathrm{T}})^e ({}^{(k)}\bar{\mathbf{Q}})^e ({}^{(k)}\mathbf{A}\,)^e \, dS$$

$$(\mathbf{K}_{AC})^e = \sum_{k=1}^{NL} \frac{1}{2} ({}^{(k+1)}z^2 - {}^{(k)}z^2) \int_{S^e} ({}^{(k)}\mathbf{A}^{\mathrm{T}})^e ({}^{(k)}\bar{\mathbf{Q}})^e ({}^{(k)}\mathbf{C})^e \, dS$$

$$(\mathbf{K}_C)^e = \sum_{k=1}^{NL} \frac{1}{3} ({}^{(k+1)}z^3 - {}^{(k)}z^3) \int_{S^e} ({}^{(k)}\mathbf{C}^{\mathrm{T}})^e ({}^{(k)}\bar{\mathbf{Q}})^e ({}^{(k)}\mathbf{C})^e \, dS$$

where $k = 1,...,NL$ and $e = 1,...,NE$ are the number of laminate layers and the number of elements. ${}^{(k)}z$ is the vertical distance from the mid-surface of the multilayer plate composite to the upper surface of the kth lamina. Any distances below the mid-surface are negative and those above the mid-surface are positive. $\bar{\mathbf{e}}$ and $\bar{\mathbf{Q}}$ are the piezoelectric and elastic constants for the composite structures, respectively.

Topology optimization of composite structures

For the orthotropic material to be considered in this work [12,16,18], instead, a simple SIMP scheme as the following is used to describe practical properties of laminated composites:

$$E_{ii}(x^{(k)}) = E_{ii,\min}^{(k)} + (x_e^{\mathrm{p}})^{(k)} (E_{ii}^{(k)} - E_{ii,\min}^{(k)}),\ (i = 1,2,3,\ \text{no sum}) \tag{16}$$

$$G_{ij}(x^{(k)}) = G_{ij,\min}^{(k)} + (x_e^{\mathrm{p}})^{(k)} (G_{ij}^{(k)} - G_{ij,\min}^{(k)}),\ (i = 1, j = 2; i = 2, j = 1; i = 3, j = 1) \tag{17}$$

$$\alpha_{T\,jj}(x^{(k)}) = \alpha_{T\ \,jj,\min}^{(k)} + (x_e^{\mathrm{p}})^{(k)} (\alpha_{T\ \,jj}^{(k)} - \alpha_{T\ \,jj,\min}^{(k)}),\ (j = 1,2,3,\ \text{no sum}) \tag{18}$$

$$\mathrm{e}_{mn}(x^{(k)}) = \mathrm{e}_{mn,\min}^{(k)} + (x_e^{\mathrm{p}})^{(k)} (\mathrm{e}_{mn}^{(k)} - \mathrm{e}_{mn,\min}^{(k)}),\ (m = 3, n = 1; m = 3, n = 2; m = 3, n = 3) \tag{19}$$

where E_{ii}, G_{ij}, α_T and e_{mn} represent the practical material Young's modulus, shear modulus, thermal expansion coefficients and piezoelectric stress constants, respectively. In the above equations, the minimum values of each constant can be determined as 0.0001 times of their solid material properties to avoid numerical singularity. Here, p=3 is used for all the material parameters including E_{ii}, G_{kl}, α_T and e_{mn}. The SIMP-based parameterization scheme is applied to both the photostrictive and host material layers to ensure a "0/1" approximation design. In addition, the standard sensitivity filtering technology [13] is included in numerical implementation to eliminate the checkerboards and mesh-dependence in the topology optimization.

The topology optimization problem can be defined as a constrained least squares problem as:

$$\begin{cases} \underset{\mathbf{x}=\{\mathbf{x}^{\mathrm{h}},\mathbf{x}^{\mathrm{p}},\mathbf{x}^{\mathrm{I}}\}^{\mathrm{T}}}{\underline{\text{Minimize}}}: J(\mathbf{u}(\mathbf{x})) \\ \text{Subject to}: \begin{cases} g_1(\mathbf{x}) = \mathbf{v}^{\mathrm{T}}\mathbf{W}\mathbf{v} - E_{\max} \le 0 \\ g_2(\mathbf{x}) = \sum_{i=1}^{N_1} (\mathrm{x}_i^{\mathrm{p}} V_i^{\mathrm{p}}) - V_0^{\mathrm{p}} \le 0 \\ g_3(\mathbf{x}) = \sum_{i=1}^{N_2} (\mathrm{x}_i^{\mathrm{h}} V_i^{\mathrm{h}}) - V_0^{\mathrm{h}} \le 0 \\ 0 < \mathrm{x}_{\min}^{\mathrm{h}} \le \mathrm{x}_i^{\mathrm{h}} \le 1,\ i = 1,2,\ldots,N_1 \\ 0 < \mathrm{x}_{\min}^{\mathrm{p}} \le \mathrm{x}_i^{\mathrm{p}} \le 1,\ i = 1,2,\ldots,N_2 \\ \mathrm{x}_{\min}^{\mathrm{I}} \le \mathrm{x}_i^{\mathrm{I}} \le \mathrm{x}_{\max}^{\mathrm{I}},\ i = 1,2,\ldots,N_3 \end{cases} \end{cases} \tag{20}$$

For the shape control problems, the objective function [9,11-12,14,] is defined as the square shape errors to minimize the difference between the desired shape $\overline{S}_i$ and the actuated shape S_i

$$J(\mathbf{u}(\mathbf{x})) = \frac{1}{2}\sum_{i=1}^{q_1}(S_i(\mathbf{x}))^2 = \frac{1}{2}\sum_{i=1}^{q_1}\left[S_i(\mathbf{x},x,y,z) - \overline{S}_i(x,y,z)\right]^2 \tag{21}$$

where the vector $\mathbf{x} = \{\mathbf{x}^{\mathrm{h}}, \mathbf{x}^{\mathrm{p}}, \mathbf{x}^{\mathrm{I}}\}^{\mathrm{T}}$ includes three different groups of design variables. $\mathbf{x}^{\mathrm{p}}$ and $\mathbf{x}^{\mathrm{h}}$, respectively, include element-wise design variables for photostricitve layers $\mathbf{x}^{\mathrm{p}} = \{\mathrm{x}_1^{\mathrm{p}}, \mathrm{x}_2^{\mathrm{p}}, \ldots, \mathrm{x}_{\mathrm{N}_1}^{\mathrm{p}}\}^{\mathrm{T}}$ and host layers $\mathbf{x}^{\mathrm{h}} = \{\mathrm{x}_1^{\mathrm{h}}, \mathrm{x}_2^{\mathrm{h}}, \ldots, \mathrm{x}_{\mathrm{N}_2}^{\mathrm{h}}\}^{\mathrm{T}}$. $\mathbf{x}^{\mathrm{I}} = \{\mathrm{x}_1^{\mathrm{I}}, \mathrm{x}_2^{\mathrm{I}}, \ldots, \mathrm{x}_{\mathrm{N}_3}^{\mathrm{I}}\}^{\mathrm{T}}$ is the vector including design variables of light intensity factors regularized from 0 to 1. $i = 1,2,\ldots,q_1$ is used to represent the number of sample points. The first constraint refers to the equilibrium equation of the system, and the second constraint is used to limit the required actuation energy with the prescribed upper limit of the control energy $E_{\max}$, where $\mathbf{W}$ is a positive definite weighting matrix [14]. The third and fourth constraints are introduced as volume constraints, where V_0^{p} and V_0^{h} are the allowable material volume to be distributed inside the design domain for piezoelectric material and host material laminas, respectively, and V_i is each element volume. The rest of the constraints are used as lateral constraints for design variables x. $\mathrm{x}_{\min}^{\mathrm{p}}$=0.0001 and $\mathrm{x}_{\min}^{\mathrm{h}}$=0.0001 are the lower limits. $\mathrm{x}_{\min}^{\mathrm{I}}$ and $\mathrm{x}_{\max}^{\mathrm{I}}$ are the lower and upper limits of practical light intensities. $N = N_1 + N_2 + N_3$ is the number of all the design variables. In this work, the least-square optimization problem can be solved using MMA method [19].

In this study, the adjoint sensitivity method (ASM) is applied to perform the design sensitivity analysis. Consider the following equilibrium equation of the composite structures bounded with optical photostrictive surface layers, we have the following sensitivity expression

$$\frac{dJ(\mathbf{u}(\mathbf{x}))}{dx_i} = \boldsymbol{\lambda}^{\mathrm{T}}\left(-\frac{\partial(\mathbf{f}(\mathbf{x}))}{\partial x_i} - \frac{\partial \mathbf{K}(\mathbf{x})}{\partial x_i}\mathbf{u}(\mathbf{x})\right) \tag{22}$$

Hence we can find the design sensitivity of the objective function with respect to the design variables according to the derivatives of load vector $\mathbf{f}$ and stiffness matrix $\mathbf{K}$.

A Numerical Example

This section is concentrated on numerical applications of photostrictive actuators to the shape control of a linear elastic cantilever composite plate. It is assumed that multilayered laminas are mid-plane symmetric laminates which are composed of *NL* (*L=1,..,N*) perfectly bonded orthotropic layers with two photostrictive layers at the top and bottom surfaces (*L=2 and N-1*). The input light is insensitive to the host layers and the input light uniformly irradiates on the optical layers. The design objective is to minimize shape errors of the transverse deflection at all FE nodes. This study uses the following bending shape, symmetrical along the longitudinal axis, as the desired shape:

$$w^d(x,y) = \frac{\cos(10x)-1}{G}, \text{ where } G = 100 \text{ in this example} \tag{23}$$

The initial value of the light intensity is given as 115 mW/(cm)^2. The design domain is 1000mm×1000mm which is meshed with 400 quadrangular 8-noded finite elements and each node has 5 degrees of freedom. The composite plate is clamped at left-side edge and subjected to a concentrated force $F = 1.25N$ at the mid-point of the opposite position (Fig. 1) and the actuators are equipped as shown in Fig. 2. The thickness of the plate is 8.0mm which is laminated by a set of single material layers as [2.0/1.0/1.0/1.0/1.0/2.0]mm and the laminates are stacked sequentially with angle directions [90/0/90/90/0/90]. There are four elastic layers and two host layers adjacent to the actuation layers are topologically unchanged. The volume fractions for actuation layers, host layers and light intensity are 50%, 50%, and 45%, and the allowable light energy limit is 3.5E+08. The widely studied PLZT (3/52/48) [20,9,11] is used, with the piezoelectric stress constant $e = 25.1 \text{ N/V·m}$, the isotropic thermal coefficient $\alpha_T = 1.75\times10^{-5}$ and the temperature change 100K.

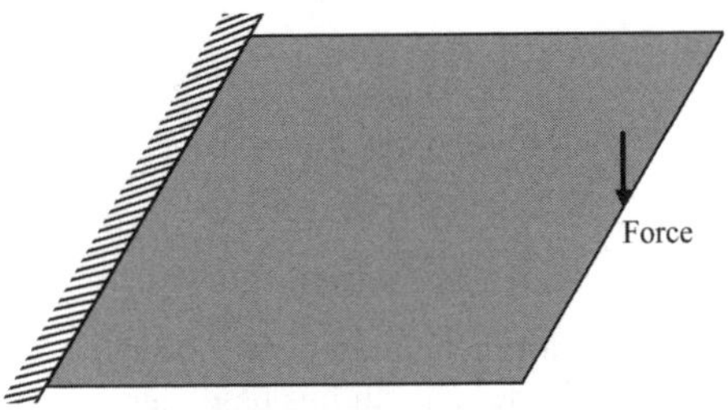

Fig. 1: Design domain

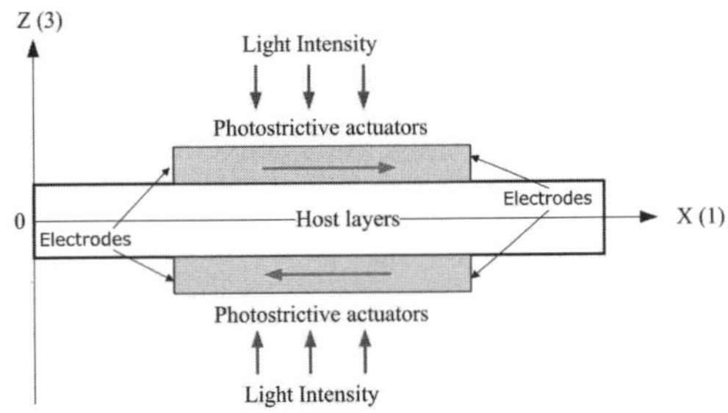

Fig. 2: 0-1 polarized PLZT composite plates

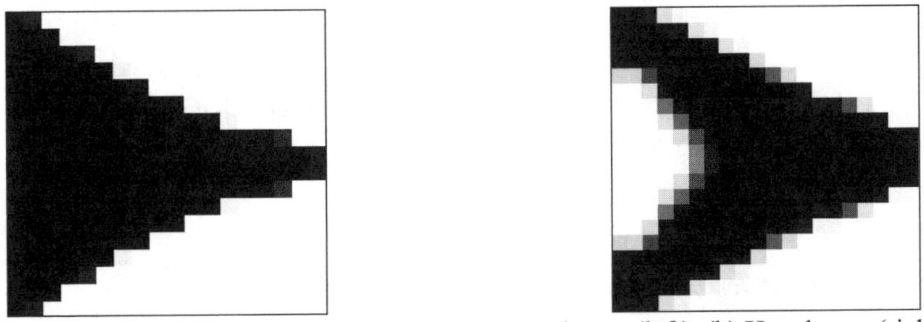

Fig. 3: Optimal topology layouts: (a) Photostrictive layers (left); (b) Host layers (right)

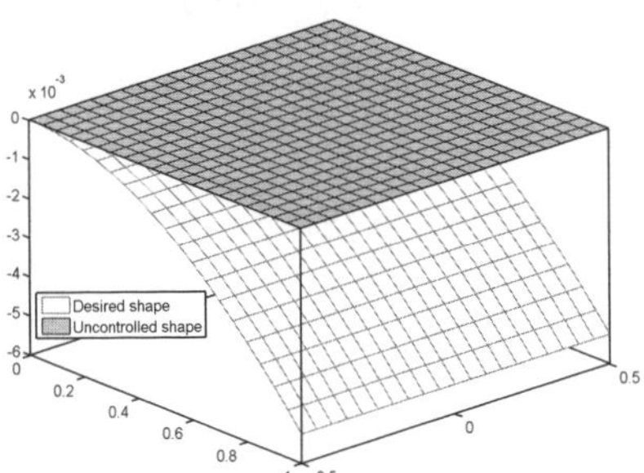

Fig. 4: Uncontrolled shape

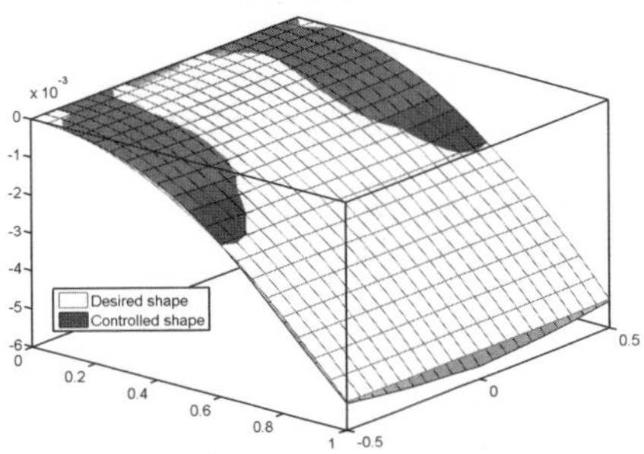

Fig. 5: Controlled shape

The design is converged after 209 iterations with the shape errors minimized from 1.7455 to 0.0068. The final topologies for photostrictive and host elastic layers are displayed respectively in Fig. 3, in which the topologies of the actuation and the host layers indicate that the materials are gradually accumulated from the left side to the centre and right side position along with the increase of the piezoelectric stress constant. Fig. 4 and 5 display the relations between structural original shape and the desired shape, and controlled shape and the desired shape, subject to the concentrated force and optical photostrictive actuation. The uncontrolled shape here refers to the undeformed shape without applying any tip-load. The iteration curve in Fig. 6 denotes that the topology changes are mainly completed in the first 60 iterations and the rest is used to implement shape evolvements until the design is at convergence. Fig. 7 shows the iteration curves of two mass conservative constraints related to the photostrictive and host structure layers.

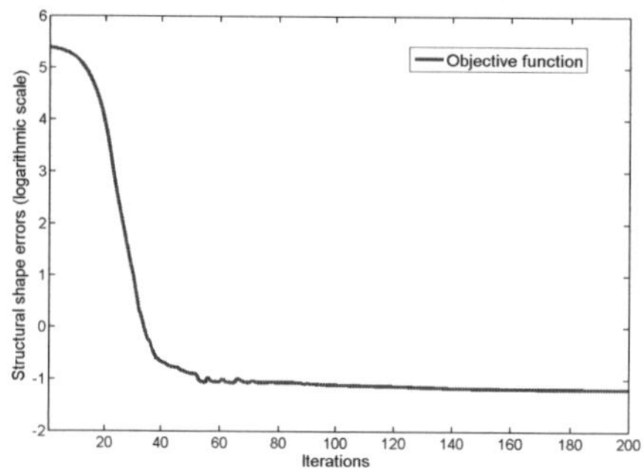

Fig. 6: Iteration of objective function

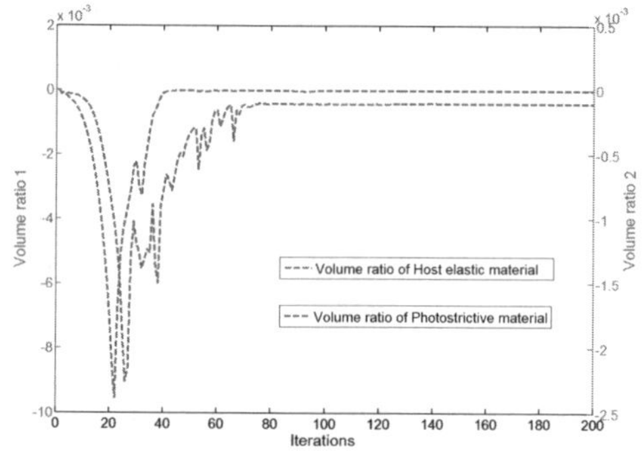

Fig. 7: Iteration of volume constraints

Summary

This paper presents a method for wireless shape control of laminated composite structures with photostrictive actuators using topology optimization methods. Numerical studies have demonstrated that the present method is effective in shape control of adaptive composite structures.

Acknowledgements

This research was sponsored in part by ARC-DP0774596, and Vice-Chancellor Research Fellowship, The University of Technlogy, Sydney, 2011.

References

[1] V.M. Fridkin: *Photoferroelectrics*. Springer, New York, (1979).

[2] K. Uchino: *Photostrictive Effect in (Pb,La)(Zr,Ti)O_3*. Ferroelectrics, 64, 1(1985), pp. 199-208.

[3] H.S. Tzou, and C.S. Chou: *Nonlinear opto-electromechanics and photodeformation of optical actuators*. Smart Mater. and Struct, 5, 2(1996), pp. 230-235.

[4] P. Poosanaas, K. Tonooka, and P. Uchino: *Photostrictive actuators*. Mechatronics 10, 4-5(2000), pp. 467-487.

[5] C.T. Lin, L. Li, J.D. Mcneill, and et al: *Photoconductivity of extrinsic ion-doped PLZT ceramics*. J. of Luminescence, 60-61, April(1994), pp. 170-174.

[6] M. Qin, K. Yao, S. Shannigrahi, and et al: *Thickness effects on photoinduced current in*

ferroelectric $(Pb_{0.97}La_{0.03})(Zr_{0.52}Ti_{0.48})O_3$ *thin films*. J. of Appl. Phys., 101, 014104(2007), pp. 1-8.

[7] H. Irschik: *A Review on Static and Dynamic Shape Control of Structures by Piezoelectric Actuation.* Eng. Struct., 24, 1(2002), pp. 5-11.

[8] D.B. Koconis, L.P. Kollar, and G.S. Springer: *Shape control of composite plates and shells with embedded actuators II: desired shape specified.* J. of Compos. Mater., 28, 5(1994), pp. 459-482.

[9] D.C. Sun and L.Y. Tong: *Modeling of wireless remote shape control for beams using nonlinear photostrictive actuators.* Int. J. Solids and Struct., 44, 1(2007), pp. 672-684.

[10] M. Ichiki, Y. Morikawa, Y. Mabune, and et al: *Preparation and photo-induced properties of lead lanthanum zirconate titanate multilayers.* J. of Phys., D: Appl. Phys., 37, 21(2004), pp. 3017-3024.

[11] Z. Luo, L.Y. Tong, J.Z. Luo, and et al: *Design of piezoelectric actuators using a multiphase level set method of piecewise constants.* J. of Comput. Phys., 228, 7(2009), pp. 2643-2659.

[12] Q.T. Luo, and L.Y. Tong: *Constitutive modeling of photostrictive materials and design optimization of microcantilevers.* J. of Intell. Mater. Syst. Struct., 20, 12 (2009), pp. 1425-1438.

[13] Z. Luo, Q.T. Luo, L.Y. Tong, and et al: *Shape morphing of laminated composite structures with photostrictive actuators via topology optimization.* Compos. Struct., 93, 2(2011), pp. 406-418.

[14] M.P. Bendsøe, O. Sigmund: *Topology Optimization: Theory, Methods, and Applications.* Springer, Berlin Heidelberg, (2003).

[15] Z. Kang, and L.Y.Tong: *Integrated optimization of material layout and control voltage for piezoelectric laminated plate.* J. of Intell. Mater. Syst. Struct., 19, 8(2008), pp. 889-903.

[16] Zhou M, and Rozvany GIN. *The COC algorithm, part II: topological, geometry and generalized shape optimization.* Comput. Methods Appl. Mech. Eng, 89, 1-3(1991), pp.197-224.

[17] M.P. Bendsøe, O.Sigmund: *Material interpolation schemes in topology optimization.* Arch. Appl. Mech,. 69, 9-10(1999), pp. 635-654.

[18] J.N.Reddy: *Mechanics of Laminated Composite Plates-Theory and Analysis.* CRC Press, Boca Raton, (2004).

[19] J. Stegman, E. Lund: *Discrete material optimization of general composite shell structures.* Int, J. Numer. Methods Eng., 62, 14(2005), pp. 2009-2027.

[20] K. Svanberg: *The method of moving asymptotes-a new method for structural optimization.* Int, J. Numer. Methods Eng., 24, 2(1987), pp. 359-373.

[21] M. Ichiki, Y. Morikawa, Y. Mabune, and et al: *Preparation and photo-induced properties of lead lanthanum zirconate titanate multi*layers. J. Phys. D: Appli. Phys., 37, 21(2004), pp. 3017-3024.

Advanced Materials Research Vol. 279 (2011) pp 194-199

doi:10.4028/www.scientific.net/AMR.279.194

A Four-node Hybrid-Trefftz Plane Elasticity Element with Fundamental Analytical Solutions

WANG Keyong[1, a]

[1]School of Mechanical Engineering, Shanghai University of Engineering Science, Shanghai 201620, China

[a]keyongw@gmail.com

Keywords: Fundamental analytical solutions, Hybrid-Trefftz element, Modified variational functional

Abstract. A new four-node hybrid-Trefftz element is developed for plane elasticity problems. From the Airy stress function, the fundamental analytical solutions of the governing equation are derived as Trefftz functions for the intra-element (Trefftz) displacement field. Together with an independent frame displacement field along the element boundary, the element formulation is then constructed following the modified variational functional. Several numerical examples indicate that the proposed element exhibits good performance.

Introduction

The original idea of Trefftz method can go back to 1926 [1] and therefore, it has been widely studied. The so-called hybrid Trefftz (HT) finite element (FE) model for plane elasticity problems, whereby the intra-element (Trefftz) displacement field is linked through auxiliary interelement displacement frame, was firstly presented by Jirousek and Teodorescu [2]. Subsequently, many versions of HT plane elasticity elements have been investigated [3-5]. In contrast to conventional FE models, the HT FE model preserves the advantages of FE approaches and boundary solutions and avoiding their drawbacks.

The construction of trail functions, i.e. Trefftz functions, plays an important role in the HT FE analysis. Among the previous works, the general way of constructing Trefftz functions usually starts at the Muskhelishvili's complex variable formulation. In this paper, the fundamental analytical solutions [6] were employed as Trefftz functions to derive HT plane elasticity element formulation by discarding rigid body motions. The accuracy of the newly developed element has been verified by comparing our results with the analytical solutions or other numerical results.

Basic Equations and Corresponding Trefftz Functions

We adopt the following matrix notation:

$$\mathbf{u}=\{u \quad v\}^T,\ \boldsymbol{\varepsilon}=\{\varepsilon_x \quad \varepsilon_y \quad \gamma_{xy}\}^T \tag{1a,b}$$

$$\boldsymbol{\sigma}=\{\sigma_x \quad \sigma_y \quad \tau_{xy}\}^T,\ \mathbf{t}=\{t_x \quad t_y\}^T \tag{2a,b}$$

If there is no existence of body forces, the differential equations of linear plane elasticity problems may be summarized as follows:

$$\mathbf{L}^T\boldsymbol{\sigma}=0,\quad \boldsymbol{\sigma}=\mathbf{D}\boldsymbol{\varepsilon},\quad \boldsymbol{\varepsilon}=\mathbf{L}\mathbf{u}\qquad \text{in } \Omega \tag{3a-c}$$

together with

$$\mathbf{u}=\bar{\mathbf{u}} \text{ on } \Gamma_u \text{ and } \mathbf{t}=\mathbf{A}\boldsymbol{\sigma}=\bar{\mathbf{t}} \text{ on } \Gamma_t \tag{4a,b}$$

where

$$\mathbf{L}=\begin{bmatrix}\frac{\partial}{\partial x} & 0 & \frac{\partial}{\partial y}\\ 0 & \frac{\partial}{\partial y} & \frac{\partial}{\partial x}\end{bmatrix}^T,\ \mathbf{D}=\frac{E}{1-\mu^2}\begin{bmatrix}1 & \mu & 0\\ \mu & 1 & 0\\ 0 & 0 & \frac{1-\mu}{2}\end{bmatrix},\ \mathbf{A}=\begin{bmatrix}n_x & 0 & n_y\\ 0 & n_y & n_x\end{bmatrix} \tag{5a-c}$$

Ω stands for the region, Γ_u and Γ_t designate the parts of the boundary $\Gamma=\Gamma_u\cup\Gamma_t$ with specified displacements and tractions, respectively, and n_x, n_y stand for unit outward normal to the boundary Γ. The Young's modulus E and the Poisson's ratio μ for plane stress state may be replaced by $E'=E/(1-\mu^2)$ and $\mu'=\mu/(1-\mu^2)$ for plane strain state.

The Trefftz functions play an important role in the derivation of the HT FE formulation. Now, the construction of Trefftz functions for the plane elasticity will be discussed in detail. The equilibrium equations, strain-displacement relations or compatibility conditions and Hooke's law in Eq. 3 can be readily reduced to a biharmonic equation with respect to the stress function $U(x,y)$

$$\nabla^4 U(x,y)=0 \tag{6}$$

where

$$\nabla^4=\frac{\partial^4}{\partial x^4}+2\frac{\partial^4}{\partial x^2\partial y^2}+\frac{\partial^4}{\partial y^4} \tag{7}$$

In order to find the fundamental analytical functions associated the plane elasticity problems, the stress function $U(x,y)$ can be readily assumed in a polynomial manner

$$U(x,y)=a_1+a_2x+a_3y+a_4x^2+a_5xy+a_6y^2+a_7x^3+a_8x^2y+a_9xy^2+a_{10}y^3 \tag{8}$$

As done in [7], the analytical solutions satisfying Eq. 6, as shown in Table 1, can be determined by the contra- solving approach.

Table 1: Fundamental analytical solutions of plane elasticity problems

term	1	2	3	4	5	6	7	8	9	10
u	1	0	y	y	μx	x	$-\mu x^2-y^2$	2xy	$-2\mu xy$	$x^2-(2+\mu)y^2$
v	0	1	-x	x	y	$-\mu y$	2xy	$-\mu y^2-x^2$	$y^2-(2+\mu)x^2$	$-2\mu xy$

HT FE Analysis

Assumed Fields. The basic idea of the HT FE model is to establish a finite element formulation whereby intra-element continuity is enforced on a non-conforming internal displacement field chosen so as to satisfy, *a priori*, the governing differential equation of the problem under consideration. With this method the solution domain is subdivided into elements Ω_e, and over each element "*e*", two groups of independent displacement fields are assumed, as shown in Fig. 1:

(a) A nonconforming intra-element displacement field

$$\mathbf{u}_e=\sum_{j=1}^{m}\mathbf{N}_{ej}c_{ej}=\mathbf{N}_e\mathbf{c}_e \quad \text{in } \Omega_e \tag{9}$$

(b) An exactly and minimally conforming auxiliary frame field

$$\tilde{\mathbf{u}}_e=\tilde{\mathbf{N}}_e\mathbf{d}_e \quad \text{on } \Gamma_e \tag{10}$$

being independently assumed along the element boundary in terms of nodal degrees of freedom (DOF) $\mathbf{d}_e$, where $\mathbf{N}_e$ are the homogeneous solutions to Eq. 6, $\mathbf{c}_e$ stands for undetermined parameters, $\tilde{\mathbf{N}}_e$ are the shape functions (frame functions) defined in the customary way as in conventional FEM. The tildes above the symbols in Eq. 10 allow the two fields to be distinguished. In general, m denotes the number of truncated Trefftz terms which is suggested as $m \geq NDOF - R$ where NDOF means the number of nodal DOF of the element and R =3 is the number of rigid body motion modes because of the rank sufficiency of the element stiffness matrix [3].

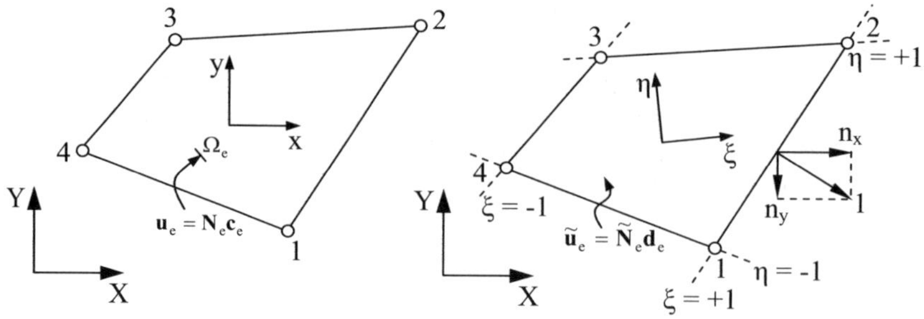

Fig. 1: Typical four-node HT element in plane elasticity

The corresponding stress field

$$\boldsymbol{\sigma}_e = \sum_{j=1}^{m} \mathbf{T}_{ej} c_{ej} = \mathbf{T}_e \mathbf{c}_e \tag{11}$$

and the boundary tractions

$$\mathbf{t}_e = \sum_{j=1}^{m} \mathbf{Q}_{ej} c_{ej} = \mathbf{Q}_e \mathbf{c}_e \tag{12}$$

can be easily deduced from $\boldsymbol{\sigma}_e = \mathbf{D}\mathbf{L}^T\mathbf{u}_e$ and $\mathbf{t}_e = \mathbf{A}\boldsymbol{\sigma}_e$ respectively.

The hybrid technique used in the HT FE formulation implies that the generated homogeneous solutions $\mathbf{N}_{ej}$ yield non-vanishing stresses $\mathbf{T}_{ej}$ and, as a consequence, the three rigid-body motion modes should be excluded. Therefore, the Trefftz functions for the four-node HT element has the following form

$$\mathbf{N}_e = \begin{bmatrix} y & \mu x & x & -\mu x^2 - y^2 & 2xy & -2\mu xy & x^2 - (2+\mu)y^2 \\ x & y & -\mu y & 2xy & -\mu y^2 - x^2 & y^2 - (2+\mu)x^2 & -2\mu xy \end{bmatrix} \tag{13}$$

Using a similar technique to that in the conventional FE model, the overall frame function matrix $\tilde{\mathbf{N}}_e$ for the four-node HT element can be directly defined as

$$\tilde{\mathbf{N}}_e = \begin{bmatrix} \mathbf{I}\tilde{N}_{e1} & \mathbf{I}\tilde{N}_{e2} & \mathbf{I}\tilde{N}_{e3} & \mathbf{I}\tilde{N}_{e4} \end{bmatrix} \text{ with } \tilde{N}_{ej} = \frac{1}{4}(1+\xi_j\xi)(1+\eta_j\eta)\,,\ j=1,2,3,4 \tag{14}$$

Element Formulation. Substituting Eq. 9, Eq. 10 and Eq. 12 into the modified variational functional below [5]

$$\Pi_m(\mathbf{u},\tilde{\mathbf{u}}) = \sum_e \Pi_{me}(\mathbf{u}_e,\tilde{\mathbf{u}}_e) = \sum_e \left[\iint_{\Omega_e} \frac{1}{2}\boldsymbol{\sigma}_e^T\mathbf{C}\boldsymbol{\sigma}_e d\Omega - \int_{\Gamma_{eu}} \mathbf{t}_e\overline{\mathbf{u}}_e d\Gamma - \int_{\Gamma_{et}} (\mathbf{t}_e - \overline{\mathbf{t}}_e)\tilde{\mathbf{u}}_e d\Gamma - \int_{\Gamma_{eI}} \mathbf{t}_e\tilde{\mathbf{u}}_e d\Gamma \right] \tag{15}$$

one can obtain the following form

$$\Pi_{me}(\mathbf{u}_e,\tilde{\mathbf{u}}_e) = -\mathbf{c}_e^T\mathbf{H}_e\mathbf{c}_e + \mathbf{c}_e^T\mathbf{G}_e\mathbf{d}_e + \mathbf{c}_e^T\mathbf{g}_e + \text{terms without } \mathbf{c}_e \text{ or } \mathbf{d}_e \tag{16}$$

where $\Pi_{me}(\mathbf{u}_e,\tilde{\mathbf{u}}_e)$ represents the element complementarity energy, $\mathbf{C}=\mathbf{D}^{-1}$, $\Gamma_e=\Gamma_{eu}\cup\Gamma_{et}\cup\Gamma_{eI}$, while $\Gamma_{eu}=\Gamma_u\cap\Gamma_e$, $\Gamma_{et}=\Gamma_t\cap\Gamma_e$, and Γ_{eI} is the inter-element boundary of element "*e*".

Taking the vanishing variation of Eq. 16 with respect to $\mathbf{c}_e$ yields

$$\frac{\partial\Pi_{me}(\mathbf{u}_e,\tilde{\mathbf{u}}_e)}{\partial\mathbf{c}_e}=0 \;\Rightarrow\; \mathbf{c}_e=\mathbf{H}_e^{-1}\mathbf{G}_e\mathbf{d}_e \tag{17}$$

and then the element stiffness equation can be obtained in a similar way with respect to $\mathbf{d}_e$:

$$\frac{\partial\Pi_{me}(\mathbf{u}_e,\tilde{\mathbf{u}}_e)}{\partial\mathbf{d}_e}=0 \;\Rightarrow\; \mathbf{K}_e\mathbf{d}_e-\mathbf{P}_e=0 \tag{18}$$

where $\mathbf{K}_e=\mathbf{G}_e^T\mathbf{H}_e^{-1}\mathbf{G}_e$ and $\mathbf{P}_e=\mathbf{G}_e^T\mathbf{H}_e\mathbf{g}_e$ are, respectively, the element stiffness matrix and the vector of the equivalent nodal forces. The auxiliary matrices $\mathbf{H}_e$, $\mathbf{G}_e$ and $\mathbf{g}_e$ are explicitly expressed as

$$\mathbf{H}_e=\int_{\Gamma_e}\mathbf{Q}_e^T\mathbf{N}_e\,d\Gamma=\int_{\Gamma_e}\mathbf{N}_e^T\mathbf{Q}_e\,d\Gamma\;,\;\mathbf{G}_e=\int_{\Gamma_{eI}+\Gamma_{et}}\mathbf{Q}_e^T\tilde{\mathbf{N}}_e\,d\Gamma\;,\;\mathbf{g}_e=-\int_{\Gamma_{et}}\tilde{\mathbf{N}}_e^T\bar{\mathbf{t}}_e\,d\Gamma \tag{19a-c}$$

Numerical Examples

To evaluate the performance of the proposed element, several numerical examples, patch test, Cook's membrane problem and thick-walled cylinder, are considered.

Patch Test. Fig. 2 describes the geometry and material properties of a patch test for the plane stress state [8]. The displacements and stresses may be analytically expressed as

$$u=10^{-3}\left(X+\frac{Y}{2}\right),\; v=10^{-3}(X+Y),\; \sigma_x=\sigma_y=1333.333,\; \tau_{xy}=400.0 \tag{20a-d}$$

In the patch test, the displacements of boundary nodes 5 to 8 are specified by Eq.(20a,b). The computational results indicate that the present element has passed the patch test because it reproduces the exact answers with any mesh configuration. Satisfaction of the patch test guarantees stability and convergence of the element when used for the solution of real problems.

Node	X	Y
1	0.04	0.02
2	0.18	0.03
3	0.16	0.08
4	0.08	0.08
5	0.00	0.00
6	0.24	0.00
7	0.24	0.12
8	0.00	0.12

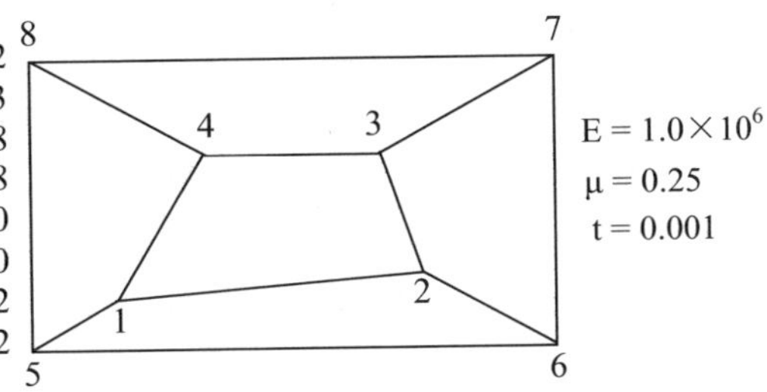

Fig. 2: Patch test in the coarse mesh

Cook's Membrane Problem. In order to assess the accuracy and the convergence rate of the present element, the Cook's membrane problem has been taken as an example. Fig. 3 shows the detailed geometry and material properties of this model. The reference vertical displacement v_c at the tip (Point C) is 23.96. The comparison displayed in Table 2 indicates that the present element has good accuracy as mesh density increase, and especially the faster convergence rate for the displacement behavior than that of conventional FEM (isoparametric quadrilateral element, Q4). There is no practical distinction between the numerical results obtained with the present element and those based on Muskhelishvili's solutions (HT)[3].

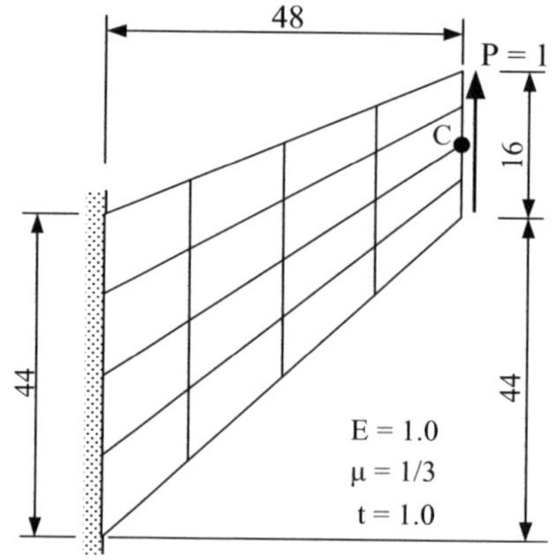

Fig. 3: Cook's membrane problem

Table 2: Vertical displacement at the tip

Mesh	Q4	HT [3]	Present
2×2	11.73082	15.03978	15.03978
4×4	17.16255	20.60818	20.60820
8×8	22.04080	22.94928	22.94923
16×16	23.40892	23.68502	23.68494
32×32	23.79947	23.88795	23.88783
64×64	23.91687	23.94485	23.94413

Thick-walled Cylinder Subjected to Internal and External Pressure. Fig. 4 shows a thick-walled cylinder subjected to internal and external pressure, which conforms to plane strain conditions. The internal and external radii are $r_a = 10$ and $r_b = 20$. Due to axisymmetric properties, one quarter of the model is considered. Hoop stress distribution is obtained and compared with the theoretical solution in Fig. 5, from which it can be seen that good agreement is achieved between the present element and theoretical solution and the HT element.

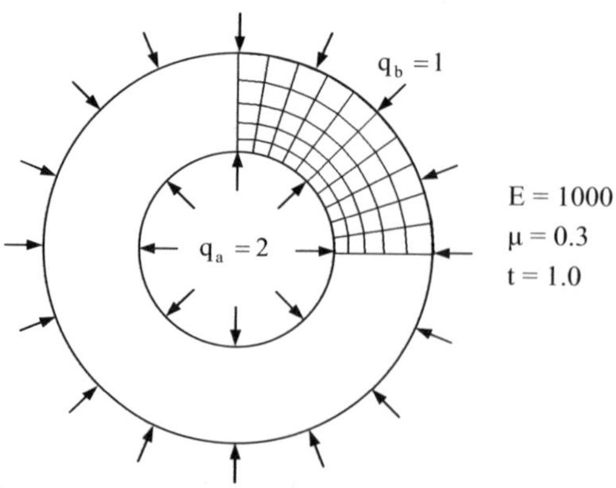

Fig. 4: Thick-walled cylinder subjected to the internal and external pressure

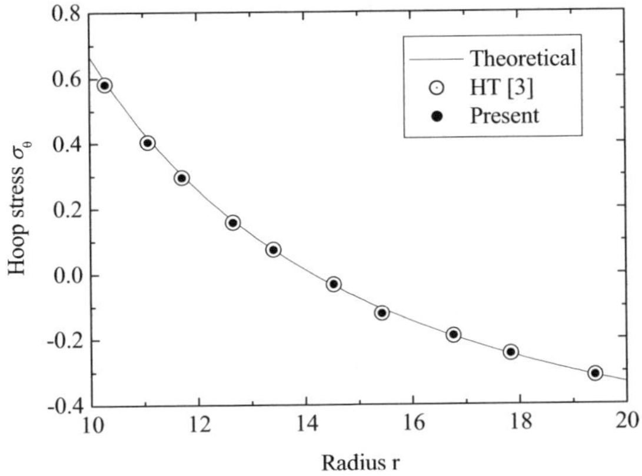

Fig. 5: Hoop stress distribution due to internal and external pressure

Concluding Remarks

A four-node hybrid Trefftz plane element has been developed based on the use of fundamental analytical solutions of the problem under consideration. Several numerical examples are considered for the assessment of the proposed element. The results indicate that the proposed element has been shown to have better performance than the conventional finite element and be similar convergence rate with the HT element based on Muskhelishbili's solutions.

Acknowledgement

The work reported in this paper has been supported by the Education Committee of Shanghai for Outstanding Young Teachers in University under the Grant No. GJD10019.

References

[1] E. Trefftz: *Ein gegenstück zum ritz'schen verfahren*. Proceedings of the Second International Congress on Applied Mechanics, (1926), pp. 131-137.

[2] J. Jirousek, P. Teodorescu: *Large finite elements method for the solution of problems in the theory of elasticity*. Computers & Structures, 15, 5(1982), pp. 575-587.

[3] J. Jirousek, A. Venkatesh: *Hybrid Trefftz plane elasticity elements with p-method capabilities*. International Journal for Numerical Methods in Engineering, 35, 7(1992), pp. 1443-1472.

[4] Y.S. Choo, N. Choi, B.C. Lee: *Quadrilateral and triangular plane elements with rotational degrees of freedom based on the hybrid Trefftz method*. Finite Elements in Analysis and Design, 42, 11(2006), pp. 1002-1008.

[5] Q.H. Qin: *Dual variational formulation for Trefftz finite element method of elastic materials*. Mechanics Research Communications, 31, 3(2004), pp. 321-330.

[6] X.R. Fu, Y.Q. Long, M.W. Yuan: *A generalized conforming plane element based on analytical trial functions*. Engineering Mechanics, 22, 3(2005), pp. 1-4.

[7] J.M. Hu, H.D. Wang: *Contra-solution of solving displacement basis formulae of incompatible element*. Journal of Harbin Institute of Technology, 36, 11(2004), pp.1547-1549.

[8] R.H. MacNeal: *A proposed standard set of problems to test finite element accuracy*. Finite Elements in Analysis and Design, 1, 1(1985), pp. 3-20.

Advanced Materials Research Vol. 279 (2011) pp 200-206
© (2011) Trans Tech Publications, Switzerland
doi:10.4028/www.scientific.net/AMR.279.200

A New Hole Filling Algorithm in Space for Triangular Model

Jie Chen[1, a], Chenghui Gao[1, b] and Bingwei He[1, c]

[1]College of Mechanical Engineering and Automation, Fuzhou University,
Fujian Fuzhou 350108, China

[a]N050220060@fzu.edu.cn, [b]gch@fzu.edu.cn, [c]bw_he@yahoo.com.cn

Keywords: Triangular mesh, Hole boundary, Least squares method, Hole filling

Abstract. Aiming at filling the holes which were generated from uncompleted point cloud data in reverse engineering, a new hole filling algorithm in space is presented. Firstly, the holes boundary was identified and pretreated, and the hole boundary feature datum was established and the boundary was projected on it to form a projection polygon. Secondly, the smallest angle of the projection polygon was found out to determine the corresponding boundary point as the mesh growing point. The original hole was covered by the new meshes covering and then filling algorithm was completed. Finally, the neighborhood points of hole boundary vertex were selected as the sampling points for the least squares fitting adjustment of new filled vertices position, which aims at the preparation hole filled result. Examples are given to prove that the method has good accuracy and stability of the hole filling.

Introduction

In reverse engineering, triangular meshes are widely used in geometric model expression for its suppleness and intuitiveness. Point cloud of models acquired from measuring equipment can be used for reconstruction of triangular mesh. However, as measurement devices functional limitation or model characteristics of its own, which will make a variety of regions can't be measured, the point cloud data is uncompleted. When the point data is used for reconstruction, it will generate holes in triangle mesh model. The existence of these holes will not only affect the visual effects of the three-dimensional model, but also cause problems in the series of follow-up treatment such as surface modeling, rapid prototyping, and finite element analysis. For this reason, many scholars have conducted in-depth research on hole filling in triangular mesh models. Zhang[1] provided a spatial mesh growing method to fill the mesh holes, which the meshes filled by this method have a optimize mesh distributed, but the increased vertex positions accuracy needs to be improved. Neural network hole repairing method, especially Radial Basis Function(RBF) neural network repairing algorithm was presented to fill holes in the triangular mesh, whose precision and speed have been made a good effect, but when it is used to deal the untrained new data, it's adaptation is rather poor. So there is an inherent learning shortage of RBF[2]. Jean-Philippe[3] filled the hole plane meshes after smoothing the boundary of the hole and then used a mechanical four-bar linkage of the minimum energy method to adjust the new added vertices. This method made a mainly goal aimed at achieving a minimum curvature variation status in global, and complicated calculation process is the main disadvantage. Yongtae Jun[4] proposed a robust hole filling method to divide complex holes in part to simple plane holes and then repaired these holes by using some mature approaches. It can solve some holes with complex plane boundary curves, but it also neglected the around meshes properties of the holes during the process, and hole filling results is not very satisfactory. Liu[5] carried out a approach on filling holes on the hole feature plane and then used least squares method to adjust the added vertices of the hole. Although this approach has a improved precision, the main filling process is carried out based on feature plane, and there is a certain precision different from hole filling in space meshes.

In this paper, a new space mesh hole filling algorithm is proposed, including holes feature extraction and preprocessing, hole feature holes building up, space triangles hole filling, vertex position adjustment based on least squares, and example is given for proving the algorithm can obtain quite a good filling effect.

Hole Boundary Feature Extraction and Hole Feature Plane Established

Triangular mesh model is composed of a series of triangular facets, and according to its topology we can see that each triangle edge belongs to two adjacent triangles. Thus, relevant definition of the boundary edges are: if an edge only belongs to a triangular, the edge is defined as a boundary edge; two vertex bonder of the boundary edge defined as boundary points; triangles contain boundary edges belong to the boundary triangles [3]. Searching the entire triangle model according to these definitions can extract all boundary edges to meet the definitions. Hole boundary is interconnected by a number of boundary edges form a closed polygon, as shown in Fig. 1. For non-closed triangular mesh model, model boundary and hole boundary are both constituted of boundary edges, and the composition of the two boundary properties are the same. In this case, it can take human-computer interaction approach to identify hole boundary for model hole filling.

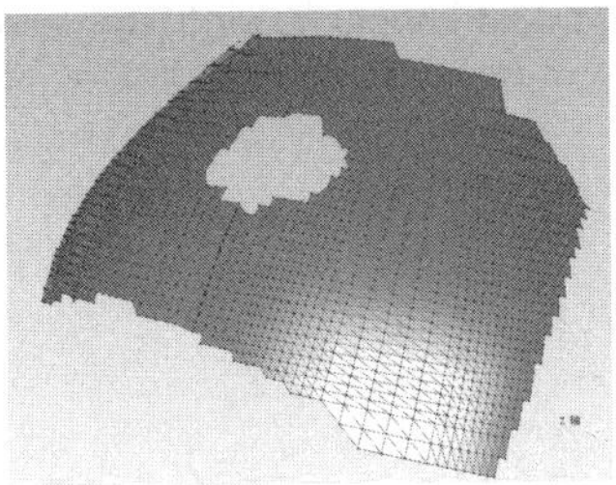

Fig. 1: Hole boundary

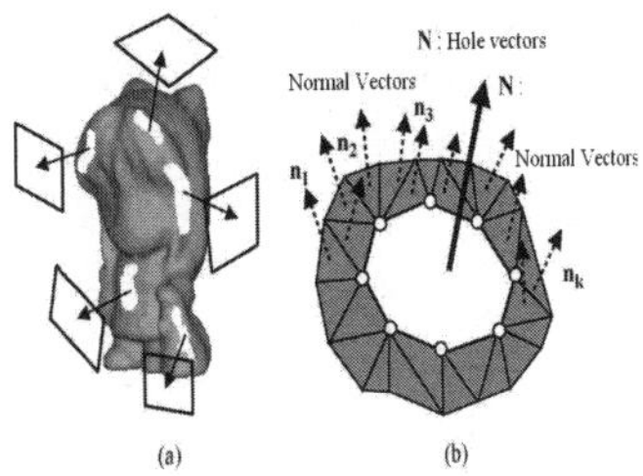

Fig. 2: Hole feature plane and hole vector

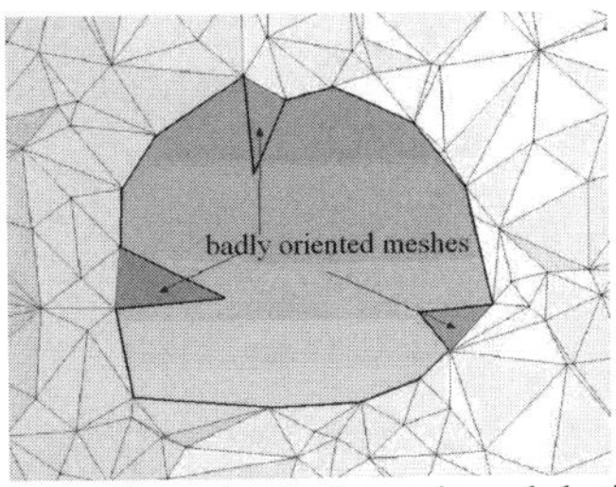

Fig. 3: Badly oriented meshes deletion

Hole feature plane established is aimed to hole filling process correctly completed in follow-up, as shown in Fig. 2(a) to find out all the feature planes corresponding to each hole. Utilize Eq. 1 to calculate out the normal vector of the hole feature plane:

$$N = n_{sum} = \sum_{k=1}^{v} n_k \quad (1)$$

Where N is normal vector of hole feature plane in Fig. 2 (b); n_{sum} is the sum of whole normal vectors of boundary triangle meshes; n_k is the normal vector of the kth boundary triangle mesh [4].

Secondly, find the based point of hoe feature plane, by Eq. 2 can be obtained:

$$O = \frac{1}{n} \sum_{i=0}^{n} P_i \quad (2)$$

Where n is the number of whole boundary vertices, P_{i} is the ith vertex coordinate value, then we can establish the necessary hole feature plane.

Hole Boundary Preprocessing

Hole boundary preprocessing concludes two main parts, badly oriented meshes deletion and hole boundary triangles optimization.

Badly Oriented Meshes Deletion

A badly oriented triangle facet which has two boundary edges in its three edges are as shown in Fig. 3, and one can notice that a triangle defined by two boundary edges can rotate around its interior edge without affecting the other triangles. If nothing is done, all these triangles may lead to undesired undulations of the inner mesh that will fill in the hole. At the same time, it also can be found that in the filling process, the existent of such triangles will lead to the boundary vertex adjacent points searching error. To overcome these drawbacks, it is necessary to clean the hole contour before hole filling processing. Therefore we delete recursively all the triangles defined by two boundary edges which is the black triangles in the example, as shown in Fig. 4(a). But by using this process, some significant ones may cause details of the contour disappearance. To overcome this limit, an additional condition should be satisfied at each step of the recursive deletion process: a triangle which has two boundary edges can be deleted only if its adjacent triangle does not have any boundary edge, as shown in Fig. 4(b). Then the process moves in the right direction to hole contour cleaning [3].

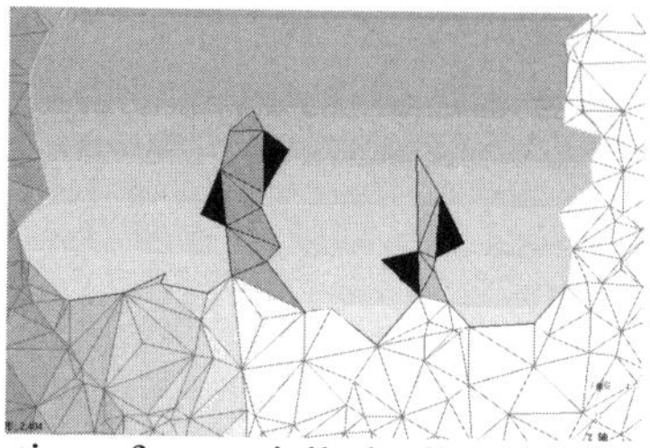

Fig. 4: Deletion of potentially badly oriented triangles

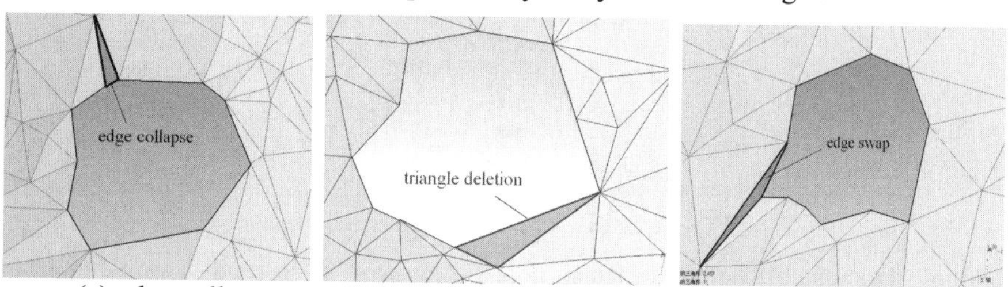

(a) edge collapse (b) triangle deletion (c) edge swap

Fig. 5: Boundary meshes optimization

Hole Boundary Triangles Optimization

In order to avoid the follow-up holes filling process appears long-narrow or irregular triangles, the boundary of the hole preprocessing is necessary, including edge collapse, triangle deletion and edge exchange. First, search out all the boundary triangles, according to the Eq. 3 to obtain area and side length factor to determine whether the triangles need treatment.

$$Q = \alpha \frac{S}{hp} \tag{3}$$

Where, $\alpha = 2\sqrt{3}$ is a standardization parameters, h is the longest edge of the triangle, p is half-perimeter, S is the area of the triangle.

According to the literature [3], we take 0.25 as the threshold for Q, triangular which it's Q <0.25 will be considered as potential triangles to optimized; otherwise, the triangles without to be dealt. To these potential triangles, if the boundary edge is the shortest side, as shown in Fig. 5(a), edge collapse is taken to it. Otherwise, if the longest edge is a boundary edge, this triangle is deleted as shown in Fig.5 (b), or the longest edge is swapped as Fig. 5(c) [3].

Space Triangle Hole Filling

In this paper, an optimization hole filling method based on boundary mesh growing hole filling theory according to literature [1], and fills the triangle hole in space. First project boundary vertices onto hole feature plane to form a hole feature curve, and then calculate the angle at each projected point using its two adjacent sides on the hole feature curve, determine the minimum angle between the projection point and find out the corresponding boundary vertex in the space. This vertex will be the fist boundary vertex for mesh growing based point. Meanwhile, according to the magnitude of the minimum angle, the mesh growing rules is divided into four types, as shown in Fig. 6. Filling the triangle model holes with this mesh growing method may appear the new added triangles that are quite different from the boundary triangles, especially in Fig. 6(a) conditions. In order to avoid such situations, a handling procedure is added: if the new edge length $|B_{i-1}B_{i+1}|$ of the added triangle has such a relationship with the average length of all boundary edges l_{avg} , when $|B_{i-1}B_{i+1}|/l_{avg}>1.5$, divide triangle $B_{i-1}B_iB_{i+1}$ into two triangles $B_{i-1}TB_i$ and $B_iT_2B_{i+1}$ by line $|B_iT|$, which vertex T is the midpoint of $|B_{i-1}B_{i+1}|$, as shown in Fig. 7; when $|B_{i-1}B_{i+1}|/l_{avg}<0.5$, take a treatment of edge swap on the new added triangles like Fig. 5(a), or to be unchanged.

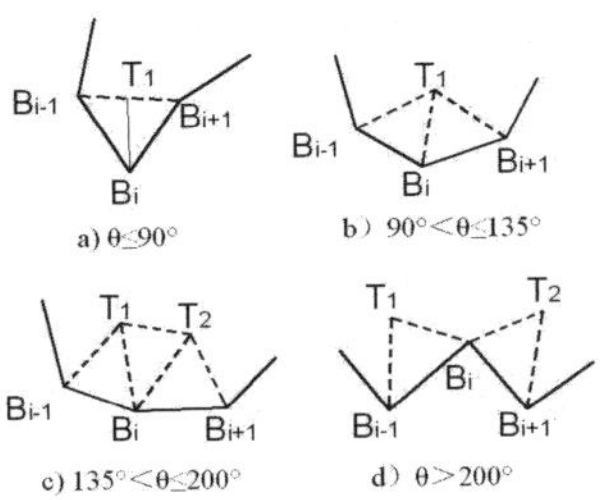

Fig. 6: Mesh growing rules

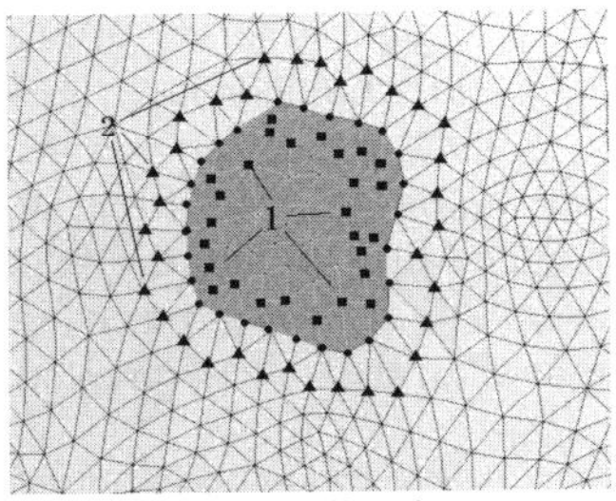

Fig. 7: Sample and adjusting vertex selection

Vertex Position Adjustment Based on Least Square Method

Aiming at the hole filling result closed to the actual model surface as much as possible, the added vertices should be dealt with position adjustment. In this paper, principles of surface fitting using least squares adjustment method will be applied, and the main approach is using the existing boundary triangle vertices to adjust the filled vertices. Surface least squares fitting is utilizing a given surface point set coordinates (x, y, z), driven by the fitting function to find a set of optimal parameters with which distance between the points on fitted surface and the original point to be least. In this paper, the least squares fitting function equation are as follows:

$$Z(x,y)=a_{00}x^0y^0+a_{01}x^0y^1+a_{02}x^0y^2+a_{10}x^1y^0+a_{11}x^1y^1+a_{12}x^1y^2+a_{20}x^2y^0+a_{21}x^2y^1+a_{22}x^2y^2 \quad (4)$$

Least square fitting is to find minimum value of S in the next formula:

$$S=\sum_{g=1}^{n}w(x_i,y_j)\left[\sum_{0}^{p-1}\sum_{0}^{q-1}a_{ij}x^iy^j-f(x_i,y_j)\right] \text{(}g\text{=0,1, …}n\text{, means vertices number)} \quad (5)$$

Where $f(x_i,y_j)$is the Z coordinate value of(x_i,y_j,z_k),$w(x_i,y_j)$ is the weight function, said the point on the contribution of the surface, here we define a_{ij} all equals 1 which is the default point on the surface has the equal contribution for the surface[6-7].

Solving the minimum value of S, a_{ij} must satisfy Eq. 6.

$$\frac{\partial S}{\partial a_{ij}}=0 \quad (ij\text{=00 01 …} \quad pq\text{)} \quad (6)$$

Take Eq. 6 into Eq. 5 and transpose after solving the partial derivative of the function, there will be

$$\begin{bmatrix} \sum x_g^0 y_g^0 \bullet x_g^0 y_g^0 & \cdots & \sum x_g^0 y_g^0 \bullet x_g^0 y_g^{q-1} & \cdots & \sum x_g^0 y_g^0 \bullet x_g^{p-1} y_g^{q-1} \\ \vdots & \ddots & \vdots & \ddots & \vdots \\ \sum x_g^{p-1} y_g^{q-1} \bullet x_g^0 y_g^0 & \cdots & \sum x_g^{p-1} y_g^{q-1} \bullet x_g^0 y_g^{q-1} & \cdots & \sum x_g^{p-1} y_g^{q-1} \bullet x_g^{p-1} y_g^{q-1} \end{bmatrix} \bullet \begin{bmatrix} a_{00} \\ \vdots \\ a_{p-1q-1} \end{bmatrix} = \begin{bmatrix} \sum x_g^0 y_g^0 \bullet z_g \\ \vdots \\ \sum x_g^{p-1} y_g^{q-1} \bullet z_g \end{bmatrix} \tag{7}$$

Use L, R, A to simplify expressions (7), there $$L*A=R \tag{8}$$
From equation(8), there will $$A=L^{-1}*R \tag{9}$$

At last we can calculate the matrix A of the least square fitted parameters which will be used for filled vertex positions adjustment in the next step.

Filled vertices adjustment algorithm is described as following. Take the variation of hole boundary curves into account and in order to get close to the original shape surface and smoothness of the new added triangles, and select 3-5 vertex rings as the least squares fitting surface sample vertices. Sample points selection which is starting from the hole boundary from inside to outside nearby boundary triangles, as shown in Fig. 7(a). In this paper we choose 4 rings, and in the figure there are 2 rings circular points and 2 rings square points. Filled vertices for adjustment are chosen by the same vertex ring from outside to inside on the filled area, as shown in Fig. 7(b). Use surface fitted parameters matrix A in the front, according to x, y coordinate value to calculate exactly z coordinate value, and then update the selected filled vertices. When a ring vertices is adjusted by least square method, this ring will be changed to be the new boundary vertices for the adjustment to the next ring until all of the hole vertices adjustment are finished.

Examples

In this section, we present a testing example to demonstrate the effectiveness and practicality of our method. A car body original triangular model (Fig. 9) is used to form several holes (Fig. 10) artificially for filling verification. The proposed hole filling algorithm in space for triangular model in this paper and the triangular mesh repair method proposed in literature [5] methods (feature plane hole filling method) were compared according to the original mesh surface data. Fig. 11 shows several triangle holes filling results used two different hole filling methods separately. We can see that the method proposed in this paper is more uniform. Table 1 - Table 3 are analysis of two different methods of triangular hole filling, including the number of filled vertices, the number of triangles ,distance between filled vertices to original surface (average error), average length and the average area of the triangles. Comparison of the data obtained from the table shows that the proposed method is more effective and practical from several aspects. The hole filling accuracy has a very good improvement, which has engineering practicability. (Note: In Table 1 to Table 3, vertices number is the number of new added filling vertices, triangles number is the number of new filled triangles, average error is the average value of distance between new added vertices to the original triangular model surface, and the average length and the average area is correspond to new added triangles.)

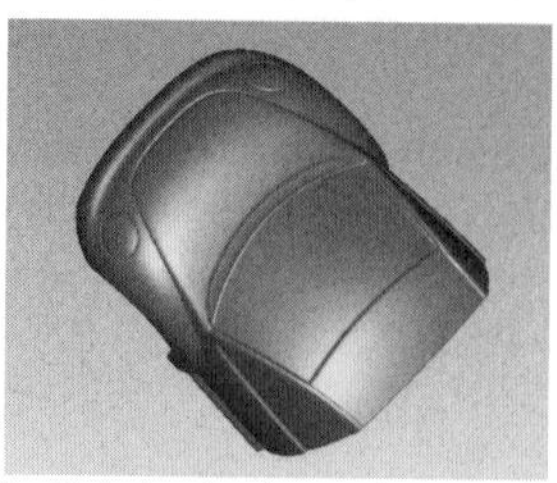

Fig. 9: Original triangular model

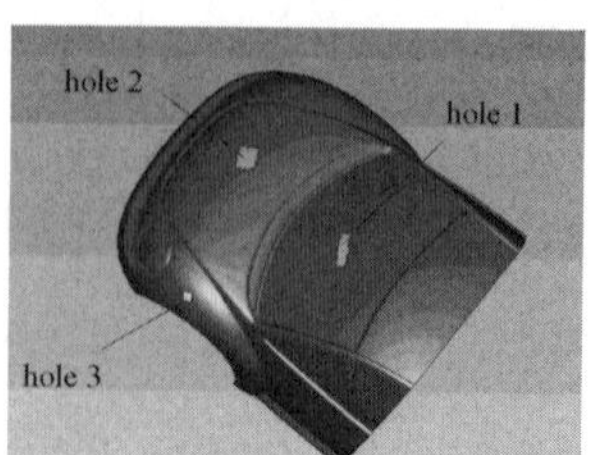

Fig. 10: Triangular model with several holes

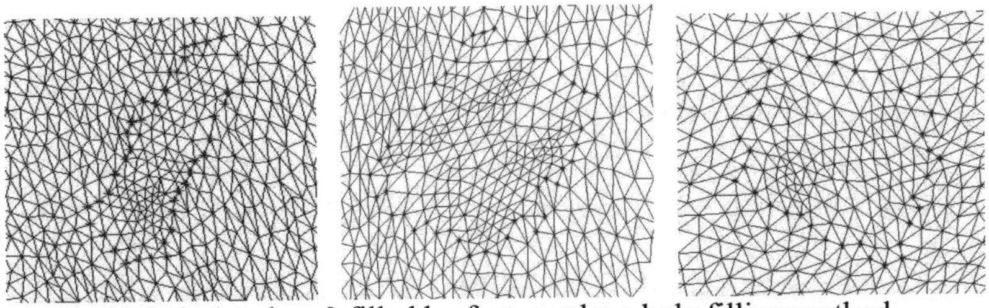

(a) hole 1 to hoe 3 filled by feature plane hole filling method

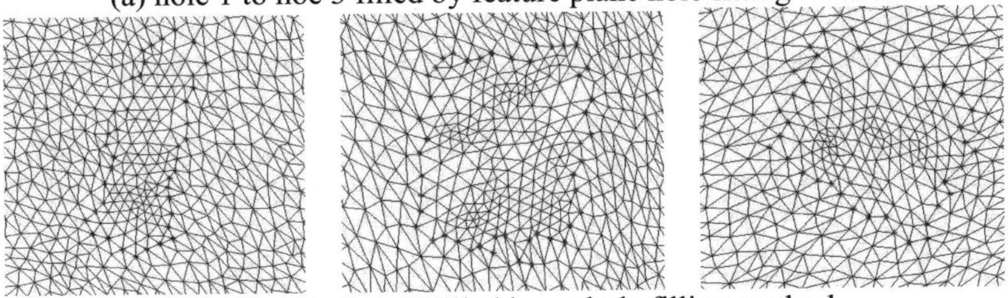

(b) hole 1 to hole 3 filled by us hole filling method

Fig. 11: Result comparison of two different hole filling method

Table 1: Comparative data of hole 1 triangles filled by two different methods

Filling method	vertices number	triangles number	average error	average length	average area
Original model	118	189	0	4.226	9.650
Feature plane hole filling method	153	340	0.729	4.560	11.054
Our hole filling method	124	205	0.619	4.331	9.820

Table 2: Comparative data of hole 2 triangles filled by two different methods

Filling method	vertices number	triangles number	average error	average length	average area
Original model	190	315	0	3.980	7.475
Feature plane hole filling method	237	578	0.630	3.520	6.524
Our hole filling method	206	364	0.512	3.782	7.014

Table 3: Comparative data of hole 3 triangles filled by two different methods

Filling method	vertices number	triangles number	average error	average length	average area
Original model	154	249	0	1.840	1.692
Feature plane hole filling method	192	399	0.272	1.029	1.038
Our hole filling method	155	271	0.258	1.805	1.611

Conclusion

In this paper, a new hole filling algorithm in space for triangular model is proposed, which fully considered the effect of filled triangles uniform and the hole boundary changeless were taken into account. While several relevant rings vertices were selected as sampling points for the adjustment of new filled vertex position by least square method, we finally found an effective way to achieve a more accurate hole filling result. Examples were given to prove the effectiveness and the processing flexibility and stability of our hole algorithm in mesh model.

Acknowledgements

This work was supported by National Natural Science Foundation of China (No.50605007); Grand Science and Technology Project Foundation of Fujian Province (No.2007H2011).

References

[1] Zhang L.Y, Zhou R.R & Zhou L.S: *Research on the Algorithm of Hole Repairing in Mesh Surfaces*. Journal of Applied Science, 20, 3(2002), p. 221.

[2] Wang H.T, Zhang L.Y, LI Z.W, Liu S.L & Zhou R.R.: *Repairing Holes in Triangular Meshes Based on Radial Basis Function Neural Network.* China Mechanical Engineering, 16, 12(2005), p. 2072.

[3] Jean-P.P, George M, Philippe V.: *Filling Holes in Meshes Using a Mechanical Model to Simulate the Curvature Variation Minimization* .Computers Graphics Vol. 30 (2006), p. 892.

[4] Yongtae J.: *A Piecewise Hole Filling Algorithm in Reverse Engineering.* Computer-aided Design Vol. 37 (2005), p. 263.

[5] Liu D.P, YU S.J, Chen J.J & Wang Y.Y.: *Hole Repairing in Triangular Meshes based on Least-squares Support Vector Machines.* Computer Integrated Manufacturing Systems, 15, 9(2009), p. 1867.

[6] Jin T, Tong S.G.: *Reverse Engineering.* China Machine Press, Beijing, (2003).

[7] Zheng Q.H & Lu D.T.: *Curve and Surface Fitting Based on Moving Least-squares Methods.* Journal of engineering Graphics, 25, 1(2003), p. 84.

[8] Wei Z. L, Zhong Y. X, Yuan C.L & Li R.J.: *Research on Smooth Filling Algorithm of Large Holes in Triangular Mesh Model.* China Mechanical Engineering, 19, 8(2008), p. 949.

Advanced Materials Research Vol. 279 (2011) pp 207-213
© (2011) Trans Tech Publications, Switzerland
doi:10.4028/www.scientific.net/AMR.279.207

Finite Element Analysis of Pipeline Steel Air-bending

Qiang Li[1, 2, a], Lifeng Fan[1, b], Ying Gao[2, c], Conghui Song[1, d], Wentao Yang[2, e]

[1]College of Mechanical Engineering, Yanshan University, Qinhuangdao, Hebei, China

[2]College of Material Science and Engineering, Hebei University of Science and Technology, Shijiazhuang, Hebei, China

[a]liqiang@hebust.edu.cn, [b]ysufanlifeng@foxmail.com, [c]gaoyingch@sina.com, [d]songconghui041@163.com, [e]yangwentaozhang@126.com

Keywords: orthogonal design, wide-punch, finite element method, JCO forming, pipeline steel

Abstract. Air bending of wide-punch large radius of curvature, as a special bending mode, has been widely applied in the production of large diameter pipeline JCO forming. The JCO forming process is accomplished by several air bending processes. The quality and shape of pipe are greatly influenced by the air bending. During the present work, the bending is investigated into using the finite element (FE) method and orthogonal design of experiment. Experiments are carried out and the experimental result is very close to the result simulated by finite element analysis. Springback, forming force and residual stress are discussed. The aim of this research is to investigate the influencing degree of parameters, such as sheet material, sheet thickness, bending angle, lubrication and punch radius; in particular the effect of springback, forming force, residual stress on air bending parameters are also discussed.

Introduction

Wide sheet air bending is applied in large diameter welding pipe JCO forming. Three main phases are individuated in large diameter welding pipe production cycle: (1)the JCO forming phase, (2) the welding phase, (3) the calibration phase; in addition, the first phase is accomplished by two steps: the crimping and the JCO forming. After a preliminary chamfering operation, the plate edges are firstly crimped into circular arcs using two involutes dies. In the JCO forming step the whole sheet profile is changed from a flat one into an O-shaped one by air bending. At first, half of the plate is bent into J-shape step by step while the punch moves down, sequently the other half is formed progressively into C-shape, and then finally the punch is actuated downwards for the final press, forcing the skelp almost into a circular shape. The seam is subsequently welded using submerged arc welding. In the final phase the pipe is circumferentially expanded to calibrate the pipe distortion and a highly circular shape using expanding machine that obtained [1].

In recent years, the wide sheet air bending about JCO forming was applied to analyze plastic mechanics [2,3]. Results of theoretical calculation are not perfect for several factors haven't been taken into consideration. According to various material types and new production items, process parameters are usually varied using a "trial and error" technique to find the optimal process to set up. But this approach is not cost effective while the final product is quite expensive and maintenance costs (machines and tools) are also high. Thus the adoption of numerical techniques is preferred. It is rare that the effect of air bending about JCO forming process parameters is evaluated by using a two-dimensional (2D) finite element model. The authors adopt a FE approach for investigating the air bending; in particular the sensitivity influenced by air bending parameters has been evaluated [4].

Air Bending Mode of Wide-punch

According to the characteristic of JCO forming process for the large diameter longitudinal-seam submerged arc welded pipes, the punch width is larger than the die opening distance and the punch displacement is less than the extreme displacement. This bending mode is named after free bending of wide-punch. In forming process, with the punch displacement increasing, the contact point O moves gradually from centre to two sides. However, the contact point A always can not arrives on point A' in two endpoints of circular arc, as shown in Fig. 1. Therefore, it is unnecessary to consider the width of the punch in the forming process.

During the bending process, the sheet wraps around the punch gradually, the section covers punch contact completely, and the uncovered is the straight line segment. The line and arc segment is naturally tangent to the die round. According to the geometric relationship, the approximate formula between the punch displacement h and bending angle α can be obtained as follows:

$$h=\frac{L}{2}\tan\frac{\alpha}{2}-(R_d+R_p+t)\left[\frac{1}{\cos\frac{\alpha}{2}}-1\right] \qquad (1)$$

Where L, R_p, R_d, t are dies span, punch radius, die radius, sheet thickness respectively. Contrasting the experimental value with the calculated bend angle with Eq. 1, the relative error is less than 0.8%. The result is shown in Fig. 2.

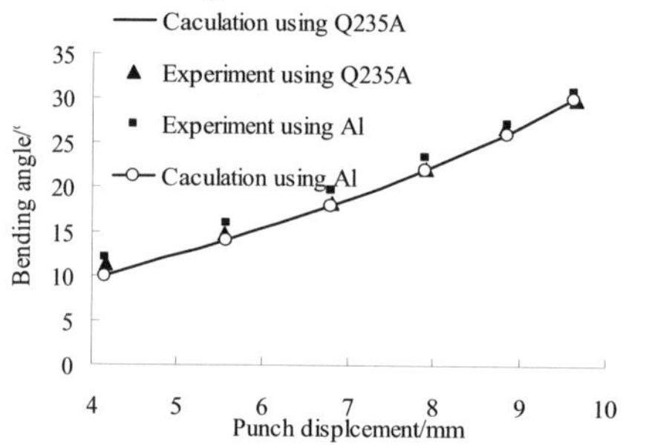

Fig. 1: Air bending schematic diagram

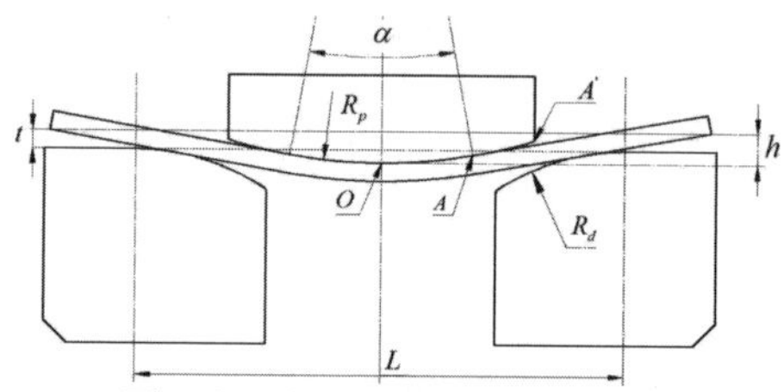

Fig. 2: Theoretical and experimental result.

Experiment Design

To obtain the sensitivity influenced by air bending parameters, orthogonal experiment method is adopted to design the air bending simulation. According to the characteristic of air bending of wide-punch large radius of curvature in production, the bending invariants are removed; in particular the interaction of thickness, punch radius and the bending angle are considered, simultaneously the material properties, thickness of sheet, punch radius, lubrication pattern, and bending angle are chosen as the factors of study, and springback angle, maximum forming force and residual stress peak value are selected as object of study. The finite element simulation is designed by the orthogonal design L_{18} table 6×36.

Finite Element Analysis

Tensile Tests. Q235A or 16Mn was adopted as main pipeline material from 1950s to 1970s in China. Before West-East gas pipeline project, X60 was adopted as the superlative degree pipeline steel. At present, X80 has been widely used. So Q235A, X60 and X80 are selected as experimental materials in uniaxial tensile. True stress-strain curve was showed in Fig. 3. The material properties parameters are showed as follows in Table 1.

Table 1: Parameters of sheet material properties

Sheet material	E [Gpa]	σ_s [Mpa]	n	K [Mpa]	v
X80	196	635	0.082	913	0.3
X60	183	475	0.1	725	0.3
Q235A	200	202	0.184	457	0.28

Here: E, σ_s, n, K, v are elastic modulus, yield strength, hardening exponent, intensity coefficients, poisson's ratio.

Finite Element Modeling. The air bending is assumed to take place under plane strain conditions and the 2D finite element model is established. The problem is solved within the nonlinear FE code ABAQUS standard solver. The sheet is discretized by linear, reducing integration plane strain continuum elements (CPE4R). There are 10 elements through the thickness. All forming moulds were modeled as analytical rigid surfaces. Whereas the sheet is deformable, the sheet surface is defined as two surfaces: the top surface and the bottom surface. ABAQUS defines the contact between two surfaces using a strict "master–slave" algorithm. Rigid surfaces constitute the master surface in a contact pair. The contact pairs are allowed "finite sliding". In this contact scheme, the nodes of the slave surface are not allowed to move normally to the surface, but they can still slide on the surface for the friction model of penalty function is defined [5].

Punch displacement is calculated by Eq. 1 for punch radius of 80mm with different bending angle. Based on the geometrical similarity theory, the bend angle is measured by portable tri-ordinate measuring instrument. The maximum error between the results of finite element analysis and the experimental datas is 0.55°, as shown in Fig. 4.

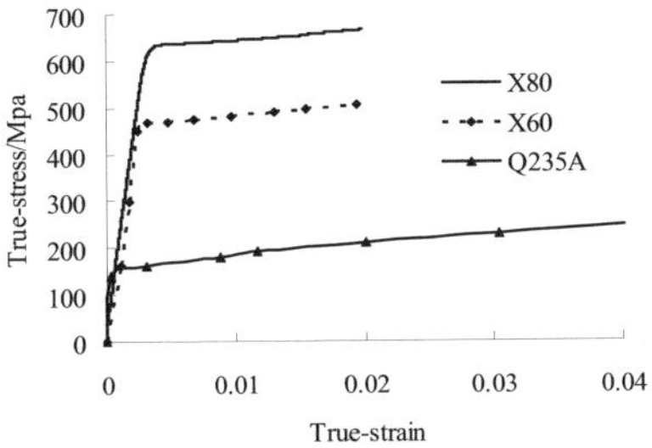

Fig. 3: True stress-true strain curve

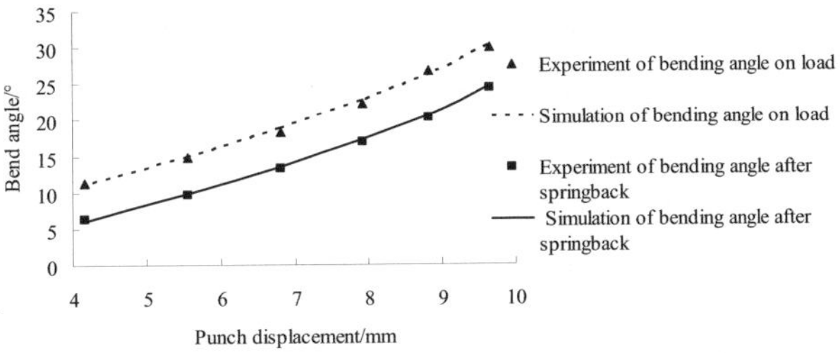

Fig. 4: Punch displacement and bending angle

Results Analysis

According to the orthogonal table, the results of FE calculation are shown in Table 2.

Table 2: Results of finite element analysis

Number	Springback angle [°]	Maximum forming force [kN]	Residual stress peak [MPa]
1	5.26	6899.04	567.5
2	4.76	14145.1	582.9
3	3.10	1168.93	239.1
4	2.19	3292.7	249
5	4.29	16649.2	577.6
6	8.61	3675.89	688.1
7	6.68	12299.8	672.4
8	1.65	7052.74	279
9	9.40	3481.83	543.9
10	6.12	11326.7	722.8
11	1.75	8783.58	262.6
12	9.30	4893.77	530.6
13	5.64	10484.7	592.4
14	4.95	29797.3	728.9
15	3.87	2928.29	218.1
16	2.68	2928.29	230.8
17	4.33	24051	621.4
18	10.13	7145.55	667.6

Springback Angle. The range of springback angle is from 1.65° to 10.13°. The sensitivity of each factor to springback angle is obtained by range analysis and variance analysis: sheet material type> sheet thickness> bend angle> punch radius> friction. Therefore the material properties and the sheet thickness are the mainly factors that influencing the springback.

Relation between springback and each factor is shown in Fig. 5, where α is bending angle, t is sheet thickness, R_p is punch type, and f is friction coefficient.

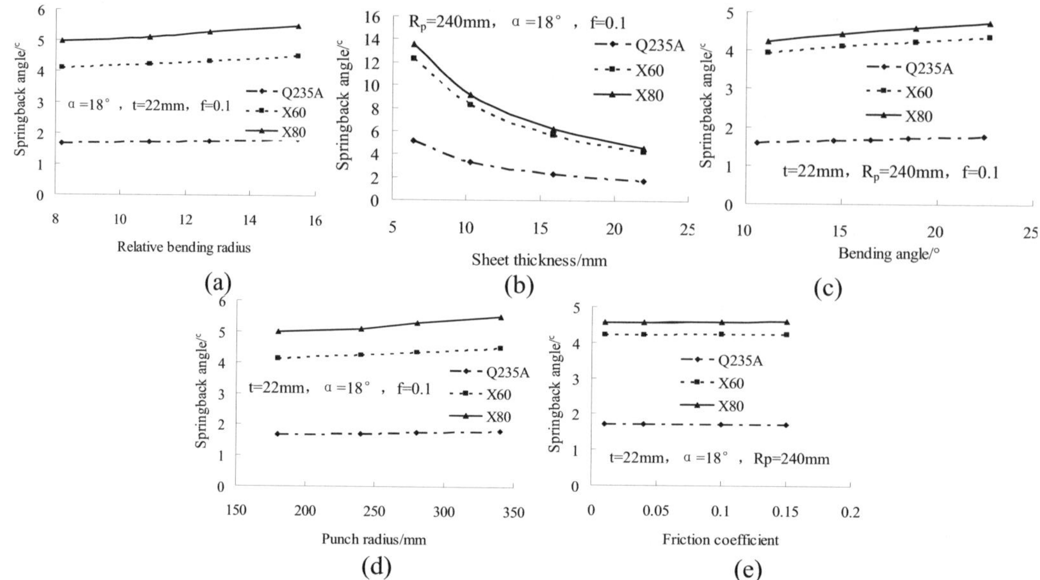

Fig. 5: (a) Relative bending radius and springback angle, (b) Sheet thickness and springback angle, (c) Bending angle on load and springback angle, (d) Punch radius and springback angle, (e) Friction coefficient and springback angle.

The springback angle of small curvature bending ranges from 10° to 24°, and the influence of friction coefficient on springback angle is negligible. The springback angle increases approximately linearly with the increasing of bend angle, punch radius and friction coefficient. The relation between sheet thickness and springback angle fits power exponent function. With the sheet material properties, such as the yield stress and the elasticity modulus varying, the springback angle changes significantly. Moreover, Fig. 5(a) shows that the ascendant trend for X80 is most obvious. The range of springback angle is only 0.5° while punch radius varies from 150mm to 340mm. The influencing degree of high-strength steel, such as X60 and X80, are stronger than Q235A. In the process of unloading, the springback angle decreases gradually with the sheet thickness increasing for the restricting of material fiber beteween each other.

Forming Force. The maximum forming force range is from 116.89kN to 29797.3kN. The sensibility sequence influencing maximum forming force is: sheet thickness> sheet material type> bending angle> punch radius> friction coefficient.

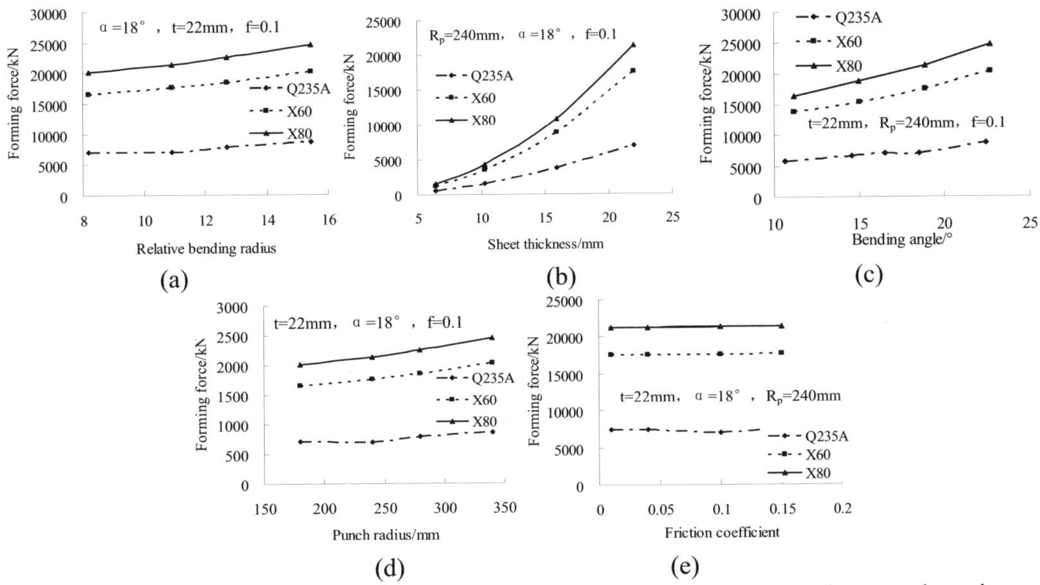

Fig. 6: (a) Relative bending radius and maximum forming force, (b) Sheet thickness and maximum forming force, (c) Bending angle and maximum forming force, (d) Punch radius and maximum forming force, (e) Friction coefficient and maximum force.

It can be concluded that the thickness and material properties are the main factors that influencing maximum forming force. Fig. 6 shows that the influence of every factor on forming force is similar to the springback angle. The friction force in the sheet tangent direction can be ignored in engineering calculation. With the further deformation, the forming force increases linearly with relative bending radius, punch radius and bend angle. The relation between sheet thickness and the maximum forming force fits power exponent function. With variation of the material properties, the deformation resistance per unit area is different and the maximum forming force changes obviously under the comprehensive effect of the material properties such as the yield stress and elasticity modulus. The influence of punch radius on the maximum forming force is small. Fig. 6(d) shows that the range of X80 maximum forming force is 4344.3 KN as the punch radius varies from 340mm to 150mm, which indicates the maximum forming force for X80 in small curvature bending mode is more than other material.

Residual Stress. After unloading, there are left some residual stress, which has great influence on the deformation of the slab and its performance. In this paper, the sensitivity of the factors on residual stress peak is investigated. The factors sequence influencing residual stress is: sheet materials type> sheet thickness> punch radius> bend angle> friction coefficient.

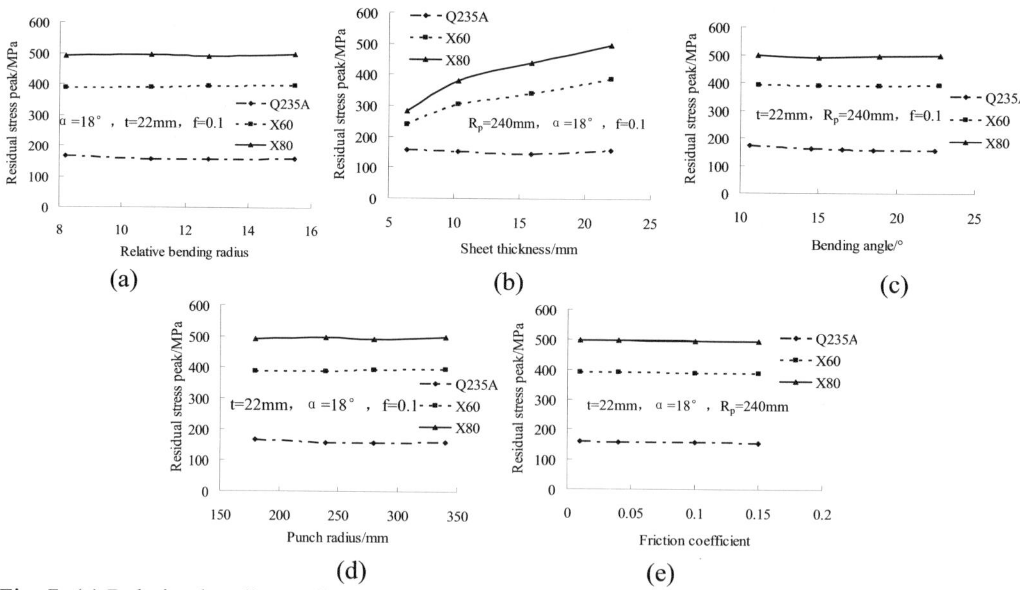

Fig. 7: (a) Relative bending radius and residual stress peak, (b) Sheet thickness and residual stress peak, (c) bending angle and residual stress peak, (d)Punch radius and residual stress peak, (e) Friction coeffcient and residual stress peak.

It also can be concluded that the material property is the main factor influencing the residual stress. Relation between residual stress and factors is shown in Fig. 7. Assuming pure bending of sheet, the inner fibers in sheet bears compressive stress, and simultaneously the outer ones bears tensile stress in bending. The moment appears in the sheet metal equilibrious to elastic stress. The residual stress is equal to the vector sum of elastic stress and the tangential stress. Residual stress σ_r in elastic part:

$$\sigma_r=\sigma_\theta-\sigma_e=E(y/\rho)-E(y/\rho-y/\rho_0)=E(y/\rho_0) \tag{2}$$

Residual stress σ_r in plastic part:

$$\sigma_r=\sigma_\theta-\sigma_e=K(y/\rho)^n-E(y/\rho-y/\rho_0) \tag{3}$$

Where σ_θ, ρ, ρ_0 is tangential stress before springback, curvature radius before springback and curvature radius after springback. From Eq. 2 and Eq. 3, the residual stress mainly depends on the material property such as yield stress and elastic modulus in same bending degree. The tangential stress in high strength steel is far away more than elastic stress. The difference between tangential stress value and elastic stress for Q235A is slight. With the sheet thickness increasing, the residual stress peak is greatly changed. And other factors have little influence on tangential stress, so the residual stress peak changes unobviously.

Conclusion

The pipeline steel air bending have been simulated under the plane strain conditions and the finite element model was verified by experiments. The simulation results were quite close to the experiments. The sensitivity sequence of factors on the springback angle, maximum forming force and residual stress peak are obtained by designing orthogonal experiment while the bending angle ranges from 10° to 24°. The quality of pipeline steel air bending influenced by lubrication slightly no matter what sheet material type was chosen. In addition, sheet material type is the main factor that influences bending quality. The relation between investigative objects and factors except sheet thickness is linear approximately; simultaneously relation of sheet thickness and each investigative object fits power exponent function. It is the firm conclusion of this study that springback, forming force and residual stress are enhanced because of the application of high grade pipeline steel such as X80. Furthermore, the analysis results can also be applied in the similar forming study with the characteristics of small curvature air bending mode.

References

[1] Ying Gao, Qiang Li, Xianchang Meng: *Numerical simulation of JCO/JCOE pipe forming*. Foundry Technology, 30, 1(2009), pp. 77-79.

[2] Jian Li: *Research on intelligent control technology for forming the large diameter longitudinal-seam submerged arc welded pipes with JCO process*. Doctoral dissertation, Yanshan University, (2009).

[3] Zhegang Ye: *Study on progressive forming theory and application*. Master's dissertation, Wuhan University of Technology, (2002).

[4] M.D. Herynk, S. Kyriakides, A. Onoufriou and H.D. Yun: *Effects of the UOE/UOC pipe manufacturing processes on pipe collapse pressure*. International Journal of Mechanical Sciences, 49, 5(2007), pp. 533-553.

[5] G. Palumbo and L. Tricarico: *Effect of forming and calibration operations on the final shape of large diameter welded tubes*. Journal of Materials Processing Technology, 164-165, 5(2005), pp. 1089-1098.

Advanced Materials Research Vol. 279 (2011) pp 214-218
© (2011) Trans Tech Publications, Switzerland
doi:10.4028/www.scientific.net/AMR.279.214

Finite Element Analysis and Optimization of Electric Field Structure of New-type Electrospinning Machine

Hongwei Duan[1, a], Jingang Jiang[2,b]

[1]School of Material Science and Engineering, Harbin University of Science and Technology, Harbin 150040, China

[2]Intelligent Machine Institute, Harbin University of Science and Technology, Harbin, 150080, China

[a]duanhw@hrbust.edu.cn, [b]jiangjingang1982@163.com

Keywords: Electrospinning, Optimization, Electric field analysis, ANSYS

Abstract. Electrospinning technique is currently one of the most important methods for preparing nanofiber. Aiming to the fact that the electric field at the receiver is relatively weak, the electric field's distribution in the typical electrospinning device was calculated and analyzed by ANSYS. It put forward the method of optimizing electric field's distribution by using an assistant electric field device. At the same boundary conditions, the electric field's distribution in the original device and improved binode device are compared. The simulation analysis provides an effective reference for the optimization design of electric field's distribution in the electro-spinning device. The experimentation results show that the distribution of fiber diameter of polyimide non-woven fabric is equal and has a certain directivity after adding assistant electric field.

Introduction

The polymer solutions or melts with charge flow and deform at strong static electric field, and solvent is evaporated or the melt is cooled and solidified, then fibroid substance is obtained. In high electric field, fluid is ejected from Taylor core and formed jet flow. Jet flow with electricity speed up and its diameter becomes tiny in the electric field [1-3]. This is the reason that electrospinning can obtain the nanofibre. Fig. 1 is schematic diagram of electrospinning device [4]. In essence, electrostatic force is unique driving force in electrospinning. So, whether the distribution of electric field is reasonable will direct affect the fineness and morphology of electrospinning fibre. The fluidic instability phenomenon exist in undesired electric field distribution, especially in sudden drop of field intensity, and this will lead to uncertainty of jet flow in the receiver. External disturbance leads to unordered diffusion in the process of electrospinning (especially in lower concentration and viscosity of solution) [5]. This not only pollute experimental environment, but also waste experimental material and reduce the efficiency of electrospinning. The fluidic instability phenomenon can be weakened by controlling the shape and the strength of macroscopic electric field based on the analysis of Reneker's electrostatic spinning process [6]. Consequently, it is important and essential to analyze the distribution of electric field in electrospinning machine.

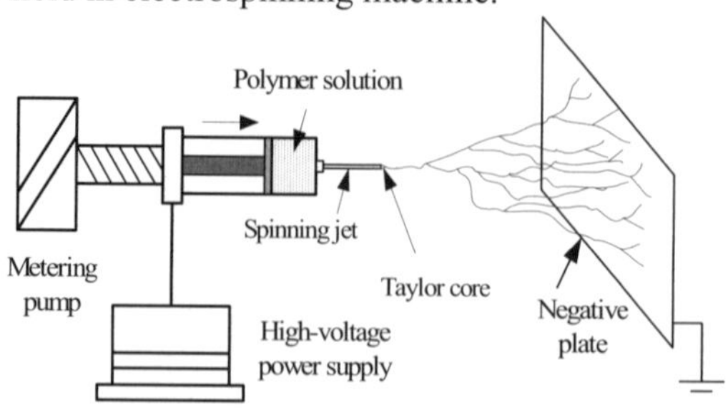

Fig. 1: Electrospinning device

Set up and simplification of electric field

In order to get better collecting result of spinning, the electric-field distribution between the spinning jet and the negative plate is the main influencing factor. Therefore, the working model of the whole electric field can be divided into two parts: one is the metal spinning jet, and the other is negative plate. The spinning jet added high voltage static electricity is the factor that produces electric field, and the generated electrostatic field has a spatial symmetry, so we can deal with it as 2D field. We analyze the model by simplify it into two dimension electric field model.

Because the aperture of metal spinning jet is very tiny (ususlly 0.4mm), in order to simplify the structure of spinning jet, we separately set up the finite-element model of the model Fig. 2(a) and model Fig. 2(b), and analyze and compare the difference of simulation results. The difference between model Fig. 2(a) and model Fig. 2(b) is that model Fig. 2(a) has pore and model Fig. 2(b) is replaced in entity. The analysis result shows that there are identical voltages distribution maps of the two types of structures, as shown in Fig. 2. It indicates that the spinning jet can be completely replaced by model Fig. 2(b). And the result from structural simplify can be controlled within allowable error.

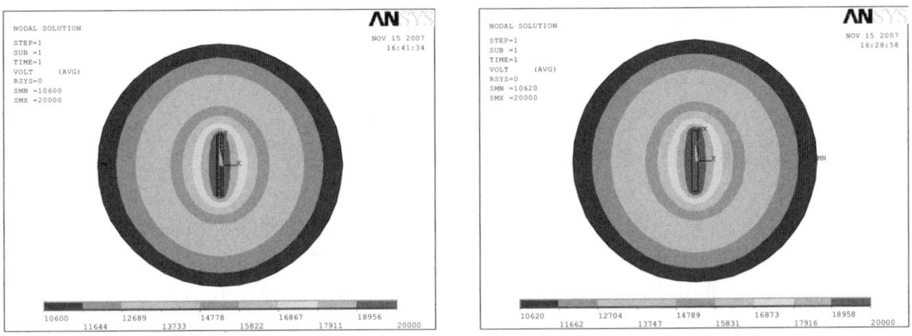

(a) Voltages distribution before simplification (b) Voltages distribution after simplification

Fig. 2: Voltages distribution maps of spinning jet

Finite-element analysis and optimization of electrospinning electric field structure based-on ANSYS

Method selection and assumption. In this paper, we adopt traditional H-method to analyze finite-element model of electric field structure based-on the following assumptions: (1) the permittivity of dielectric is constant and independent of electric field. (2) the electric charge volume density $\rho=0$.

Set up of model and material property. The model should consider the following three characteristic parameters: (1) the size of metal spinning tube, (2) the width of negative plate, (3) the distant between spinning jet and negative plate. The size of model is shown in Fig. 3. In the model, there are two unit types. One is the analysis unit of tetrahedron electric field- Plane121, which is matched with two-dimensional structure of electrostatic field. Air, spinning jet and negative plate adopt this unit type. The other is distant field unit INFIN110, which is specially set to describe the distant field dissipation problem of electrostatic field. The air distant field adopts this unit. The dielectric is air in the model. The relative dielectric constant of air is 1 and the permittivity of spinning jet and negative plate is 2.

Setting boundary conditions and load. This analysis does not involve in charge density, so we just add a voltage of 20KV at the spinning jet and a voltage of 0V at negative plate. Due to the use of distant field unit, we need to add distant field sign to the external boundary of distant field region. We solve this problem using static analysis.

Result analysis. Fig. 4 shows electric-field vector distribution between spinning jet and negative plate. From this figure, we can see the following characteristics of the distribution of electric field intensity: electric field intensity vector distributes symmetrically, the maximum value of electric field intensity appears at jet of spinning jet, and the direction of electric field intensity is perpendicular to negative plate. The maximum value of field intensity is $E = 1.0\times10^6\ V/m$, which is identical with the truth. The value of electric field intensity mainly keeps the same in this region, which has a distance of 0 to 2 centimeters far from spinning jet. It provides stable electric field force for the production of spinning. However, the closer to negative plate, the faster of the decline of the electric field intensity. We also can see the tendency that the vector lines diffuse outward gradually. These factors make it easy for spinning to diffuse all around at the collecting end, which results in a bad collecting result, as shown in Fig. 5.

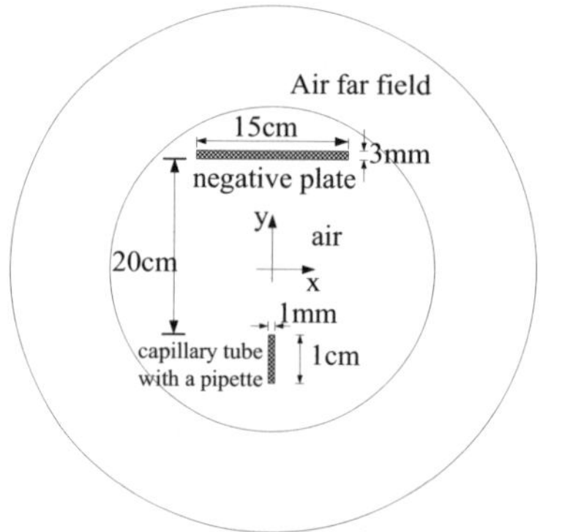

Fig. 3: Finite element model of electrospinning

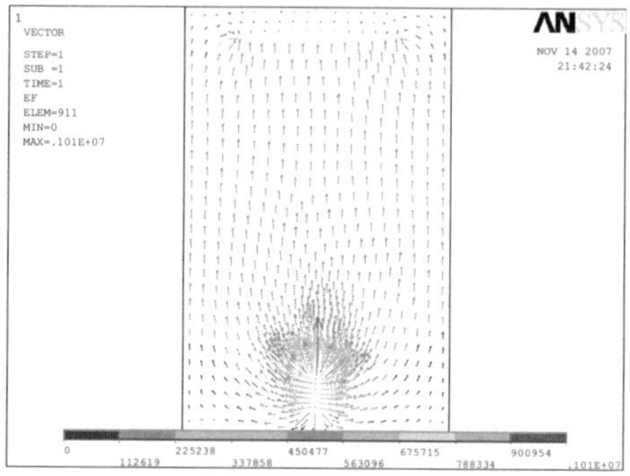

Fig. 4: Vectors distribution of electric-field strength

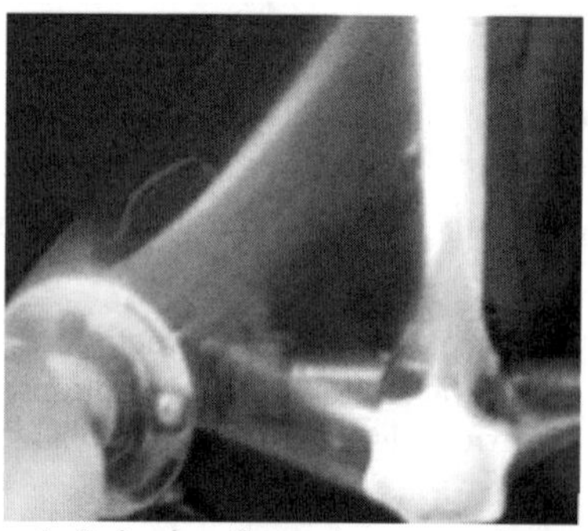

Fig. 5: Spinning floated on experimental device

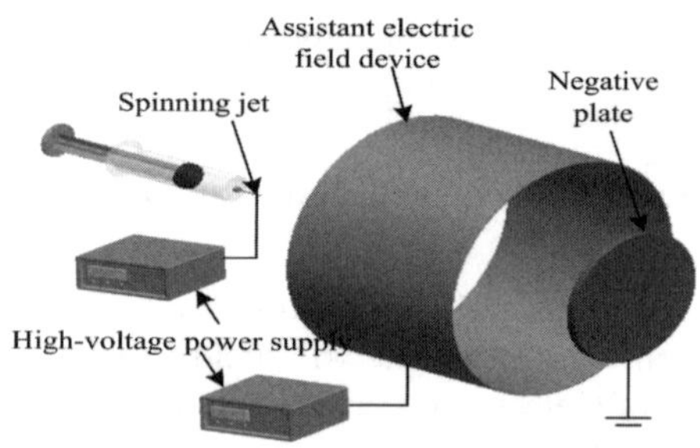

Fig. 6: Electrospinning device with assitant electric field

Optimization of the electric field. In order to collect much more spinning at the negative plate, we need to optimize the distribution of electric field intensify between spinning jet and negative plate. Based on the analysis of the results, we know that the value of electric field intensify becomes relatively small near the front of negative plate. The value falls to $1.13\times10^5\ V/m \sim 2.25\times10^5\ V/m$ and vector lines gradually diffuse. In order to get relatively strong and inwards convergent electric-field distribution at the region next to the negative plate end, we can add assistant electric field device. On the basis of the distribution theory of electrostatic field power line, electric field can optimized through adding circinate electric field. We design double-anode device, as shown in Fig. 6. That is to add a metal cylinder with a certain voltage to intrinsical device. Finite-element model is analyzed by setting the same boundary conditions and loading (adding a voltage of 5KV to assistant electric field device). Fig. 7 shows the distribution of electric-field vector after adding assitant electric field. From Fig. 7, we see that the improved electric field is obviously convergent compared with the previous and electric intensity increase notablely in the region next to negative plate. Improved electric-field distribution will make it spinning with electric charge to get relatively strong electric field force at the collecting end and the forced direction presents convergent tendency toward the centre area of negative plate. It is effective to solve the diffusing problem of spinning at the collecting end. But the noteworthy is that the maximum value of field intensity somewhat decrease at the jet of spinning jet due to assitant device with a certain voltage. We go on with a couple of simulations and find that: while the voltage of assitant device improves gradually and keeps the voltage of spinning jet constant, the maximum value of field intensity decreases quickly at the jet. So, the voltage of assitant device should change in a certain range (or properly increase voltage at the spinning jet), or else this may affect the regular production of spinning. In addition, strong electric intensity appears at the tip of assistant cylinder device and negative plate. In order to reduce the possibility of marginal discharge caused by voltage rise, we can substitute the edges of assistant cylinder assembly and negative plate for fillet.

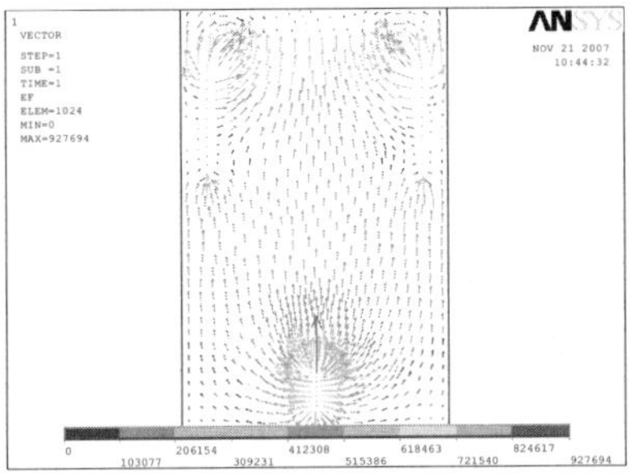

Fig. 7: Distribution of electric-field vector after adding assitant electric field

Electrospinning experiment

Experimental system of the electrospinning machine is shown in Fig. 8, which is consistent of following components: high-voltage power supply, injection unit, metal receiver, two-dimension workbench and shield mesh. High-voltage power supply, which is made of electrical department of Tsinghua University, can realize the continuous adjustment of the voltage in the range from 0V to 100KV. Polyimide is selected as solution electrospinnability.

Fig. 9(a) and Fig. 9(b) are the SEM image of polyimide non-woven fabric obtained in the structure of needle-plate and assistant electric field respectively. We can see obviously that dispersity of fiber diameter in Fig. 9(a) is larger than in Fig. 9(b), and fiber orientation in Fig. 9(a) is not better than in Fig. 9(b). From Fig. 9, after adding assistant electric field, the distribution of fiber diameter of polyimide non-woven fabric is equal and has a certain directivity.

Fig. 8: Experimental system of the electrospinning machine

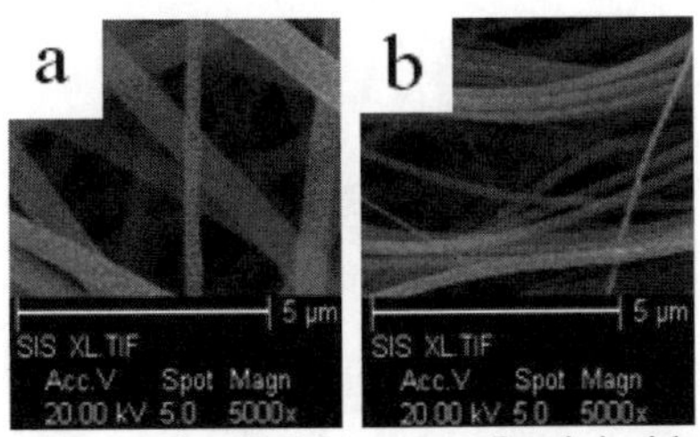

Fig. 9: The SEM image of polyimide non-woven fabric

Conclusion

Based on ANSYS, the distribution of electric field of new-type electrospinning machine is calculated, and the feature of the electric field distribution is analyzed. The simulation analysis provides an effective reference for the optimization design of electric field's distribution in the electro-spinning device. Aiming to the fact that the electric field at the receiver is relatively weak, the method of optimizing electric field's distribution by using an assistant electric field device is put forward. Double-anode device is designed to strengthen the electric intensity of the receiver. This solves the problem of spinning diffusion and improves the morphology of spinning fibres. The fineness of fiber is 270nm~400nm. The experimentation results show that the distribution of fiber diameter of polyimide non-woven fabric is equal and has a certain directivity after adding assistant electric field.

Acknowledgements

This research was supported by the Heilongjiang Postdoctoral Foundation (Grant No. LBH-Z09098).

References

[1] Dacheng Wu, ZZhongliang Du, Xushan Gao: *Nanofibre*. Chemical industry press, Beijing, (2004).

[2] Na Liu, Jianzhong Yang: *Progress in research and application of electrospun nannfibers*. China synthetic fiber industry, 29, 3(2006), pp. 46-49.

[3] S. A. Theron, E. Zussman, A. L. Yarin: *Experimental investigation of the governing parameters in the electrospinning of polymer solutions*. Polymer, 45, 6(2004), pp. 2017-2030.

[4] Jihuan He, Yong Liu, Lan Xu, et al: *Micro sphere with nanoporosity by electrospinging* . Chaos Solitons and Fractals, 32, 3(2007), pp. 1096-1100.

[5] H. Reneke, L. Y. Alexander, Fong Hao, et al: *Bending instability of electrically charged liquid jets of polymer solutions in eletrospinning*. Journal of Applied physics, 87, 9(2000), pp. 4531-4547.

[6] Reneker, I. Chen: *Nanometer diameter fiber of polymer produced by electrospinning*. Nanotechnology, 18, 6(1996), pp. 216-223.

Advanced Materials Research Vol. 279 (2011) pp 219-223
© (2011) Trans Tech Publications, Switzerland
doi:10.4028/www.scientific.net/AMR.279.219

Numerical Analysis of Co-cured Composite Single Lap Joint with Spew Fillets

Guohua Zhao[1, a], Qinglian Shu[1,b] and Bosheng Huang[1,c]

[1] School of Civil Engineering and Architecture, University of Jinan, Jinan 250022, P.R. China

cea_zhaogh@ujn.edu.cn, [b]cea_shuql@ujn.edu.cn, [c]cea_huangbs@ujn.edu.cn

Keywords: co-cured composite single lap joint, spew fillet, peel stress, shear stress, submodeling

Abstract. Spew fillet is common in co-cured composite single lap joint. It plays an important role in the safety of the joint. But few attentions were paid to the subject in former investigations. In this paper, co-cured composite single lap joints with spew fillets under static tensile load were investigated numerically using the commercial finite element code ANSYS. The submodeling technique was employed to overcome the problem caused by extra aspect ratio of elements. The computation results indicated that the existence of spew fillets will not change the possible failure location, but will improve the anti-failure performance of single lap joint.

Introduction

Investigations on single lap joints have been launched for more than half a century [1]. But the composite single lap joint received the attention of researchers just within recent 30 years [2]. Some investigations about stress distribution and failure modes of composite single lap joints have been conducted [3-7] in recent years. These studies indicated that two ends of single lap joint are the zones where the stress concentration happens.

The spew fillet at joint end is common in bonded and co-cured single lap joint, which was not mentioned in the above literatures, but may effect the stress distribution of the joint. Up to now, there were some investigations concerning the spew fillet of adhesively bonded single lap composite joint: In literature [8], a numerical model of adhesive bonded joint with a spew fillet was employed to study the effect of it on stress distribution. In literature [9] the authors presented eight spew geometries and discussed the effect of them on stress distributions of bonded single lap joints.

To our knowledge, few works were conducted on spew fillet of co-cured single lap joint. Actually, literature [10] was the only reference we could find, in which the authors described the natural fillets at co-cured single lap joint ends would be formed by squeezed resin in curing process and when applying fatigue load, the failure would be initiated from the edge of the fillets. But the impact of the fillets on the performance of joint was not involved in this paper. Even though, the authors reminded us spew resin fillets at the end of the joint should be taken into account when building the finite element model of co-cured single lap joints.

In this paper, a finite element model of the co-cured single lap composite joint with spew fillet was built to investigate the effects of spew fillet on static performance of the joint. For co-cured joint, the joining method requires no adhesive. So there is no adhesive layer between two adherends, which is the main difference between our model and the model in [8]. Meanwhile, unlike the spew fillets used in literature [9], which seems to be designed by authors for their study, our model was built according to the spew fillets of a real specimen.

Numerical simulation procedures

Finite element model. The finite element model developed in this paper was based on a real specimen of co-cured composite single lap joint (Fig. 1). The photograph of a spew fillet of this specimen is shown in Fig. 2. The two adherends of the specimen are made from HS160 RME Textipreg® (SEAL, Italy) graphite/epoxy prepreg unidirectional tapes. Prepreg layers were first laid up with a $[0_2/90]_s$ cross-ply sequence for use as adherends and joined along the overlap length of 30mm without the insertion of any adhesive layer between the adherends. Then joined panels were bonded with a co-curing process in an autoclave. Finally, coupon specimens were cut from the cured panels with desired dimension.

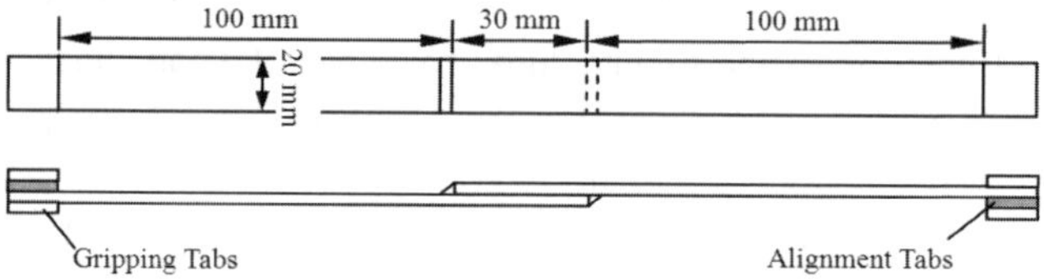

Fig. 1: Sketch of co-cured single lap joint specimen

Fig. 2: Photograph of a typical spew fillet

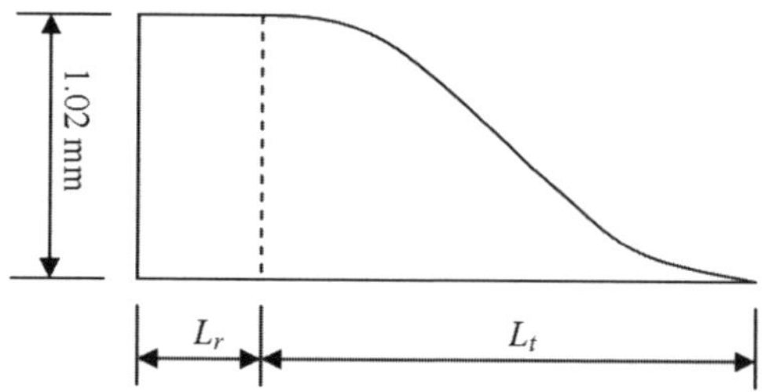

Fig. 3: Profile of spew fillet

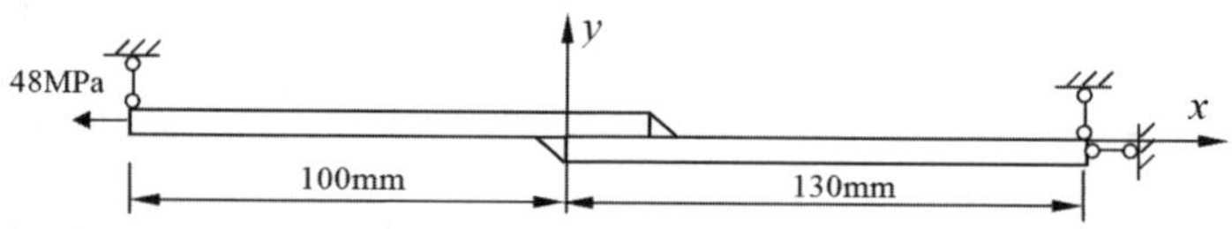

Fig. 4: The Coordinate system and boundary conditions of the finite element model

The properties and dimensions of our sample are as follows: the two adherends of joint are carbon/epoxy laminated plates, and the stacking sequence of plies of each adherend is $[0_2/90]_S$. Each ply has the thickness of 0.17mm, and the mechanical properties of each ply are shown in Table 1. The mechanical properties of spew fillet, which is made of resin of epoxy and taken as isotropic material, are also shown in table 1.

Table 1: Mechanical Properties of Graphite/Epoxy Lamina and Epoxy

	E_1 [GPa]	E_2 [GPa]	E_3 [GPa]	G_{12} [GPa]	G_{23} [GPa]	G_{31} [GPa]	v_{12}	v_{23}	v_{31}
Lamina	93.7	7.45	7.45	3.97	3.97	3.97	0.261	0.261	0.261
Epoxy	3.4						0.3		

In order to reflect the 'real' distribution of the stress and strain in spew fillet, as mentioned above, the actual dimension of spewed resin fillet was adopted in our model, which means the dimension of fillet in finite element model was determined by that of a real spew fillet of the above specimen. Based on the photograph (Fig. 2), the standard shape of spew fillet in our model is shown in Fig. 3, where the fillet is divided into two parts: one rectangle and one curve-side triangle. The heights of the two parts are all 1.02mm (equal to the thickness of one adherend) and the lengths of them are denoted by L_r and L_t, respectively. According to the real specimen size, L_r =0.3mm and L_t =1.5mm. The curve-side is formed by using spline curves to fit the curve edge of the real fillet.

Boundary conditions. The coordinate system adopting for finite element model is shown in Fig. 4. On the plane of x=130mm, the fixed boundary condition is applied. On the plane of x= -100mm, the y-direction displacement is set to zero, and the x-direction displacement is free. Referred to reference [10], a uniform tensile stress of 48MPa is applied on this plane.

Submodeling. The calculation was performed using the general finite element code ANSYS 12.0. Considering the thickness (about 1mm) of the adherend is far less its width (20mm), the model was meshed using quadrilateral plane strain element (element number: 42). Because each adherend is consist of six plies and every ply has different mechanical properties due to the orientation of fibers, there must be adequate elements along the thickness of the adherend. On the other hand, the thickness (about 1mm) of the adherend is also far less than its length (130mm). So if we want to get accurate results, more elements must be used along adherend length to keep the good aspect ratio of element (the ratio of element length to width), which means this ratio shouldn't be too big or small. But under the condition, too much computation time and memory were consumed, especially when using personal computer.

To overcome the above obstacle, a technique named submodeling in ANSYS 12.0 was used. Submodeling is a finite element technique used to get more accurate results in a region of the big or high aspect ratio models. Often in finite element analysis, the finite element meshes in these models are too coarse to produce satisfactory results in the region of interest. The results away from this region, however, may be adequate. To obtain more accurate results in such a region, an independent, more finely meshed model of only the region of interest can be generated and analyzed. Submodeling is also known as the cut-boundary displacement method or the specified boundary displacement method. The cut boundary is the boundary of the submodel which represents a cut through the coarse model. Displacements calculated on the cut boundary of the coarse model are specified as boundary conditions for the submodel.

For our model, the whole was coarsely meshed to get the first step result. Then the overlap portion and fillets were finely meshed to achieve the accurate solution.

Results and discussion

The stress and strain distribution of the spew fillet. For a single lap joint model without spew fillet, two end corners are the stress concentration zones and failure always initiates here. But when spew fillet is considered, we wonder whether the failure would change its location.

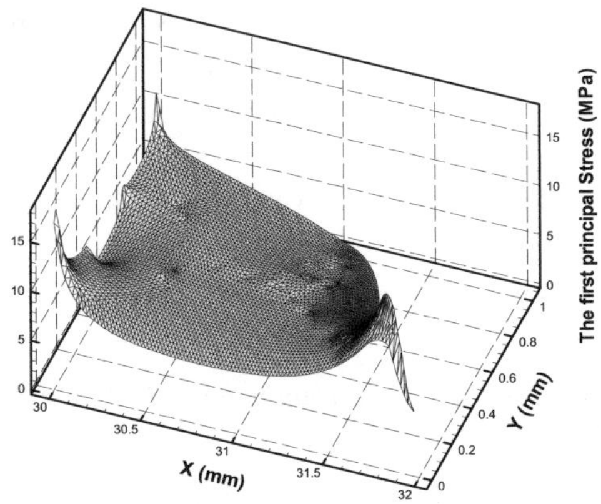

Fig. 5: The first principal stress distribution

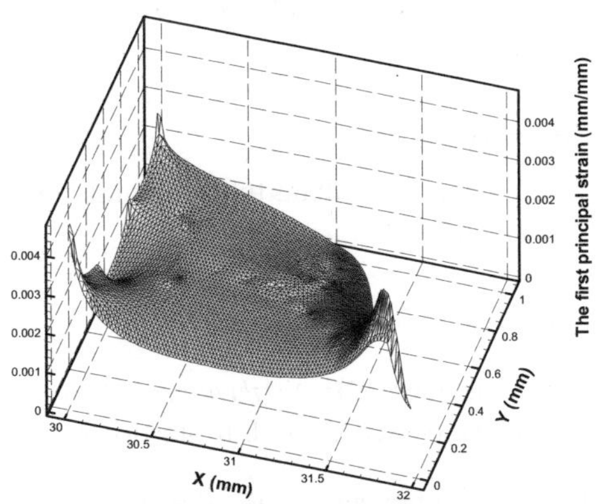

Fig. 6: The first principal strain distribution

To determine the possible failure location, a proper failure criterion must be employed: The fillets of our model are made of epoxy, which is a representative brittle material, so Maximum Strain Failure Criterion [7] and Maximum Stress Failure Criterion [11] can be used to identify the failure point. Thus the first principal strain and the first principal stress are employed here to indicate the possible failure location. Fig. 5 and Fig. 6 are the distributions of the first principal stress and the first principal strain for the above model. From the figures we can see the fact: no matter adopting the first principal stress or strain, the maximum values are always located at the corner of the joint end. This indicates the corner is a possible failure location, which is right next to the dangerous point of the joint interface. So even considering the spew fillet, the failure still initiates at the corner of joint end.

The influence of spew fillet on the stress concentration of the joint ends. To evaluate the influence of spew fillet on the stress concentration at the two joint end corners, a model without spew fillet was also employed as comparison.

From the failure micrograph of the joint in literature [12] we can see that separation between the adherends was mainly dominated by a peel or mode I fracture mechanism, and partly by a shear or mode II fracture mechanism. So the comparisons of the peel stress (the normal stress of the interface) and shear stress along joint interface were made.

The calculation result of model with spew fillet suggests: the distributions of peel stress and shear stress along joint interface are very alike as that of model without fillet, that is, no matter the model with fillet or without fillet, the peak values of the peel and shear stresses along joint interface are always at the two ends.

The result also indicates: the spew fillet can greatly lower the peak values of the peel and shear stresses at two joint ends, but it has little effect on the distribution at the middle part of interface.

To make it clear, the maximum peel stress and maximum shear stress of the two models are listed in Table 2.

Table 2: The maximum stresses at joint end corner (Units: MPa)

No Fillet		**With Fillet**	
Max. Peel Stress	Max. Shear Stress	Max. Peel Stress	Max. Shear Stress
46.51644	28.28474	6.17034	12.59018

Conclusions

In this paper, a numerical model of co-cured composite single lap joint with natural spew fillets was developed to investigate the influences of spew fillet on failure of joint. The following conclusions can be drawn from the calculation results:

1. The possible failure point of spew fillet is located at the corner of the joint end, so the existence of spew fillet will not change the initial failure location of co-cured single lap joint.
2. Spew fillet can greatly cut down the peak values of peel stress and shear stress of adherend interface.

Acknowledgement

The authors were thankful for the funding of Natural Foundation of University of Jinan (Granting Number: XBS1013).

References

[1] O. Volkersen: *Die Nietkraftverteilung in Zugbeanspruchten Nietverbindungen Mit Konstanten Laschenquerschnitten*. Luftfahrtforschung, Vol.15 (1938), pp.41-47.

[2] J. W. Renton and J. R. Vinson: *On the Behavior of Bonded Joints in Composite Material Structures*. Engineering Fracture Mechanics, Vol.7 (1975), pp.41-52.

[3] K. C. Kairouz and F. L. Matthews: *Strength and Failure Modes of Bonded Single Lap Joints between Cross-ply Adherends*. Composites, Vol. 24 (1993), pp.475-484.

[4] G. Wu and A.D. Crocombef: *Simplified Finite Element Modeling of Structural Adhesive Joints*, Computers and Structures. Vol.61 (1996), pp.385-391.

[5] K.S. Kim, J.S. Yoo, Y.M. Yi and C.G. Kim: *Failure mode and strength of uni-directional composite single lap bonded joints with different bonding methods*. Composite Structures, Vol.72 (2006), pp.477-485.

[6] M.S. Seonga, T.H. Kima, K.H. Nguyena, J.H. Kweon and J.H. Choi: *A parametric study on the failure of bonded single-lap joints of carbon composite and aluminum*. Composite Structures, Vol.86, 1-3(2008), pp.135-145.

[7] M.G. Songa, J.H. Kweona, J.H. Choia, J.H. Byunb, M.H. Songc, S.J. Shinc and T.J. Lee: *Effect of manufacturing methods on the shear strength of composite single-lap bonded joints*. Composite Structures, Vol.92, 9(2010), pp.2194-2202.

[8] M.Y. Tsai and J. Morton: *The Effect of a Spew Fillet on Adhesive Stress Distributions in Laminated Composite Single-lap Joints*. Composite Structure, Vol.32 (1995), pp.123-131.

[9] T.P. Lang and P.K. Mallick: *Effect of Spew Geometry on Stresses in Single Lap Adhesive Joints*. International Journal of Adhesion & Adhesive, Vol.18 (1998), pp.167-177.

[10] F. Aymerich, R. Onnis, and P. Priolo: *Analysis of the effect of stitching on the fatigue strength of single-lap composite joints*. Composites Science and Technology, Vol.66 (2006), pp.166-175.

[11] B.L. Kemp and E.R. Johnson: *Response and Failure Analysis of a Graphite-epoxy Laminate Containing Terminating Internal Plies*. 26th Structures, Structural Dynamics, and Materials Conference, Orlando, FL, United States, (1985), pp.13-24.

[12] L. Tong, L.K. Jain, K.H. Leong, D. Kelly and I. Herszberg: *Failure of Transversely Stitched RTM Lap Joints*. Composites Science and Technology, Vol.58 (1998), pp.221-227.

Advanced Materials Research Vol. 279 (2011) pp 224-229
doi:10.4028/www.scientific.net/AMR.279.224

Analysis of shear stress of two elasto-plastic random suraces

Lianfeng Lai[1, a], Chenghui Gao[2,b] and Jianmeng Huang[3,c]

[1]School of Mechanical Engineering and Automation, Fuzhou University, Fuzhou, Fujian, 350108, China

[a]llf2531@yahoo.com.cn, [b]gch@fzu.edu.cn, [c]huangjianmeng@fzu.edu.cn

Keyword: two fractal rough surfaces, sliding model, shear stress, delaminating theory of wear

Abstract. A FEA model of two-dimensional (2D) sliding between two interfering fractal rough solids was build, and the results was presented. The trends in shear stress between the rough solids were provided by the FEA. Combined with the delaminating theory of wear, the Maximum shear stresses and its distance are presented when loading and sliding. The result showed that the Maximum shear stresses was located in the subsurface which range about 2 to 8 um from the surface. In addition, with different plastic characteristics, the location of Maximum shear stresses were different. With modeling the contact and sliding between rough surfaces, the friction, sub-surface crack initiation and propagation and wear mechanisms can be understood deeply.

Introduction

Contact mechanics, as essential part of solid mechanics, deals with the stresses and deformation within two bodies loaded together. It is a powerful tool for investigation of basic problems in tribology, which is the science of the mechanisms of friction, lubrication, and wear of interacting surfaces in relative motion[1]. Finite element modeling is the most commonly used numerical method for solving contact problems with friction between deformable bodies. Most previous studies were static contact model, which did not consider the plastic deformation of rough body and the problem of friction and wear during sliding process[2-6]. Actually, friction is a dynamic process. Therefore, to establish a model of sliding contact between two rough surfaces can give a better understanding of friction and wear mechanisms, and has important application value. For example: Chen Guoan, Ge Shirong and Liu Jinlong[7] developed the relationship between sliding friction force and contacting condition of rough surfaces according to kinematics and dynamics analysis of two contacting asperities. They derived a fractal prediction model of sliding friction forces based on the fractal geometry theory and verified theoretically the validity of the model.

Gong and Komvopoulos[8] modeled the elasto-plastic sliding contact of a sphere sliding on a periodically patterned surface. They used the plane strain assumption and considered surface adhesion. The results provided useful descriptions of the stresses within the surfaces during contact.

In 2000, Faulkner and Arnell[9] developed an elasto-plastic finite element asperity model, which included sliding friction. This was performed by modeling the three-dimensional spherical asperities at various distances apart while still in sliding contact. The data was then statistically used to model an entire surface of asperities. The major problem with this model was that it was computationally intense and took many hours to execute. It was also limited by the specific materials simulated in that work.

Vijaywargiya and Green[10] presented the results of a finite element analysis (FEA) used to simulate two-dimensional (2D) sliding between two interfering elasto-plastic cylinders. The material for the cylinders is modeled as elastic-perfectly plastic and follows the von Mises yield criterion. The FEA provided trends in the deformations, reaction forces, stresses, and net energy losses as a function of the interference and sliding distance between the cylinders. Results are presented for both frictionless and frictional sliding and comparisons are drawn. The effects of plasticity and friction on energy loss during sliding are isolated. They also presented empirical equations that relate the net energy loss due to sliding under an elasto-plastic deformation as a function of the sliding distance.

What's more, they presented contour plots of the Von Mises stresses to show the formation and distribution of stresses with increasing plastic deformation as sliding progresses. They showed that for the plastic loading cases the ratio of the horizontal force to the vertical reaction force is non-zero at the point where the cylinders are perfectly aligned about the vertical axis. In addition, a "load ratio" of the horizontal tugging force to the vertical reaction force is defined. Although this is analogous to the common definition of the coefficient of friction between sliding surfaces, it just contains the effect of energy loss in plasticity. The values of the contact half-width are obtained for different vertical interferences as sliding progresses.

More recently, Jackson et al[11] sought for characterize the component of friction, which arises from energy being dissipated via elasto-plastic deformation during sliding contact (as apposed to adhesive mechanisms). They analyzed the sliding interaction between spheres using two approaches (a semi-analytical and finite element simulation). These analyses were used to formulate empirical equations, which describe the average tangential and normal forces resulting from the sliding interaction. A parametric study of the properties of typical metals was applied to verify the effectiveness of the empirical equations. The study showed that the effective friction coefficient between spherical asperities increases with the elastic modulus, decreases with yield strength, and increases with the interference between the contacts (dependant on the normal load).

Previous models were analyzed the force or deformation of the rough surface, during the process of sliding. However, these rough surfaces were built by regular shapes. Since the force works when contact and sliding, deformation, fatigue, wear, adhesion and groove etc were happened in the rough surfaces. This is the nature of friction and wear. Friction and wear are great relationship with the surface characterization and the contact property. Objective and accurate to the surface characterization is the prerequisite for contact modeling. Contact analysis was advantageous for understanding the tribological behaviors (including wear, friction, and lubrication). The contact model system, which takes the fractal characteristics in both rough surfaces into account, was more practical for studying the nature of friction and wear. So, a two rough surfaces model was established in this paper. The fractal rough surface was modeled by Weierstrass-Mandelbrot function and the maximum shear stress was discussed by using the finite element analysis combined with the delaminating theory of wear[12].

Model of Friction Sliding and Calculation

Since the geometric structures of rough surface affect engineering, especially through contact problems, a realistic representation of rough surface, other than the traditional statistical methods, is strongly necessary for obtaining accurate information about the stresses and strains at contact interfaces. In this study, the rough surface is simulated based on the two-dimensional W-M fractal[13]. Fig. 1 shows a model of the two dimension rough solids A and B with a W-M fractal rough surface. In Fig. 1, L_1 and L_2 are the length of the solid A and B, respectively. h_1 and h_2 are the thickness of the solid A and B, respectively.

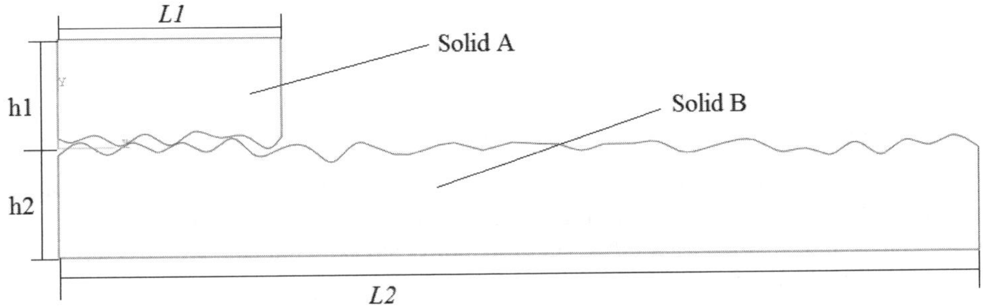

Fig. 1: model of the two dimension with fractal surface

In order to simplify calculation, it was assumed that:

(1) Consistent with Coulomb's law of friction, the friction coefficient remains unchanged in the process of sliding;

(2)Without consider the impact of adhesion, wear failure;

(3) Material is isotropy.

In the process of contact, the uniform pressure *(P)* is progressive applied on the up surface of the solid A, and then the pressure was kept as a constant in the sliding process. The amplitude of load was:

$$\begin{cases} P(t_1) = P_0 + \dfrac{P_{\max}}{t_{load}} \times t_1 & 0 < t_1 \le t_{load} \\ P(t_2) = P_{\max} & t_{load} < t_2 \le t_{slide} \end{cases} \tag{1}$$

Here, P_0 is the initial load (this model, $P_0=0$), P_{max} is the final laod, t_{load} is the time of loading, t_{slide} is the time of sliding.

On the process of sliding, the relative sliding velocity *(v)* between the solid A and B

$$v(t_2) = v_{\max} \quad t_{load} < t_2 \le t_{slide} \tag{2}$$

For the three-dimensional stress space, Mises yield criterion is expressed as:

$$\frac{1}{6}\left[(\sigma_1 - \sigma_2)^2 + (\sigma_2 - \sigma_3)^2 + (\sigma_3 - \sigma_1)^2\right] - \frac{1}{3}\sigma_y{}^2 = 0 \tag{3}$$

On the plane of a Cartesian coordinate system, $\sigma_3 = 0$, so the Eq. 3 can be change as:

$$\sigma_1{}^2 - \sigma_1\sigma_2 + \sigma_2{}^2 = \sigma_y{}^2 \tag{4}$$

Taken into account the influence of different contact characteristics, the paper established the model *I*(A is elasto-plastic and B is rigid)and model *II*((A and B are elasto-plastic) to analyzed. The boundary of structure is:

solid A: $\begin{cases} \sigma_y = P, & 0 \le x \le L_1, y = h_1 \\ u_x = 0, & x = 0, 0 \le y \le h_1 \\ u_x = 0, & x = L_1, 0 \le y \le h_1 \end{cases}$

solid B: model *I* $\begin{cases} u_y = 0, & RP \\ u_x = s_x, & RP \\ M_x = M_y = 0, & RP \end{cases}$ model *II* $\begin{cases} u_y = 0, & s_x \le x \le L_2, y = -h_2 \\ u_x = s_x, & s_x \le x \le L_2, -h_2 \le y \le 0 \end{cases}$

RP is the reference point of rigid body.

Simulation and Analysis

Parameters. In the model, the material parameter referenced from the document[14].

Analysis of shear stress. There are only a few asperities coming into contact when two rough solid contact under the nominal loading which makes the contact pressure on the contact region hugeness. When two bodies slide against each other, the irregularity stresses and contact pressure distribution will be generated in the bodies.

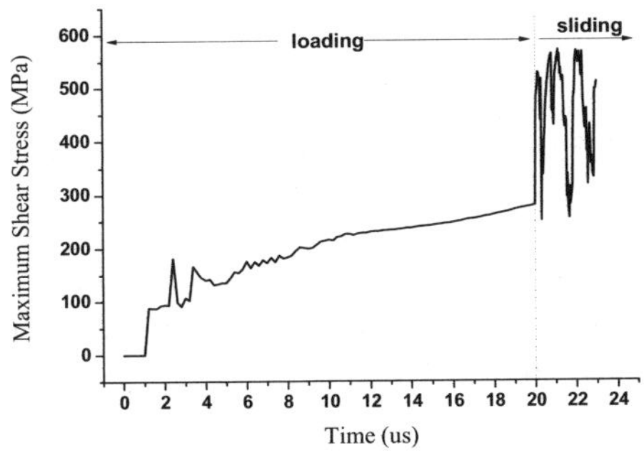

(a) model *I*

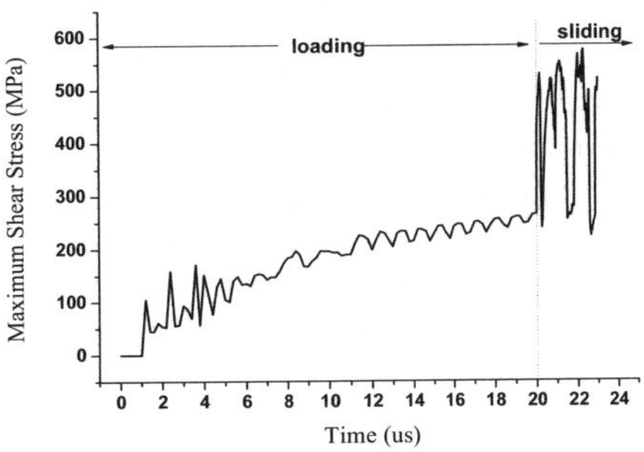

(b) model *II*

Fig. 2: The maximum shear stress versus the loading and sliding time of the rough solid A

Fig. 2 shows the maximum shear stress versus the loading and sliding time of the rough solid A. From Fig. 2, it can be seen that the fluctuation of the maximum shear stress is larger at the beginning of normal loading in both the model *I* and model *II*. This was because the load was small, and only a spot of asperities begins to contact and produce elastic deformation. With the increment of the loading, the number of yielding asperities was increase and the maximum shear stress rose gradually. At the beginning of sliding, with the combined effects of normal and tangential force, the maximum shear stress rose quickly, and the soft rough solid overcame the frictional resistance and became plastic deformation or fracture failure. So the maximum shear stress declined evidently and the fluctuation was more severe than the loading.

Fig. 3 shows the maximum shear stress's distance versus the loading and sliding time of the rough solid A. From Fig. 3, it can be found that lay a foundation to further research on. The reason is that with the process of loading and sliding, normal and tangential force will pass through the asperities when two different hardness surfaces contact, and the asperities on the soft rough solid will get into plastic deformation. During many asperities have passed, they will produced plastic deformation accumulated in the soft surface and formed a layer of plastic deformation, and then a large number of dislocations accumulated in the layer below the soft surface because of the dislocation motion. Finally, a crack in the subsurface would form. The results of simulation closed with the delaminating theory of wear: the traction at the sliding contact deforms the subsurface layer, generating dislocations and vacancies. Continued plastic deformation leads to the formation of voids in the subsurface layer.

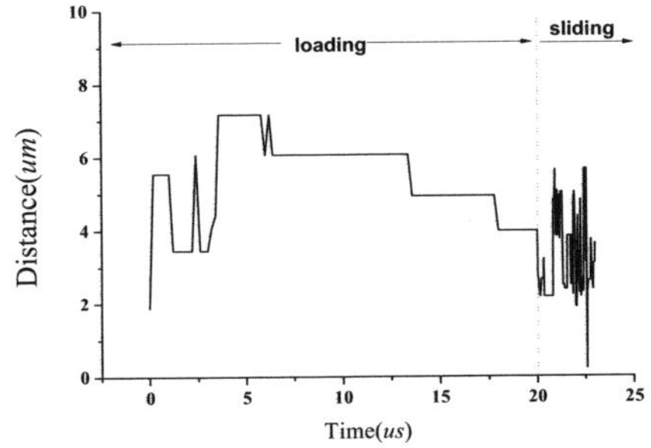

(a) model I

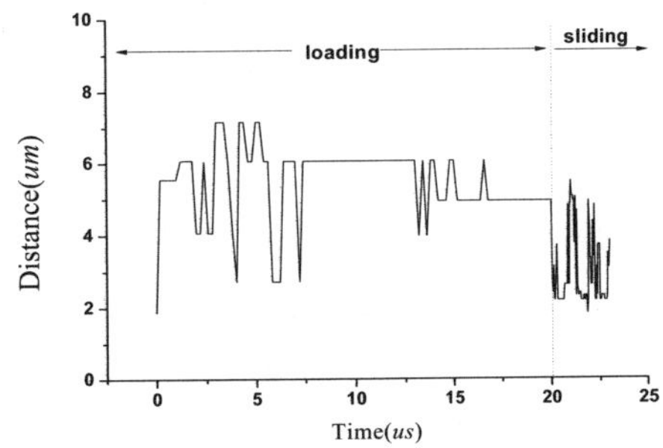

(b) model II

Fig. 3: The maximum shear stress's distance versus the loading and sliding time of the rough solid

From the maximum shear stress and the depth from the surface of maximum shear stress, it can be found that: the greater the maximum shear stress, the more shallow depth from the surface. The Delaminating Theory of Wear was introduced in 1973 by N.P.Suh[12], which provided an explanation for the wear phenomenon based on the mechanical behavior of metals and their surface and metallurgical characteristics. It explained wear of metals as a process of dislocation motion below the surface, sub-surface crack initiation and propagation, the subsequent joining of cracks, and finally the loss of material through plate-like wear debris. The simulation results keep with Suh's views.

From these two models it can be seen that: when the solid A just under the normal load, two solids didn't move relatively, and the value of load was small, so the maximum shear stress was small, and most of the depth was arrange 5-6*um* from the surface; In the process of slide, with the combined effects of normal and tangential force, the Maximum shear stress generated in the shear band when the asperities in contact with the side, and the depth was arrange 2-4*um* from the surface.

Model *II* considered the material deformation of rough solid B, and the situation of contact was more complex, so the fluctuations of the maximum shear stress and distance were more intense than that of Model *I*.

Conclusion

In this paper, a two-dimensional slide model was established. The fractal rough surface was modeled by Weierstrass -Mandelbrot function and the Maximum shear stress was discussed using the finite element analysis. Combined with the delaminating theory of wear, the maximum shear stress and its distance were discussed. The analysis and simulation identified that there was complex relation between the maximum shear stress, the distance of maximum shear stress and the different elasto-plastic characteristics:

(1) The maximum shear stresses was in the subsurface layer which range about 2 to 8 um from the surface.

(2) The fluctuations of the maximum shear stress and distance of Model *II* (both of were elasto-plastic) were more intense than that of Model *I* (One was elasto-plastic, and the other was rigid body).

Analysis of the mechanics problems with the static contact and sliding contact will lay a foundation to further research on the substance of the process of friction and wear, and can helped to explain wear of metals as a process of dislocation motion below the surface, sub-surface crack initiation and propagation, the subsequent joining of cracks, and finally the loss of material through plate-like wear debris.

Acknowledgment

This research is supported by the National Natural Science Foundation of China (NSFC) under grant No: 50775039, the Science Development Foundation of Fuzhou university(2010-XQ-13)

References

[1] Z.zhongrong:*The development of tribology*. Sciences Press, Beijing, (2006).

[2] Z. Lisowski, T. Stolarski: *An analysis of contact between a pair of surface asperities during sliding*. J. Appl. Mech. 48, 3(1981), pp.493-499.

[3] Y. Xie, J.A. Williams: *The prediction of friction and wear when a soft surface slides against a harder rough surface*. Wear, 196, 1-2(1996), pp.21-34.

[4] J.Jamari, M.B.Rooij, D.J.Schipper: *Plastic deterministic contact of rough surfaces*. ASME Journal of Tribology, 129, 4(2007), pp.957-962.

[5] L.Kogut,I.Etsion: *A static friction model for elastic -plastic contacting rough surfaces.* ASME Journal of Tribology, 126, 1(2004), pp.34-40.

[6] J. Yang, K. Komvopoulos: *A mechanics approach to static friction of elastic-plastic fractal surfaces.* ASME Journal of Tribology, 127, 2(2005), pp.315-324.

[7] C Guoan, G Shirong and LJinlong: *Fractal prediction model of sliding friction force.* Journal of China University of Mining & Technology, 29, 5(2000), pp.492-495.

[8] Z.Q.Gong, K.Komvopoulos: *Effect of surface patterning on contact deformation of elastic-plastic layered media.* ASME Joumal of' Tribology, 125, 1(2003), pp.16-25.

[9] A.Faulkner,R.D.Arnell: *Development of a finite element model to simulate the sliding interaction between two, three-dimensional ,elasto-plastic, hemispherical asperities.* Wear, 242, 1(2000), pp.114-122.

[10] R.Vijaywargiya, I.Green: *A finite element study of the deformations, forces, stress formations, and energy losses in sliding cylindrical contacts.* International Journal of Non-Linear Mechanics, 42, 7(2007), pp.914-927.

[11] R.Jackson, R.Duvvuru, H.Meghani, M.Mahajan: *An analysis of elasto-plastic sliding spherical asperity interaction.* Wear, 262, 1-2(2007), pp.210-219.

[12] N.P. Suh: *The delamination theory of wear.* Wear, 25, 1(1973), pp.111-124.

[13] W Yan, K Komvopoulos: *Contact analysis of elastic-plastic fractal surfaces.* Journal of Applied Physics, 84, 7(1998), pp.3617-3624.

[14] G.Chenghui, H.Jianmeng and D.Lei: *Influence of different sliding velocity on the thermo-mechanical coupling of the rough surface contact.* Applied Mechanics and Materials Vol. 29-32 (2010), pp.332-336.

Advanced Materials Research Vol. 279 (2011) pp 230-234

doi:10.4028/www.scientific.net/AMR.279.230

Investigation Of TIG Dressing And Chilling Treatment For Improving The Fatigue Strength Of Welded Joints

Ping Zhu[1, a], ChunYuan Shi[1, b], Cheng Jin[1, c] and XiuJuan Zhang[2, d]

[1]the department of material sciences and engineering, Dalian Jiaotong University, Dalian, Liaoning Province,china, 116028

[2]the department of mechanical engineering, Dalian Jiaotong University, Dalian, Liaoning Province,china, 116028

[a]warten@163.com, [b]shicy@sina.com ,[c]jincheng@126.com, [d]zhangxiuj@djtu.edu.cn

Keywords: TIG dressing, chilling treatment, residual stress, fatigue strength

Abstract. The negative temperature field with a certain strength is exerted on the TIG dressing zone in welding toes, which could decrease and offset the residual stress. When the temperature of TIG dressing zone is or above 450 ℃, the chilling treatment could produce the stress field of biaxial compression. The chilling treatment is performed when the temperature of dressing zone is 550 ℃. The fatigue strength of joints can be increased more than twice comparing with those after welding.

Introduction

According to statistics, about 90% accident damages of welded structures are the fatigue damage and the fatigue cracks are usually originated from welded toes of joints. TIG dressing treatment for welding toes is the simple and effective treatment process after the welding [1-3]. But the surface of welding toes still has the residual tensile stress after TIG arc melting revision [4]. Thus, the further improvement of fatigue strength is limited. The welding process is proposed in which TIG dressing of welding toes is firstly performed and then the emergency cooling of welding toes is implemented. The negative temperature field is exerted on the welding toes in order to decrease and offset residual tensile stress. When the cold source is strong enough, the biaxial residual compressive stress field is generated. Obviously, the compressive stress of the welding toes is conducive to further improve the fatigue performance of welding joints.

Experiment material and method

Experiment Material. The experiment material is Q345C low-alloy structural steel which is selected by the welded frame of the passenger train bogies. The thickness of the steel specimen is 14mm. The chemical composition of the steel specimen is described as follows: C≤0.20%, Mn:1.00~1.60%, Si≤0.55%, P≤0.035, S≤0.035%,V:0.02~0.15%, Nb:0.015~0.060%, and Ti:0.02~0.20%. The mechanical properties of the steel specimen are described as follows:$\sigma_s \geq 345$MPa, σ_b:470-630MPa, and $\delta \geq 22\%$. The Akv is 34J at 20℃,0℃, and-20℃.

The forms of welded joint is T-fillet joint. The multi-MAG welding process is used. The welding wire is the CHW-50C6 (φ1.2mm) solid wire and the shielding gas is 80% Ar +20% CO_2. The welding process parameters are listed in Table 1.

Table 1: Welding parameters

Welding level	electric current of welding [A]	Arc Voltage [V]	Welding velocity [cm/min]	Air flow [l/min]
Bottom layer	130	18	8	19
Other layers	240	26	12	

Experiment Methods. Before the TIG arc dressing, the rusts, slag, and spatters of large particles on the surface of the welding toes are cleaned up at first, since too much slag could form the inclusions and porosities after TIG dressing. The spatters of large particles could bring the adverse affect on not only the arc dressing process and its stability, but also the quality of dressing zone [5-6]. Dressing power is DCSP with the current of 180A, the voltage of 17V, the speed of 15cm/min, and the line energy of 12.5KJ/cm. The diameter of TIG tungsten electrode is φ3.2mm, the nozzle diameter is φ12mm, and the flow rate of Argon gas is 10L/min.

After TIG dressing process, the chilling treatment is carried out and the rapid cooling is also performed to the room temperature when the dressing zone of welding toes reaches a certain temperature.

The blind hole method is used to test the residual stress. BE120-2CA-K three-resistance strain gauge (the sensitivity coefficient is 216) is pasted on the welding toes. CCZ-1 drilling equipment is used to drill a blind hole with the diameter of φ2mm and the depth of 2mm on the welding toes. YJ-25 strain gauge is used to measure the release strain and then the value of residual stress is calculated in the light of the elasticity theory.

The three-point bending fatigue experiment is performed according to the standard JB/T7716-95. The fatigue experiment is performed on the PLG-100 high-frequency fatigue testing machine controlled by the computer. The fulcrum span is 100mm, the vibration frequency of 154-168Hz, the ratio of cyclic stresses is 0.1, and the specified cycle life is 5×10^6N.

Measurement results and analysis

Residual Stress for the Welding Toe Surfaces. The measurement results of residual stress for welded state, TIG dressing of welding toes, and the chilling treatment of welding toes after TIG dressing for different temperatures are given in Table2.

Table 2: Measurement results of residual stress for the welding toe surfaces of butt joints

Processing state of welding toes	Vertical stress σ_x [Mpa]		Horizontal stress σ_y [Mpa]	
	Stress value	Average value	Stress value	Average value
Welded state	163.90, 154.14, 145.41, 133.92, 110.98	141.67	28.93, -22.01, -69.63, 67.58, 51.50	11.27
TIG remelting treatment	162.99, 160.65, 144.15, 222.37, 135.00	165.03	-61.20, 66.86, -64.03, 130.86, -54.88	3.36
TIG remelting zone after the chilling treatment under 350°C	128.37, 140.85, 128.87, 80.45, 119.63	119.63	-56.93, 99.67, 89.98, -95.71, -57.98	-4.19
TIG remelting zone after the chilling treatment under 450 °C	-36.18, -59.18, -47.68, -30.40, -52.01	-45.09	-104.81, -159.84, -132.32, -153.93, -136.66	-137.51
TIG remelting zone after the chilling treatment under 550 °C	-214.56, -200.77, -142.62, -64.30, -257.38	-175.93	-52.14, -26.91, -119.75, -176.39, -239.08	-122.85

Table 2 shows that the longitudinal residual stresses of welding toes are the tensile stresses for both the welded state and the TIG arc dressing. The transverse residual stresses are both the tensile stresses and compressive tresses, because the test plate can be freely contracted in the horizontal direction. It can be seen that TIG dressing has not changed the residual tensile stress state of the welding toes, which is consistent to the literature [4]. The chilling treatment is implemented under different temperatures after the TIG dressing of welding toes. With the increase of dressing zone temperature for the chilling treatment, the residual tensile stress of welding toes is gradually decreased and is gradually changed into the residual compressive stress. When the temperature of chilling treatment reaches 450 °C, the vertical residual tensile stress for the welding toe surface of joints is close to zero. While the vertical residual tensile stress for the welding toe surface of T-joint is less than zero. Thus, when performing the chilling treatment and the dressing zone temperature of TIG welding toes is higher than 450 °C, the area can remove the residual tensile stress or even achieve the two-way residual compressive stress state.

When the chilling treatment for the dressing layer of welding toes is performed under the high temperature, the compressive residual stress can be formed on the surface of welding toes. The basic principle is that the action of shrinkage is generated on the surface metal in the dressing zone of welding toes after the high-temperature chilling treatment and the tensile plastic deformation is generated due to the binding effect of surrounding metal. When the temperature difference is large enough, the tensile plastic deformation can not only offset the compressive plastic deformation produced during the welding process, and produce the residual tensile plastic deformation. The biaxial compression residual stress is then formed on the surface of the welding toes.

Fatigue property of the welded joints. The results of three-point bending fatigue experiment for the welding joint under different processing states is shown in Table3.

Table 3: Fatigue test results of T-joints

Welded state			TIG remelting treatment of welding toes			TIG remelting zone after the chilling treatment		
Maximum stress S_{max}[MPa]	Stress range ΔS [MPa]	Fatigue life [N_f]	Maximum stress S_{max} [MPa]	Stress range ΔS [MPa]	Fatigue life [N_f]	Maximum stress S_{max} [MPa]	Stress range ΔS [MPa]	Fatigue life [N_f]
373	335.7	175900	448	403.2	138300	550	495	66200
299	269.1	380400	373	335.7	283200	500	450	616200
261	234.9	805400	299	269.1	3017900	450	405	5000000
246	221.4	613900	261	234.9	905900	525	472.5	263800
224	201.6	1129500	259	233.1	2577600	475	427.5	590500
187	168.3	1480000	252	226.8	5000000	462.5	416.25	1011500
168	151.2	2853800				450	405	5000000
149	134.1	5000000				462.5	416.25	5000000

The S-N curves of T-joints under three states, which are welding, TIG dressing of welding toes, and chilling treatment on the welding toes after TIG dressing, are given in Figure1. The comparison of fatigue strength for these three states is given in Table4.

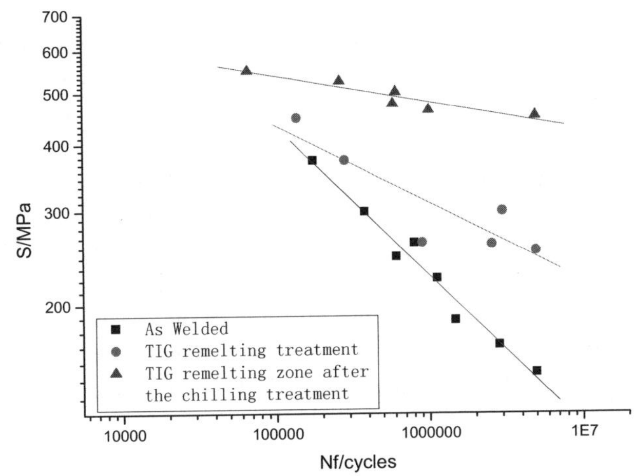

Fig. 1: S-N curve of T-joints in three states

Table 4: Fatigue strength of T-joints in three states

Welding process	$\lg \Delta S = a + b \lg N$	Fatigue strength [Mpa]	Fatigue strength improving level [%]
Welded state	$\lg \Delta S = 3.99181 - 0.28029 \lg N$	130	—
TIG remelting treatment	$\lg \Delta S = 3.292 - 0.14504 \lg N$	209	61
TIG remelting zone after the chilling treatment	$\lg \Delta S = 2.98283 - 0.0505 \lg N$	382	294

It can be seen from the experiment results that the fatigue strength of joints is increased by 61% after TIG dressing comparing with the T-joint after welding. The fatigue strength of joints is increased by 294% after the chilling treatment in TIG dressing zone under 550℃. The fatigue performance of the T-joints is further improved. The obvious improvement of fatigue strength is relevant with the extent of stress concentration and the state of residual stress, etc.

TIG dressing of welding toes improves the weld forming, generates the smooth transition between the welds and basic metals, and decreases the tiny nonmetallic slag in this field. Thus, the fatigue strength is improved. But, the dressing treatment does not change the tensile state of the residual stress in the welding toes, which limits the effect of TIG dressing of welding toes on the fatigue strength. The chilling treatment is performed under 550°C after TIG dressing. The peak value of longitudinal residual stresses in the welding toes is only -175.93Mpa. The vertical residual tensile stress is also the compression stress. Thus, the chilling treatment under 550°Cafter TIG dressing can offset the residual tensile stress of welding toes and further improve the fatigue strength.

Conclusions

The longitudinal residual stresses of welding toes are the tensile stresses for the welded state, the transverse residual stresses are both the tensile stresses and compressive tresses, after TIG dressing, the residual stresses of the surface of welding toes don't decreased obviously, still are tensile stresses.

With the increase of dressing zone temperature for the chilling treatment, the vertical residual tensile stress of welding toes is gradually decreased and is gradually changed into the residual compressive stress. The lower limit of temperature for the vertical residual tensile stress for the welding toe is 450 °C.

Comparing the welding, the fatigue strength of joints in the welding toes is increased by 61% after TIG dressing. While the fatigue strength of joints in the welding toes is increased by 294% when they are performed the chilling treatment under 550°Cafter TIG dressing.

References

[1] Cheng Bing et al.: *Overview on Railway Passenger Cars of Technology Development in China.* rolling stock, 2005(11).

[2] Huo Lixing et al.: *the latest advances and technology on Improving the fatigue performance of welded joints*. welding technology, 2002.

[3] li xu et al.: *study on improving TIG dressing process for welded bogie of high-speed vehicles.* machinist, 2002(5).

[4] A. M. Horn, I. Huther, H. P. Lieurade: *Fatigue behaviour of T-joints improved by TIG dressing.* Weld in the World Vol.41 (1998), pp. 273-280.

[5] Zhao Jianming et al.: *Application of TIG dressing of welding toes in the manufacturing of bogie frame in high speed passenger train.* Railway Locomotive & CAR, 1995(3).

[6] T. Dahle: *Design fatigue strength of TIG-dressed welded joints in high-strength steels subjected to spectrum loading.* Int. J. Fatigue, 20, 9(1998), pp. 677-681.

Advanced Materials Research Vol. 279 (2011) pp 235-238

doi:10.4028/www.scientific.net/AMR.279.235

Housing Die-Casting and Its process Based on Numerical Simulation

Yong Huang[1, a], Jibin Guo[1, b], Qiong Wu[1, c], Xiaoming Du[1, d]

[1]School of Materials Science and Engineering, Shenyang ligong University, Shenyang 110159, China

[a]hy195953@163.com, [b]guojibin19851016@sina.com, [c]wuandqiong@126.com, [d]du511@163.com

Keywords: Casting Process, Numerical Simulation, Housing

Abstract. Mould filling and solidification of aluminum alloy housing in die casting were simulated by the casting simulation software. Distribution and change of velocity field and temperature field were visualized. The optimized processing parameters were obtained with pouring temperature at 600°C and pre-heated temperature of mould at 180°C as well as injected velocity at 2.5m/s. The qualified aluminum alloy housing and the dense microstructure were successfully produced by adopting the optimized processing parameters, which is proved by the actual production.

Introduction

The control, design and process of die casting depend on experience to verify, so the quality of castings is difficult to guarantee, which cause a high rejection rate. Casting simulation technology to solve these problems has provided favorable conditions, which can help designers to give effective prediction for a variety of defects and their size, location and timing to occur, so as to take countermeasures before the casting mold and the product may be manufactured. So casting quality is ensured, and the cost of production is reduced [1, 2].

Filter housing is an important component of automotive engines, and its quality is critical to the performance of the car. In order to resolve the problem of the big shrinkage and high waste ratio of production, three groups of process solution is designed. Using ProCAST software, numerical simulation of mould filling and solidification process of housing to predict the shrinkage were done to predict the location of shrinkage. By optimizing the casting process parameters, and controlling the quantity of shrinkage, we can search an optimal process to supply the reliable theoretical basis for the actual production.

Mathematical model

The k-ε model is employed to describe the turbulent characteristics on numerical simulation of filling process at present, which is solved by SOLA-VOF method [3]. For the viscous layer near the wall, the turbulent dissipation rate and kinetic energy on the grid points near the wall are corrected by wall function method. Free surface is treated by using the improved VOF method. In a Cartesian coordinate system, an integrated equation is created by means of the following one:

$$\frac{\partial}{\partial t}(\rho\phi)+\frac{\partial}{\partial x_j}(\rho u_j\phi)=\frac{\partial}{\partial x_j}\left(\Gamma_\phi\frac{\partial\phi}{\partial x_j}\right)+S_\phi \tag{1}$$

where the parameter x_j corresponds to the coordinate component, $j=1,2,3$; u_j is the velocity component of the x_j direction; ϕ is the flux; Γ_ϕ represents the transmission coefficient; S_ϕ denotes the source term. The varied Γ_ϕ and S_ϕ are determined by the specific parameter ϕ.

In this paper, the $G/\sqrt{R}$ method used to predict casting shrinkage is proposed by E. Niyama [4]. The $G/\sqrt{R}$ method proposed by Professor Eisuke Bahru is used to predict casting shrinkage comprehensively. G represents the temperature gradient, R is the cooling gradient [5]. G is defined specifically as the following one:

$$G=\left[\left(\frac{\partial T}{\partial x}\right)^2+\left(\frac{\partial T}{\partial y}\right)^2+\left(\frac{\partial T}{\partial z}\right)^2\right]^{1/2} \quad (2)$$

The cooling rate is defined as:

$$R=\left|\frac{T_{upper}-T_{lower}}{t_{upper}-t_{lower}}\right| \quad (3)$$

where T_{upper} is the temperature of liquidus, $^{\circ}$C; T_{lower} denotes the temperature of solidus, $^{\circ}$C; t_{upper} represents the time of liquidus temperature, s; t_{lower} is the time of solidus temperature, s.

Simulation of die casting process

The three-dimensional die casting mold is built by using PRO/E software. Fig. 1 is the casting 3D mould. The die casting and mould are put into ProCAST software, respectively. The grid mesh and tetrahedron mesh is generated (Fig. 2).

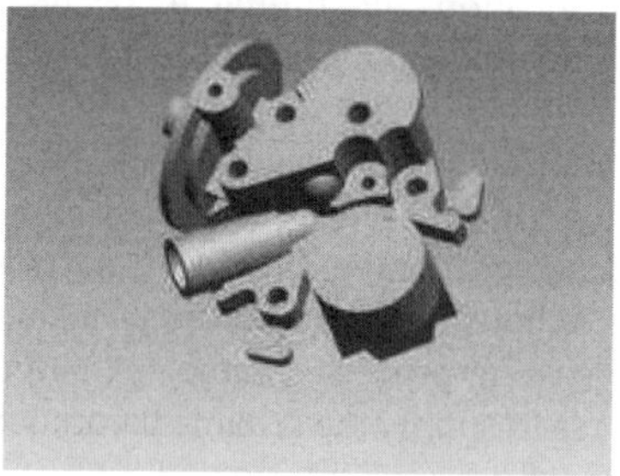

Fig. 1: The casting 3D mould

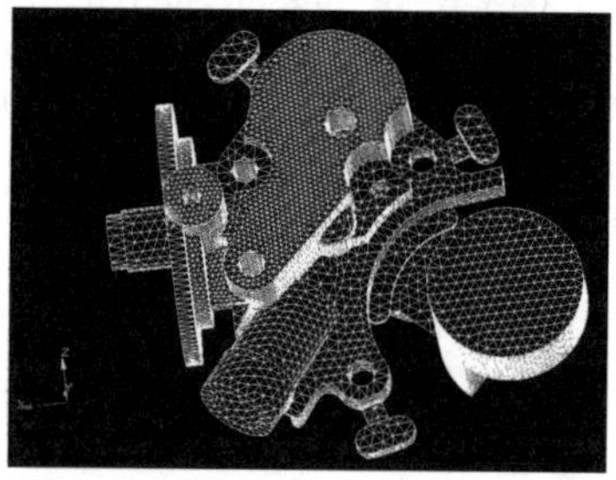

Fig. 2: The finite element model

The Aluminum alloy ADC12 is selected in this numerical simulation. Its freezing point is 522°C, the melting point is 580°C. The material of mold is H13 steel. The environmental temperature is 20°C [6]. The outside surface temperature of the mold is the pre-heating temperature of the mold. The heat transfer coefficient is 25W/(m^2·k), 1000W/(m^2·k) and 1500W/(m^2·k) for the mold and environmental, casting and mold and mold and mold, respectively.

Results and Discussion

In this paper, there are three groups of process parameters as follows:

Table 1: Simulation program

	Injection speed [m/s]	Pouring temperature [$^{\circ}$C]	Pre-heating temperature of mould [$^{\circ}$C]
Sol.1	1	590	160
Sol.2	2.5	600	180
Sol.3	4	610	200

The results of mould filling and solidification in solution 1 are shown in Fig. 3:

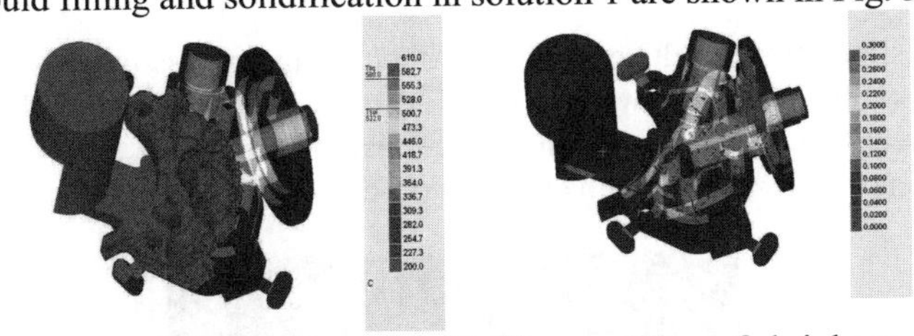

(a) Filling process (b) The condition of shrinkage and porosity

Fig. 3: Solution 1

The volume of shrinkage and porosity was so big(>0.01) that the macro-defect appeared. The reason is that the parameters in solution1 are relatively small, the alloy liquid is super-cool easily and the liquidity is poor in the filling and solidification process. In the mould filling process, the front metal was mixed with a certain amount of liquid oxides, coating and other impurities. This forms shrinkage and porosity easily in the solidification process. In particular, the ingate have a lower pouring temperature and mold preheating temperature. It likely causes the solidification of the gate early. The internal castings are not feeding, leading to shrinkage, porosity defects eventually.

The results of mould filling and solidification in solution 2 are shown in Fig.4:

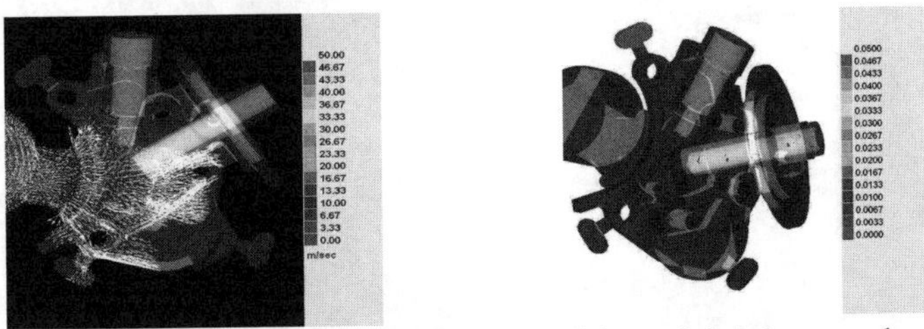

(a) Filling process (b) The condition of shrinkage and porosity

Fig. 4: Solution 2

It can be seen that there exist no volume gas and sputtering phenomenon. This indicates that the filling process is smooth under the group process parameters. The volume fraction of shrinkage is very small, about as 0.0033, which is at a very low micro-porosity, shrinkage level. In the actual production, this level of shrinkage and porosity is negligible.

The results of mould filling and solidification in solution 3 are shown in Fig. 5:

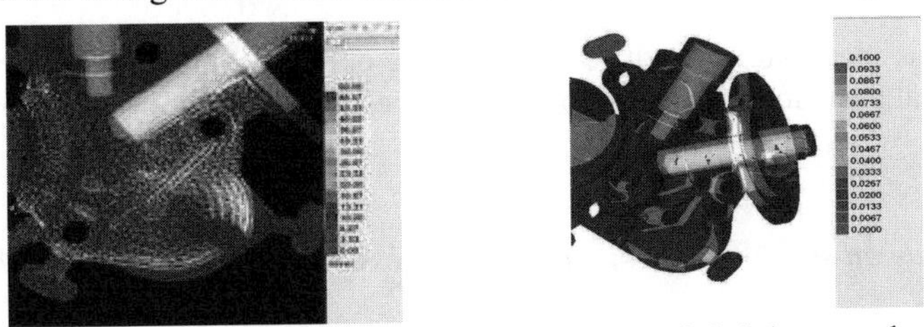

(a) Filling process (b) The condition of shrinkage and porosity

Fig. 5: Solution 3

Using casting process parameters at high pouring temperature and injection speed in solution 3, the premature solidification occur easily and proper feeding is not supplied to casting, which were prevented. With higher pouring temperature, the solubility of gas and degree of oxidation varied rapidly from pouring temperature. The higher injection speed cause possibly to wrap gas and to create other casting issues. The porosity and shrinkage rate is about 0.01 in this group casting parameters, which either contained micro-shrinkage or macro-shrinkage.

Based on the above three solutions of the simulation results and analysis, the second set of parameters obtained the desirable result of the filling and solidification process.

Experimental Verification

According to the simulation results obtained by die casting process parameters of production, the actual verification is good, as is shown in Fig. 6. Then, Fig. 7 and Fig. 8 show the difference in microstructure between the change of casting process parameters.

It can be seen from Fig. 6, after optimization of process parameters, the quality and dense of die casting is good.

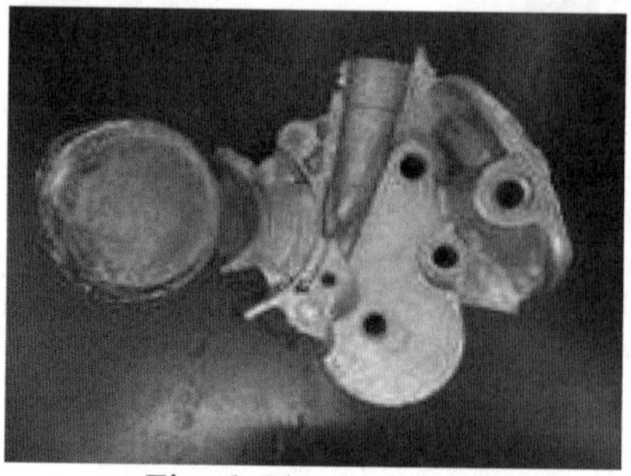

Fig. 6: Physical map

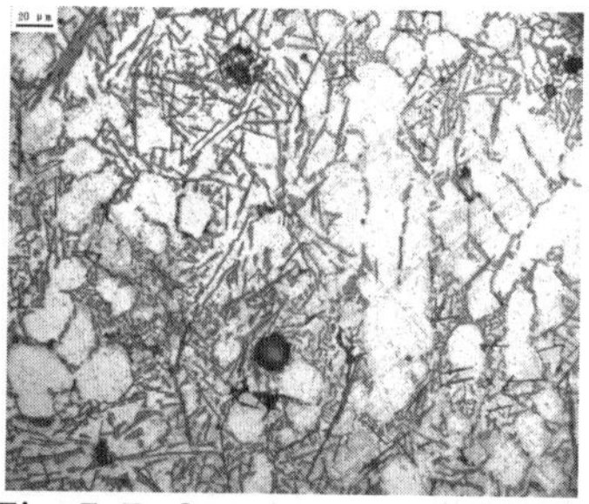

Fig. 7: Before the optimization of process parameters × 200

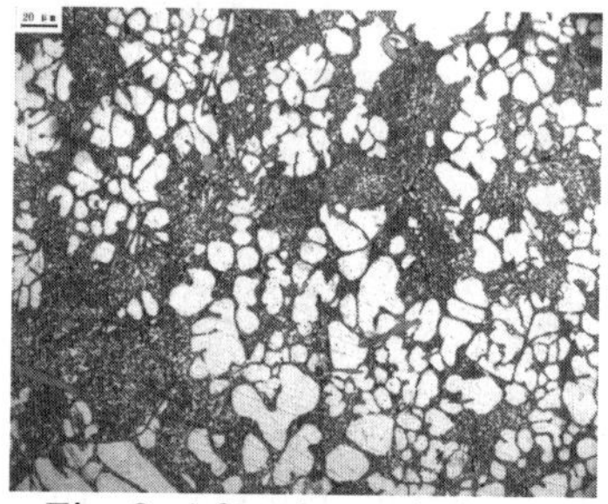

Fig. 8: After the optimization of process parameters × 200

Summary

(1) Simulation results show that the desirable processing parameters can be obtained with pouring temperature at 600°C and pre-heated temperature of mould at 180°C as well as filling velocity at 2.5m/s.

(2) The housing was successfully produced by adopting optimized processing parameters. It supplies a reliable theoretical basis for the production and verifies the correctness of the simulation results.

References

[1] Fang L. G: *Study on solidification simulation process of steel castings technology.* Forging. Vol. 96 (2002), p. 1.

[2] B.C. Liu, T. Jin: *Casting simulation engineering and quality control.* CHINA MACHINE PRESS, Beijing, (2002).

[3] Q. Ma, Y.Q. Yu, K.P. Zhang: *Study on filling simulation process of magnesium die casting.*.Journal of Harbin university of science and technology. Vol. 9 (2004), p.28.

[4] Y. M. Lai, C.L. Hu, Q.K. Liu: *Comparative study on die casting of aluminum chassis*. Mould Industry. Vol. 32 (2006), p. 4.

[5] I. Imafuku, K.A. Chijiiwa: *Mathematical model for shrinkage cavity predication in steel castings*. AFS Transactions. Vol. 10 (1983), p. 527.

[6] Y. Huang, D.W. Pan and X.M. Du: *Numerical simulation and die casting process for shell.* Special Casting & Nonferrous Alloys Vol. 30 (2010), p. 59.

Advanced Materials Research Vol. 279 (2011) pp 239-246
© (2011) Trans Tech Publications, Switzerland
doi:10.4028/www.scientific.net/AMR.279.239

Analysis and Numerical Simulation about the Influence of the Flux Ratio of Methane and Air on Combustion

Weiyang Cao[1,2,a], Bin Meng[2,3,b], Jinhui Peng[1,2,c] and Runze Fu[1,2,d]

[1]Faculty of Metallurgy and Energy Engineering, Kunming University of Science and Technology, Kunming 650093, Yunnan, China

[2]Key Laboratory of Unconventional Metallurgy of Kunming University of Science and Technology, Ministry of Education, Kunming 650093, Yunnan, China

[3]Faculty of Materials Science and Engineering, Kunming University of Science and Technology, Kunming 650093, Yunnan, China

[a]kmcwyrex@163.com, [b]hitmengbin@163.com, [c]jhpeng@kmust.edu.cn, [d] frzmargot@163.com

Keywords: Temperature Field, Numerical Simulation, Concentration Field, Combustion, Nitrogen Oxide

Abstract. The combustion technologies of natural gas and air have been studied by numerical simulation using the software Fluent. The factors are mainly analyzed which affect the temperature distribution and the concentration distribution in the firebox. The results show that different ratios of methane and air greatly affect the temperature distribution, the concentration distribution and the formation of NOx. By selecting a reasonable proportion, not only can it improve fuel combustion efficiency, thereby increasing fuel efficiency and reducing the waste of natural gas, but also can lessen NOx emissions, reduce fuel combustion pollutants, and lessen the damage to the environment because of the combustion of natural gas and air.

Introduction

The survival and development of human and the energy access are always inseparable. The consumption is declining among the OECD countries. The proportion of the natural gas consumption has exceeded. With the wide range of environmental issues of the international community and the extensive progress of the energy technology, the clean energy has developed rapidly, which can replace coal and oil. The proportion of coal and oil in the total requirements of primary energy has dropped and the ratio of natural gas has increased [1].

The natural gas is not only green but also has high calorific value, which is used widely in industry. However, the ratio will affect the combustion temperature and the resultant concentration when natural gas is burned with air. If marched irrationally, it may result in lower combustion efficiency, industry costs rise, energy waste and even a large amount of nitrogen oxides, causing environmental pollution. Therefore, this paper applies the software Fluent to study the influence of the ratio of natural gas and air on combustion, making the mainly components of natural gas-methane as the main object of study.

The numerical simulation using the software Fluent is to build mathematical models firstly; then choose the solving model, set the boundary and initial conditions, control and iterative parameters and do the calculation; and then, a conclusion is made after post-processing and analyzing, or adjust the parameters on the basis of experimental results and re-calculated till an acceptable result is got; at last, compare with the actual industrial production and the experimental data.

Combustion Mechanism of Methane

Hydrocarbons can be cracked into carbon and hydrogen at high temperatures. Gao Zhichong[2] proposed that the hydrocarbons issue light in the combustion process, and it takes part of the energy, which is not used for heating. Therefore, a number of photons emits and its wavelength during the combustion of hydrocarbons will directly affect the flame temperature. On the contrary, the flame temperature can be used to issue the number of photons in the combustion process, and to determine the mechanism of combustion reaction. According to the relationship of the combustion reaction enthalpyΔH, the number of photon and the wavelength: $\Delta H=W=-0.1196n/\lambda$(the first law of thermodynamics), the methane combustion flame temperature is calculated. The calculated temperature 3134K is very close to the actual temperature 3120K, which shows the combustion reaction mechanism [3-8], and the proposed mechanism is as follows.

1: $O_2+hv\rightarrow 2O\cdot$ (1)

2: $CpH_2q\rightarrow pC+qH_2$ (2)

3: $H_2+O\cdot\rightarrow CO+hv$ (3)

4: $C+O\cdot\rightarrow CO+ hv$ (4)

5: $2CO+O_2\rightarrow 2CO_2$ (5)

According to the theory above, the equation of the combustion reaction of methane is:

$CH_4+2O_2\rightarrow CO_2+H_2O$ (g) +1.5hv (6)

Where "h" is Planck constant (6.626×10^{-31}J·s),"v" is velocity. The following calculation is made by adopting the theory above.

Physical model and work conditions

Physical model. As the size of the furnace mentioned in this paper is big, and the velocity of the fuel flow at the entrance is high, the fuel flow is often turbulent and the physical quantities are all random variables of space and time. Even so, the turbulent flow still follows the general movement law of the continuum, having a regular statistical characteristic; its instantaneous flow condition still meets the viscous fluid flow equations. The equations describing all the instantaneous flow field of combustion are as follow: [9]

Continuity equation:

$$\frac{\partial \rho}{\partial t}+\frac{\partial}{\partial x_i}(\rho u_i)=0\ . \quad (7)$$

Momentum equation:

$$\frac{\partial}{\partial t}(\rho u_i)+\frac{\partial}{\partial x_J}(\rho u_i u_j)=-\frac{\partial p}{\partial x_i}+\frac{\partial}{\partial x_j}(\tau_{ij})\ . \quad (8)$$

Components equation:

$$\frac{\partial}{\partial t}(\rho m_s)+\frac{\partial}{\partial x_J}(\rho u_j m_s)=\frac{\partial}{\partial x_J}\left(\frac{\mu}{s_c}\frac{\partial}{\partial x_J}m_s\right)-\bar{w}_s\ . \quad (9)$$

Energy equation:

$$\frac{\partial(\rho C_P T)}{\partial t}+\frac{\partial}{\partial x_j}(\rho u_j C_P T)=\frac{\partial}{\partial x_j}\left(\frac{\mu}{P_r}\frac{\partial C_P T}{\partial x_L}\right)+\bar{w}_s Q_s\ . \quad (10)$$

Where τ_{ij} is the viscous stress:

$$\tau_{ij} = \mu\left(\frac{\partial u_i}{\partial x_j} + \frac{\partial u_j}{\partial x_i}\right) - \frac{2}{3}\mu\delta_{ij} \tag{11}$$

The model adopted in this paper is the $k-\varepsilon$ turbulence model; the $k-\varepsilon$ model is two-dimensional model and also called the standard $k-\varepsilon$ model. The viscosity coefficient of the turbulence in this model is defined as:

$$\mu_t = \rho C_u \frac{k^2}{\varepsilon} \tag{12}$$

The governing equations of "k" and "ε" are respectively expressed as :

$$\frac{\partial(\rho k)}{\partial t} + \bar{u}_i \frac{\partial(\rho k)}{\partial x_j} = \rho p - \rho\varepsilon + \frac{\partial}{\partial x_j}\left[\left(\mu + \frac{\mu_t}{\sigma_k}\right)\frac{\partial k}{\partial x_j}\right] \tag{13}$$

$$\frac{\partial}{\partial t}(\rho\varepsilon) + \bar{u}_j \frac{\partial}{\partial x_j}(\rho\varepsilon) = C_{\varepsilon 1}\rho\frac{\varepsilon}{k}p - C_{\varepsilon 2}\rho\frac{\varepsilon^2}{k} + \frac{\partial}{\partial x_j}\left[\left(\mu + \frac{\mu_t}{\sigma_\varepsilon}\right)\frac{\partial \varepsilon}{\partial x_j}\right] \tag{14}$$

"p" in the Turbulence Kinetic Energy is:

$$p = 2\frac{\mu_t}{\rho}\overline{S_{ij}S_{ij}} \quad ; \quad \overline{S_{ij}} = \frac{1}{2}\left(\frac{\partial \bar{u}_j}{\partial x_i} + \frac{\partial \bar{u}_t}{\partial x_i}\right). \tag{15}$$

In the equations above, "ρ" is the fluid density, kg/m3; "t" is time, s; "p" is pressure, Pa; "i" "j" represent 1,2,3. "μj" is the velocity vector along j in the rectangular coordinate system, m/s; "xj" is the coordinate along j, "μi" is the viscosity coefficient of the turbulence, Pa·s; "Cu" is one coefficient, Cu=0.09; "μ" is the laminar viscosity coefficient, Pa·s; σ_ε and σ_k are the Planck numbers in correspondence to the turbulence energy and the turbulence dissipation ratio. In the software Fluent, in default σ_ε=1.0, σ_k=1.3 ; $C_{\varepsilon 1}$, $C_{\varepsilon 2}$ and $C_{\varepsilon 3}$ are empirical constants which are respectively 1.44, 1.92, 0.09; "k" is the turbulence kinetic energy, $k = \frac{1}{2}\overline{u_i u_j}$; $\varepsilon = C_D \frac{k^{\frac{3}{2}}}{l}$, C_D is a empirical constant, C_D=0.08~0.09 ; l is the Planck mixing length of the turbulence.

Work conditions. The regenerative furnace in one factory was selected as the subject investigated. The effective length of the combustion space is 4000 mm, and the width 3400mm, the height 2500mm. A burner is arranged on each side of the furnace body. Both burners jet at the same time, and gas vents are arranged in front of the burner body. Outer equipment is installed in which gas discharge ports connect with outside. The effective flow area of the burner is a circle section. Industrial natural gas is used as the fuel, whose major component is methane, and whose average calorific value is 35 000 kJ/m3. In order to avoid the air vent being too big as the theoretical air consumption of natural gas is 9.4, three vents are designed in the burner. The middle vent jets natural gas with a diameter of 100mm; the vents on both sides jet air, both with a diameter of 300mm. The air vents and the gas vents can be into a certain angle. Relying on the transverse momentum of the gas, the turbulence is produced so as to make an easier combustion in the furnace.

Results and Analysis

Temperature distribution at different ratios (Methane: Air).

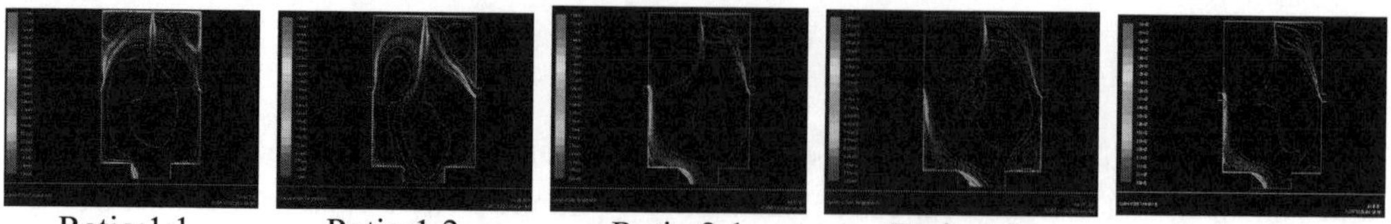

Ratio:1:1 Ratio:1:2 Ratio:2:1 Ratio:3:2 Ratio: 3:1

Fig. 1: Temperature contours at different work conditions

The properties are dependent on temperature in some degree, so the property data which change with temperature are used for calculating and simulation. A section of the furnace was selected as the subject to investigate to observe the temperature and the concentration distribution in the furnace. The temperature conditions in the furnace under different ratios of methane and air are shown as follows. When the ratio of the methane and the air increases, the combustion value increases. However, as the ratio increases to a certain value, the combustion value begins to reduce. This shows the existence of the critical value of the ratio of methane and air which affects the temperature.

Table 1: The distribution of the furnace temperature under different work condition

Work condition	1: 1	1: 2	2: 1	3: 1	3: 2
Maximum furnace temperature[K]	1108.907	1056.644	1642.400	1519.729	1799.538
Average furnace temperature[K]	965.204	826.121	1435.261	1304.542	1605.674
Average exit temperature[K]	580.141	556.448	617.678	625.397	742.896

As is shown in Table 1, the temperature is the maximum with the ratio of methane and air 3:2, and the minimum with the ratio 1:2. This shows that the combustion of gas is not only related with the circumstances, but also with the ratio of fuels.

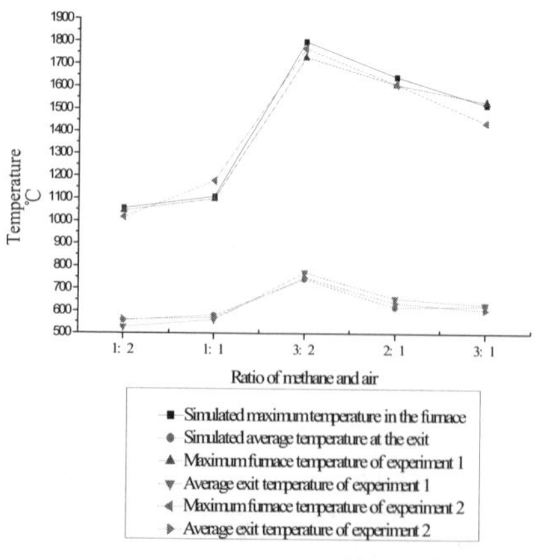

Fig. 2: Furnace temperature distribution under different ratios.

It can be seen from Fig. 2 that to a certain extent, the higher ratio of methane and air, the higher the maximum furnace temperature. However, when the ratio reaches a certain value, the average temperature tends to fall. This apparently declares a major change have taken place. This is to say, the temperature of the furnace is more non-uniform, which is consistent with the actual results of industrial experiments. Visibly, the ratio of natural gas and air is not the bigger the better but only at a value with a more uniform temperature.

Concentration Distribution of Furnace gas. The distribution of the furnace gas concentration is an important basis for determining the burner performance and the thermal characteristics of mill furnace. As the working process of the mill furnace is a process which is almost closed at high temperatures, it is difficult for the sampling and analysis. The normal method is to do the extraction of samples in the flue, to understand if the emission of pollutants and the gas combustion is complete. But this is difficult to identify and seize the combustion situation in the furnace. Therefore, the approach of simulation analysis was adopted to analysis the distribution of the furnace concentrations under various conditions.

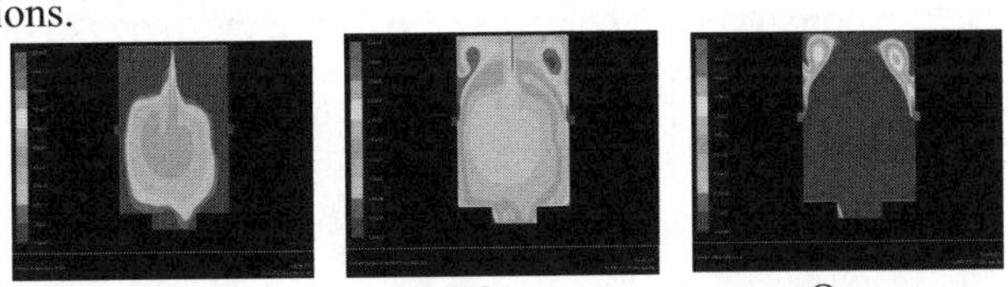

CH_4 CO_2 O_2

Fig. 3: Contours of mass fraction of CH_4, CO_2 and O_2 with the ratio of methane and air 1:1

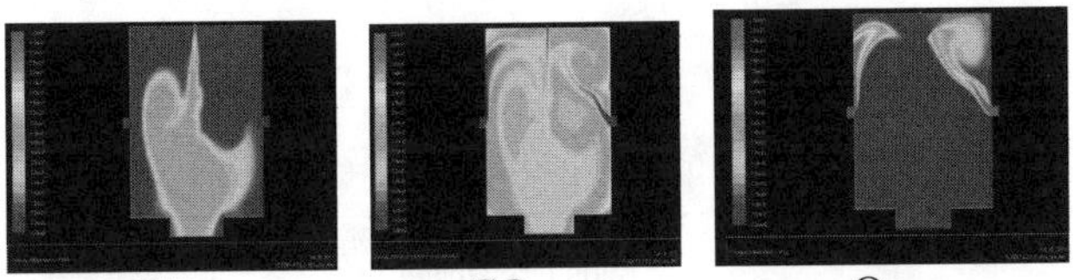

CH_4 CO_2 O_2

Fig. 4: Contours of mass fraction of CH_4, CO_2 and $O_2$with the ratio of methane and air 1:2

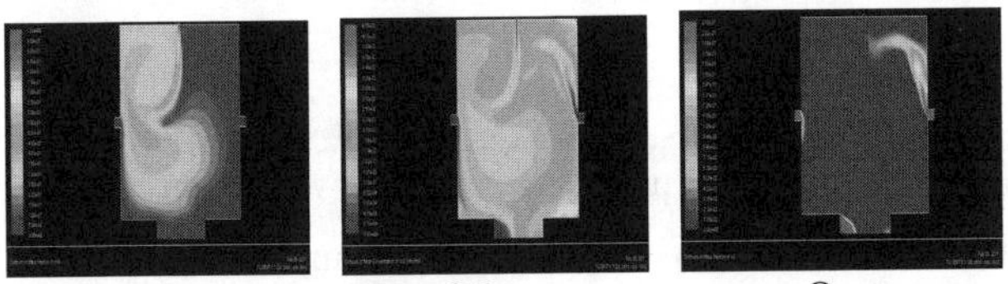

CH_4 CO_2 O_2

Fig. 5: Contours of mass fraction of CH_4, CO_2 and $O_2$with the ratio of methane and air 2:1

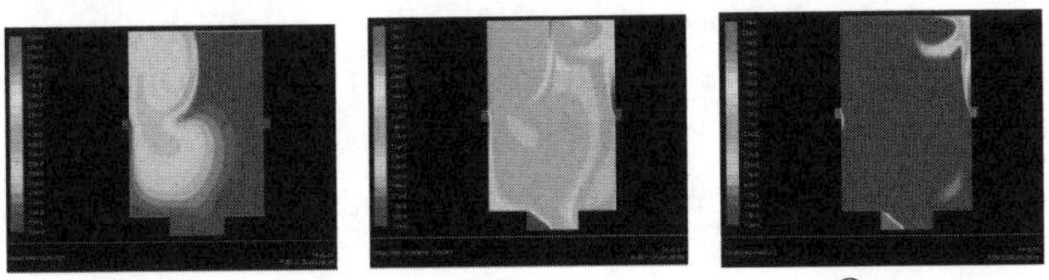

CH_4 CO_2 O_2

Fig. 6: Contours of mass fraction of CH_4, CO_2 and $O_2$with the ratio of methane and air 3:1

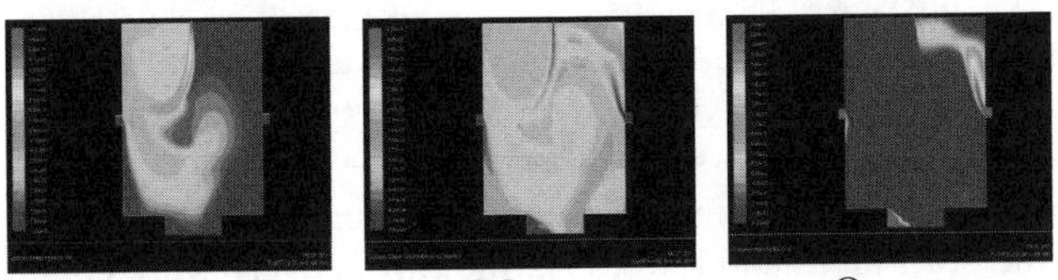

CH_4 CO_2 O_2

Fig. 7: Contours of mass fraction of CH_4, CO_2 and $O_2$with the ratio of methane and air 3:2

The figures of contours of mass fraction 3-7 show that the concentration of CH_4 and O_2 is lower but that of the CO_2 is higher at the exit while the condition of the entrance is on the contrary. Fig. 8 shows that the concentration of CH_4 at the exit tends to fall with a higher flux ratio, however, when the ratio increases to a certain value, the concentration turns to enhance instead. This is contrary to the change of O_2. The certain value at the work conditions of this paper is 3:1. These indicate that the combustion efficiency is different with different flux ratio. Taking a reasonable ratio of natural gas and air in industrial production can lessen the waste of natural gas and can improve the combustion efficiency at the same time. The concentration of O_2 was controlled to make a low oxygen environment so as to reduce the amount of oxidation loss and to avoid secondary combustion and fuel waste.

Table 2: Concentration of each component at the exit [kmol/m3]

Work condition	1: 2	1: 1	3: 2	2: 1	3: 1
Concentration of CH_4	0.31282973	0.15669545	0.1232256	0.00717916	0.10342359
Concentration of O_2	0.00060638	0.00478454	0.0131432	0.02734240	0.01164144
Concentration of CO_2	0.12882622	0.17952135	0.1366381	0.17657034	0.15691781

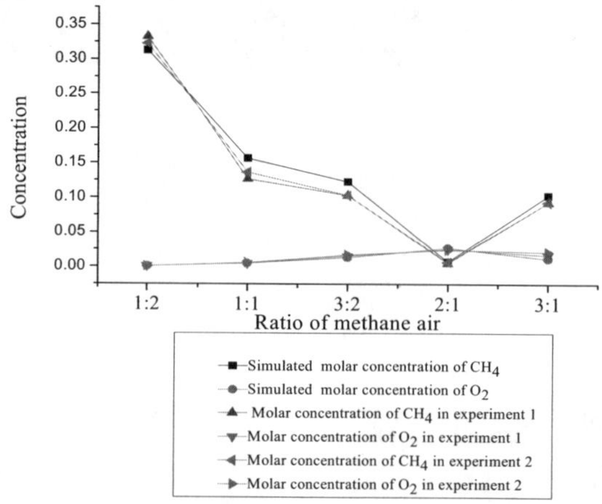

Fig. 8: Changes of mass concentration with different flux ratio

The impact of different ratios of NOx [11]. The main component of natural gas is methane, while combustion products mainly consist of water, carbon dioxide, oxygen, nitrogen and unburned methane. The formation of nitric oxide mainly depends on the oxidation of nitrogen and oxygen at high temperature. As a kind of atmospheric pollutants, it greatly damaged the environment. Therefore, the model of nitrogen oxide of software Fluent was used in this paper to calculate the concentration of nitric oxide in various work conditions.

Table 3: Mass fraction of nitric oxide in the furnace and at the exit

Work condition	1: 1	1: 2	2: 1	3: 1	3: 2
Mass fraction of NO in the furnace	3.460055e-6	5.354095e-10	5.298334e-8	3.950328e-8	7.432367e-09
Mass fraction of NO at the exit	4.205844e-7	6.558421e-11	4.354552e-9	2.881516e-9	8.833599e-10

The contrast between Fig. 3-7 and Fig. 9 shows that the concentration of fuel and oxygen is high at the entrance, low at the other areas and lower near the exit in every work condition. The contours of mass fraction (Fig. 9) and the contours of static temperature (Fig. 1) show that the maximum of NO both appeared at the high temperature region, which proves the formation amount of nitric oxide has a great relationship with temperature; however they are not one-to-one in correspondence. The formation of NO is not just associated with temperature but also depends on the ratio of constituent concentration of nitrogen and oxygen of local air. Certain ratio can get a high temperature and a high reaction rate, but high temperature not always can get a high nitric oxide concentration, and the concentration of nitric oxide also hinges on the oxygen content before the formation of nitric oxide. A reasonable ratio of methane and air can realize high combustion efficiency and low amount of nitric oxide. Work condition 2 has the lowest concentration of nitric oxide, where the flux ratio of methane and air is 2:1 as showed in the Table 3, which indicates the amount of air obviously restrain the formation of nitric oxide.

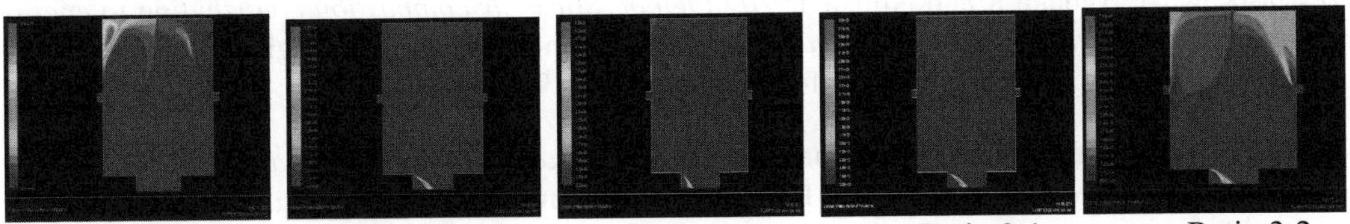

Ratio:1:1 Ratio:1:2 Ratio:2:1 Ratio:3:1 Ratio:3:2

Fig. 9: Contours of mass fraction of NO with different flux ratio of methane and air

Conclusions

The pretreatment software Gambit was adopted to proceed geometric modeling and meshing, and software Fluent was used to calculate mathematical model and fitting the results with images, which all visually show the mass fraction and temperature distribution, realizing result visualization.

I : The flux ratio of methane and air greatly affects the furnace temperature distribution during the combustion, and the temperature decreases at both sides of critical values. Therefore, the ratio of methane and air should be chosen according to the specific work conditions to reach the best fuel efficiency and combustion efficiency.

II : Different ratios of methane and air have a great influence on the combustion products. Therefore, the selection of fuel should follow a principle of high-temperature in correspondence to low-oxygen to get the best fuel efficiency and to reduce the loss of methane.

III : Different ratios of methane and air have different amounts of nitric oxide. The amount of air blasted in the furnace and the temperature of the furnace also impacts the produce of nitric oxide, and these factors are also influenced by the ratio of methane and air. Therefore, the selection of the ratio must pursue the lowest emission of nitric oxide in order to reduce the pollution of the environment during combustion.

References

[1] Zemin Jiang: *Reflections on China's energy issues*. Journal of Shanghai Jiaotong University, 42, 3(2008), pp. 345-359.

[2] Zhizong Gao: *Reaction mechanism of combustion of hydrocarbon*. Journal of Liaoning university (natural sciences edition), 29, 3(2002), pp. 266-271.

[3] Yujiao Liao, Hanpin Huang and An Fu: *Physics*. Higher Education Press, Beijing, (1994).

[4] Zhizong Gao: *Discussion of flame temperature of combustion*. Journal of Liaocheng Teachers University (natural science edition), 14, 1(2001), pp. 57-60.

[5] Information on http://www.me.berkeley.edu/gri_mech/

[6] Coppens, E.H.V, De Ruyek, J. and Konnov, A.A: *The effects of composition on burning velocity and nitric oxide formation in laminar premixed flames of CH4+H2+02+N2*. Combustion and Flame, 149, 4(2007), pp. 409-417.

[7] Konnov, A.A:*Detailed reaction mechanism for small hydrocarbons combustion*. Release 0.5. (2000).

[8] Li, S.C. and Williams, E.A: *NOx formation in two-stage methane-air flames.* Combustion and Flame, 118, 3(1999), pp. 399-414.

[9] Hasegawa T, Tanaka R and Niioka T: *High temperature air combustion contributing to energy saving and pollutant reduction in industrial furnace.* The Federation of Engineering Societies of China Association for Science and Technology, 35, 2(1999), pp.102-114.

[10] Hongjun Zhu, Yuanhua Lin and Longhan Xie: *Practical course of fluid analysis and simulation of software Fluent.* China Press of Posts and Telecommunications, Beijing, (2010).

[11] Yuan J, Naruse I: *Modeling of combustion characteristics and NOx emission in highly preheated and diluted air combustion.* International Journal of Energy Research, 22(1998), pp.1217-1234.

Advanced Materials Research Vol. 279 (2011) pp 247-252
© (2011) Trans Tech Publications, Switzerland
doi:10.4028/www.scientific.net/AMR.279.247

Dynamic Analysis of Angular Position Simulator

YANG Xu-dong[1, a], JIN Liang-liang[1, b]

[1]School of Mechanical Engineering, Guizhou Univ., Guiyang 550003, China

[a]yxd_gzu@163.com, [b]jll2009@foxmail.com

Keywords: angular position simulator, dynamic analysis, explicit method

Abstract. In order to study the dynamic characteristics of angular position simulator, a finite element model was established in ABAQUS environment. After getting the reasonable static analysis results, the results of dynamic analysis where the model suffered impact load were obtained and presented in this paper using explicit method. The maximum displacement of this model in analysis was 0.03484mm. This indicates that the maximum displacement meets the expected requirements. Meanwhile, it means this model is feasible in application and of engineering application value. Also, it provides another idea and an access to study and assess dynamic characteristics in virtual environment before manufacture is carried out.

Introduction

Angular position simulator is mainly used in the field of aviation simulation or in other spatial area. When working, it tracks and receives corresponding signals by changing the angular position of the antenna (though linear motor's movement and servo motors' rotation). The antenna is mounted on rotation arm (see part 1 in Fig. 1). Each moving part is driven by linear motor or servo motors in order to improve the accuracy of this system and reduce response time. In designing three dimensional movement simulator, Zheng Xinglin and Ma Long have studied three dimensional movement devices and their functions [1]. In fact, relative studies are less compared to other subjects.

The designing process of angular position simulator is so important that it provides support for the antenna and rotation arm and ensures motion accuracy. It also has strict requirements in its own weight and size. It is known to us all that reducing weight will affect the stiffness of the simulator but the maximum deformation must be assured still. The simulator suffers impact of movement caused by the motor's accelerations and decelerations. This makes dynamic characteristics bad and position accuracy reduced compared to static analysis. Therefore, use finite element method to study its dynamic responses [2].

The dynamic characteristics with the motor's accelerations imposed on the simulator were studied using finite element software ABAQUS combined with explicit method [3, 4].

Working Principles and Precision Requirements

The rotation arm driven by the servo motor (see part 2 in Fig. 1) mounted on the center of the simulator can rotate along the circular rail (see part 3 in Fig. 1) which is also fixed on the simulator. A set of siemens linear motor is installed on the rotation arm and motor's secondary coils are fixed on motor holder (see part 4 in Fig. 1) where two little servo motors (see part 5 in Fig. 1) are mounted. These two little servo motors are used to make fine adjustment of angular position for the antenna while linear motor and large servo motor is used to make coarse adjustment. The antenna's range of motion is a circular domain with radius 2220mm and this domain covers required range of motion.

When working, the rotation arm driven by large servo motor rotates to the correct position with low uniform speed, and it is locked by electromagnetic brakes (see part 6 in Fig. 1) immediately. Then, linear motor moves to suitable location. Finally, two little servo motors rotate respectively to help the antenna receive signals. The system's positioning accuracy should be better than 0.1mm while motion accuracy better than 0.1mm/s. The linear motor's velocity ranges from 1mm/s to 8m/s and it should not jitter in uniform motion. Because of foundation can only withstand 350Kg weight per square meter, steel profiles can be used after weight calculation.

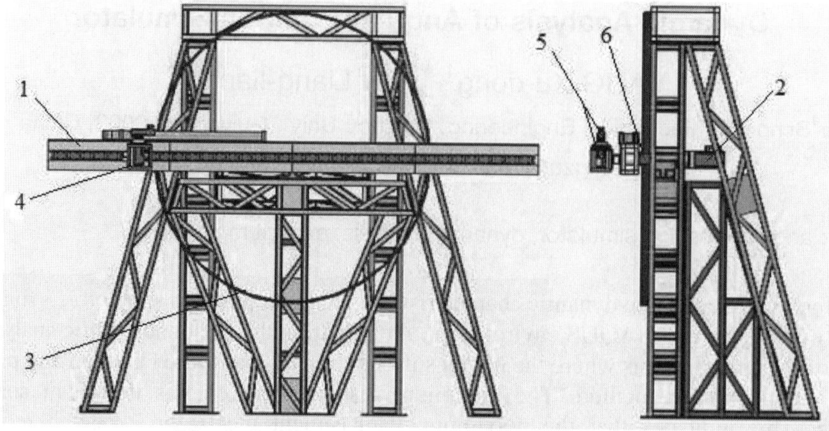

Fig. 1: Structure of angular position simulator

Analysis Process

Analysis Method. In analysis, structure weight is considered after the density of material was defined. The gravity of linear motor is simplified and loaded on the beam in the form of distribution through coupling. According to working principles, the force caused by arm's rotation may not be considered. The worst case is when the rotation arm moves to horizontal position and locked by electromagnetic brakes, and then the linear motor starts to move with maximum acceleration. The simulator suffers transverse impact load in that time.

The analysis process is shown in Fig. 2. After the static analysis results meet the requirements, maximum circular frequency is extracted in order to define time increment and dynamic analysis is carried out finally. The dashed box in Fig. 2 means last 2 analyses are completed in 2 continuous steps.

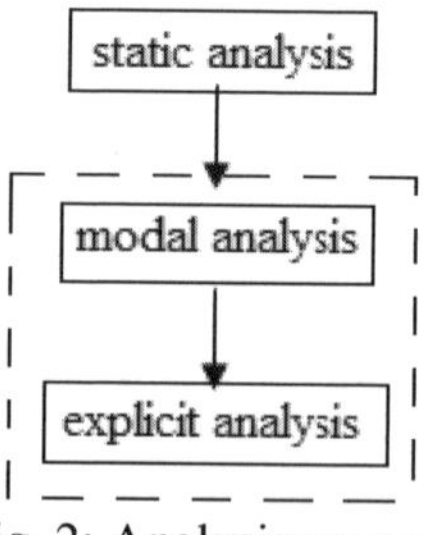

Fig. 2: Analysis process

In ABAQUS environment, the model's material property is given; and an assembly is defined. And coupling effect between the arm and the main part was set up. The interaction between the point of applied load and the relative part was also defined. Motor's flat secondary coils were regarded as uniform load. The finite element model was obtained after it was meshed. Fig. 3 shows the finite element model. The main body and the rotation arm use B32 elements. Thus, high precision results can be obtained. According to static analysis results, the maximum deformation is about 0.0251mm; and it met preset requirement (Fig. 4).

The maximum linear motor-driven acceleration with load is 10m/s^2. An acceleration chart of linear motor driven by siemens 840D ranging from 0 to maximum is shown in Fig. 5.

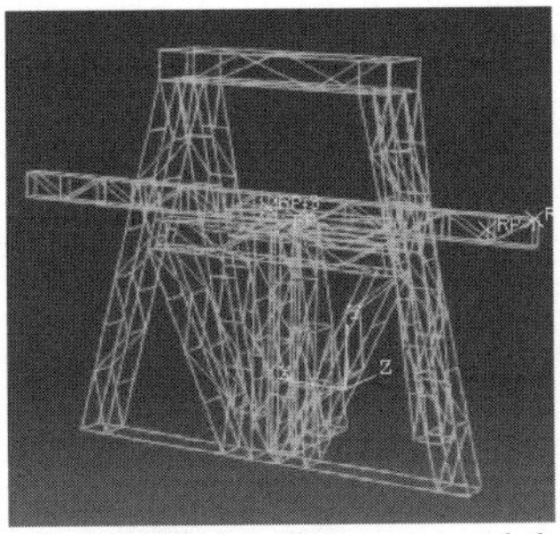

Fig. 3: Finite element model

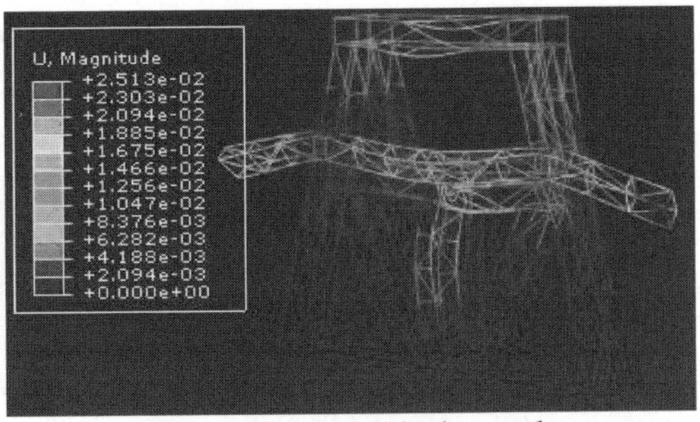

Fig. 4: Static analysis result

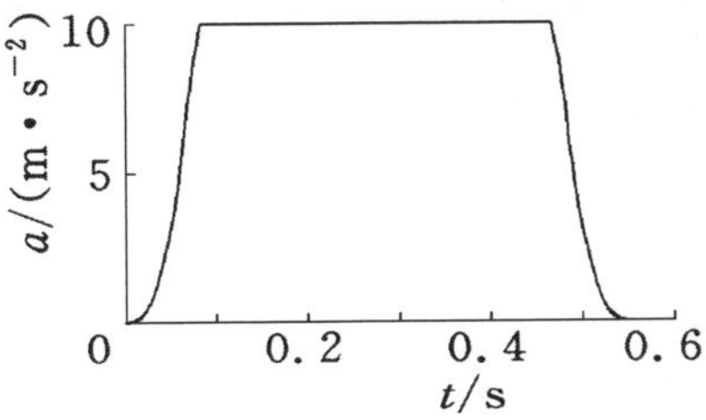

Fig. 5: Acceleration-time curve of linear motor

Apparently, transverse impact load caused by linear motor's acceleration would have influence on dynamic characteristics. According to Fig. 5, at first, the acceleration ranges from 0 to 10m/s^2 within 0.1s; thus, impact loading appears. Form 0.1s to 0.4s, acceleration remains 10m/s^2 and it reduces to 0 after 0.55s. This indicates that after 0.1s, the load can be treated as static load. At the time 0 to 0.1s, it is considered impact load.

In the time of t+Δt, dynamic equilibrium equation is known as:

$$M\ddot{u}_{t+\Delta t}+C\dot{u}_{t+\Delta t}+Ku_{t+\Delta t}=P_{t+\Delta t} \quad (1)$$

In Eq. 1, M is mass matrix; C is damping matrix; K is stiffness matrix; P is external incentives. u denotes nodal displacement, nodal velocity, and $\ddot{u}$ nodal acceleration. The displacement and velocity in the time of t+Δt are obtained using newmark integration method and they can be stated as:

$$u_{t+\Delta t}=u_t+\Delta t\dot{u}_t+\Delta t^2[(0.5-\beta)\ddot{u}_t^2+\beta\ddot{u}_{t+\Delta t}] \quad (2)$$

$$\dot{u}_{t+\Delta t}=\dot{u}_t+\Delta t[(1-\gamma)\ddot{u}_t+\gamma\ddot{u}_{t+\Delta t}] \quad (3)$$

Through this method, differential equation is reformulated. The displacement and velocity are obtained by explicit method. Using central difference integration, they can be expressed as

$$\dot{u}^{(i+\frac{1}{2})}=\dot{u}^{(i-\frac{1}{2})}+\frac{\Delta t^{(i+1)}+\Delta t^{(i)}}{2}\ddot{u}^{(i)} \quad (4)$$

$$u^{(i+1)}=u^{(i)}+\Delta t^{(i+1)}\dot{u}^{(i+\frac{1}{2})} \quad (5)$$

Where, i is increment number. Because $\dot{u}^{(i-\frac{1}{2})}$ and $\ddot{u}^{(i)}$ are known, so it is called explicit method. Similarly, the following formula is established:

$$C\dot{u}_t+Ku_t\approx C\dot{u}_{t-\frac{\Delta t}{2}}+Ku_{t-\frac{\Delta t}{2}} \quad (6)$$

If not, explicit method can not be used because there has unknown value. In addition, time increment with damping is defined by the following formula [5]:

$$\Delta t \le \frac{2}{\omega_{max}}(\sqrt{1+\xi^2}-\xi) \tag{7}$$

In formula, ξ is the fraction of critical damping in the highest mode, and ω_{max} is the maximum circular frequency which is extracted from previous step.

Analysis Results. Fig. 6 shows the deformation of rotation arm suffering impact loading in several time increments. Fig. 6 (a) shows the deformation in the first time increment; Fig. 6 (b) shows the deformation in the fifth time increment; Fig. 6 (c) shows the maximum deformation.

From Fig. 6, the maximum deformation is 0.03484mm, which meets the requirement. Compared with static analysis results, it indicates that the dynamic analysis result is 1.39 times as large as static analysis results. Fig. 7 shows the displacement of a point in one end of the arm where deformation came to the maximum alone three coordinate directions. From Fig. 7, when the step comes to the end, the deformation reduces progressively in all three coordinate directions. This means the solution is convergent and the simulator especially the rotation arm finally came to balance. Note that in Fig. 7(a-c), the displacements do not come to 0 finally, because these final displacements in three dimensions are caused by static load. According to final static displacements in Fig. 7, the analysis results are consistent with the actual situation.

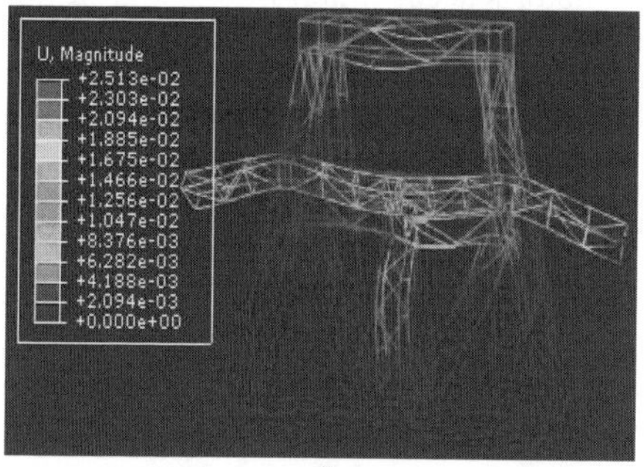

a)

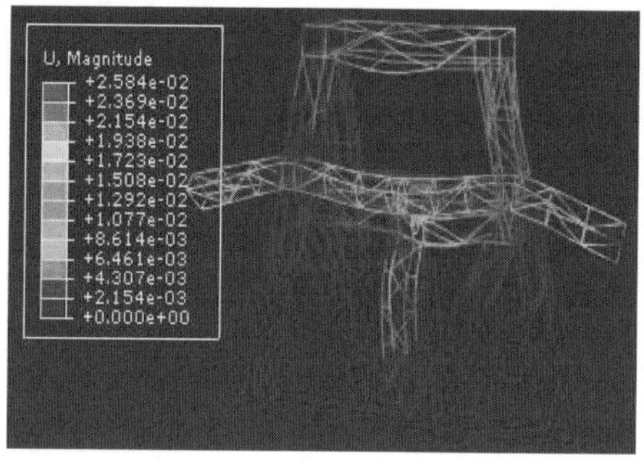

b)

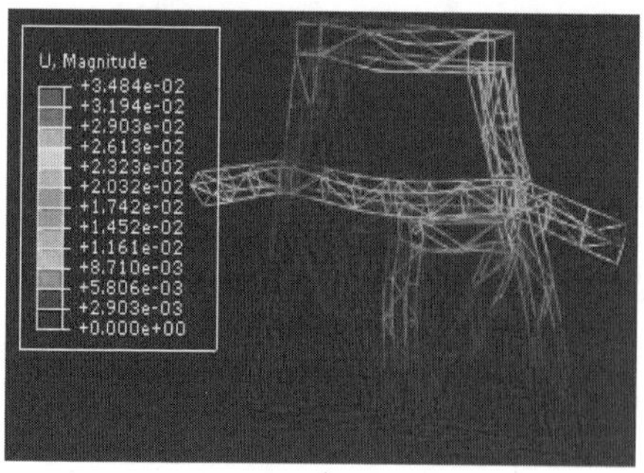

c)

Fig. 6: Deformation of the simulator

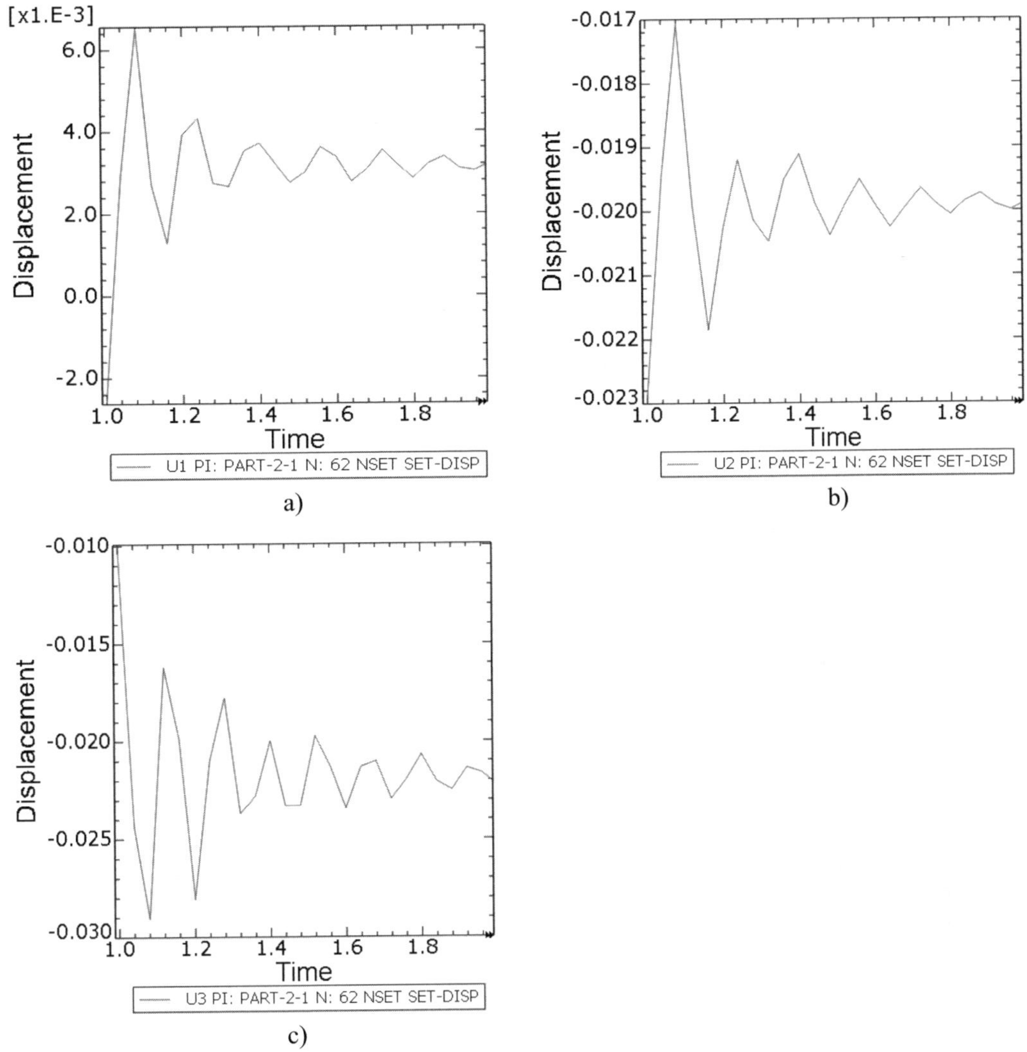

Fig. 7: Displacement of a point in one end of the beam in three dimensions

Conclusion

The dynamic responses of the simulator with the motors' acceleration were analyzed using ABAQUS. The deformations of all time increments were obtained by explicit method. Compared the results between static and dynamic analysis, larger impact loading makes the dynamic characteristics worse but the maximum deformation still meets the requirement. This means the model is feasible in application. It provides an idea to study and assess dynamic characteristics in virtual environment before manufacture is carried out.

Acknowledgments

The authors are grateful to Foundation of Guizhou Science and Technology Department (No. [2008]2210 and No. [2009]3009) and Foundation of Guizhou University (No. [2007]25) for funding this research.

References

[1] Zheng Xinglin, Ma Long: *Design and application of three-dimensional movement simulator under space environment*. Spacecraft Environment Engineering, 26, 5(2009), pp. 442-446, (in Chinese).

[2] GUO Zeying, LI Qingning: *An explicit integral method and the stability of dynamic response analysis*. Journal of Earthquake Engineering and Engineering Vibration, 28, 2(2008), pp. 15-19, (in Chinese).

[3] LIU Heng, LIAO Zhenpeng: *An explicit time integration method for structural dynamic equations*. Journal of Earthquake Engineering and Engineering Vibration, 29, 1(2009), pp. 32-42, (in Chinese).

[4] Zhuang Zhuo: *ABAQUS Nonlinear Finite Element Analysis and Examples*. Science Press, Beijing, (2005).

[5] Dassault Systemes Simulia Corp.: *ABAQUS Theory Manual* (version6.8), (2008).

Advanced Materials Research Vol. 279 (2011) pp 253-257
© (2011) Trans Tech Publications, Switzerland
doi:10.4028/www.scientific.net/AMR.279.253

Acoustic Performance Analysis of Impedance Compound Muffler with Aluminum Foam

Bao Huaiqian[1, a], Wang Changtian[1, b] and LI Jun[1, c]

[1]College of Mechanical and Electronic Engineering
Shandong University of Science and Technology Qingdao, Shandong China, 266510

[a]bhqian@163.com.com, [b]w6255@tom, [c]Li Jun87111@163.com

Keywords: muffler, aluminum foam, transmission loss, acoustic performance

Abstract. The common reactive muffler has poor acoustic properties in high frequency, so aluminum foam is applied to the reactive muffler and the impendence compound muffler is designed. The acoustic performance of the muffler is analyzed; the interior sound field of the muffler is modeled and meshed in the ANSYS, then they are imported into SYSNOISE, and imposed reasonable boundary conditions to carry out the analysis of acoustic performance of the muffler. The results show that, compared to reactive muffler, the muffler with aluminum foam has a higher amount of noise reduction and a wider frequency band in middle and high frequency.

Introduction

Aluminum foam is a kind of new materials developed in recent years. As light weighted materials it has high strength characteristic; as functional material, it has porous, vibration, damping, absorption and other physical properties[1,2]. It is widely used in the general industrial and high technology filed. This thesis designs the new structure of impedance compound muffler with the use of the aluminum foam and carries out the sound filed simulation and studies the acoustic performance. It establishes the foundation for the engineering applications of aluminum foam material.

Acoustic model of the muffler

Muffler's model design. The designed muffler has two cavities which were separated by the elliptical cross section, and were wrapped by the rigid aluminum foa. The length of the tube was determined [3] as follow, and the inlet of the tube has the half length of the expansion cavity. The outlet tube takes 1/4 length of the expansion cavity. The model of the aluminum foam muffler is shown in Fig. 1.The left was inlet, and the right is outlet. The inlet, outlet and middle tide have the same diameter of 30mm, and the thickness of aluminum foam is 10mm. The diameter of long axle of internal cavity's elliptical cross section is 180mm, and the short cavity is 140mm, and the single expansion chamber is 200mm. The thickness of intermediate partition is 10mm. The inlet, outlet and middle tube get the same material of aluminum foam, the outer sphere of expansion cavity is aluminum foam with rigged backing added to it, and the others are steel plate.

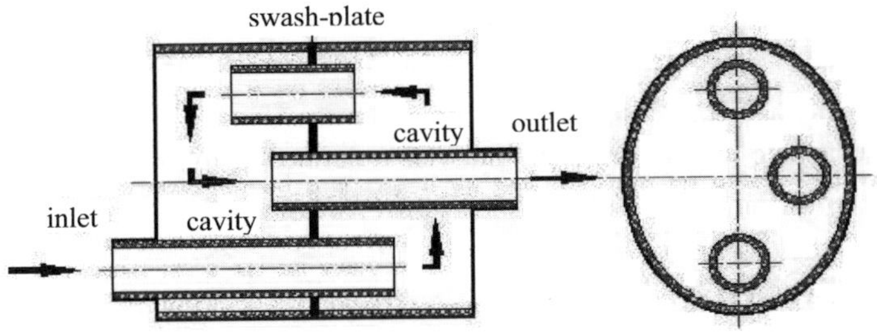

Fig. 1: The schematic diagram of muffler's structure

Modeling the muffler's interior sound field. The muffler's interior sound field is modeled in ANSYS, and the frequency cave rage for calculation is 20-3200 HZ. One wavelength contains at least six unit at the max frequency[4]. Sound field model's grid size is determined by the following formula,

$$L_{\max} \leq \frac{1}{6}\frac{c}{f_{\max}} \tag{1}$$

In the formula, c depends the speed of sound , m/s ; $f_{\max}$ is the max calculation frequency, HZ.

The max size of grid is 18.8mm according to Eq. 1. This thesis chooses 15mm, and uses eight-node hexahedral element. This kind of element can not only get high accuracy but also reduce the number of units. The size of model is 84964 units, 16941 nodes.

Numerical analysis of acoustic field model

Foam aluminum parameters. The selected aluminum foam material parameters are as follows: structure factor is 3, porosity is 0.8, the aperture of the aluminum foam is 1mm, and frequency cave rage is 20-3200Hz.

Sound Field Analysis. The acoustic properties of sound-absorbing material are often reflected through the acoustic impedance of the material. We solve the acoustic impedance by the setting parameters, then obtain admittance table and exert impedance boundary condition to solve it, and analyze acoustic performance of impedance compound muffler with aluminum foam. We transform aluminum foam to the steel material for comparing it with the traditional reactive muffler without aluminum foam. The compound muffler's detailed boundary conditions are set as follows. The velocity boundary conditions at the selection set of sound field's entrance are defined. Real part is -1, and the imaginary part is 0; the boundary conditions of admittance is set at the export's selection; the relevant boundary conditions is defined at the surface contacted with aluminum foam; the rigid wall takes the default conditions.

The radiation specific acoustic impedance of muffler's tail tube can calculate the muffler's outlet boundary condition[5]. The radiation specific acoustic impedance can be calculated by approximation analog line,

While $ka \leq 0.5$,

$$Z_{out} = \rho_0 c_0 \left[0.25(ka)^2 + j \cdot 0.6ka\right] \tag{2}$$

While $0.5 < ka \leq 5$,

$$Z_{out} = \pi \rho_0 c_0 \frac{5.7 + j(9.9ka - 11.4)}{86.7 + j\left(31.2ka - 87.9/ka\right)} \tag{3}$$

While $ka > 5$,

$$Z_{out} = \rho_0 c_0 \tag{4}$$

The acoustic impedance of aluminum foam material is defined as follow,

$$Z_{foam} = \rho c \tag{5}$$

The acoustic impedance of foam aluminum after exerting rigid backing:

$$Z_{bc} = -j\rho c \cot kl \tag{6}$$

Where: ρ_0 is air density; c_0 is propagation speed of sound wave in the air; k is the number of the wave; a is the radius of the sound field's exit; ρ is the effective air density inside the foam; c is the Propagation speed of sound wave in the foam aluminum; l is the thickness of foam aluminium material; j represents imginary part, $j=\sqrt{-1}$.

The results analysis

Analysis of sound pressure distribution. Taking the radiation resistance of exerting tail pipe at the end of the type for example, we can get the sound pressure distribution inside of the muffler at the different frequencies. Fig. 2 and Fig. 3 are the sound pressure distribution of impendance compound muffler's with aluminium foam at frequencies of 500Hz and 2500Hz respectively, and the color represents the value of sound pressure.

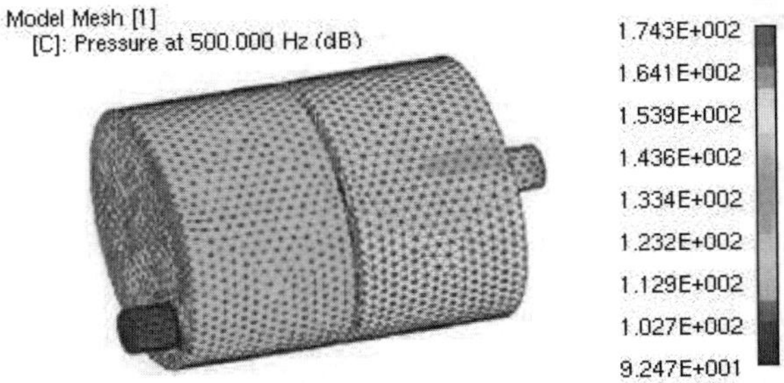

Fig. 2: Contours of sound pressure at the frequency of 500Hz

It can be seen from Fig. 2 that sound wave enters it from the left inlet tube, and then spread through the intermediate pipe expansion chamber and outlet pipe expansion chamber. The contours of the sound pressure inside the muffler at the low frequency correspond with one-dimensional plane wave theory.

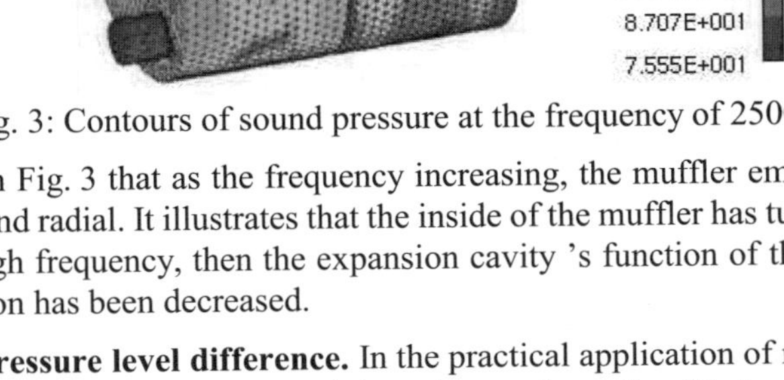

Fig. 3: Contours of sound pressure at the frequency of 2500Hz

It can be seen from Fig. 3 that as the frequency increasing, the muffler emerges high mode wave along with the axial and radial. It illustrates that the inside of the muffler has turned up radial pressure fluctuations at the high frequency, then the expansion cavity 's function of the muffler has dropped and the noise reduction has been decreased.

Analysis of sound pressure level difference. In the practical application of muffler, we often apply sound pressure level between the entrance and the exit to evaluate the effects of muffler, which can be shown as below,

$$NR = 20\lg\left|\frac{p_1}{p_2}\right| \tag{7}$$

Where, p_1 and p_2 are the muffler's inlet and outlet sound pressure respectively, Pa.

In the software SYSNOISE we read the middle node's sound pressure of inlet and outlet pipe respectively, and draw sound pressure level difference curve. It can be seen in Fig. 4 that curve 1 shows the pure reactive muffler's sound pressure level difference, and curve 2 shows the sound pressure level difference of impedance compound muffler with aluminum foam.

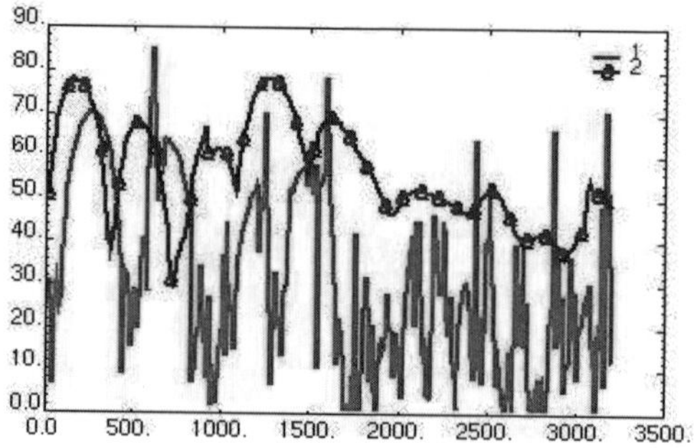

Fig. 4: The curve Sound pressure level difference

The exerting aluminum foam material can be seen from Fig. 4. Sound pressure level difference has a significant increase, and noise elimination band becomes wider. The max noise reduction can be got 76dB and 81 dB; while the frequency exceed 1500Hz, the noise reduction is weaken, but it was still more stranger than the pure reactive muffler without using aluminum foam.

Analysis of transmission loss. Transmission loss is often used to assess siliencing effect of a singal silencer element or noise reduction performance, which can be shown as follows[6],

$$TL = 20\lg\left|\frac{p_1 + \rho_0 c_0 v}{2p_2}\right| \tag{8}$$

In the formula, v represents the particle velocity at the entrance, m/s .

When using transmission loss to evaluate the muffler's noise elimination performance, because the impedance of the outlet pipe is not constant, the outlet pipe will generate reflection. In other words, the sound pressure of the outlet pipe is actually the composition of transmission pressure and reflection pressure of the outlet. Thus the transmission sound power at this place must be a total absorption outlet without reflection. In this case there is no distribution of reflection pressure, and we exert air characteristics impedance at the outlet of the model.

To calculate the muffler's transmission loss, we need to read the middle node's pressure of inlet and outlet respectively, calculate it following Eq. 8, and obtain the relationship between the transmission loss and frequency as Fig. 5. From the figure we solve the transmission loss curve before and after exerting aluminum foam.

The curve1 represents the transmission loss of pure reactive muffler before exerting aluminum foam, while the curve 2 represents the transmission loss of impedance compound muffler with aluminum foam.

From this figure we can see that, the muffler's noise reduction increases apparently and the noise elimination band becomes wider after exerting aluminum foam material. The transmission loss could get 48dB at 230Hz, while the transmission loss could get 72dB at 1270Hz. At the muffler's outlet (Radiation resistance of the tail pipe) we can draw that the acoustic properties have a significant effect to the muffler's noise effect, and it is particularly evident to low frequency noise reduction performance.

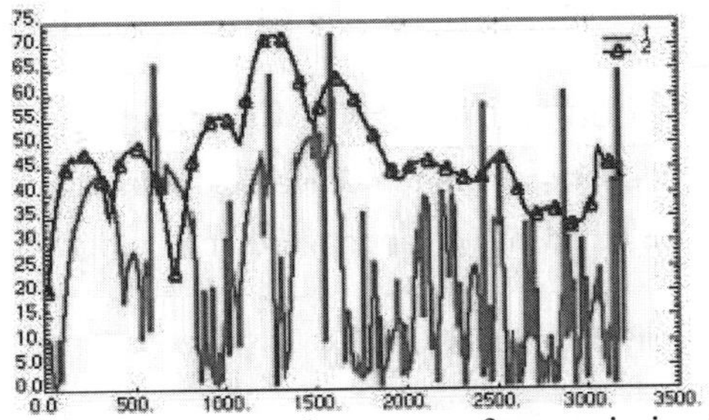

Fig. 5: The constractive curve of transmission loss

Conclusions

By analyzing acoustic performance of impedance compound muffler with aluminum foam, we can draw the conclusions.

The reactive muffler without using aluminum foam has a good noise reduction at low frequency, but appears the apparent axial resonance at high frequency. The noise elimination band becomes the narrow and turns up peak obviously, and the noise elimination becomes not ideal.

Comparing with pure reactive muffler, impedance compound muffler with aluminum foam has higher noise reduction and wider frequency band, which indicates that aluminum foam material has better acoustic performance.

Both mufflers have good acoustic performance at low frequency, and the noise reduction decrease at high frequency, which describes that acoustic performance is influenced by the muffler's structure, the length of expansion chamber and the insertion tube.

In practical application, the muffler's acoustic performance will be influenced by airflow. In this thesis we haven't taken the impact of the airflow into consideration. The future study will still need experiment test to verify the accuracy of simulation results.

Acknowledgment

This work was financially supported by the China National Coal Association Science Foundation (MTKJ-2010-342), the support of a research grant from the scientific and technological development plan project of Qingdao Municipal (KZJ-50).

References

[1] Michael F. Ashby, LU Tianjian: *Metal foams:a survey*. Seienee in China(B), 46, 6(2003), pp. 521-532.

[2] F. D. DENIA, J. ALBELDA, F. J. FUENMAYOR and A. J. TORREGROSA: *Acoustic behaviour of elliptical chamber mufflers*. Journal of Sound and Vibration, 241, 3(2001), pp. 401-421.

[3] Fan Qianwang: *Acoustic Performance Simulation Analysis and Experiment Study for Exhaust Muffler of Four-Cylinder Diesel*. Kunming University of Science and Technology, Kunming, (2007), pp. 89-102.

[4] Li Zenggang: *SYSNOISE Rev5.6*. Nation Defense Industry Press, Beijing, (2005), pp. 17-56.

[5] H.F: *Calculation of Acoustic Characteristics of Automobile Exhaust Muffler and its Optimum Design*. Hefei University of Technology, Hefei, (2004), pp. 18-34.

[6] M.L.Munjal: *Acoustics of Ducts and Mufflers*. Wiley-Interscience, New York, (1987), pp. 30-67.

Advanced Materials Research Vol. 279 (2011) pp 258-261
© (2011) Trans Tech Publications, Switzerland
doi:10.4028/www.scientific.net/AMR.279.258

An Alkaline SiO_2 Slurry for Fine Atomizing CMP

Jing Zhai[a], Zifeng Ni[b], Qingzhong Li[c]
School of Mechanical Engineering, Jiangnan University, Wuxi Jiangsu, 214122, PR China
[a]zhaijing1234@163.com, [b]nizf@jiangnan.edu.cn, [c]qingzhongli@163.com

Keywords: fine atomization, polishing slurry, CMP

Abstract. A kind of slurry which is applicable for fine atomizing CMP was made and the optimal results were obtained through orthogonal experiments by comparing fine atomizing CMP and traditional CMP. The research results show that the material removal rate of fine atomizing CMP is 52.23% of traditional CMP, and the dosage of the slurry used in fine atomizing CMP only accounts for 10 vol% compared to traditional CMP. The surface roughness after the fine atomizing CMP is 2.5nm which is better than that of the traditional CMP (3.0nm).

Introduction

At present, with the rapid development of Communication and Network, there is an increasing demand for Integrated Circuit (IC). Scaling-down of the chip feature size and increasing of the wafer size promote the IC industry to develop rapidly. And the chip feature size of Ultra Large Scale Integration (ULSI) has reached the deep sub-micron level. Semiconductor Industry Association (SIA) predicts that the feature line width will reach 0.05μm in 2012, which demands the flatness of chip reach the nanometer level. Chemical mechanical polishing (CMP) is recognized as the only global flattening practical technology [1-8].

The equipment and consumables for CMP technology include CMP equipment, polishing slurry, polishing pad, cleaning equipment, detection equipment, etc [9]. The flow velocity of polishing slurry has important influence on the polishing rate and polishing quality. Too low flow velocity will increase friction, make the temperature uneven distribution and reduce the wafer surface roughness. High flow velocity can ensure the consistency of the wafer surface[10]. However, the polishing slurry utilization only reached 20%, and the treatment of the massive slurry is quite troublesome, which will cause environmental pollution after the emission to nature. In order to improve the polishing utilization, we proposed fine atomizing chemical mechanical polishing technology. The special polishing slurry is atomized to uniform micron-grade droplet whose Sauter Mean Diameter (DSM) is 5~15μm. Then the polishing fog is imported to the polishing pad because of the pressure difference.

Experiment

The preparation of the polishing slurry was as follows: Add DI water to White carbon black first, and then mix them with magnetic blender. Meanwhile, add suitable amount of organic alkali and surfactant into the mixture, and then add proper Silicon sol and oxidant. At last, we can get good dispersion polishing slurry which do not form the stratification or precipitation after quietly being placed for 48 hours.

Experiment instrument: Millipore water purification systems; UNIPOL-1502 Polishing Machine; Mettler Toledo XS205DU precision electronic balance; IKA Magnetic blender; CSPM5000 scanning probe microscopy.

We compared the material removal rates of the traditional CMP and the fine atomizing CMP. The necessary conditions for polishing were as follows: The pressure of polishing is 7PSI; the speed of polishing pad is 55r/min; the time of polishing is 5min; the flow velocity of polishing slurry in the traditional CMP is 100ml/min and in the fine atomizing CMP is 10ml/min; the materials are monocrystalline silicon chips (20mm×20mm).

Experimental results and analysis

Table 1 indicates the material removal rates of different conditions through the traditional CMP and fine atomizing CMP.

Table 1: Test conditions and results

level	SiO_2 [wt%] A	Oxidant [wt%] B	organic alkali [pH] C	Surfactant [wt%] D	The MRR of the traditional CMP [nm/min]	The MRR of the fine atomizing CMP [nm/min]
1	28	1	9.5	1.5	150.86	68.84
2	28	2	10.5	2.5	188.84	72.96
3	28	3	11.5	3.5	215.64	95.67
4	30	1	10.5	3.5	171.67	84.12
5	30	2	11.5	1.5	225.21	104.94
6	30	3	9.5	2.5	170.60	83.64
7	32	1	11.5	2.5	240.28	118.03
8	32	2	9.5	3.5	194.64	91.74
9	32	3	10.5	1.5	216.74	113.20

From the experimental conditions, it is known that the material removal rate is 52.23% of traditional CMP's. And the velocity of the fog is 10ml/min in the fine atomizing CMP which only accounts for 10% of that in the traditional CMP. Table 2 indicates the analysis aiming at material removal rate after fine atomizing polishing. And $\overline{y}$ shows the mean value of n times test results; *Kij* says the sum of the first *i* (*i*=1,2,3) level in the first *j* column; ω_{ij} says the effect of the first *i* (*i*=1,2,3) level in the first *j* column, the following type said: $\omega_{ij} = \frac{K_{ij}}{n} - \overline{y}$, n=the times of the first *j* listed first *i* level; R_j says the range. $R_j = (\omega_{ij})_{\max} - (\omega_{ij})_{\min}$.

Table 2: Analysis and calculation that aiming at the material removal rate after fine atomizing polishing

Factor	SiO_2 A	Oxidant B	organic alkali C	Surfactant D	
K_{1j}	237.47	265.84	244.22	286.98	$\overline{y} = \frac{1}{n}\sum_{i=1}^{n} y_i$
K_{2j}	272.70	269.64	270.28	269.48	=92.00
K_{3j}	317.82	292.51	313.49	271.53	
ω_{1j}	-12.84	-3.39	-10.59	3.66	
ω_{2j}	-1.10	-2.12	-1.91	-2.17	
ω_{3j}	13.94	5.51	12.50	-1.49	
R_j	26.78	8.90	23.09	5.83	

We can get the following information from the ω_{ij}:

(a)The effect is increased with the increase of A, B and C, respectively. But the effect of factor D first decreased and then increased. The max appears when the content of surfactants is 1.5%.

(b)The size of R_j can judge the primary factors and secondary factors to indexes. Aiming to the effect of material removal rate, the factor A plays an ultimate role and factor D does the minimum role.

The wafer surface roughness is 9.51nm before polishing. The surface roughness is 2.5nm after the fine atomizing CMP. And in the same conditions, the surface roughness is 3.0nm after the traditional CMP. Fig. 1 shows the surface morphology of the monocrystalline silicon chips after the traditional polishing. Fig. 2 shows the surface morphology of the monocrystalline silicon chips after the fine atomizing polishing.

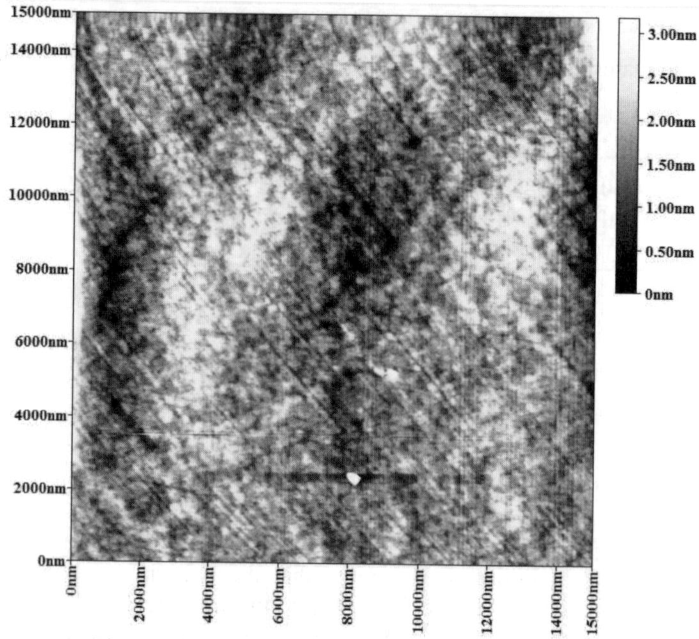

Fig. 1: The surface after the traditional polishing

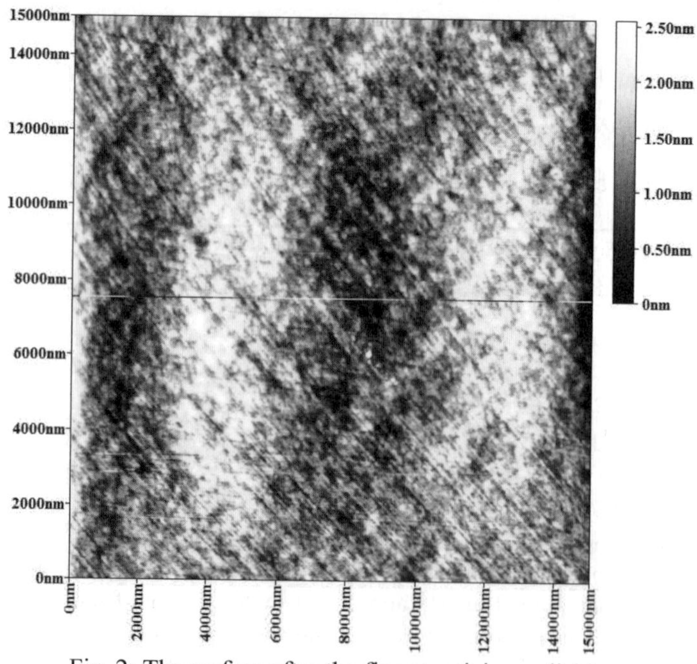

Fig. 2: The surface after the fine atomizing polishing

The surface quality of the wafers through the fine atomizing CMP is better than the one through the traditional CMP and the reasons may have two points, as follows:

(a) In the fine atomizing CMP, the special slurry is atomized to the micron-grade liquid grains which cluster less abrasive particles than the traditional slurry. So the scratch of the polishing surface caused by a lot abrasive particles clustered can be avoided;

(b) At the same time, in the fine atomizing CMP, the rapid absorption function of the micron-grade liquid grains in polishing interface and highly active chemical reactions lead the speed and strength of the chemical reactions to far outweigh the traditional CMP of importing the slurry in the fluid. These alleviate the contradictions of surface roughness and the material removal rate to a certain extent.

Conclusion

The material removal rate through fine atomizing CMP is 52.23% of that through traditional CMP, and the dosage of the slurry used in fine atomizing CMP only accounts for 10 vol% compared to traditional CMP. The surface roughness after the fine atomizing CMP is 2.5nm which is better than that of the traditional CMP (3.0nm). The components ratio of the best material removal rate (118.03nm/min) is the following: SiO_2 accounts for 32 wt %, PH is 11.5, surfactant accounts for 2.5 wt % and oxidant accounts for 1 wt %.

Acknowledgements

This research work was provided by the National Natural Science Foundation of Jiangsu Province (BK20080605) and China Postdoctoral Science Foundation funded project (No.20060390984).

References

[1] P.B. Zantye, A.Kumar and A.K. Sikder: *Chemical mechanical planarization for microelectronics applications*. Mater. Sci. Forum Vol. 45 (1992), p. 89.

[2] K.L. Zhang and Z.T.Song: *Research and Prospects of Chemical Mechanical Polishing for ULSI*. Microelectronics Vol. 35 (2005), p. 226.

[3] A F. Michael: *The early days of CMP*. Solid State Technol Vol. 40 (1997), p. 81.

[4] T. Minoru and Oxide: *CMP mechanisms*. Solid State Technol Vol. 40(1997), p. 1701.

[5] T.Ohmi, et al: *Tatal room temperature wet cleaning for Si substrate surface*. Electrochem Soc Vol. 43(1996), p.749 .

[6] Y.L.Liu, B.M.Tan and K.L.Zhang: *Substrate material properties and processing test technology project in ULSI*. Metallurgical Industry Press (2002).

[7] J.X.Su, R.I.Kang and D.M.Guo: *Technology analysis of wafer chemical mechanical polishing in the manufacture of ULSI*. Semiconductor technology Vol.28(2003), p.18.

[8] G.J.Pietsch, Y.J.Chabal and G.S.Higashi: *The atomic-scale removal mechanism during chemo-mechanical polishing of Si(100) and Si(111)*. Surface Science Vol.331-333(1995), p.395.

[9] L.M.Cao, Y.H.Hu, et al.: *Chemical-Mechanical Polishing Technique in the Semiconductor Industry*. Hubei chemical Vol.4(2000), p.7.

[10] G.Tan, J.L.Wu: *Chemic-mechanical Polishing of Silicon Wafer*. Chinese Journal of Science Instrument Vol. 26(2005), p.104.

Advanced Materials Research Vol. 279 (2011) pp 262-265
© (2011) Trans Tech Publications, Switzerland
doi:10.4028/www.scientific.net/AMR.279.262

Synthesis and Simplification of Fractal Profile

Chao Zhou[1, a], Chenghui Gao[1,b], Jie Chen[1,c] and Lianfeng Lai[1,d]

[1]School of Mechanical Engineering and Automation, Fuzhou University, Fujian Fuzhou 350108, China

[a]zandz1978@126.com, [b]gch@fzu.edu.cn, [c]chen830608@163.com, [d]llf2531@yahoo.com.cn

Keywords: Fractal geometry, fractional Brownian motions, Discrete Fourier transform, Chord deviation algorithm

Abstract. In order to import a synthesized fractal profile into finite element software, the profile synthesized by discrete Fourier transform was studied. The synthesized profile was filtered in frequency domain in order to filter its high frequency components and to make it smooth, and then the chord deviation algorithm was used to reduce its redundant data in space domain. It was found that: after filtering, the profile is smooth but with lots of redundant data; the chord deviation algorithm can simplify the profile which is redundant in space domain; the time needed in the process of importing a profile into finite element software can be reduced greatly after profile simplification.

Introduction

With the miniaturization of mechanical parts, their surface geometries which influence a machine's engineering performance become very remarkable. Real surfaces are often called rough surfaces in tribology, which are filled with irregular micro asperities. Tribological phenomena such as friction and wear are, in essence, the interaction between these asperities. Tribological phenomena occurs in contacting surfaces, and they are quickly happened, hard to be observed and tested, so the computer simulation methods, mostly the finite element method(FEM), are often extensively used[1,2]. Reasonable surface geometry model ensure a reliable simulation result. Only the three dimensional surface model contains the whole information, but because of the complexity of the surface geometries, if the three dimensional model is used in FEM, the computing time will be very long and the simulation will be hard to converge. Therefore, if surfaces are isotropic or have a specific direction, their two dimensional profiles are often used instead[3,4].

The fractal geometry developed by Mandelbrot is the best modeling method for nature phenomena. A lot of experiments have proven that the geometries of rough surfaces are fractal in certain scale at least, so fractal surfaces are extensively used in tribology research. Fractal Brownian motions(FBMs)[5] is one of the most important fractal models, and discrete Fourier transform(DFT)[6] is a useful synthesis method for generating FBMs. Comparing with other methods, such as random midpoint displacement method or two dimensional Weierstrass-Mandelbrot function, the profile synthesized by DFT has the advantages of stable fractal parameters and explicit meaning of input arguments.

To synthesize fractal profile with fine structures and to computer the tribology conventional parameters which are strongly related to the tribological performance, the sampling points are needed as many as possible. To use the synthesized profiles in FEM, the profile's high frequency components must be filtered. These make the profiles containing a lot of redundancy data, and much unnecessary time is needed when they are imported into FEM software. To solve this problem, a profile synthesized by DFT and filtered in frequency domain was simplified using the chord deviation algorithm[7] in space domain. It was found that only a small part of data were needed to express the filtered fractal profile entirely, and the importing time were much more reduced if the simplified profile were used.

Fractal Profile Synthesis

The Fourier transform builds a relationship between space domain and frequency domain of any function. Since it is discrete in both space domain and frequency domain, and has fast algorithm which is called fast Fourier transform(FFT), DFT is appropriate for computer realization. If let the space domain expression of a profile be x[n], and its corresponding frequency domain Fourier coefficient is X[k], its DFT can be written as[8]:

$$X[k]=\sum_{k=0}^{N-1} x[n]W_N^{nk}\ . \tag{1}$$

$$x[n]=\frac{1}{N}\sum_{n=0}^{N-1} X[k]W_N^{-nk}\ . \tag{2}$$

$$W_N=e^{-j\frac{2\pi}{N}}\ . \tag{3}$$

Eq. 1 is DFT, and Eq. 2 is inverse DFT; N is the total number of sampled points, for sake of FFT, N often equals an integer power of 2; n, k is the sample number of space domain and frequency domain respectively. For a real sequence, X[k] is ofter a complex number, and can be expressed as:

$$X[k]=|X[k]|e^{j\phi(k)}\ . \tag{4}$$

|X[k]| is called the magnitude of DFT, and φ(k) is the phase; X[k] is circular conjugate symmetry, and it means |X[k]| is even symmetry,φ(k) is odd.

For FBMs, its discrete power spectrum density is[6]:

$$E\left(|X[k]|^2\right)\propto\frac{1}{k^{2H+1}}=\frac{C^2}{k^{2H+1}},\quad k=1,2,\cdots,N/2-1\ . \tag{5}$$

E() means mathematical expectation; H is Hurst exponent which is related to fractal dimension D as D=2-H, 1<D<2; X[k] is the Fourier coefficient; C is a constant; N is the total number of sampling points; k is the sample number of frequency domain. Because of the circular conjugate symmetry of X[k], when k counts from N/2+1 to N-1, the value of $|X[k]|^2$ is just the same as $|X[k]|^2$ in Eq. 5, and we let the value of |X[0]| and |X[N/2]| equal 0.

Just like the Weierstrass-Mandelbrot function for fractal profile synthesis, an assumption was used in fractal synthesis in this paper, which assumed that the power spectrum of the synthesized profile was met the power law strictly, and it meant that Eq. 5 could be simplified as:

$$|X[k]|^2=\frac{C^2}{k^{5-2D}},\quad k=1,2,\cdots,N/2-1\ . \tag{6}$$

So with the given fractal dimension D, constant coefficient C, the total sample numbers N, and let φ(k) be a random number which was uniform distributed in the range of [0,2π), a fractal profile could be synthesized using Eq. 2. Fig. 1 showed a fractal profile with D=1.5, C=7.9346×10^3, N=1024.

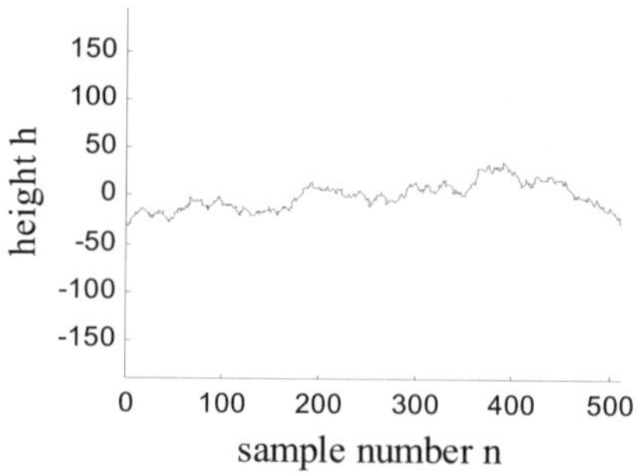

Fig. 1: a profile synthesized by DFT

Profile Filter in Frequency Domain

As shown in Fig. 1, the synthesized profile has a lot of fine structures, and its high frequency components should be filtered before FEM. Because there are not extensively accepted filter rule, an appropriate range of scale is selected according to some specific research needs.

The power spectrum of the profile is shown in Fig. 1 is Fig. 2(a). The main energy of the profile is contained in its low frequency parts. If only the 14 low frequency components which possess high energy are preserved, and the rest components are made equal to zero, the profile became smooth and was suit for FEM, which is shown in Fig. 2(b).

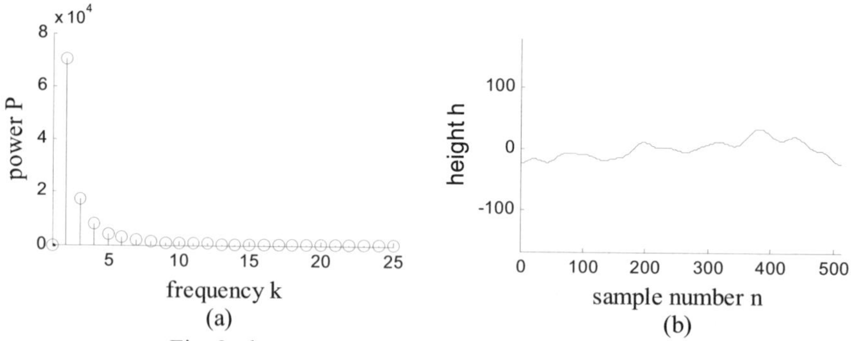

Fig. 2: the power spectrum of the profile and the filtered profile

Profile Simplification in Space Domain

The number of profile's sample points was not reduced after frequency domain filtering, which meant although the profile looked smooth but the data contained in a filtered profile was as many as the origin profile. It needed a lot of time to import the profile into FEM software. Fig. 3(a) shows a filtered profile and Fig. 3(b) is a partial enlarged drawing of Fig. 3(a) at position A.

Fig. 3: the profile before simplification in space domain

It was found in Fig. 3 that the dense sampling of smooth profile caused unnecessary data redundancy. So the chord deviation algorithm[7] was used to simplify the profile in space domain. The fundamental theory of chord deviation algorithm is: for three consecutive points, for example P_{i-1}, P_i, P_{i+1}, in the profile shown in Fig. 4, if e_i which is the distance from P_i to P_{i+1} is smaller than a prescribe value, the point P_i should be deleted.

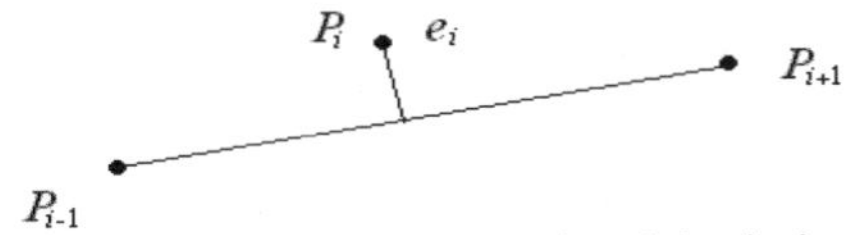

Fig. 4: a brief description of the chord deviation algorithm

The simplified profile was shown in Fig. 5. The number of sampling points was 61 which only occupied 6% of the original 1024 points.

(a) (b)

Fig. 5: the fractal profile after simplification

Conclusion

A fractal profile was synthesized using DFT, and it was filtered in frequency domain and simplified in space domain consecutively. The results found are as follows.

1. The frequency filter could filter the high frequency components of a synthesized profile, which made the profile smooth and be suit for FEM, but can not reduce the data contained in the profile.

2. Because the number of sampling points was not changed after frequency domain filter, there are a lot of redundancy data in the profile. After simplifying in space domain using the chord deviation algorithm, a large amount of redundancy data were deleted and the time of importing data into FEM software are reduced greatly.

References

[1] Chenghui Gao, Jianmeng Huang, Lei Dai: *Influence of different sliding velocity on the thermo-mechanical coupling of the rough surface contact*. Applied Mechanics and Materials Vol. 29-32(2010), pp. 332-336.

[2] Jianmeng Huang, Chenghui Gao: *Research of contact pressure of an elastoplastic rough body in contact with a smooth rigid plane*. Advanced Materials Research Vol. 97-101(2010), pp. 1248-1252.

[3] A. Ozer, H. Sofuoglu: *Thermo-mechanical analysis of the magnetic head-disk interface with a fractal surface description*. Wear, 266, 11-12(2009), pp. 1185-1197.

[4] B. Buchner, M. Buchner, B. Buchmayr: *Determination of the real contact area for numerical simulation*. Tribology International, 42, 6(2009), pp. 897-901.

[5] B.B. Mandelbrot, Van Ness: *Fractional Brownian motions, fractional noises and applications*. SIAM Review, 10, 4(1968), pp. 422-437.

[6] Heinz-Otto Peitgen, Dietmar Saupe: *The Science of Fractal Images*. Springer-Verlag, New York, (1988).

[7] Jin Tao, Tong Shuiguang: *Reverse Engineering*. China Machine Press, Beijing, (2003).

[8] Oppenheim A.V., Schafer R.W., Back J.R.: *Discrete-time Signal Processing*. Prentice Hall, NJ, (1999).

Advanced Materials Research Vol. 279 (2011) pp 266-270
© (2011) Trans Tech Publications, Switzerland
doi:10.4028/www.scientific.net/AMR.279.266

Computer Aided Design of Technology and Die for the Shoe Toe

Xiaohong Fan[1, a], Zhengjun Yuan[1, b], Jing Li[1, c], Bin Xu[1, d]

[1]School of Material Science and Engineering, Shandong Jianzhu University, Jinan 250101,China

[a]xiaotong324@163.com, [b]yuangtt@163.com, [c]lijing-ck@163.com, [d]xubin@sdjzu.edu.cn

Keywords: 3D model, DYNAFORM, technology program, die construction

Abstract. A three-dimensional(3D) model of the shoe toe is established by the CACX and UG software in this paper. Then the process analysis and calculations of the product are done with DYNAFORM software. The most reasonable technology scheme and the related die construction are determined in the end.

Introduction

The technological design and the die design of the shoe covers of special forces are carried out in this paper. Since there is only the workpiece with irregular curved surface and with no related blueprint and datas, first, the 3D precise model of the workpiece must be obtained with the CACX software, and then the technology and the die design are discussed [1-4].

Three Dimensional Modeling

First, spatial points of the work are got with the numerical control machine tool, and then the CACX software is used to obtain the parameter curve. Next, 3D surfaces are constructed using curves and surfaces in CACX. In the end, the 3D model is acquired in UG software.

The specific process is shown as follows. First, the workpiece is fixed on the workbench of the numerical control machine tool, and a point on the workpiece surface is chosen randomly as the zero point. Then, as much as possible points are selected, three-dimensional space coordinates of every points are recorded, and the points of high saltus are deleted. Next, the optimized points are imported into the CACX software·to acquire a parameter curve and further the rounded up curve is obtained at the same time. By the same rule, multiple curves are acquired, as shown in Fig. 1. Many curves further constitute a grid surface and then the surface is built, as shown in Fig. 2. More surfaces are obtained in the same way. Through rounding up all the surfaces, the complete and accurate surfaces (as shown in Fig. 3) are finally obtained. At last, the surfaces generated in CACX are imported into the UG software in which the 3D model of the shoe toe is finally established, and the 3D model of the shoe toe is shown in Fig. 4.

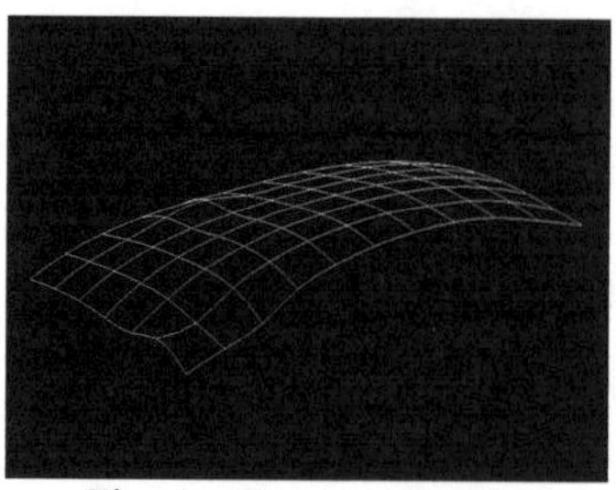

Fig. 1: Curve networks

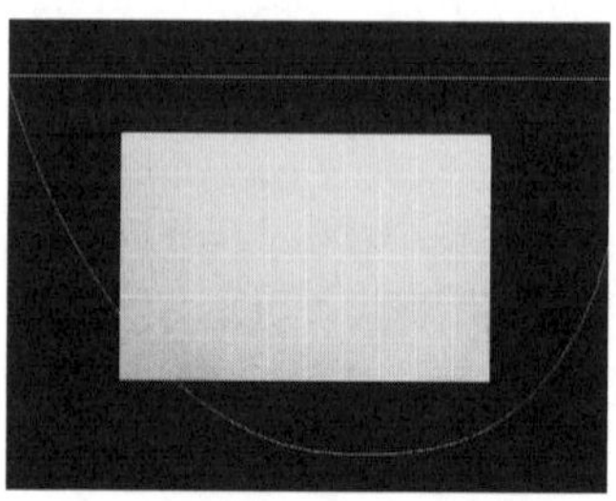

Fig. 2: Mesh surface

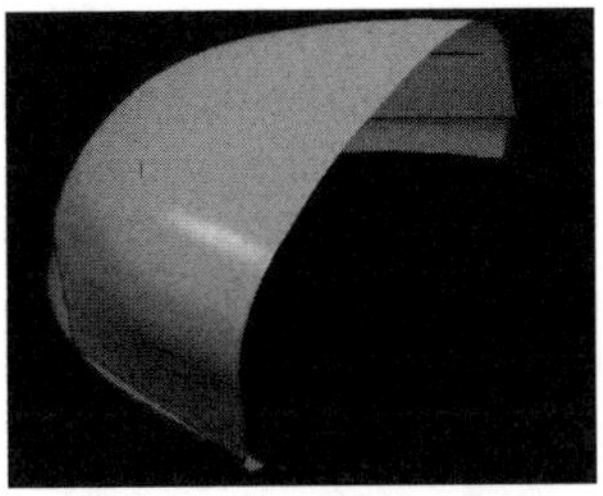

Fig. 3: The 3D surfaces

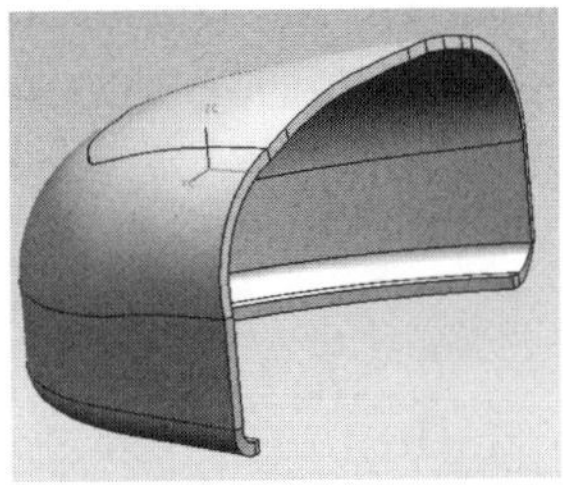

Fig. 4: The 3D model of the shoe toe

Technology Program

The workpiece studied in this paper is shoe cover for special forces. The shoe cover is a half opening drawing part, and the right shoe cover is enantiomorphic with the left one. Therefore, the two shoe covers should be built-up drawing, and then be cut open. Since the oral area of the drawn part is radon transform, trimming is necessary after drawing, and then flanging to form the final work.

Since the workpiece is irregular, DYNAFORM software is used to determine the figure and size of the blank. First, neutral surface (as shown in Fig. 5) generated in UG is imported into the DYNAFORM, in which the surfaces are optimized, the blank figure of the work is obtained (as shown in Fig. 6), and the final size of the blank is showed in Fig. 7.

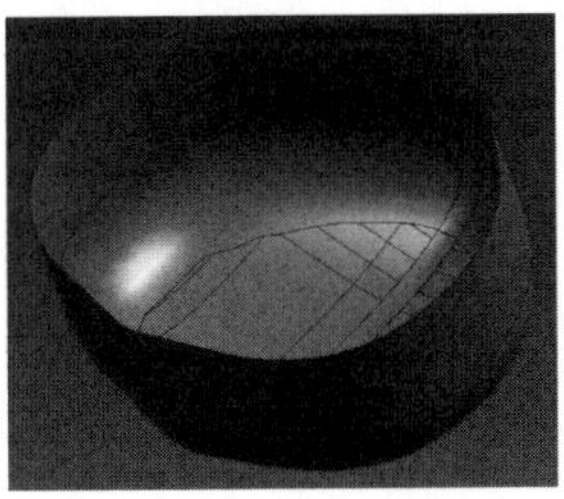

Fig. 5: Neutral surface

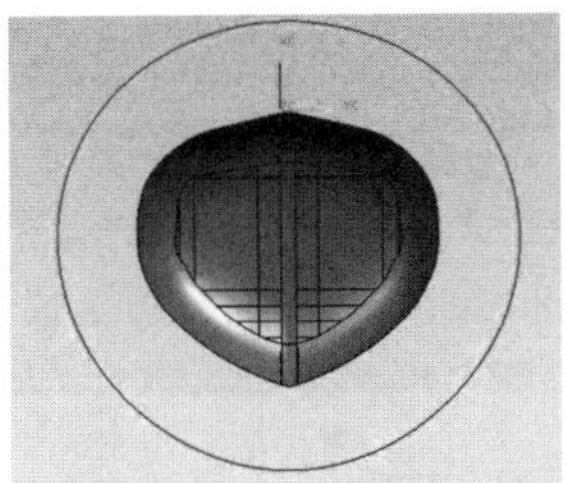

Fig. 6: The blank figure

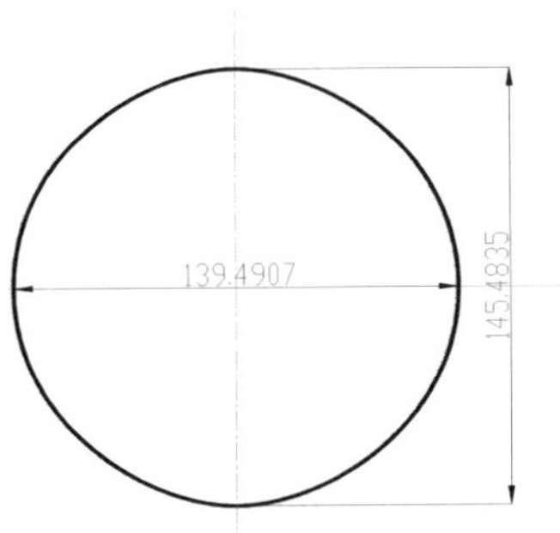

Fig. 7: The final blank size

The size of the workpiece: d_1=97mm, d_2=86.81mm;

The size of the blank: D_1=145.48mmm, D_2=139.49mm;

Drawing coefficient: $m_1=d_1/D_1$=97/145.48=0.67; $m_2=d_2/D_2$=86.81/139.49=0.62;

Relative thickness of the blank:

$$\frac{t}{D}\times 100=\frac{1.8}{145.48}\times 100=1.24$$

Critical drawing coefficient: K_1=0.56[5]

$$m_2>K_1$$

Consequently, the drawing coefficient of the workpiece is greater than the first critical drawing coefficient, so the workpiece can be drawn in one process.

In conclusion, the most reasonable technology program used is compound process of blanking and drawing, slitting ,cutting edge and flanging.

Die Construction

Compound Die for Blanking and Drawing. The die construction for blanking and drawing is shown in Fig. 8. The punch progression is as follows: the blank passes through the stock guide rail 8, and then blanking is carried out by the punch-die15 and the blanking die 16. The upper plate 5 continues going down, and drawing is completed by the punch-die15 and the drawing punch 11. The workpiece in punch-die 15 after drawing is ejected by 12, and the waste is removed by the stripper plate 17.

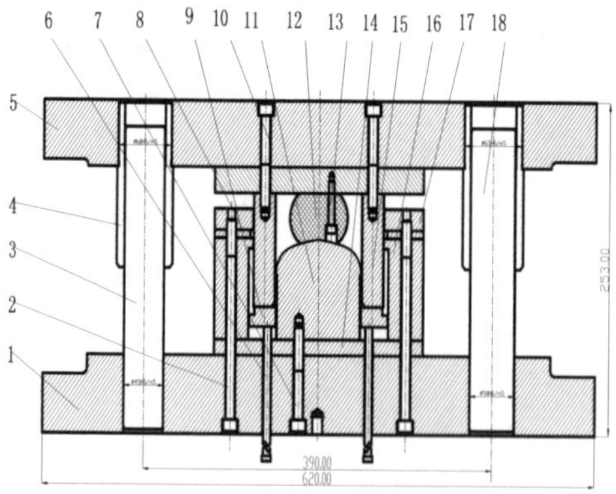

Fig. 8 Compound Die for Blanking and Drawing
1- lower plate 2, 7, 10, 13-screw 3, 18-guide pillar 4-guide bushing 5- upper plate 6-knockout-pin 8- stock guide rail 9-blank holder 11- drawing punch 12-rubber 14-bolster 15- punch-die 16-blanking die 17- stripper plate

Slitting Die. The blanking and drawing die is shown in Fig. 9. The semi-product after drawing is placed in die 4. The upper plate 5 descends, and the rubber 13 holds down the workpiece. The die12 touches and then cut off the stand wall of the workpiece. The upper plate continues going down, until the workpiece is completely parted off.

Cutting Edge Die. The cutting edge die is shown in Fig. 10. The semi-product after slitting is placed on die 1, and the press ram descends and holds down the workpiece. The press ram continues going down, and the die sets to wok, until the cutting is completed.

Flanging Die. The flanging die is shown in Fig. 11. The punch progression is as follows: The semi-product after cutting edge is placed on flanging punch whose shape is similar to the internal shape of the shoe cover. The press ram descends. First, the stripper plate 3 presses the workpiece up against the flanging punch, with the press ram going down. The workpiece gradually enters the flanging die, when the press ram gets to the bottom dead center, and the flanging course is over.

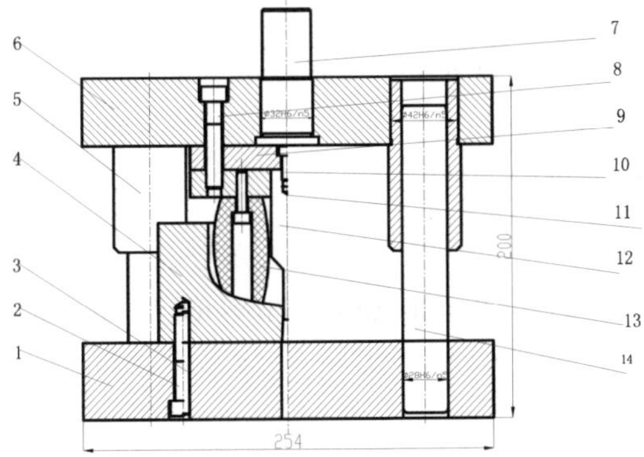

Fig 9 Slitting Die

1- lower plate 2- screw 3-dowel 4-die 5- guide bushing 6-upper plate 7-punch shank 8, 10-screw 9- bolster 11-punch-hlder 12-punch 13-rubber 14- guide pillar

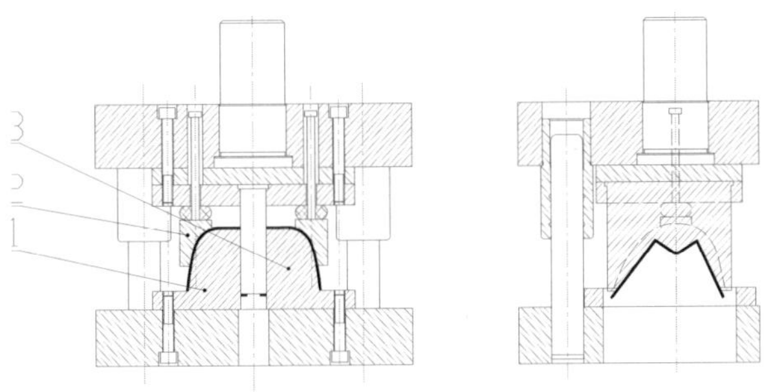

Fig.10 cutting edge die

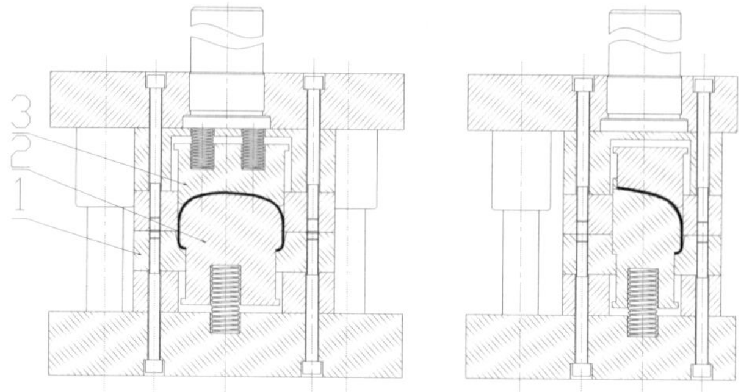

Fig.11 flanging die

Summary

1) Spatial points of the work are got with the numerical control machine tool, and then the CACX software is used to obtain the parameter curve and the 3D surfaces. In the end, the 3D model of the workpiece is acquired in UG software.

2) DYNAFORM software is used to determine the figure and size of the blank. The most reasonable technology scheme finally determined is compound process of blanking and drawing, slitting ,cutting edge and flanging.

3) The related die construction is determined.

Acknowledgements

This work is financially supported by the Shandong Province Young and Middle-Aged Scientists Research Awards Fund (2007BS04020).

References

[1] M. Kegl, B. J. Butinar, M. M. Oblak: *Shape optimal design of elastic planar frames with non-linear response.* International Journal for Numerical Methods in Engineering, 38, 19(1995), pp. 3227-3242.

[2] Zhang Zhaotao, Liang Bingwen: *Determination of blank shapes for drawing irregular cups using an electrical analogue method.* International Journal of Mechanical Sciences, 28, 8(1986), pp. 499-503.

[3] Guo Y. Q., Batoz J. L., et al: *Finite procedures for strain estimation of sheet metal forming parts.* Int.J.N um. Methods Eng, 30, 8(1990), pp. 1385- 1401.

[4] LEE W. B., TAI W. H., TANG C. Y.: *Damage evolution and forming limit predictions of an Al2024-T3 aluminium alloy.* Journal of Materials Processing Technology, 63, 1-3(1997), pp. 100-104.

[5] Q.X.Xia: *sheet metal forming technology and die design.* South China University of Tecnology Press, Guangzhou, (2004).

Advanced Materials Research Vol. 279 (2011) pp 271-274
© (2011) Trans Tech Publications, Switzerland
doi:10.4028/www.scientific.net/AMR.279.271

A Fine Atomization CMP Slurry for Copper

Hui Zhang[a], Zifeng Ni[b], Qingzhong Li[c]

School of Mechanical Engineering, Jiangnan University, Wuxi Jiangsu, 214122, PR China

[a]349855293@qq.com, [b]nizf@jiangnan.edu.cn, [c]qingzhongli@163.com

Keywords: Chemical mechanical polishing (CMP), copper, slurry, atomization

Abstract. In this paper, a kind of alkaline slurry was introduced, in which silica was used as the abrasive, H_2O_2 was used as the oxidize, glycine was used as the complexing agent, azimidobenzene was used as the surfactant, and borax was used as the pH regulator. The atomization polishing method was used, and the effects of the traditional polishing and atomization polishing were compared. After the atomization polishing, the surface roughness of copper was 7.61 nm and the material removal rate was 188 nm/min; After the traditional polishing, the surface roughness was 15.22 nm and the material removal rate was 236 nm/min. The dosage of polishing slurry used in the atomization polishing is dozens of times less than that in the traditional polishing.

Introduction

Chemical mechanical polishing process (CMP) is the most effective method which is currently used to implement local and global material planarization, and it is widely applied in flattened surface treatment of hard brittle materials and IC process [1-3]. With the stringent requirements of microelectronics devices, IC chips are developing to high density, high integration and high performance. These make the feature size of integrated circuits more and smaller, with the size decreasing from 0.25 μm to 0.13 μm or smaller. The interconnected metal layers of chip are developing to multi-layers [4,5]. These changes will put forward more requirements on the layer planarization of IC process. Copper polishing slurries are the key factors of copper CMP global flattening and they determine the overall polishing crafts good or not.

Alkaline slurry has some advantages, such as little corrosive, high selectivity (forming automatically a layer of oxide film in the copper surfaces to enhance selectivity) etc. Domestic research level has developed a lot, but there is a big gap compared with international famous companies [6]. Therefore, the alkaline slurry study which is suitable for sub-micron and deep submicron process is necessary. In the alkaline slurries, arasive, oxidant, pH regulator and complexing agent are the main components, and they almostly determine material removal rate and surface quality in the CMP process. The method of fine atomization CMP is that the components are atomized with controlling ultrasound frequency to form SMD5-15um uniform droplets, and then it is transported to the polishing interface. They can form uniform film which has low shear strength on the surface of workpiece through the strong adsorption and the high-performance uniform chemistry between interface, and then the film is removed by mechanical function. The method can significantly reduce the dosage of slurry, and it is helpful in getting better workpiece surface quality. Therefore, there is far-reaching significance in researching the polishing slurry which is suitable to fine atomization CMP. This article mainly studied on the preparation sequence of slurry, polishing efficiency and polishing surface quality in order to get good atomizing slurry which is suitable for the IC process.

CMP Experiment

Composition choice: Slurries affect both chemical and mechanical process of CMP [7]. In order to form a soft and crispy oxidation film in copper surface to improve removal efficiency and flatness, alkaline slurry is introduced. H_2O_2 was used as the oxidize in this paper, and white carbon black was used as the abrasive in order to avoid surface damage and Al^{3+} pollution[8]. Glycine was used as complexing agent in alkaline slurry in order to form copper-amine soluble ion, improve the copper removal rate, control the pollution of metal ions and avoid gel phenomenon of slurry. BTA was used as surfactant, and it can guarantee the slurry's dispersion and reduce the formation of corrosion pit in copper surface. In addition, borax was used as pH regulator. Borax in water has great solubility and can make the pH value keep stable at 9.5-10.5. According to the past experiences and the analysis of experimental phenomena, we got a compounding order for stability test. Firstly, pH regulator was put in the beaker, and surfactant was added. Then a certain volume of deionized water was put in to adjust the pH value, silica (particle size is 12nm) was put in with electromagnetic stirring, and complexing agent, oxidants were put in by turns. Finally, deionized water was added into reach predetermined amount. It can get stable slurry with this method.

Polishing technology: UNIPOL-1502 experiment machine was used in the experiments. The rotate speed of polishing pad was 50 r/min; The polishing pressure was 5 psi; Environmental temperature was room temperature; The time of traditional polishing was 5min and the atomization polishing was 9min. Specimens were Φ25 mm×1 mm, and they were wire-cut and burnished copper slices. The material removal rate is measured with electronic balance whose accuracy was 0.01mg, and the removal rate was calculated with the average weight of four times weighing. Surface topography and roughness were watched with atomic force microscopy (CSPM5000).

Through researching and analysising on documentary about the copper alloys CMP and repeating test, we have obtained a low pollution and good effect polishing slurry.

Table 1: Slurry components and main parameters

Slurry	I	II	III
Abradant material (quality)/wt%	1	1.5	3
Oxygenant(hydrogen peroxide) /wt%	8	5	5
Surfactant (BTA) /wt%	1	0.5	1
Complexant(glycine) /wt%	1.5	1.2	0.7

Experimental Results and Discussion

From different kinds of test, we choosed three kinds of slurries to compare their polishing effects, respectively. Slurry I and slurry II were used in the traditional CMP, and slurry III was used in the atomization CMP.

Table 2: polishing effects contrast

Slurry	I	II	III
Quality changes after polishing Δm [g]	3.95×10^{-3}	4.54×10^{-3}	6.50×10^{-3}
Removal rate [nm/min]	205	236	188

Table 2 shows that there is no great difference in removal rate between traditional and atomization CMP, and it was found that the time of atomization CMP was longer than that of the traditional CMP, which is good for the mist fully diffused to achieve ideal removal rate. Different components' adding sequence and their proportions also influenced the removal rate. The surface of workfiece polished with polishing slurry I produced corrosion pit obviously, and it produced many scratches which was polished with polishing slurry II, but the corrosion pit and scratches significantly reduced after polishing with slurry III. After the experiment, the consumption dosage of polishing slurry III was dozens of times less than that of the slurry I and II.

The surface morphology of workfieces were observed by AFM (Fig. 1-3).

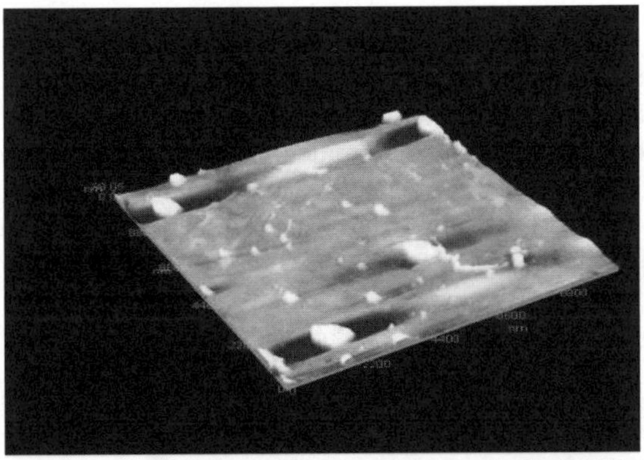

Fig. 1: The surface morphology of copper after polished with the slurry I

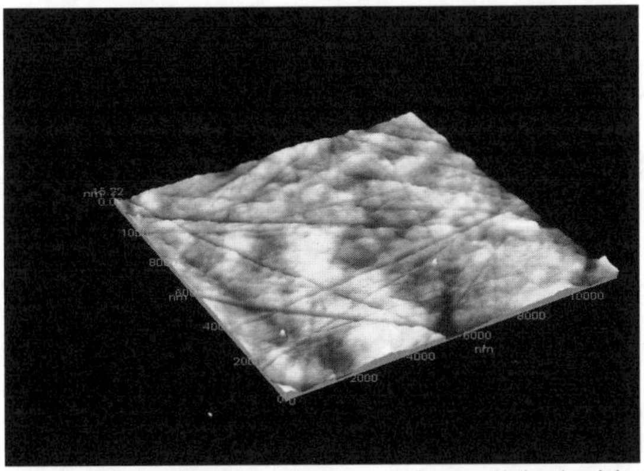

Fig. 2: The surface morphology of copper after polished with the slurryII

Fig. 3: The surface morphology of copperafter polished with the slurry III

Both the traditional CMP or the atomization CMP method can obtain higher surface planeness of the copper. However, the traditional CMP method causes the imbalance of chemical and mechanical action easily, such as surface pollution, scratches and corrosion etc. After the experiments, slurry I reduced the surface roughness of copper significantly from 50nm to 38.05 nm, slurre II reduced the surface roughness of copper from 25 nm to 15.22 nm, and slurry III used in the atomization CMP reduced the surface roughness of copper from 12nm to 7.61 nm. It is obvious that CMP with the atomization slurry can not only greatly reduce the consumption of slurry, but also achieve better surface quality of workpiece.

Conclusion

From the experimental results can be found:

a) The prepared polishing slurry III is applicable to fine atomization CMP, and it has no excitant odour and has good stability to satisfy the requirements of copper CMP. The material removal rate is 188 nm/min and the surface roughness of copper reach 7.61 nm.

b) The consumption dosage of polishing slurry in atomization CMP is dozens of times less than that in the traditional polishing.

Acknowledgements

This research work was provided by the National Natural Science Foundation of Jiangsu Province (BK20080605) and China Postdoctoral Science Foundation funded project (No.20060390984).

References

[1] D.M.Guo, R.K.Kang, J.X.Su, et al.: *Future development on wafer planarization technology in ULSI fabrication*. Journal of Mechanical Engineering Vol.39 (2003), p. 100.

[2] J.X.Su, D.M.Guo, R.K.Kang, et al.: *Modeling and analyzing on nonuniformity of material removal in chemical mechanical polishing of silicon wafer*. Mater Sci Forum Vol.471~472 (2004), p. 26.

[3] J.X.Su, D.M.Guo, R.K.Kang, et al.: *Kinematic mechanism of chemical polishing in ULSI manufacturing*. Chinese Journal of Semiconductors Vol. 26 (2005), p. 606.

[4] A.Michael, Fury: *Emerging developments in CMP for semiconductor planarization*. Solid state technology Vol. 38 (1995), p. 81.

[5] B. L.Gehman: *In the age of 300mm silicon,tech standards are even more crucial*. SolidState Technology Vol. 44 (2001), p. 128.

[6] J.C.Li, Z.X.Li, S.Z.Jin: *Combination content of polishing* .Chinese patent zL99111114.1. (2000).

[7] P.Wrschka, J.Hernandez: *Chemical mechanecal planarization of copper damascene structures*. Journal of the Electrochemical Society Vol. 147 (2000), p. 706.

[8] H.Y. Wang, Y.L. Liu, D.C. Zhang: *A new type of copper CMP slurry in ULSI*. Semiconductor journals and technology Vol.23 (2003), p. 217.

[9] Q.Z.Li, Z.J.Jin, R.Zhang, et al.: *Experimental study on influences of dispersant on material removal rate and the surface roughness of Cu CMP*. Lubrication and sealing Vol.32 (2007), p. 71.

[10]Y.L. Liu, J.X. Li, et al.: *CMP study of multilayer wiring conductor copper in ULSI manufacturing*. Electronic industrial specialized equipment Vol.153 (2007).

[11]X.H.Chen: *The study on the slurry of the copper CMP of 300mm silicon wafer* (2007).

[12]R. Zhang: *Study on CMP Slurry of ULSI Copper Interconnection Layer* (2007).

[13]M. Zhong, J.L.Zhang, et al.: *Research on CMP slurry for fine polishing of Si substrate*. Semiconductor journal Vol.27 (2006), p. 400.

Advanced Materials Research Vol. 279 (2011) pp 275-279
© (2011) Trans Tech Publications, Switzerland
doi:10.4028/www.scientific.net/AMR.279.275

Research on the ICP Acceleration Sensors Detecting Roadbed Compaction Degree Test

CAO Yuan-wen[1,a], JIAO Jian-hua[2,b], MA Li-Ying[1,c] and GUI Shao-xiong[1,d]

[1]School of Mechatronics and Automotive Engineering, Chongqing Jiaotong University, Chongqing 400074, China

[2]CCC First Highway Survey and Design Institute, Xi'an 710075, China

[a]caoyw@cquc.edu.cn, [b]xajjh2008@126.com, [c]maliying801@yahoo.com.cn, [d]guishaoxiong@yahoo.com.cn

Keywords: acceleration sensors, vibratory acceleration, compaction degree, field test, dependence relation

Abstract. Compaction is an important influencing factor that makes various structural materials of road embankment and surface sufficiently dense, and it becomes more and more important that effective detecting methods of the compaction degree are employed to monitor the quality of compaction. In this paper, the testing methods and means that the compaction degree of sub grade soils was detected by the acceleration sensors during the compaction process of vibratory roller were presented, with which the field tests were performed to measure the acceleration of vibratory drum of the vibratory roller and the compaction degree of compacted soils, and then the relational expression between the vibratory acceleration and the compaction degree of soils were obtained. Results indicate that there exists such a significant positive correlation relationship between the virtual value of vibration acceleration and the soil compaction degree that the former can be used to characterize the latter accurately. Therefore, it provides a technical support for the development of nondestructive and live vibratory roller compaction degree detection technology.

Introduction

In highway base construction, whether the compaction degree is qualified or not is an important criterion of judging the quality of construction engineering . Compaction makes the soil base reach the required compaction degree standard so as to improve the bearing capacity of the road. The previous research indicated that the compaction degree of roadbed increased 1% with a foundation bearing capacity increase of 10%. Improving the roadbed compaction degree helps to prevent roadbed settlement, reduce the penetration of moisture in roadbed, and promote the roadbed durability [1,2].

With the development of the highway construction, more and more attention has been paid to the construction quality. The insufficient compaction degree is one of the important reasons for the early damage of the road. Compaction operation is an important link that makes various structural materials of the road embankment and surface obtain sufficient density. Adopting effective methods of detecting compaction degree is an important means of monitoring it. At present, the traditional methods such as sand cone method, cutting ring method and so on have still been mainly applied to inspect the compaction degree of highway sub grade, the disadvantages of which included the low efficiency, the bad representative, and the destruction brought to the pavement structure layer. However, using the internal circuit piezoelectric (ICP) acceleration sensor to detect the roadbed compaction degree made it possible for the staff to lively examine compaction degree throughout the compacting process, to control the compaction quality, and thus to ensure sufficient compaction of roadbed with the least compactor pass, which effectively avoided under-compaction or over-compaction. It can not only save manpower and shorten the construction period but also speed up the construction progress based on the construction quality. So, it is a new kind of nondestructive testing technology which can achieve quick, accurate, real-time consecutive detection for the compaction degree of roadbed.

Test Method

Test Plan. The test plan that ICP acceleration transducers were adopted to measure the compaction degree of vibratory roller was as follows:

(1) ICP acceleration transducers must be inspected and calibrated prior to test in order to assure the stability and accuracy of the testing system;

(2) Field tests were performed for the acquisition, storage, and treatment of the vibratory roller acceleration signals;

(3) One of the traditional methods, sand cone method was applied to test the compaction degree on site;

(4) The dependence relation was obtained based on the analysis of the relationships between the compaction degree in conventional test and acceleration tested from ICP sensors.

Installation of Acceleration Sensors. How and where was the sensor installed would directly affect the measurement results. Magnetic holdfast was applied since it is quite convenient to measure different measurement points rapidly and simultaneously. First, the acceleration sensor was fixed to a magnetic seat and the mounting surface was grinded smooth, rub-up, and then the insulation slices on the magnetic seat of the sensor was removed. At the same time the acceleration sensor was installed to vibrating shaft or the position where it can best reflect the position of the vibratory drum, as shown in Fig. 1. Second, one end of the data transmission line was linked to the sensor, and the other end was linked to the input interface of the acquisition instrument. Finally, the cables were fixed in the frame with adhesive tape to avoid the relative motion between the vibrating plane and the position where the cables were fixed, which would reduce measurement error of the frequency response [3,4].

Fig. 1: Installation position of acceleration sensors

Fig. 2: The site of data acquisition

Parameter Setting and Data Acquisition. Presently the working frequency of the vibratory roller used on the construction site is general in 20-50Hz. In addition, the acceleration of vibratory drum is no more than 15g when the vibratory roller is running. Therefore, the proper choice of the examining instruments were as follows: the KD1010L type acceleration sensors, the YE6263type acquisition instrument (dynamic data acquisition channel: 8, the channel bandwidth: 0-200kHz, A/D resolution: 12Bit, maximum static error:≤±0.3%), and the YE7600 type acquisition software [5].

Fig. 2 shows the situation on the data acquisition site. The length of test section was 20m, and the vibratory roller model selected was YZ16 which run at speed of 1.7km/h. During the tests the soil base was compacted repeatedly. After every compactor pass sand cone method was adopted to measure the compaction degree of roadbed.

Test Results of Base Soil Compaction Degree

Traditional Testing Results of Compactness. In this paper, three different kinds of soil sub grade were collected for the compaction degree consecutive detection. All soils tested were cohesive soil whose granular compositions met the condition of Cu>5, 1<Cc<3, so the size gradation was obviously good. The particle composition of three kinds of soil were given in Table 1, and in vibratory roller compacting process the compaction degree tested results from sand cone method were shown in Table 2 [6].

Table 1: Particle composition of the 3 kinds of the testing soil

Sieve hole dimension[mm]	quality percentage of the soil passing the sieve hole [%]		
	No.1 soil	No.2 soil	No.3 soil
20	100	100	100
10	100	84	86.7
5	80	73.4	72
2	62	60	48.8
1	54.5	48.3	35
0.5	40.6	36.9	25.6
0.25	26	23.1	17.2
0.074	14.4	10.9	11.3
<0.074	14.4	10.9	11.3

Table 2 Compaction degree tested results from sand cone method

soil coding	Compactor pass									
	1	2	3	4	5	6	7	8	9	maximum test error
No.1 soil	46.0	55.0	69.5	78.5	89.4	94.3	95.2	90.6	92.4	11.5%
No.2 soil	67.4	76.5	84.8	91.2	94.2	92.6	91.5	-	-	9.6%
No.3 soil	68.5	78.2	86.5	90.9	96.2	91.5	90.5	-	-	7.1%

Compaction degree tested results from ICP acceleration sensor. With treatment of filtering, elimination of polynomial tendency item, and smooth processing to the collected signals, the effective value of acceleration after every compactor pass were calculated, as shown in Table 3.

Table 3: The effective value of acceleration (m/s^2)

soil coding	Compactor pass							The maximum error [%]
	1	2	3	4	5	6	7	
No.1 soil	25.8	33.3	30.5	36.7	40.5	45.4	49.6	2.5
No.2 soil	20.5	35.9	44.1	42.2	44.7	-	-	3.6
No.3 soil	31.2	33.2	37.5	42.3	48.3	-	-	1.8

Table 3 indicated that the effective value of acceleration did improve with the increase of compactor passes, which also proved that the effective value of acceleration may be used to reflect the compaction degree of soil.

Test Analysis of the Relationship between Compactness and Acceleration

Fig. 3 and Fig. 4 showed the relationship between the effective value of vibratory drum acceleration and compaction degree and compactor pass for No.1 soil. The corresponding regressive relation and the correlation coefficient of each kind of three soils were provided in Table 4.

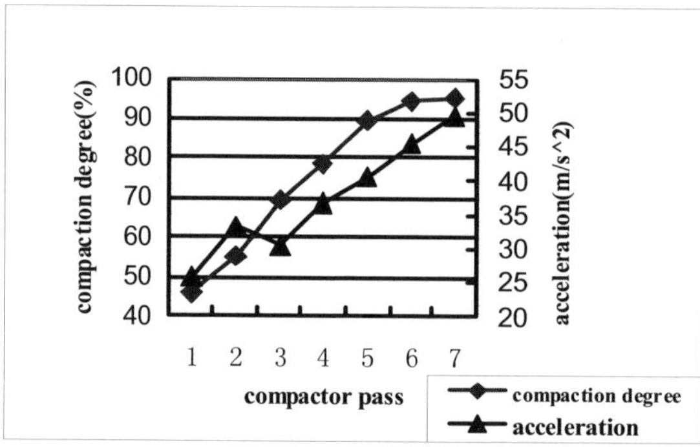

Fig. 3: Relationship existing in the effective value of acceleration, compaction degree and compactor pass for No.1 soil

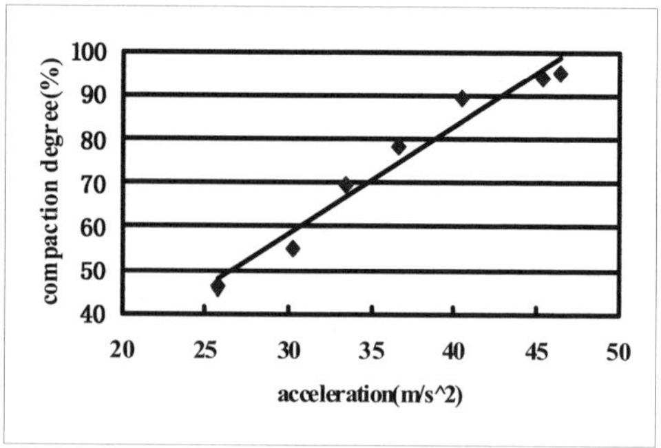

Fig. 4: Relationship between compaction degree and the acceleration for No.1 soil

Table 4: regression relationship between compaction degree and effective value of acceleration

soil coding	regression relation between compaction degree and effective value of acceleration	
	Regression equation	Correlation coefficient R^2
No.1soil	y=2.1a-4.3	0.84
No.2soil	y=a+45.6	0.84
No.3soil	y=1.5a+26.8	0.90

Note: y represents the compaction degree; x represents the effective value of acceleration

Fig. 3-4 and Table 4 implied that there was a good linear relationship between the acceleration and the compaction degree. The vibration roller compaction degree real-time consecutive detection system designed and manufactured according to this principle can accurately characterize the compaction degree of roadbed so as to effectively prevent "under compaction" and "over compaction", which for the roadbed compaction provided a continuous rapid nondestructive testing technology whose measurement data was better at reflecting the base bearing capacity.

Conclusions

(1) The changes of vibratory drum acceleration may reflect the compaction degree of soil, so it is a correct and feasible method to describe the actual compaction with the processed and analyzed vibratory acceleration signal.

(2) The test results prove that there is a significant linear relationship between soil base compaction degree and the effective value of the vibratory roller vibration acceleration.

(3) By means of regression analysis the expression of dependence relationship has been observed, which indicates that it is possible for the effective value of vibratory acceleration to accurately characterize the compaction degree of soil compacted

(4) The vibration roller compaction degree real-time consecutive detection system composed of the acceleration sensor, the data acquisition instrument, and the computer data processing and analysis system provides a continuous, rapid and nondestructive testing technology for the roadbed compacting.

References

[1] Yang Ren-feng: *Impact-vibration Composite Compaction Technologies and Equipment.* China science and technology press, Beijing, (2003).

[2] Kundt, M: *Zur Optimierung der veranderlichen Maschinenparameter und Vibrationswalzen.* Dissertation, TU Magdeburg, (1997).

[3] Lu Zhao-feng, Cao Yuan-wen, Qin Min: *The application of the piezoelectric accelerometer in engineering test.* Journal of Chongqing Jiaotong University Natural science , 4 (2008), pp. 643-646.

[4] Wen Bang-chun: *Mechanical Vibration Study.* Zhejiang University Press, Zhejiang, (2006).

[5] Huang Xian-wu, Zheng Xiao-Xia: *Sensor Principle and Application.* Electronic Science and Technology University Press, Chengdu, (2005).

[6] Du Shan-qi: *Compaction degree real-time measurement and control acquisition signal analysis of vibration roller.* Road Machinery & Construction Mechanization, 21, 8(2004), pp. 47-49.

Advanced Materials Research Vol. 279 (2011) pp 280-286
© (2011) Trans Tech Publications, Switzerland
doi:10.4028/www.scientific.net/AMR.279.280

A New Hydraulic Excitation Mode with Liquid-filled Pipe

Ziming Kou[1, a], Chunyue Lu[1, b], Juan Wu[1, c] and Huixian Zhang[1, d]
[1]College of Mechanical Engineering, Taiyuan University of Technology, Taiyuan, Shanxi, China
[a]zmkou@163.com, [b]luchunyue@nuc.edu.cn, [c]wujuanz@163.com, [d]zhxsky2008@163.com

Keywords: High-pressured Pipe, Pressure Pulsation, Excitation Source, Vibration Utilization

Abstract. The current vibration modes, whether inertial vibration or hydraulic vibration, have some problems such as large power consumption and low energy utilization rate. A new low energy hydraulic excitation mode is presented. This mode takes the high-pressured pipe as the new excitation source. Periodic pressure pulsation generated by wave exciter makes the pipe vibrating. The vibration mode of pipe with fluid-structure interaction is established and solved with numerical method. The modal analysis and the simulation of the vibration have been carried out. An experimental platform is built to verify that the new hydraulic vibration can be exploited.

Introduction

Vibrating machines are widely used in industrial production. Most of them are still work by inertial excitation. Therefore the energy consumption is large. The hydraulic vibration is gradually replacing the inertia vibration for its high power density, power transmission and control convenient. At present although the hydraulic vibration forms are various, it can be divided into two categories: one is valve control [1-3] and the other is pump-control[4-6]. Because of the vibration of the valve control excitation relaying on the valve's fast reversing, there are two shortcomings: throttling loss and interact between the frequency and amplitude. While the vibration of pump-control system is produced through hydraulic motor driving inertia exciter ultimately, whose essence is still inertia vibration, it has problems of heavy vibration quality and large power consumption. This paper puts forward a new excitation way to solve the above problems. That is the pipe that can vibrate under the effect of pressure transient flow. The high-pressure pipe or liquid cavity vibrated in the incentive of the periodic pulsation is taken as a new vibration source. This vibration whose frequency and amplitude can be adjusted is able to be controlled artificially and applied to industrial production to realize the new vibration mode of low energy consumption.

The Principle of the Hydraulic Pipe Vibration

The key to make the pipe vibrating is that it must be able to produce pressure transients in the hydraulic system. This is achieved by a wave exciter and its driving device, and the principle is shown in Fig. 1. The high-pressure oil output from the hydraulic pump station enters the shock device after flowing through the pipe, and then flow back into the tank after the flow control valve. The wave exciter is a component producing pressure pulse, which can make high-speed motor-driven rotary, and it is working in the full open state as a fast moving two-position & two-port valve. When interface of the wave exciter alters position from left to right, flow valve port is from the all-pass flow to zero flow, and positive water hammer will appear in the pipe. When interface of the wave exciter alters position from right to left, flow valve port is from the zero flow to all-pass flow, and negative water hammer will appear in the pipe. As the wave exciter rotating continuously and shocking wave device rotating continuously, waves will appear in alternating high and low pressure to make the pipe vibrating. The rotating speed of the wave exciter can be controlled by the frequency conversion unit.

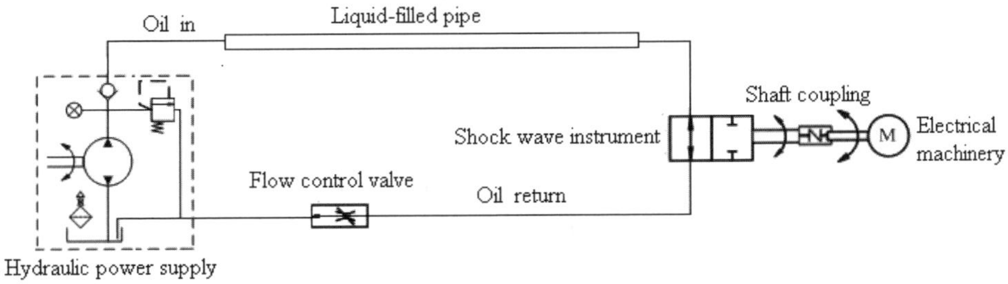

Fig. 1: Hydraulic power supply

Vibration Model and Modal Simulation of Pipe

Establishment of Vibration Model. The vibration of hydraulic pipe is a spatial multidimensional vibration. Because most vibration mechanism is using up-down vibration to work in actual industrial production, the vibration characteristics of pipe in longitudinal direction are only studied in this paper. The pipe's vibration in the action of fluid is a typical fluid-structural interaction problem. We suppose that the pipe is put horizontally, the pipe wall's material is linear, uniform and isotropic, the y axis is coincident with the axis of pipe before distortion, and the x axis and z axis are separately two mutually perpendicular directions for transverse vibration. The pipe's force is shown in Fig. 2. According to the existing documents [7-9], considering Poisson coupling, the coupling vibration equations in vertical direction of pipe is as follows.

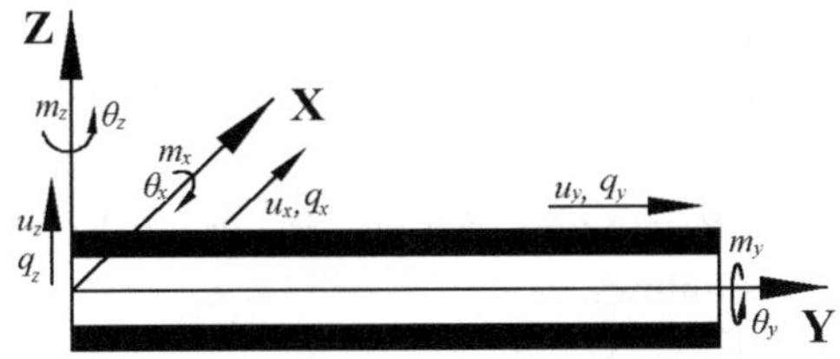

Fig. 2: The pipe's force diagram

$$\begin{cases} (\rho_p A_p + \rho_f A_f)\dfrac{\partial^2 u_z}{\partial t^2} + \dfrac{\partial q_z}{\partial y} = 0 \\ \rho_P I_P \dfrac{\partial^2 \theta_x}{\partial t^2} - q_z - \dfrac{\partial m_x}{\partial y} = 0 \\ EI_P \dfrac{\partial^2 \theta_x}{\partial y^2} + \dfrac{\partial m_x}{\partial t} = 0 \\ \dfrac{\partial^2 u_z}{\partial y \partial t} + \dfrac{\partial \theta_x}{\partial t} + \dfrac{4+3\mu}{2(1+\mu)A_p G} \cdot \dfrac{\partial q_z}{\partial t} = 0 \end{cases} \tag{1}$$

Where u_z is the pipe's longitudinal displacement, q_z is shear force of pipe in the Z direction, m_x is torque that pipe is around the X axis, θ_x is pipe's angle, ρ_P is pipe's density, A_p is pipe's cross-sectional area, ρ_f is liquid density, A_f is liquid cross-sectional area, E is pipe's elastic modulus, I_P is the inertia moment of section, μ is poisson's ratio for pipe, G is the shear modulus for pipe.

After transforming above four equations, the variable wave equation about u_z is obtained

$$EI_p\frac{\partial^4 u_z}{\partial y^4}-m\frac{\partial^2 u_z}{\partial t^2}-\frac{mI_p\rho_p}{\kappa GA_p}\cdot\frac{\partial^4 u_z}{\partial t^4}+[\rho_p I_p-\frac{mEI_p}{\kappa GA_p}]\frac{\partial^4 u_z}{\partial^2 t\partial y^2}=0 \tag{2}$$

In this equation, $m=\rho_p A_p+\rho_f A_f$ is the mass per unit length of hydraulic pipe, $\kappa=\frac{2(1+\mu)}{3\mu+4}$ is the shear distribution coefficient. Supposing that pipe's longitudinal vibration is a simple harmonic form, let $u_z(y,t)=U_z(y)e^{j\omega t}$, Eq. 2 can be arranged as follows:

$$\frac{d^4U_z}{dy^4}-(\frac{L^2\rho_p}{E}+\frac{mL^2}{\kappa A_p G})\frac{\omega^2}{L^2}\cdot\frac{d^2U_z}{dy^2}+\frac{1}{L^4}(\frac{m\omega^2L^4}{EI_p}-\frac{m\rho_p\omega^4L^4}{\kappa A_p GE})U_z=0 \tag{3}$$

Then the secular equation is

$$\lambda^4-(\alpha+\beta)\lambda^2+(\alpha\beta-\gamma)=0 \tag{4}$$

The solution of Eq. 4 is $\lambda=\pm\lambda_1,\ \pm j\lambda_2$

$$\lambda_1^2=\frac{\alpha+\beta}{2}+\sqrt{\frac{(\alpha-\beta)^2}{4}+\gamma},\ \lambda_2^2=\frac{\alpha+\beta}{2}-\sqrt{\frac{(\alpha-\beta)^2}{4}+\gamma}$$

Then the general solution of Eq. 4 is

$U_z(y)=C_1\cosh(\lambda_1\eta)+C_2\sinh(\lambda_1\eta)+C_3\cos(\lambda_2\eta)+C_4\sin(\lambda_2\eta)$, in which $\eta={}^{y}/_{L}$, L is pipe's length.

Let $q_z=Q_ze^{j\omega t}$, $m_x=M_xe^{j\omega t}$, $\theta_z=\Theta_ze^{j\omega t}$, then Q_z, M_x and Θ_z have the same form as U_z. Solve and write them in matrix form as follows:

$(U_z\ \Theta_z\ M_x\ Q_z)^{\mathrm{T}}=[\mathrm{D}(\eta)]\ [\mathrm{C}_1\ \mathrm{C}_2\ \mathrm{C}_3\ \mathrm{C}_4]^{\mathrm{T}}$

$$[\mathrm{D}(\eta)]=\begin{pmatrix}\cosh(\lambda_1\eta) & \sin(\lambda_1\eta) & \cos(\lambda_2\eta) & \sin(\lambda_2\eta)\\ D_1\sinh(\lambda_1\eta) & D_1\cosh(\lambda_1\eta) & D_2\sin(\lambda_2\eta) & D_2\cos(\lambda_2\eta)\\ D_3\cosh(\lambda_1\eta) & D_3\sinh(\lambda_1\eta) & D_4\cos(\lambda_2\eta) & D_4\sin(\lambda_2\eta)\\ D_5\sinh(\lambda_1\eta) & D_5\cosh(\lambda_1\eta) & D_6\sin(\lambda_2\eta) & D_6\cos(\lambda_2\eta)\end{pmatrix} \tag{5}$$

$D_1=-\frac{\lambda_1^2L^3}{\gamma EI_p}$, $D_2=-\frac{\lambda_2^2L^3}{\gamma EI_p}$, $D_3=-\frac{(\lambda_1^2+\alpha)L^2\lambda_1^2}{\gamma}$, $D_4=\frac{(\lambda_2^2+\alpha)L^2\lambda_2^2}{\gamma}$, $D_5=-\frac{(\lambda_1^2+\alpha)L^2}{\gamma EI_p}$, $D_6=-\frac{(\lambda_2^2+\alpha)L^2}{\gamma EI_p}$

The pipe's left node satisfies η=0, and then $(C_1\ \ C_2\ \ C_3\ \ C_4)^T=[D(0)]^{-1}(U_z\ \ \Theta_z\ \ M_x\ \ Q_z)_L^T$. The right node satisfies η=1, and then

$$(U_z\ \ \Theta_z\ \ M_x\ \ Q_z)_R^T=[D(1)](C_1\ \ C_2\ \ C_3\ \ C_4)^T=[D(1)][D(0)]^{-1}(U_z\ \ \Theta_z\ \ M_x\ \ Q_z)_L^T \tag{6}$$

So the pipe's transfer matrix is

$$T=[D(1)][D(0)]^{-1}=[D(1)]\begin{pmatrix}1&0&1&0\\0&D_1&0&D_2\\D_3&0&D_4&0\\0&D_5&0&D_6\end{pmatrix}^{-1}$$

$$=(\lambda_1^2+\lambda_2^2)^{-1}\begin{pmatrix}A_1-\alpha A_3 & A_2-(\alpha+\beta)A_4 & A_3 & -\gamma^{-1}\left[-\alpha A_2+(\gamma+\alpha^2)A_4\right]\\ \gamma A_4 & A_1-\beta A_3 & A_2-\beta A_4 & -A_3\\ \gamma A_3 & -\beta A_2+(\beta^2+\gamma)A_4 & A_1-\beta A_3 & -A_2+(\alpha+\beta)A_4\\ -\gamma(A_2-\alpha A_4) & -\gamma A_3 & -\gamma A_4 & A_1-\alpha A_3\end{pmatrix} \tag{7}$$

in which, $A_1=\lambda_1^2\cos\lambda_2+\lambda_2^2\cosh\lambda_1$, $A_2=\frac{\lambda_1^2}{\lambda_2}\sin\lambda_2+\frac{\lambda_2^2}{\lambda_1}\sinh\lambda_1$, $A_3=\cosh\lambda_1+\cos\lambda_2$, $A_4=\frac{\sinh\lambda_1}{\lambda_1}-\frac{\sin\lambda_2}{\lambda_2}$.

Modal Analysis of Longitudinal Vibration for Pipe. For the system shown in Fig. 1, the initial conditions are that system pressure is 3MPa, internal velocity is 5m / s, the pipe is fixed at left, and sliding bearing is at right end, so velocity is zero in the pipe's left, the final speed of pipe's right and the speed at the end of the fluid are equivalent. The boundary conditions are as follows:

Left side: P=3Mpa, u_z=0, θ=0.

Right side: $v_f=u_y'$, m_x=0, q=0.

Adding the boundary conditions into Eq.6, then

$$(U_z\quad \Theta_z\quad 0\quad 0)_R^T=T\cdot(0\quad 0\quad M_x\quad Q_z)_L^T,$$

that is $\begin{pmatrix}0\\0\end{pmatrix}=\begin{pmatrix}T_{33}&T_{34}\\T_{43}&T_{44}\end{pmatrix}\begin{pmatrix}M_x\\Q_z\end{pmatrix}_L$. (8)

Because $M_x\neq 0$, $Q_z\neq 0$, if the above equation has the solution, the determinant of coefficient must be zero, that is $\begin{vmatrix}T_{33}&T_{34}\\T_{43}&T_{44}\end{vmatrix}=0$. (9)

When the structure parameters of the system have been known, each element in the determinant is a function of the modal frequency ω. The frequency ω_1, ω_2 ...... ω_n of each mode can be calculated from the above equation. Adding ω_1, ω_2 ...... ω_n values to Eq. 6, the output state vector for each frequency vibration in the valve can be calculated.

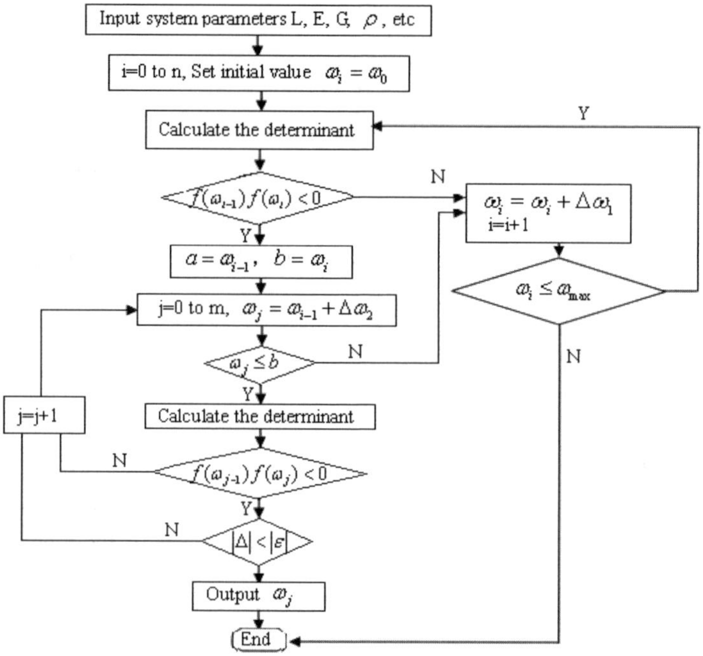

Fig. 3: Iterative flow

The analytical solution of the Eq. 9 is difficult to be calculated, so the approximate solution got with numerical methods substitutes for the precise one, which can be solved by iteration through computer. The iteration process is shown in Fig. 3. In this paper, the initial value of ω is 10Hz, ranging from 10Hz to 300Hz. The initial step size is to take $\Delta\omega$=10Hz. Eq. 9 is a continuous function, and according to its nature, if the two successive determinant values obtained satisfy $f(\omega_i)f(\omega_{i+1})<0$, a solution of the equation exists between ω_i and ω_{i+1}, which is modal frequency. Then the interval is taken as $[\omega_i, \omega_{i+1}]$, and step size reduced an order of magnitude, $\omega_i = \omega_{i-1} + \Delta\omega_i$, so again, until reaches the desired accuracy.

The pipe is a 20 # steel fluid-conveying pipe, whose length is 8m, diameter is 34mm, wall thickness is 5mm, density is 7850 kg/m^3, elastic modulus is 2.13×1011N/m^2, and Poisson's ratio is 0.286. The density of hydraulic oil is 900 kg/m^3. The five step modal frequencies calculated by computer programs are shown in Table 1. The displacement's amplitudes can be obtained, and then each step displacement's vibration model will be calculated. Finally, its vibration form will be got after superimposing various steps mode of vibration. The emulation oscillogram is shown in Fig. 4.

Table 1: The modal frequency

ω_i [Hz]	10	20	40	50	90	100	160	170	230	240
The determinant value	-4600	26600	2.07×10^6	-406000	-1.98×10^8	1.82×10^8	1.12×10^9	-3.81×10^{10}	-5.91×10^{11}	1.80×10^{11}
The modal frequency[Hz]	17.679		49.507		97.034		160.466		238.409	

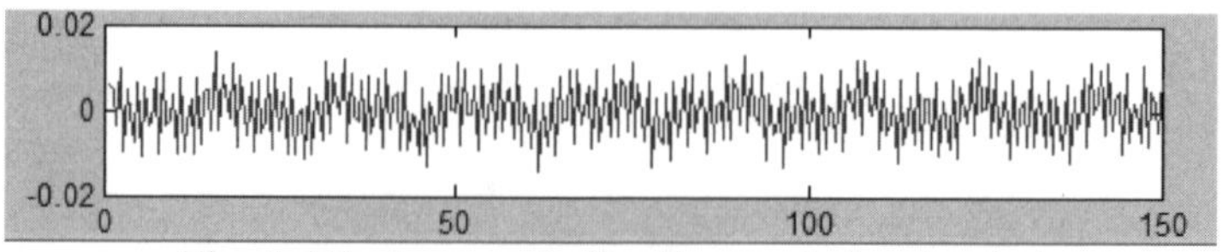

Fig. 4: Displacement vibrating mode by simulation

Experimental Verification

The experimental platform is built according to the excitation principle of pipe. The experimental system is composed of three parts, the hydraulic vibration system, the frequency control system and the data acquisition and processing system, as shown in Fig. 5. The pumping station, wave exciter [10], pipe, hydraulic cylinder and the throttle valve constitute the hydraulic vibration system. The frequency converter, signal motor and the wave exciter constitute the frequency control system. The acceleration sensor, PCI data acquisition card, the data processing software and the computer constitute the data acquisition and processing system.

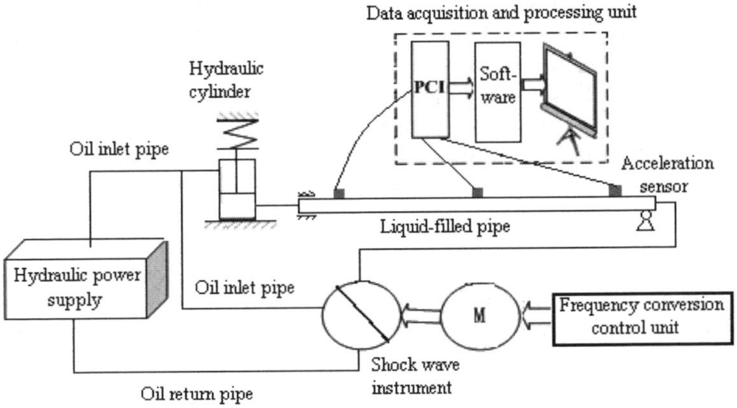

Fig. 5: Experimental system

The conversion among the converter's frequency (f_1), motor's rated speed (n_0) and the wave exciter's frequency (f) is $f = \frac{n_0 f_1}{3000} \cdot n$, in which n is the number of entries for wave exciter. Let f_1=30Hz, n_0=910r/min and n=2, then f =18.2Hz. The shape obtained from experiment is shown in Fig. 6. Comparing it with the shape in Fig. 4, the experimental shape agrees well with the simulation one, that is the high frequency wave components superimposing on the low frequency wave. It is shown that the mathematical model adopted and the calculation methods are correct. It also indicates the pipe's vibrating regulation can be artificially controlled. Therefore, the vibration can be utilized. However, the vibration wave in Fig. 4 does not consist with the one in Fig. 6 fully, which is because a hydraulic cylinder is connected in the system in order to increase system pressure in actual experiment while modeling without considering its influence.

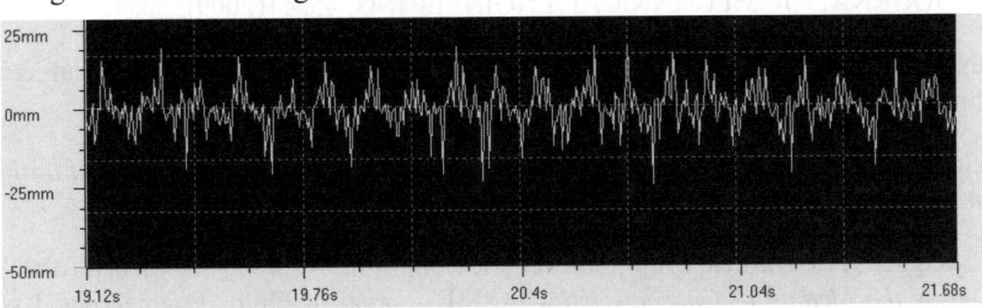

Fig. 6: Displacement wave form obtained from experiment

Conclusions

In this paper, the deficiencies of existing hydraulic excitation vibration were analyzed, and a new mode of excitation was presented, whose mathematical model was established to simulate, and finally the experiment was conducted to verify the vibration. Conclusions are as follows.

(1) Based on the principle of the hydraulic pipe vibrating for pressure transient flow, this new type of excitation made the wave exciter as a pulsating flow generator and the high-pressure steal pipe as a new vibration source.

(2) A mathematical model of pipe vibration was established, solved by numerical methods, and simulated with MATLAB to go the displacement vibration waveform.

(3) The experimental platform was built. The experimental vibration waveform agreed well with the simulation one, which proved the controllability of this excitation mode. Therefore, this paper provided a new reference method for hydraulic excitation.

This paper only initially discussed the hydraulic excitation of the single straight pipe. The theory and methods are not enough thorough and comprehensive. The influence on pipe's excition caused by wave exciter and hydraulic cylinder should be considered in further research.

Acknowledgements

The authors express their gratitude to the project supported by the National Natural Science Foundation of China (No.50775154), for its supports and the facilities to carrying out this work.

References

[1] V. K. Astashev and A. N. Tresvyatskij: *Hydraulic self-sustained vibrations exciter for vibrating technological devices*. Problemy Prochnosti i Nadezhnos'ti Mashin Vol. 6 (1993), pp. 52-59.

[2] A. F. El'muratova, D. N. Eshutkin and A. K. Kozhakhmetova: *Dynamics and stability of a hydraulic vibrator*. Journal of Mining Science, 28, 3(1992), pp. 270-272.

[3] T. Wada, M. Suzuki and T. Mori: *High speed casting of 3.0 m/min at NKK Fukuyama. No. 5 slab caster, Steelmaking Conference Proceedings*, ISS Steelmaking Conf. Proc. (1987), pp. 197-206.

[4] Information on http://www.bomag.com/worldwide

[5] RUAN Jian, LI Sheng, PEI Xiang, et al: *Electrohydraulic Vibration Exciter Controlled by 2D Valve*. JOURNAL OF MECHANICAL ENGINEERING, 45, 11(2009), pp. 125-132.

[6] WANG Fengshan, KOU Ziming and YANG Guiyuan: Colliery Mechanical & Electrical Technology. Issue 1 (2005), p. 81.

[7] Valentin R. A., Phillips J. W. and Walker J. S.: *Reflection and Transmission of Fluid Transients at an Elbow*. Trans. of SMIRT5, Berlin, August 1979, paper B2/6.

[8] D. C. Wiggert, F. J. Hatfield and S. Stuckenbrick: *Analysis of Liquid and Structural Transients in Piping by the Method of Characteristics*. ASME Journal of Fluids Engineering, 109, 2(1987), pp. 161-165.

[9] YANG Chao: *Modal Analysis of Axial Vibration of Straight Pipes of Hydraulic System*. Machine Tool & Hydraulics, 36, 1(2008), pp. 84-86.

[10] B. Q. Xu, China Patent, ZL00102009.9 (2003).

Advanced Materials Research Vol. 279 (2011) pp 287-290

doi:10.4028/www.scientific.net/AMR.279.287

Principle and Experiment of Ultrasonic Subtle Atomization in CMP

Xiaopeng Liu[a], Xiaochun Chen[b], Qingzhong Li[c]

School of Mechanical Engineering, Jiangnan University, Wuxi Jiangsu, 214122, PR China

[a]liuxiaopeng18@sina.com, [b]xchen@jiangnan.edu.cn, [c]qingzhongli@163.com

Keywords: chemical mechanical polishing (CMP), ultrasonic, subtle atomization, material removal rate

Abstract. The method of chemical mechanical polishing (CMP) using slurry which was ultrasonic subtle atomized was researched, and the system of Ultrasonic Subtle Atomization—Chemical Mechanical Polishing was established. The effects of polish parameters on polishing were also investigated. The results show that the experimental system can fully realize the expected function of polishing, the use of slurry is about one-tenth of the amount of traditional CMP, material removal rate can reach 113.734nm/min and the surface roughness is similar to the surface roughness in the traditional way.

Introduction

Wafer planarization by chemical mechanical polishing (CMP) usually obtains super-smooth surface without damaging. In today's semiconductor industry, it requires a higher level silicon manufacturing technology with the increasing of wafer size, the getting smaller of the groove width, and the more and more high degree of integration [1]. When IC feature size is smaller than 22nm, this traditional method will be no longer applicable in theory. It needs to find cyber-shot technologies of other principles, some of which originate from the traditional CMP technology such as abrasive-free CMP (AF-CMP), fixed abrasives (FA-CMP), electrochemical mechanical deposition (ECMD), plasma assisted chemical etching (PACE), stress-free polishing(SFP), etc[2-4]. The development of AF-CMP is based on the traditional CMP by removing the abrasive from slurry, and there are almost no scratches and other defects on silicon surface because no abrasive. FA-CMP condense abrasives into small abrasive blocks with specific three-dimensional structure by using resin binder and forming a polish pad with composite structure by pasting or inlaying the blocks on the surface of base material with thin organic film according to certain rules. Slurry is demonized water or aqueous solution only containing basic chemical composition. Effect of the initial application of this technology has raised concerns [5].The research results show that the cost of slurry accounts for 40% ~ 60% in traditional CMP process [6-7] and the utilization rate of slurry is low, resulting in poor economic and social benefit.

In this study, the slurry which is ultrasonic subtle atomized is used in polishing, so it could improve the utilization rate of slurry and reduce the emission of waste and harmful chemicals. The method of Ultrasonic Subtle Atomization polish uses specific frequency ultrasonic to atomize special slurry and get uniform micron-sized droplets whose diameters are 5~15μm. The droplets are imported to polished/polishing interface by negative pressure and converted to a uniform film of low shear strength on the surface after chemical reaction. The surface is super-smooth in nano-scale without damaging after the film is removed by mechanical action.

Technical principles and equipment

In this system, high-frequency ultrasonic atomizer is used to atomize slurry by controlled frequency, and micron-sized droplets which obtain uniform chemical composition are produced. Then the droplets are imported to polishing interface to join chemical reaction by negative pressure. The scratches on polished surface made by a large number of abrasive particles' reunion can be avoided both in theory and in practice because of the ultrasonic vibration effect. Slurry can be dispersed well-proportioned. Micron-sized slurry that is atomized has high specific surface area, so it can be adsorbed rapidly in the polished interface. Formation of super-smooth surface without damaging lies on the uniform chemistry on the interface and the mechanical effect of abrasive and polish pad with the reaction film. The polish experimental system is shown in Fig. 1. Wafer is nipped by the clamp and rotates around its axis. The pressure depends on the pressure parts which are added to the clamp. Clamp swings around the swing arm. The droplets are imported to polishing interface by negative pressure. They are adsorbed on the surface of polish pad and demonized water film, then join the polishing reaction, and start polishing the wafer.

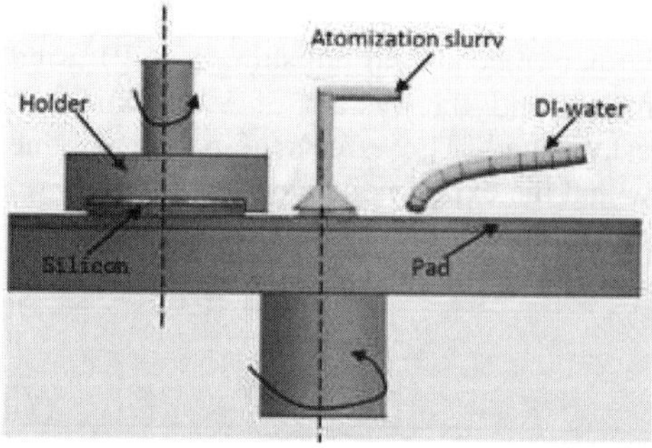

Fig. 1: Ultrasonic subtle atomization polishing system

Experiment

Pressure, nozzle angle and radius which between polish pad center and nozzle center were analyzed respectively. Pressure, nozzle height, nozzle angle and radius were orthogonal experimented under the same condition. Experimental conditions were as follows: ambient temperature was 25°C, and polishing time was 5minute. Deionized water was used to polish for 1minute after polishing was finished, and then the silicon was washed with deionized water[8]. The speed of swing arm was 9 times/min. Deionized water flow was 100ml/min. Polish pad speed was 40r/min. The size of silicon wafer was 20mm×20mm×0.6mm. The polish pad was made of polyurethane (Φ381mm). Slurry was made of silica suspension, the size was 50nm and the concentration was 15%. The wafer was weighted (n=3) by Mettler-Toledo XS205-DU precision electronic balance (accuracy: 0.01mg). Surface roughness was measured by CSPM5000s Atomic Force Microscope System.

Experimental results and analysis

Fig. 2 shows the silicon surface morphology after conventional CMP. Fig. 3 shows the silicon surface morphology after ultrasonic subtle atomization CMP. The surface roughness of silicon wafer is 5nm after conventional CMP while 9nm after ultrasonic subtle atomization CMP, and it can reach the same magnitude as conventional CMP, but the amount of slurry in ultrasonic subtle atomization CMP is only about 1/10 compared to traditional CMP.

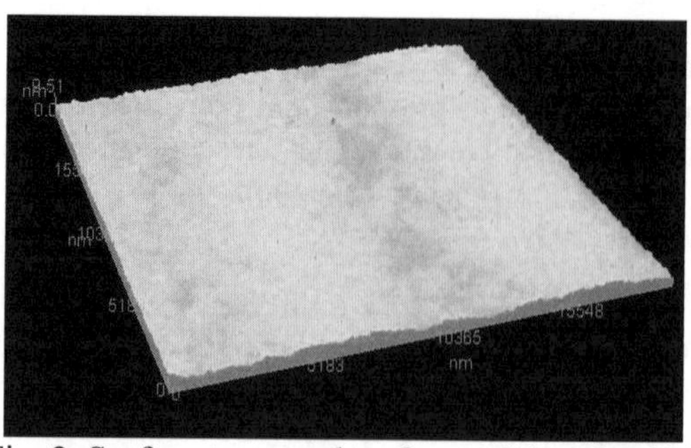
Fig. 2: Surface topography after traditional polish

Fig. 3: Surface topography after ultrasonic polish

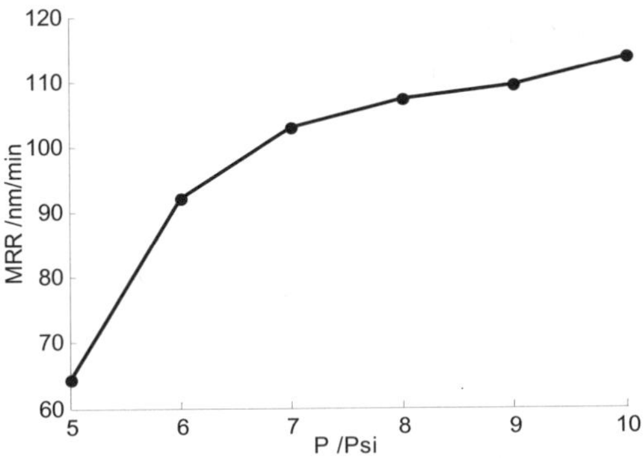

Fig. 4: MMR's change with pressure

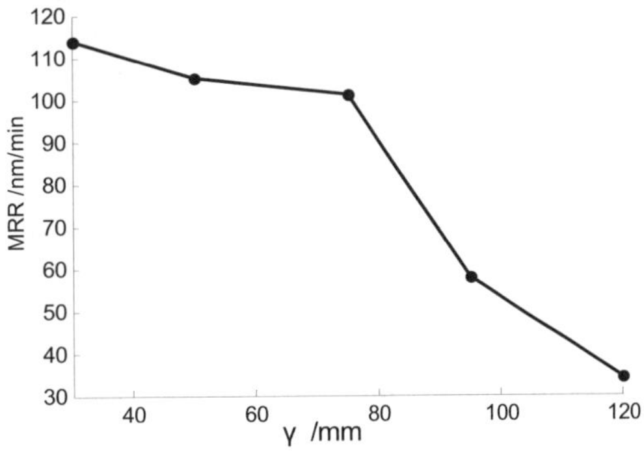

Fig. 5: MMR's change with radius

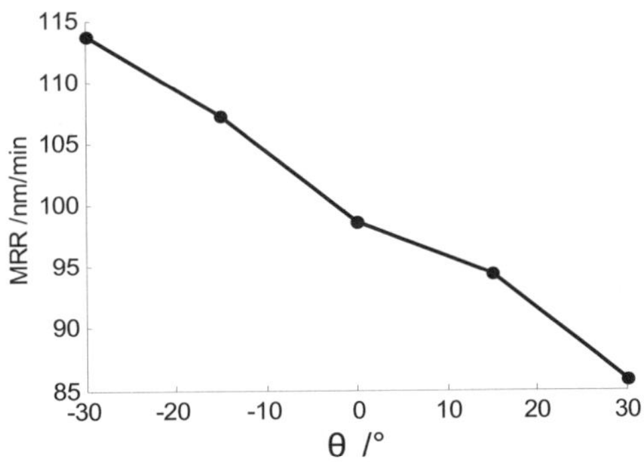

Fig. 6: MMR's change with nozzle angle

As the pressure increased, MMR of silicon wafer increased (shown in Fig. 4). When the pressure increased to a certain level, the change trend of MMR slowed down. And high pressure would lead to defects such as the accelerated loss of polishing pad, too high temperature on polishing interface and prone to debris and scratches and so on. It can be seen from Fig. 5 that as the radius increased, the MMR decreased. And when the radius increased to a certain level, the decrease trend of MMR increased sharply. As the angle increased, MMR decreased (shown in Fig. 5). And there was almost an inversely proportional relationship between MMR and angle.

Radius was the most important parameter among the four parameters from this orthogonal experiment. The optimal parameter combination in this experiment were: pressure 10Psi, nozzle height 5mm, radius 30mm and nozzle angle -30°. The experiment results also proved that MMR of this optimal parameter combination was 113.734nm/min, and it was the maximum.

Abrasives are dispersed because of the ultrasonic vibration, so it can avoid making the scratches on polished surface by a large number of abrasive particles' reunion. Meanwhile, micro-sized slurry can absorb fast and put up chemical reactions for high-activity because of its' high specific surface area.

This character makes the strength and speed of reaction much higher than that of traditional CMP, and it can alleviate or even solve the conflicts in surface roughness, surface scratches and MMR. Ultrasonic subtle atomization polish improves the utilization rate of slurry and reduces the amount of slurry. Thereby, it can reduce the initiate and expense of cracks from surface /subsurface with the mechanical action and improve the surface roughness. Reduction in the amount of slurry reduces the number of abrasive which remains on the surface of polish pad and reduces the surface scratches made by the remained abrasives, so it can improve the surface roughness.

Conclusions

(1) The ultrasonic subtle atomization experiment system can realize to ultrasonic subtle atomize slurry and can get ideal surface roughness.
(2) Amount of slurry in ultrasonic subtle atomization CMP was only about 1/10 compared to traditional CMP. Therefore, this system can be used to research ultrasonic subtle atomization CMP.
(3) MMR increased with pressure and decreased while nozzle angle and radius increased. In this experiment the maximum of MMR was 113.734nm/min while pressure was 10Psi, nozzle height was 5mm, the distance between polish pad center and nozzle center was 30mm and nozzle angle was -30°. The distance between polish pad center and nozzle center has the most significant effect to MMR among the four selected parameters.

Acknowledgements

This research work was provided by the National Natural Science Foundation of Jiangsu Province (BK20080605), China Postdoctoral Science Foundation funded project (20060390984).

References

[1] Yan ZR, Lu JJ, Li YD, et al.: *Technology Analysis of 300mm Wafer CMP*. Semiconductor Technology, Vol. 31(2006), p.561.

[2] Economikos L.: *STI Planatization Using Fixed Abrasive Technology*. Future Fab Vol.12, (2002).

[3] Li SJ, Sun LZ: *A Low Cost and Residue-Free Abrasive-Free Copper CMP Process with Low Dishing*. Erosion and Oxide Loss, IITC 2001/IEEE.

[4] Muratov VA, Fischer TE: *Tribochemical polishing*. Annual Review of Materials Science Vol.30 (2000), p. 27.

[5] Gagliardi J, Buley T: *3M SlurryFree™ CMP Technical Brief: Fixed Abrasives For Direct HDP STI CMP*. 3M Internal Pub, Jun 2001.

[6] Hahn PO: *The 300mm silicon wafer-A cost and technology challenge*. Microelectronic Engineering, Vol. 56(2001), p. 3.

[7] Braun AE: *Slurries and Pads face 2001 Challenges*. Semiconductor International. Vol.21 (1998), p.65.

[8] Li QZ, Jin ZJ, Zhang R, et al.: *Experimental Study on Influences of Dispersant on Material Removal Rate and the Surface Roughness of Cu CMP*. Lubrication Engineering, Vol. 32(2007), p.70.

Advanced Materials Research Vol. 279 (2011) pp 291-295
© (2011) Trans Tech Publications, Switzerland
doi:10.4028/www.scientific.net/AMR.279.291

A New Method for Hardened Gears Machining——Ausform Finishing Process

Li Yanbin[a], Jiang Tong[b] and Zheng Peng[c]

School of Mechanical Engineering, Shenyang University of Technology,

Liaoning Shenyang, 110023, China

[a]Liyanbin66@163.com, [b]Jangtong68@126.com, [c]Zhengpeng64@163.com

Keyword: Gear finishing, Hardened gears, Ausform finishing

Abstract. Ausform finishing process plastically deforms the surface of a steel component, generates compressive residual stresses in the surface layer, and imparts added strength and durability to the component. It integrates heat treatment, roll finishing and hardening of steel components into a single in-line manufacturing operation. The gear steel 20CrMo is introduced to determine various operation parameter of this process. The experiment results shown that ausform finished gears have the advantages of improved surface finish and profile accuracy.

Introduction

The demands of many high-performance vehicles are pushing the limits of their transmission gears. This is particularly crux of combat aircraft for which speed and payload requirement are steadily increasing and demanding more output from the gear transmission which must stay within present size and weight limitations.

In the mass production of vehicles transmission gears at home and abroad, the processes of hobbing teeth and gear shaving are still widely adopted. The process method that gear tooth surface after heat treatment were generally not finishing is difficult to achieve high accuracy requirements. Grinding teeth, the main processing means for high-precision gears, has been limited due to production efficiency, manufacturing cost and other reasons. In recent years, although there appears hardened precision hobbing technology, fine shaving technology and hardened precision machine tools, the price of equipment and tools is too expensive to our current situation. Based on the above situation, this paper proposes a new machining process for high-accuracy gear called ausform finishing and this process integrates heat treatment, roll finishing, and hardening of steel component into a single in-line manufacturing operation. The hot rolling process plastically deforms the surface of a steel component, generates compressive residual stresses in the surface layer, and imparts added strength and durability to the component. The hot rolling also creates the final net shape and surface finish of the part without the more costly traditional techniques of metal removal, such as grinding. Simulations results show that the hardness of gear after processing increased by 25%, and the accuracy increased by grade of 1 ~ 2. Based on 20CrMo carburizing gear which was widely applied in automobile industry, this article describes its processes and methods.

Gear Material and Machining Process

The gear made of 20CrMo steel materials is tooth face hardened, wear-resisting, internal low hardness ,and can withstand greater exposure to stress and bending stress after cementite and quenching, and has the advantages of small deformation after quenching. This material is widely used in automotive and high-precision machine tool manufacturing internal gear transmission in China. According to the present situation, this experiment uses steel 20CrMo as experimental material. Process route: Tooth roughing (hobbing, slotting)——Heat treatment(carburizing and quenching)——Ausform-finishing (insulation)——Cooling. Specific process is as follows:

Induction heating of the gear teeth, followed by rapid oil quenching at 150 to 250°C to obtain a metastable austenite (soft outer layer) of the teeth. This requires subjecting the gear to a specific temperature and time-usually less than a minute to soften the out layer without transforming it to martensite.

Precision roll finishing (in the hot oil) in which roller dies plastically deform the surfaces of the gear teeth into the final dimensions.

Final quenching at 50 to 65°C to harden the surface layer (by transforming austenite to martensite).

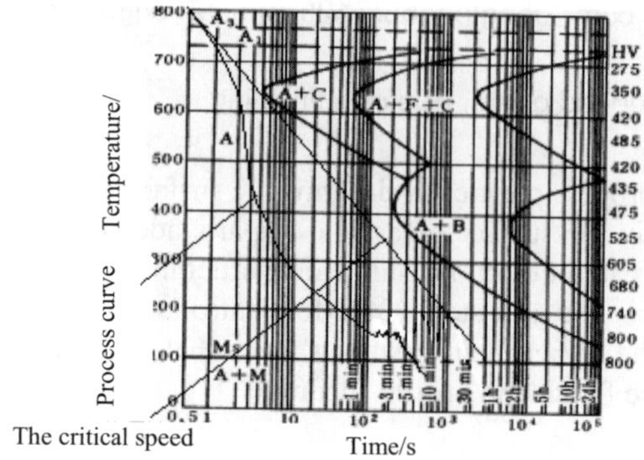

Fig. 1: Ausform finishing process

Process Parameters

Austenitizing Temperature. Since the transformation and homogenization of austenite during the process of heat treatment have a significant effect on transformation products of microstructure and properties of austenite in the subsequent cooling, the steel must be entire austenitic during heating. Entire austenitizing temperature of steel 20CrMo is 875°C [2].

Deformation Temperature. Generally, ausforming is heating the steel to the austenite temperature and thermal homogenization then cooling rapidly, and crossing the C curve of the shortest nose at the breeding area. The temperature of ausform finishing is in the temperature range of metastable austenite above the Ms point, usually in the C curve in the area[3]. The Ms temperature of carburized steel 20CrMo is 100°C, so the deformation temperature must be above this temperature. The ausform finishing must be carried out in the hot oil and maintain a certain time in a certain temperature to prevent the formation of martensite. If the temperature of ausform is too high, such as up to temperature 400°C ~ 500°C , the oil will have a break, and tooth bonding of gear and many other negative factors will appear. Therefore, the temperature of ausform should be close to the Ms point. The experiments show that for the steel 20CrMo deformation temperature of 150°C ~ 250°C is more appropriate. In this temperature gear tooth obtained metastable austenite(The organization is characterized by a soft outer layer).

Rate of Deformation. Deformation has a greater impact on the gears accuracy after hot roll finishing. The size of deformation depends directly on the reserved amount for roll finishing, if the reserved amount is too small to reach the purpose of tooth profile correction, to effectively improve properties such as tooth surface strength, and to get enough of the surface hardened layer. If the reserved amount is too large, the roller will be broken, the burr will appear on top of tooth and the bulge will occur on root of tooth. As a result, the tooth flank errors will increase, the diameter of top circle will be augment and the phenomenon of concave in mid of tooth profile aggravate. Referring to cold extrusion processed, reserved amount from 0.11mm to 0.22mm is more appropriate, and the relationship between deformation and reserved amount is as follows:

$$\Delta s - e_W - e_D = \Delta s_0 \tag{1}$$

Δs—— Deformation;
e_W—— Rough gear tooth profile error;
e_D—— Squeeze roller gear tooth profile error;
Δs_0—— the reserved space of squeeze roller gear.

The Critical Cooling Speed. The isothermal cooling curve shows the critical cooling speed should be

$$v_c = \frac{A_1 - T_m}{\tau_m} \tag{2}$$

When the continuous cooling transformation, CCT curve over to the right than the TTT curve, therefore, a correction coefficient k should be added into Eq. 2.

$$v_c = \frac{A_1 - T_m}{k\tau_m} \quad (k>1) \tag{3}$$

When T_m=600℃, τ_m=6s, to make the steel austenitize completely, making A_1=775 °C, then the critical quenching rate can be calculated. Therefore, the actual quenching rate must be greater than 30 °C / s. As the nose part of 20CrMo TTT cooling curve is at between 600 °C and 700 °C, the cooling rate of 52# cylinder oil in this range is about 90 °C / s, which is fully able to meet the requirements. 52 # cylinder oil cooling curve is shown in Fig. 2.

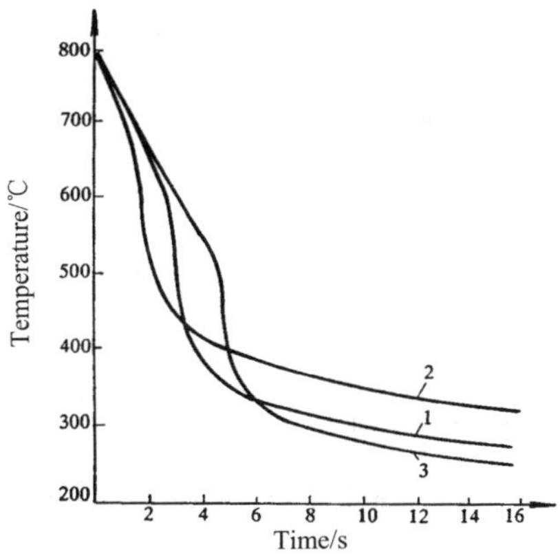

1——Rating quenching oil, 2——52 # cylinder oil, 3——Blended oil
Fig. 2: the cooling curves of several oil

Determination of Quenching Medium. The quenching oil in the process is used as quenching medium and as heat preserving and lubricating oil in ausform finishing, and the temperature required is about 150 °C ~ 250°C. In this high temperature, the selection for the appropriate technology quenching oil is one of the difficulties. Quenching oil should have a high cooling rate and flash point. Usually at room temperature, the cooling capacity of low viscosity oil is higher than that of high viscosity oil. When the temperature is rising, the cooling capacity of the low viscosity oil drops, and the viscosity of the high viscosity oil decreases, liquidity increases, cooling ability enhances. And increasing the oil temperature will reduce the stress in the workpiece, which help to reduce deformation and cracking in the workpiece. Therefore, the use of hot oil quenching is favorable. However, to ensure safety, the flash point must be higher than the oil temperature 60°C [4]. It should select quenching oil with high viscosity, flash point higher than 310°C. In view of this, we use the 52

cylinder oil (flash point 310°C) as the quenching medium. In addition, using temperature of isothermal quenching medium grade oil is usually between 150 °C and 250 °C. The extreme pressure additive should be added to prevent agglutination.

Determine for the Ausform Time. Ausform time should be less than the incubation period of ausform temperature, otherwise the austenite will be transformed into bainite, without transforming it to martensite, so that the tooth surface hardness will decrease. From the ausform finishing process curve (Fig. 1), it can be seen that ausform time should be less than 2 minutes, and ausform time should be less than 1 minute if the related operation time has been considered.

Performance Validation

The test results for 20CrMo gear whose module is 2.5 and teeth number is 49 show following advantages, as shown in Fig. 3 and Fig. 4.

(1) Enhanced tooth surface finish on 3 or 6 μ(Ra).

(2) Stronger surface layers and finer microstructure.

(3) Increased compressive residual stress: 40% higher than for conventionally hardened gears.

(4) Elimination of hardened gear-teeth finishing operations such as grinding, honing, and burnishing, as well as off-line processes such as nitaletch inspection, refrigeration, and shor peening. Such processes are often required for high-performance aircraft and automotive transmission gears.

(5) The finishing process time has decreased to less than a minute, compared to grinding, which can take several minutes.

The ausform finished gears showed significantly improve surface finish and profile accuracy.

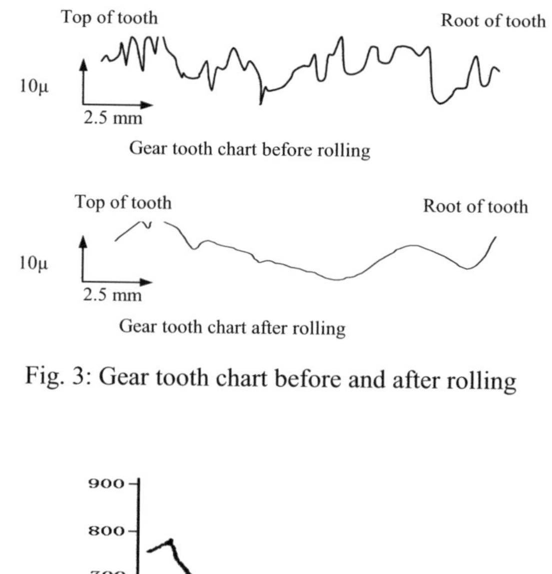

Fig. 3: Gear tooth chart before and after rolling

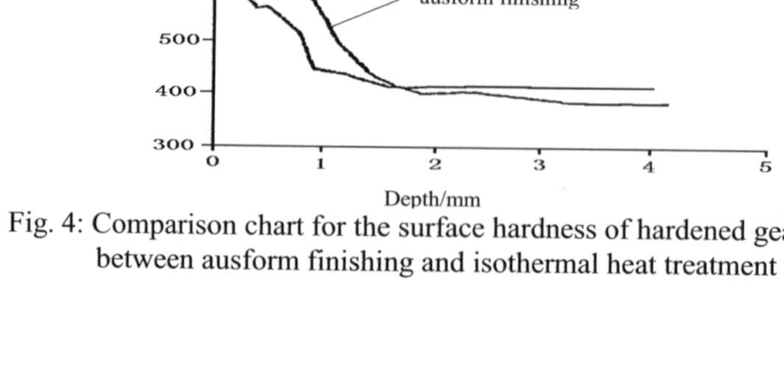

Fig. 4: Comparison chart for the surface hardness of hardened gear between ausform finishing and isothermal heat treatment

Conclusion

This paper presents a ausform finishing process which integrates heat treatment, roll finishing, and hardening of steel component into a single in-line manufacturing operation.

The hot rolling process plastically deforms the surface of a steel component, generates compressive residual stresses in the surface layer, and imparts added strength and durability to the component. Rolling also creates the final net shape and surface finish of the part without the more costly traditional techniques of metal removal, such as grinding.

With 20CrMo gear that is currently widely used in China's current precision machine tool and automotive industries, the processing temperature, deformation, critical cooling rate and other parameters of the process is determined. Experiment results show that tooth surface finish is enhanced on 3 or 6 μ(Ra), stronger surface layers and finer microstructure, significantly improved surface finish and profile accuracy.

Compared to a typical processing sequence used for automotive gears, ausforming(after case-hardening) replace the gear shaving or roll finishing process prior to heat treatment, and it eliminates burnishing and shot blasting processes.

Ausforming promises to substantially increase power-to-weight ratios of aircraft and rotorcraft transmissions so they can transmit higher load without expanding the transmission envelop. It also has the potential of producing high dimensional accuracy and improved surface finish, while eliminating several offline processing steps associated with heat treatment and grinding.

Ausforming offers a faster, less costly way to finish automotive gears. It can deliver high throughout, up to four gears per minute for a typical planetary helical pinion.

Acknowledgement

This work was supported by Science Research Plan of Liaoning Province Education Ministry Grant 2009A552.

References

[1] Joe Brown: *Future gear pack more power*. Power Transmission Design, 1998, 4, pp. 38~41.

[2] Cui Zhongxi: *Theory for metal and heat treatment*. China Machine PRESS, Beijing (2000)

[3] Fan Dongli: *Manual for data of heat treatment technology*. China Machine PRESS, Beijing (2006).

[4] Wang Yundi:*Quenching medium*. science and technology PRESS, Beijing (2001).

Advanced Materials Research Vol. 279 (2011) pp 296-300
© (2011) Trans Tech Publications, Switzerland
doi:10.4028/www.scientific.net/AMR.279.296

Development for the comprehensive optimization technology of process lubrication regime in double cold reduction

Zhen-Hua Bai[1,a], Hong-Xin Si[1,b], Qing-Tian Zhou[1,c], Xiu-Jun Li[2,d], Shou-Min Wu[2,e] and Xiao-Dong Shi[1,f]

[1]School of Mechanical Engineering, State Key Laboratory of Metastable Materials Science and Technology, YanShan University, QinHuangdao 066004, People's Republic of China

[2]Thin Strip Co.Ltd., Baoshan Iron & Steel Co.Ltd., Shanghai 200431, People's Republic of China

[a]bai_zhenhua@yahoo.com.cn, [b]sihongxin@126.com, [c]zhou_qingtian@yahoo.com.cn, [d]li_xiujun_2000@ yahoo.com.cn, [e]wu_shoumin@yahoo.com.cn, [f]xiaodong_shi@126.com

Keywords: Double cold reduction, Process lubrication, Strip shape, Surface quality, Fuel consumption

Abstract. The setting goals of process lubrication were not all-inclusive in the previous manufacturer of double cold reduction. Strip shape, surface quality, the fuel consumption of unit and rolling stability could not be considered comprehensively. Aimed at the equipment and production craft characters of double cold reduction unit, at the basis of putting forward the comprehensive control index of a strip-shaped fuel consumption cleanliness for the first time, ensuring the shape quality of strip, lowering the oil residue on strip surface, improving the cleanliness of strip surface and reducing fuel consumption were defined as control targets; besides, the control of slip and heat-scratches were defined as constraints. A complete comprehensive optimization technology of process lubrication regime was established and has been applied to a 1220 double cold reduction unit belonging to Baosteel in China. Rolling speed, surface quality and strip shape quality are improved effectively, and oil consumption is reduced, meanwhile, the productivity and rolling yield of unit are ensured. This technology has created economic benefits for domestic steel makers and has further extended application value.

Introduction

Double cold reduction (DCR) refers to the further reduction after once cold rolling, and it will increase the hardness and strength of materials. During the double cold reduction, in order to reduce the surface temperature of rollers and strip, minish the friction coefficient and friction force of the contact surface of deformation zone, prevent metal attaching to the rollers' surface and reduce the abrasion of rollers as well, the process lubrication of spraying emulsion to the surface of rollers and strip is needed. The performance of process lubrication will directly affect the rolling stability and product quality. When the kind of emulsion has been determined, process lubrication regime includes three parts: (1) the flow of emulsion; (2) the concentration of emulsion; (3) the setting of initial temperature. So, under the premise of ensuring rolling stability and developing the potential of rolling mill adequately, how to conduct integrated optimal setting of the above-mentioned three parts to improve product quality and reduce fuel consumption has become the emphasis of field breakthrough. Previously, for the setting of the above-mentioned three parameters, people often selected the method that form combines with operators' experience. This method has the features of less targeted, arbitrarily large, and it will lead to the stability of product quality is poorer. Therefore, fully combined the equipment and production craft characters of double cold reduction unit, at the basis of putting forward the comprehensive control index of a strip-shaped fuel consumption cleanliness for the first time, ensuring the shape quality of strip, lowering oil residue on strip surface, improving the strip surface cleanliness and reducing fuel consumption were defined as control targets; besides, the control of slip and heat-scratches were defined as constraints. A complete comprehensive optimization technology of process lubrication regime was established and applied to production practice, and good effects of use are achieved.

Impact Analysis of Process Lubrication Regime on the Important Parameters of Double Cold Reduction

The influence of the emulsion flow, the emulsion concentration and the setting of initial temperature on friction coefficient is caused mainly by changing the dynamic viscosity of basic oil. The oil film thickness of deformation zone is affected firstly, and finally the friction coefficient is changed. Based on [1], under the premise that rolling parameters have been determined, the friction coefficient can be written to the function of process lubrication regime:

$$\mu = f_1(C, w, T_c) \tag{1}$$

where C is the flow of emulsion, w is the concentration of emulsion, and T_c is the initial temperature of emulsion.

Impact Analysis of Process Lubrication Regime on Strip Shape. The influence of process lubrication regime on strip shape mainly manifests in the following two aspects: (1) the change of process lubrication regime will influence friction coefficient, and the change of friction coefficient will influence rolling pressure as well. The change of rolling pressure will influence on-load roll gap, and the change of on-load roll gap will directly affect the exit shape; (2) the change of process lubrication regime will influence the thermal crown of working roll. The change of working roll's thermal crown will influence on-load roll gap, and finally affect the strip shape through the change of on-load roll gap as well. According to the above analysis, the impact model of process lubrication regime on strip shape is as follows: according to the model of shape control [3], it can be known that when front and back tension, the shape of incoming material, the cross-section shape of incoming material, the width of incoming material, reduction, roller bending value and roller shifting value etc. have been determined, the exit shape of mill can be expressed as follows:

$$\sigma_{1i} = f_2(P, \Delta D_{wi}, \Delta D_{bi}, \Delta D_{mi}) \tag{2}$$

where T_1 and T_0 are respectively the front and back tension; P is the total rolling force; $\Delta D_{wi}, \Delta D_{mi}, \Delta D_{bi}$ are respectively the roll contour's distribution value of working roll, middle roll, back-up roll.

In Eq. 2, the roll contour's distribution value of working roll can be divided into two parts. One part is the superposed roll contour which is formed by the original roll contour and abrasion, and the other part is the thermal crown of working roll, expressed as follows:

$$\Delta D_{wi} = \Delta D_{wyi} + \Delta D_{wti} \tag{3}$$

where ΔD_{wyi} is the superposed roll contour which is formed by the original working roll contour and abrasion; ΔD_{wti} is the thermal contour of working roll.

Considering that compared with working roll, process lubrication regime has a minimal effect to the thermal contour of middle roll and back-up roll and can be ignored basically. Thinking namely:

$$\begin{cases} \Delta D_{mi} = \Delta D_{myi} \\ \Delta D_{bi} = \Delta D_{byi} \end{cases} \tag{4}$$

where ΔD_{myi} is the superposed roll contour which is formed by original middle roll contour and abrasion, ΔD_{byi} is the superposed roll contour which is formed by original back-up roll contour and abrasion.

Actually, during the double cold reduction, when the parameters of cooling liquid and rolling technological parameters have been determined, the thermal contour of working roll mainly depends on the process lubrication regime, [4], which can be expressed as followings:

$$\Delta D_{wi} = f_3(C, w, T_c) \tag{5}$$

Meanwhile, according to the rolling theory [2], it can be known that during the double cold reduction, for the specific steel grade, after front and back tension, the width of incoming material, reduction and working roll diameters etc. have been determined, rolling pressure mainly depends on the friction coefficient and it can be expressed with the following Eq. (6).

$$P = f_4(\mu) \tag{6}$$

Eq. 1 and Eq. 3-6 are substituted into Eq. 2, and the following Eq. 7 can be derived:

$$\sigma_{1i} = f_5(C, w, T_c, \Delta D_{wyi}, \Delta D_{byi}, \Delta D_{myi}) \tag{7}$$

As can be seen clearly from Eq. 7, for a specific rolling process, after its steel grade, front and back tension, entry and exit gauge, the thickness of incoming material, the width of incoming material, the distribution of strip shape, the bending force of working roll and middle roll, middle roll shifting value, the original roll contour and the abrasion value of working rolling, middle roll, back-up roll have been determined respectively, and the exit shape of strip depends on the initial temperature of emulsion, the emulsion concentration and the emulsion flow, which can be expressed with the following Eq. (8):

$$\sigma_{1i} = f_6(C, w, T_c) \tag{8}$$

Impact Analysis of Process Lubrication Regime on Rolling Stability. During the double cold reduction, when the setting of process lubrication regime is unreasonable, with the increase of rolling speed, the phenomenon of slipping will appear. According to [5], slipping factor ψ can be expressed the function of process lubrication regime parameters.

$$\psi = f_7(C, w, T_c) \tag{9}$$

Impact Analysis of Process Lubrication Regime on Heat-scratches. Heat-scratches refers to the phenomenon that due to the lacking of lubrication, localized rupture appeared in lubricating oil film, and the local bonding of strip and roller's surface cause damages. Heat-scratches are also called hot scratch, friction roller marks, surface crack and fishlike fringe etc. Heat-scratches are mainly determined by oil film thickness and rolling temperature. According to [6], under the premise that rolling parameters have been determined, slipping index φ which is used to express the probability and severity of heat-scratches' appearance can be written to the function of process lubrication regime:

$$\varphi = f_8(C, w, T_c) \tag{10}$$

Establishment of the Comprehensive Control Indexs of Strip Shape Fuel Consumption Cleanliness

During the double cold reduction, the index of fuel consumption can be expressed as follows:

$$k_y = C \cdot w \tag{11}$$

Similarly, during the double cold reduction, from the perspective of process lubrication regime, the surface cleanliness (residual oil) mainly depends on the concentration of emulsion, the higher of concentration, and the larger of residual oil. Based on the field tests, tracking and statistics regression, the index of fuel consumption can be appropriately expressed by the following Eq. (12):

$$k_c = \alpha_{c_1} C^{\alpha_{c_2}} \tag{12}$$

where α_{c_1} and α_{c_2} are respectively the residual oil coefficient, which can be obtained through the regression analysis of in-situ data.

During the double cold reduction, people generally determine the performance of strip shape often through the transverse distribution of front tension; that is to say, the target of shape control is to minimize the difference of transverse front tension. Based on this, the objective function of shape control can be defined as:

$$F(X) = ((\max(\sigma_{1i}) - \min(\sigma_{1i}))/T_1) \tag{13}$$

Finally, in order to ensure optimal exit shape, high surface cleanliness, low oil consumption, combined the equipment and production craft characters of double cold reduction unit, the comprehensive control index of a strip-shaped fuel consumption cleanliness is put forward for the first time, which is expressed as follows:

$$G(X) = (k_y)^{\alpha} \cdot (k_c)^{\beta} ((\max(\sigma_{1i}) - \min(\sigma_{1i}))/T_1) \tag{14}$$

where α, and β respectively are the coefficient of oil consumption and residual oil

Development for Comprehensive Optimization Technology of Process Lubrication Regime

In recent years, with the continuous improvement for double cold reduction products quality from customers, present double cold reduction not simply pursue production efficiency and output, but should pay more attention to the product quality. Based on the above considerations, during double cold reduction, there are four targets of comprehensive optimization technology of process lubrication: (1) stable rolling. Without slipping phenomenon, a certain rolling speed should be ensured, and full play to the potential of mill should be given to improve the production efficiency and output; (2) without heat-scratches. The emulsion must keep a certain concentration and flow, or there will be a localized rupture of lubricating oil film and lead to heat-scratches; (3) good exit flatness; (4) high surface cleanliness and low oil consumption. The concentration of emulsion should be reduced as much as possible. In order to achieve the above goals, the following functions of the comprehensive optimization technology of process lubrication are put forward especially:

$$\begin{cases} G(X) = (k_y)^{\alpha} \cdot (k_c)^{\beta} ((\max(\sigma_{1i}) - \min(\sigma_{1i}))/T_1) \\ \psi(X) \le \psi * \\ \varphi(X) \prec \varphi * \\ T_{\min} \le T_c \le T_{\max} \\ C \le C_{\max} \\ w \le w_{\max} \\ X = \{C, w, T_c\} \end{cases} \tag{15}$$

where $T_{\min}$ is the needed minimum temperature for keeping the activity of emulsion; $T_{\max}$ is the permitted maximum temperature for emulsion; $C_{\max}$ is the allowable maximum concentration; $w_{\max}$ is the allowable maximum flow; $\psi *$ is the critical slipping factor of DCR; $\varphi *$ is the critical slipping index of DCR.

So, the comprehensive optimizing technology of process lubrication regime of the whole double cold reduction can be described as follows: finding a suitable $X = \{T_c, C, w\}$ which will minimize the $G(X)$ under the premise of meeting constraint conditions.

Field Application of Models

In order to improve rolling stability and product quality, the comprehensive optimization technology of process lubrication regime which is mentioned in the article has been adopted by a domestic certain 1220 double cold reduction mill. A software about the comprehensive optimization technology of process lubrication regime of the 1220DCR under rolling mode has been worked out. The comprehensive optimizing of process lubrication regime of the 1220DCR has been realized. After relevant technologies were put into operation, under the premise of that slip and the defect of heat-scratches don't occur, the actual average rolling speed increases 19.85%. Among these, the steady state maximum rolling speed of BA material increases from 508m/min to 609m/min, increasing 19.90%; the steady state maximum rolling speed of CA increases from 1000m/min to 1198m/min, increasing 19.8%; the surface oil consumption decreases from 90-140 mg/m2 to 48-65 mg/m2; at the stable state, and the wave height of strip is less than 3mm and reaches to 88%.

Conclusion

Aimed at the problem that the setting goals were not all-inclusive in the previous manufacturing process of double cold reduction, strip shape, surface quality, fuel consumption of unit and rolling stability etc. could not be considered comprehensively. Based on lots of field experiments and theoretical research, fully combined the equipment and production craft characters of double cold reduction unit, at the basis of putting forward the comprehensive control index of a strip-shaped fuel consumption cleanliness for the first time, ensuring the shape quality of strip, lowering the oil residue on strip surface, improving strip surface cleanliness and reducing fuel consumption were defined as control targets; besides, the control of slip and heat-scratches were defined as constraints. Corresponding comprehensive optimizing technologies of process lubrication regime in high speed rolling process of double cold reduction unit were given and used to a 1220 double cold reduction mill. Through the comprehensive optimization settings of emulsion flow, density and initial temperature, good effects were obtained, stability rolling in high speed was realized under the premise of that slip and the defect of heat-scratches don't occur. At the same time, the effects of good flatness in mill outlet, high surface cleanliness and less fuel consumption were gained, and a larger economic benefit for production was created.

References

[1] Bai Zhenhua: *The core technology mathematical model in high-speed tandem cold rolling process*. China Machine Press, Beijing, (2009).

[2] Cao Hongde: *The mechanical foundation of plastic deformation and rolling principles*. China Machine Press, Beijing, (1981).

[3] Lian Jiachuang, Liu Hongmin: *Guage and shape control*. Weapon Industry Press, Beijing, (1995).

[4] W.L.Roberts: *Cold rolling strip production*. China Machine Press, Beijing, (1991).

[5] Bai Zhenhua, Wang Junfei: *The condition of judgment about slip in cold tandem mill and the prevention measures*. Shanghai Metals, 26, 4(2004), pp. 15-20.

[6] Bai Zhenhua, Li Xingdong, Wang Junfei: *Research of Screw-down Schedule Optimization for Preventing and Curing Hot Slip of IF Steel*. Steel Vol. 7 (2005), pp. 40-45.

Advanced Materials Research Vol. 279 (2011) pp 301-306
© (2011) Trans Tech Publications, Switzerland
doi:10.4028/www.scientific.net/AMR.279.301

Research on six DOF vertical vibration model of a four roller Hot Strip rolling mill stand

Weijin Ma[1,a], Wang Junyuan[2,b], Fenglan Li[3,c], Zeng Zhiqiang[4,d], Xiao Denghong[5,e] and Xiong Shibo[6,f]

[1, 2, 4, 5]College of mechanical engineering and automatization, North University of China, Taiyuan, China 030051

[3, 6]Mechano-electronic Institut, Taiyuan university of science and technology, Taiyuan, China 030024

[a]maweijin1957@yahoo.com.cn, [b]wjy@nuc.edu.cn, [c]lifenglan1957@yahoo.cn, [d]zzhqzhy@nuc.edu.cn, [e]denghong_xiao@hotmail.com, [f]sbxiong@tyut.edu.cn

Keywords: Hot strip rolling mill stand, vertical vibration model, modal frequency, numerical computing

Abstract. A lumped mass model with six degrees of freedom for the vertical vibration of four roller hot strip rolling mill stand was proposed, while its natural frequencies were calculated by MATLAB software. The numerical calculation results were proved more accurate through field vibration test. This vertical vibration model can be applied for dynamic simulation and structural dynamic modification with important theoretical and practical significance in mastering the vibration characteristics of four-roll mill, avoiding vertical self-excited vibration and improving production efficiency.

Introduction

As one of the most important components in rolling mill, frame was used to install the roller systems and roller adjustment devices, whose vibration characteristics and deformation affect the equipments reliability and rolling precision [1]. Natural frequency of the rolling mill vertical vibration system was an important dynamic characteristic, which was the absolutely necessary parameter for mill vibration type or nature judgment and vibration inhibition measures [2,4]. The four-roll mill's inherent vertical vibration characteristics were also important for dynamic design and modification of modern rolling mill. Mill stand's self-excited vibrations were closely related to its natural frequencies and mode shapes. Therefore, comprehensive understand of natural frequencies and vibration characteristics was the basic condition of rolling mill vibration analysis [5].

Recently lots of scholars at home and abroad have being researched the four-roll mill's inherent vertical vibration calculation with model proposed mainly on two types as follows: the finite element model and lumped mass model [6,9]. Although the finite element analysis was much more accurate, the model description was too complex. As study research contents were different, several simplified models were used in the vertical vibration analysis. Usually the models were simplified as six degrees of freedom (6 DOF), four degrees of freedom (4 DOF), two degrees of freedom (2 DOF), or a single degree of freedom(1 DOF) system, according to the research purpose and precision. 2 DOF model and 1 DOF model were simplified seriously, making the model less accurate, and lots of vibration characteristics couldn't be expressed sufficiently [10]. In this paper, the mill's vertical vibration system was simplified as a non-symmetric 6 DOF model with appropriate simplicity, which could be further used in the overall vibration controlling simulation model. The numerical calculation results were accurate and credible according to field vibration test results.

The 6 DOF vertical vibration model for rolling mill stand

An hot mill work base's simplified diagram with descriptions was shown in Fig. 1. The mill's simplified model as 6 DOF spring mass systems was shown in Fig. 2.

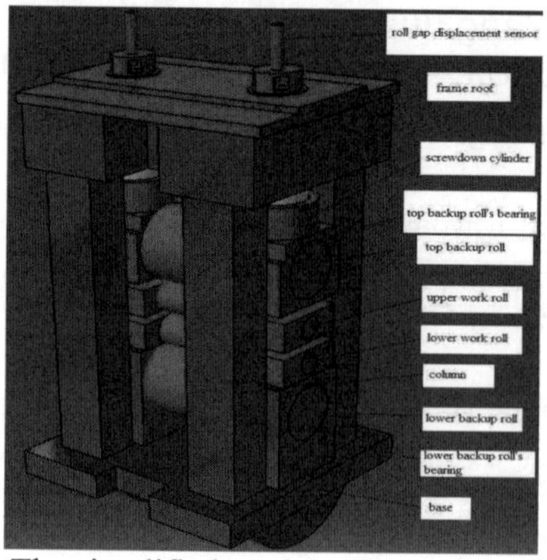

Fig. 1: The simplified model of hot strip mill stand

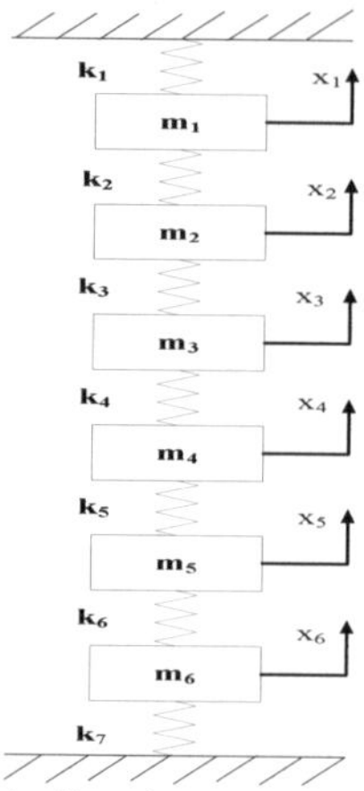

Fig. 2: Six DOF vertical vibration model of hot strip mill stand

Graphic symbols were explained as follows:

M_1—equivalent mass of column and upper beam (including fuel tanks);

M_2—equivalent mass of top backup roll with its bearings and bearing supports;

M_3—equivalent mass of roll system's upper work roll;

M_4—equivalent mass of roll system's lower work roll;

M_5—equivalent mass of lower backup roll with its bearings and bearing supports;

M_6—equivalent mass of lower beam;

k_1—equivalent stiffness of frame column and upper beam;

k_2—equivalent stiffness of mid top backup roll and mid upper beam, including cylinder oil film stiffness, bearings and bearing supports stiffness of the top backup roll and bending rigidity of the top backup roll;

k_3—elastic contact stiffness between the upper work roll and upper bearing roll;

k_4—equivalent stiffness between upper and lower work rolls and the strip when affected by the rolling force;

k_5—elastic contact stiffness between lower work roll and lower backup roll;

k_6—equivalent stiffness between the mid lower backup roll and mid lower beam;

k_7—equivalent bending stiffness of the lower beam;

x_i, i=1,2,…,6—each element's displacement.

Rolling mill's natural frequencies were determined by equivalent mass and equivalent stiffness of the system, which was directly affected by the theoretical calculation accuracy.

Equivalent mass and stiffness were calculated with the principle of conservation of energy as shown in Table 1 according to the factory's hot strip mill fifth frame structure.

Table 1: Equivalent mass and stiffness of the six DOF vibration mill model

Numbers	1	2	3	4	5	6	7
m_i [t]	186	87.6	30	30	87.6	30	
k_i [GN/m]	97	6.57	121	9.43	121	112	119

Natural frequencies calculation for mill's vertical vibration systems

From Fig. 1 the 6 DOF mass-spring systems' motion differential equations without damping could be deduced as follow:

$$\boldsymbol{M}\ddot{\boldsymbol{x}} + \boldsymbol{K}\boldsymbol{x} = 0. \tag{1}$$

$$\boldsymbol{M} = diag(m_1, m_2, m_3, m_4, m_5, m_6) \tag{2}$$

$$\boldsymbol{K} = \begin{bmatrix} k_1+k_2 & -k_2 & & & & \\ -k_2 & k_2+k_3 & -k_3 & & & \\ & -k_3 & k_3+k_4 & -k_4 & & \\ & & -k_k & k_4+k_5 & -k_5 & \\ & & & -k_5 & k_5+k_6 & -k_7 \\ & & & & -k_7 & k_6+k_7 \end{bmatrix} \tag{3}$$

$\boldsymbol{M}$ ——system mass matrix;
$\boldsymbol{K}$ ——system stiffness matrix;
$\ddot{\boldsymbol{x}}, \boldsymbol{x}$ ——system acceleration vector and displacement vector;
Eq. 1 could be expressed as the following standard eigenvalue problem.

$$\boldsymbol{K}\boldsymbol{v} = \boldsymbol{p}^2\boldsymbol{M}\boldsymbol{v}\text{, or } \boldsymbol{K}\boldsymbol{v} = \lambda\boldsymbol{M}\boldsymbol{v} \tag{4}$$

$$f_i = \frac{p_i}{2\pi} = \frac{\sqrt{\lambda_i}}{2\pi} \tag{5}$$

In the formula, eigenvalue expression $p_i = \sqrt{\lambda_i}$ was the frequency of the system; feature vector v was the vibration mode of the system. This eigenvalue problem was solved by MATLAB, and the natural frequencies calculation results f_i were shown in Table 2. Each modal shape was shown in Fig. 3. From Table 2 and Fig. 3, it can be found that the hot rolling mill vibration contained first three components, and vibration energy was concentrated in the second order (114Hz) and third order (119Hz) particularly. And the vibration modes corresponding to these two frequencies were exactly reverse movement of the two work rolls, being consistent with the mill's self-excited vibration characteristics.

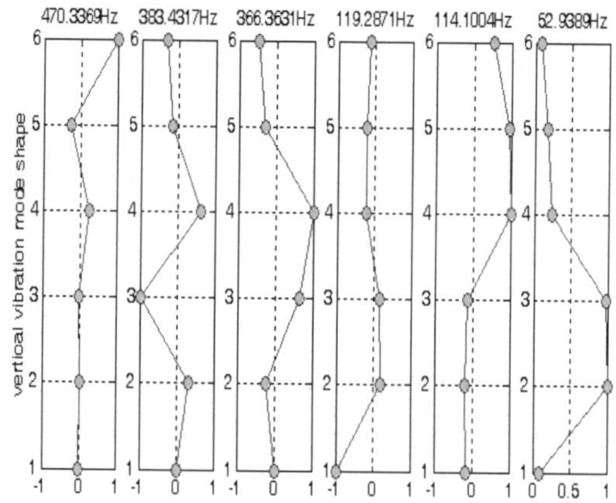

Fig. 3: Mode shapes of the hot strip mill stand

Field vibration test and model parameters verification

The measuring points' distribution of field vibration test was shown in Fig. 4. Vertical vibration system's natural frequencies were measured through the vibration waveform inspired when bit into a flat strip during test process. Measuring points were arranged in the top of the rack as well as on the upper backup roll bearing, reflecting roll systems' dynamic characteristics. Sampling rate and analysis frequency was 2000Hz and 1000Hz respectively. The record length was 141 seconds, which was the rolling slab band process. Slab band thickness from the fifth frame of precision rolling was 3mm; the velocity of slab band was 4.9m/s and 5.3m/s when steel was bit and casted respectively. The work roll diameter was 650 mm.

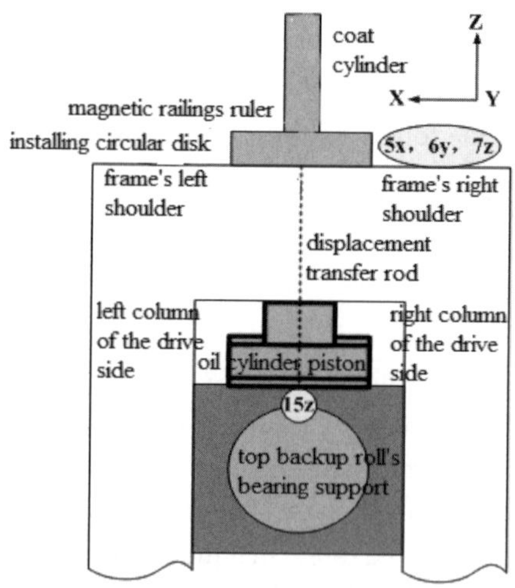

Fig. 4: Vibration test point distribution

The spectral analysis results using MATLAB were shown in Fig. 5 and Fig. 6. During the analysis process, the interception length of analysis signal was 3000 samples per 1.5 seconds, while analysis frequency was 1000Hz.

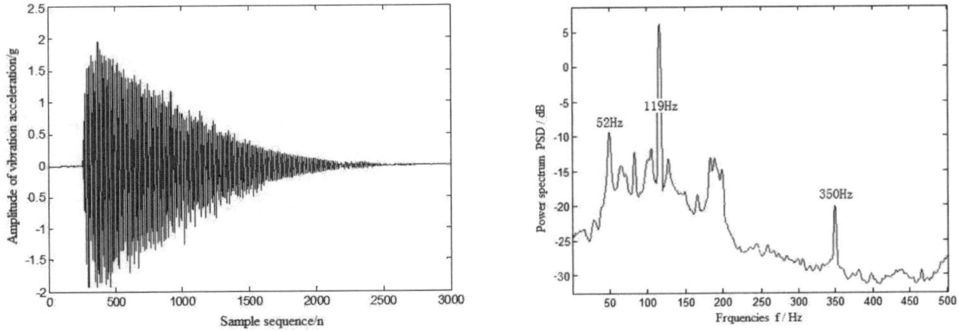

(a) Original signal (b) Power spectrum

Fig. 5: Acceleration signal and its power spectrum of the 15th measuring points on upper support roller bearing

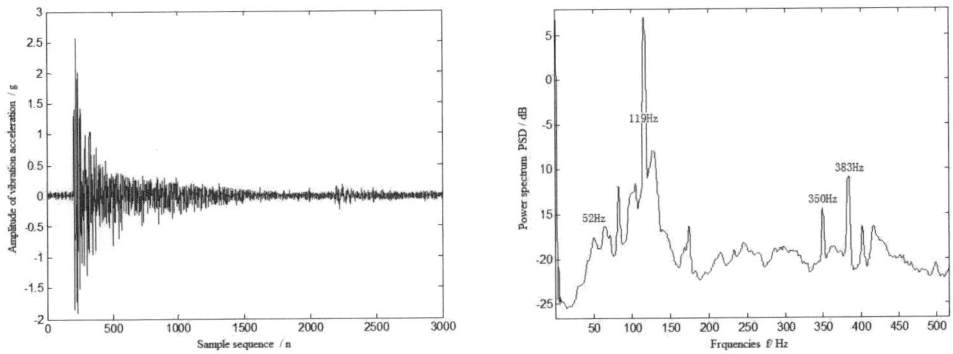

(a) Original signal (b) Power spectrum

Fig. 6: Acceleration signal and its power spectrum of the 7th measuring points

From original signal in Fig. 5 and Fig. 6, it can be found that the frame vibrated severely when steel was bit, then the amplitude was awaken, and vibration was disappeared after about 1 second, entering into the steady-state rolling process.

Table 2: Natural frequencies' theoretical value calculation and actual test value

Modal number	1	2	3	4	5	6
Calculated value f_i [Hz]	52.9389	114.1004	119.2871	366.3631	383.4317	470.3369
Tested value f_i [Hz]	52.0		119.0	350.0	353.5	

The support rolls upper part's vertical vibration mainly contained components of the first order modal frequency (52Hz), the third order modal frequency (119Hz) and the forth order modal frequency (350Hz), which was easily found from Fig. 5; while the top of the frame contained the third order modal frequency (119Hz), the forth order modal frequency (350Hz) and the fifth order modal frequency (383Hz), which was shown in Fig. 6.

The test signal in Table 2 mainly contained the first order modal frequency (52Hz), the third order modal frequency (119Hz), and the fifth order modal frequency (383Hz). Compared with previous theoretical calculation results, the error was less than 1Hz; then if frequency components 350Hz were considered as the fourth-order modal frequency, the error was still less than 5% compared with the theoretical results (366HZ), indicating that the computational model proposed in this paper was of higher accuracy, and natural frequencies' calculation results were well in accordance with test value.

Conclusions

In this paper the first sixth order natural frequencies and modal shapes were computed theoretically through the four roller hot strip rolling mill stand's simplified vertical vibration systems as a non-symmetric six DOF model. Scene vibration test results proved that the system's actual vibration signals included four modal frequencies. The calculated result of natural frequency consisted with measured values very well, and calculation errors were less than 5%, revealing that the proposed computational model was of higher precision, and the appropriate simplicity can be used in the overall vibration control of the frame's simulation model.

Acknowledgments

This research project was supported by the National Natural Science Foundation of China (Grant No.50975188) and Shanxi Province Foundation for Returned Scholars, China (Grant No. 20101477).

References

[1] Lian Jiachuang, etc: *Research on natural frequency of four roller mill horizontal vibration*, Chinese Journal of Mechanical Engineering, 19, 3(1983), pp. 1-10.

[2] Yang Qijun, Lian Jiachuang: *Stability analysis of self-excited vertical vibration of high speed cold strip mills by numerical method.* Journal of Vibration and Shock, 15, 3(1996), pp. 16-21.

[3] Cai Ganwei, Duan Jian, Yi Youping, Zhong Jue, Chen Peilin: *A finite element model for dynamic analysis of rolling mill.* Chinese Journal of Mechanical Engineering, 36, 7(2000), pp. 66-68.

[4] Tang Honghai, Li Yourong, Liu Anzhong: *Finite element analysis of four-high mill stand.* Journal of Wuhan University of Science and Technology, 32, 2(2009), pp. 149-153.

[5] Xiao Yunping, Liu Yilun, Zhou Xian: *Simulation and analysis on the vertical vibration of four-high rolling mill stand.* Machinery, 35, 1(2008), pp. 17-18.

[6] Ma Weijin, Li Fenglan, Xiong Shibo: *Analysis and Diagnosis of Self -Excitation-Vibration of Hot Rolling Mill, Journal of Vibration.* Measurement & Diagnosis, 26, 4(2006), pp. 261-264.

[7] Ma Weijin, Wang Junyuan, Li Fenglan etc: *Experimental modal analysis of roller gap sensor structure on hot rolling mill.* Advanced Materials Research Vol. 97-101 (2010), pp. 4328-4331.

[8] Ma Weijin, Xiong Xiaoyan, Li Fenglan, etc: *Modal parameter testing of roller gap sensor structure on hot rolling mill.* Chinese Journal of Mechanical Engineering, 45, 7(2009), pp. 259-264.

[9] Kazushige Ishino, Kazuhisa Kabeya, Takao Yoshikawa: *Vibration analysis of chatter in cold rolling mill (Theoretical study on self-excited vibration by a Two-DOF mass-spring-damper model).* Transactions of the Japan Society of Mechanical Engineers, code C, 687, 69(2003), pp. 2975-2982.

[10] I.S. Yun, W.R.D. Wilson, K. F. Ehmann: *Review of chatter studies in cold rolling.* International Journal of Machine Tools & Manufacture Vol. 38(1998), pp. 499-1530.

Advanced Materials Research Vol. 279 (2011) pp 307-312
© (2011) Trans Tech Publications, Switzerland
doi:10.4028/www.scientific.net/AMR.279.307

Analysis on transmission of UV radiant energy in Rapid Prototyping

Dianhong Yu[1, a], Lin Li[1,b] and Kai Zhao[1,c]
[1]School of Mechanical & Instrumental Engineering, Xi'an University of Technology, China
[a]dhyu0401@163.com, [b]lil2305@xaut.edu.cn, [c]zhaokaiii@163.com

Keywords: UV radiation, rapid prototyping, energy transmission, optical fiber

Abstract. Based on the optical-wave's theory, focusing energy transmission of UV radiation in rapid prototyping technology has been analyzed in depth. The reflecting focusing theory that continuous spherical wave radiated by point light source is reflected from concave ellipsoidal surface has been enucleated and the axial and lateral optical energy distribution of the reflecting optical wave field has been showed up. The analysis indicates that distribution of the intensity in optical axis is approximately Gauss distribution. And by means of experimentation, the research educes that the UV radiant energy being coupled by optical fiber and being focused by lens is approximately Gauss distribution, too. For effectively utilizing the UV radiant energy and improving the prototyping precision in rapid prototyping, the research supplies sufficient academic basis and has important guidance on the optical design of rapid prototyping.

Introduction

Rapid Prototyping manufacture, which is the technology to achieve part-making by accumulate thought, is the key breakthroughs in the manufacturing technology field. It can change the design idea from CAD model to certain functional prototype or parts automatically, directly, quickly, and rapidly [1,2]. Therefore, the technology can evaluate and perfect the design rapidly. Ultraviolet rapid prototyping technology is one of the RP technologies. Using UV light source instead of He-Cd Laser Light Source is core of the technology, and it was popularized and developed because of the obvious advantage on the equipment cost [3]. One of the key technologies of Ultraviolet rapid prototyping is perfecting the Coupled transmission of the light energy and enhancing the energy density of the scanning point. In UV Rapid Prototyping system, we adopt concave ellipsoid reflecting technique to achieve the reflecting and the convergence of light energy, whereafter the light energy was coupled and was transferred to fiber. Utilizing the flexile fiber, the light energy is transmitted expediently, and focused. Then, it can achieve the effective supply of light energy in the Rapid Prototyping. This paper mainly studies light energy distribution problem in RP including UV light source reflecting by surface of an concave ellipsoid, fiber coupling and transmission, and optical lens focusing[4].

Rapid Prototyping equipment based on ultraviolet and its energy transmission

One typical equipment of rapid prototyping is Compact Prototyping System (CPS). This system, as shown in Fig. 1 can manufacture parts utilizing photo-polymerization of liquid resin in certain wavelength UV irradiation.

The light source used in CPS prototyping belongs to point light source without directivity. So the light source lacks the focusing function, only combines with some focusing device can enhance the light energy density greatly, and can achieve the curing and molding of the liquid resin. Firstly, the CPS system uses concave ellipsoidal mirror to reflect and focalize, and makes most of the light energy of the UV source(UV lamp) placed on a focus of the ellipsoid gathered in another focus. Secondly, the most energy was transmitted to a coupler by cylindrical fibers (glass rod). Finally, the light energy was transferred to the convergent lens by means of optical fiber, then focus on the resin surface. In this way, on the basis of sectional layer information of CAD model (part), focusing lens can be driven to achieve two-dimensional scanning, and to complete curing of the layer. And then a three-dimensional part entity has been created by piling up the curing layers one by one.

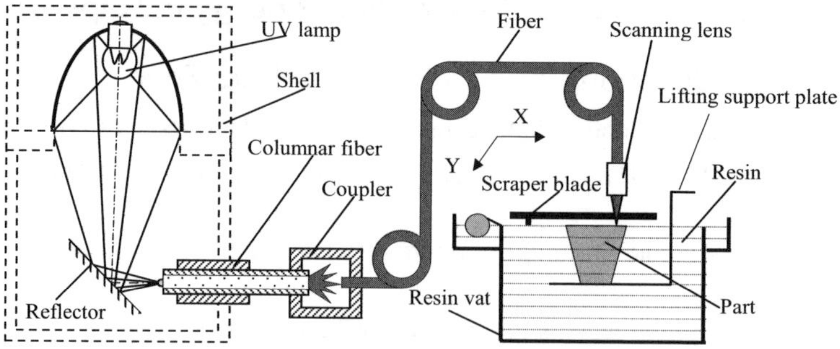

Fig. 1: Manufacturing Principle of the CPS

Optical system is an important component of the CPS. It directly decides the molding quality and efficiency. To solidify the liquid resin in a faster scanning speed, the light energy transferred to surface of the liquid resin must exceed certain threshold. On the other hand, the energy distribution of a light beam coupling to the liquid resin will affect the shape of scanning curing line and the quality of the curing parts. Therefore, analysis and research on CPS optical system in theory and practice will make great influence to complete the UV rapid prototyping technology and enhance its competitiveness.

Analysis on the focusing of light energy reflected by concave ellipsoid reflector

The schematic diagram of UV light energy reflected by the ellipsoidal reflector is shown in Fig. 2.

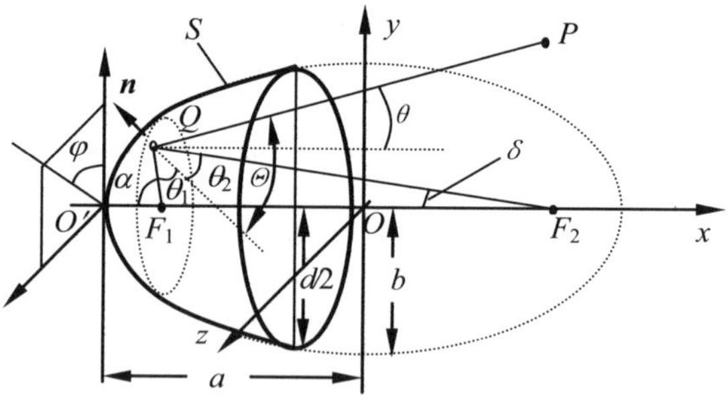

Fig. 2: Geometrical Structure of Concave Ellipsoidal Surface

The light energy from point light source located in one focus F_1 of ellipsoidal reflector is focused to the other focus F_2. In order to simplify theoretic analysis, we suppose that the point light source is monochromatic. The complex amplitude function (light wave motion) of any point P in the point source radiate field is shown as following[5]:

$$\phi(P)=\phi_i+\phi_s=\frac{A}{r_F}e^{-jkr_F}+\frac{1}{4\pi}\iint_s\{\phi_S'\frac{\partial}{\partial n}(\frac{e^{-jkr}}{r})-\frac{1}{r}e^{-jkr}\frac{\partial\phi_S'}{\partial n}\}dS \tag{1}$$

where ϕ_i is light wave motion at point P directly from light source, ϕ_s is light wave motion at point P from light source after being reflected, and ϕ'_s is any Green's Function, A is the amplitude of point source' spherical wave, r_F is the distance between P and F_1, k is wave number, $k=\omega/v$, ω is angular frequency, v is the speed of light, r is the distance between point P and point Q located in micro area dS, ρ is the distance between point F_1 and point Q, $\partial/\partial n$ is partial derivatives of light wave motion function on normal vector of the curved surface S.

In cartesian coordinates (Fig. 2), the function equation of ellipsoidal surface is[6]:

$$\frac{x^2}{a^2}+\frac{y^2+z^2}{b^2}=1 \tag{2}$$

Its polar coordination equation is:

$$\rho=\frac{(1+\varepsilon)F}{1+\varepsilon\cos\alpha} \tag{3}$$

Where ε is the eccentricity of the ellipsoid, $\varepsilon=\frac{c}{a}=\frac{(a^2+b^2)^{1/2}}{a}$, α is the angle between vector ρ and $-x$ axis, F is the distance between F_1 and the apex O' of ellipsoid (if α=0, then ρ=F). If ε<1 Eq. 3 is a ellipsoid equation, if ε=1, it will be a paraboloid equation, and when ε=0 , it will be a sphere equation.

Based on geometry relations as shown in Fig. 2, we can know: $\frac{\partial\rho}{\partial n}=\cos\theta_1=\cos\frac{\theta'}{2}$

$$\begin{aligned}\frac{\partial r}{\partial n}&=\cos\Theta=\cos(\alpha-\frac{\theta'}{2}+\theta)=\\&=\cos\theta(\cos\alpha\cos\frac{\theta'}{2}+\sin\alpha\sin\frac{\theta'}{2})-\sin\theta(\sin\alpha\cos\frac{\theta'}{2}+\cos\alpha\sin\frac{\theta'}{2})\end{aligned} \tag{4}$$

In the light wave field, monochromatic light (its frequency equals to f) intensity can be expressed as:

$$\begin{aligned}p_f&=\frac{r_1p_1}{r_F}e^{-jkr}-\frac{jkr_1p_1}{4\pi}\iint_s\frac{1}{r}e^{-jk(r+\rho)}\{1+\cos\theta[\cos\alpha+\sin\alpha\tan\frac{\theta'}{2}]-\\&-\sin\theta[\sin\alpha-\cos\alpha\tan\frac{\theta'}{2}]\}\rho\sin\alpha d\alpha d\varphi\end{aligned} \tag{5}$$

where p_1 is the light intensity of a point that is at a distance r_1 from the light source.

If the observation point falls on the x axis, angle θ and angle α and angle θ' are situated a same plane including the x axis. The reflected light intensity in the ellipsoid is:

$$\begin{aligned}p_s&=-jkF\frac{(1+\varepsilon)r_1p_1}{2\varepsilon}\int_1^{\rho_m/F}\exp\frac{\{-jk[Fu(1+\cos(\alpha-\theta))+x_F\cos\theta]\}}{[Fu\cos(\alpha-\theta)+x_F\cos\theta]}\cdot\\&\cdot\{1+\cos\theta[\cos\alpha+\sin\alpha\tan\frac{\theta'}{2}]-\sin\theta[\sin\alpha-\cos\alpha\tan\frac{\theta'}{2}]\}\frac{du}{u}\end{aligned} \tag{6}$$

where $u=\rho/F$, $x_F=r_F$, and $\theta=\tan^{-1}[u\sin\alpha/(x_F/F+u\cos\alpha)]$, $\alpha=\cos^{-1}\{[(1+\varepsilon)/u-1]/\varepsilon\}$,

$$\rho_m=[(c-a[1-(d/(2b))^2]^{1/2})^2+(d/2)^2]^{1/2}$$

Where d is the calibre diameter of the concave ellipsoid reflector. Based on Eq. 4 axial light intensity distribution can be calculated. It is similar to a Gaussian curve. Eq. 6 expresses the relation between the light intensity and the real part as well as imaginary part of the complex function p_S: $|p_s|=[\mathrm{Re}(p_s)^2+\mathrm{Im}(p_S)^2]^{1/2}$

The analysis of the light intensity near the focal point

The reflected light intensity is $p_{sF} = p_s(x_F = 2c) = \frac{-jkr_1 p_1}{2\pi} \iint_S \frac{1}{\rho r} e^{-jk(\rho+r)} \cos(\theta'/2) dS$ in focus F_2.

Taking into account $\rho+r = 2a$, $\alpha=0\sim\alpha_{max}$, we can obtain $\cos\alpha_{max} = \frac{c - a[1-(d/(2b))^2]^{1/2}}{a - c[1-(d/(2b))^2]^{1/2}}$.

Make the transformation from cartesian coordinates to polar coordinates: $y = \xi\cos\eta$, $z = \xi\sin\eta$. The light intensity in focus F_2 can be described as:

$$p_{sF} = \frac{-jkr_1 p_1}{2\pi a} e^{-j2ka} \int_0^{\pi} \int_0^{d/2} \frac{\xi d\xi d\eta}{[a^2 - c^2(1-\xi^2/b^2)](1-\xi^2/b^2)^{1/2}} \tag{7}$$

Given again: $(b^2 - \xi^2)^{1/2} = v, v = b \sim [b^2 - (d/2)^2]^{1/2}$, Then the integration result of Eq. 7 is:

$$p_{sF} = -jkr_1 p_1 \frac{1-\varepsilon^2}{2\varepsilon} \ln\{\frac{1+\varepsilon}{1-\varepsilon} \frac{1-\varepsilon[1-[\varepsilon d/(2c)]^2/(1-\varepsilon^2)]^{1/2}}{1+\varepsilon[1-[\varepsilon d/(2c)]^2/(1-\varepsilon^2)]^{1/2}}\} e^{-j2ak} \tag{8}$$

Eq. 8 shows that, as long as $1>[\varepsilon d/(2c)]^2/(1-\varepsilon^2)$, optical intensity of the focus increases with the increase in eccentricity, with the increase in the value of d/c. When an ellipsoidal reflector system is designed, the fact should be fully taken. It will enhance the energy density of the UV light as fully as possible to satisfy requirements of rapid prototyping.

In the vicinity of the focus F_2, the relation that the micro unit Q of the concave ellipsoid surface and the observation point P relate to the location of F_1 can be expressed as:

$$\overline{F_1Q} = \vec{\rho} = (\rho, \alpha, \varphi), \overline{F_1P} = \vec{r_F} = (x_0, r_0, \varphi_0)$$

The distance between Q and P is

$$r = \overline{QP} = R[1 - 2\rho r_0 \sin\alpha\cos(\varphi - \varphi_0)/R^2]^{1/2} \tag{9}$$

$$R = (r_0^2 + \rho^2 + x_0^2 + 2\rho x_0 \cos\alpha)^{1/2} \tag{10}$$

So in near-axis ($r_0 \ll x_0$) area the reflecting light intensity can be expressed as

$$p_s = -jkFr_1 p_1 \frac{1+\varepsilon}{2\varepsilon} \int_1^{\rho_m/F} \exp[-jk(Fu+R)]\{\mathbf{J}_0(kFur_0 \sin\alpha/R)/(Fu+R)\} \cdot$$

$$\cdot \{1 + [\cos\alpha + \sin\alpha\tan(\theta'/2)](Fu\cos\alpha + x_0)/R\} \frac{du}{u} \tag{11}$$

Where $\mathbf{J}_0(\cdot)$ is the Zeroth-order *Bessel* function. It can be obtained the lateral optical wave field distribution of the near ($x_F<2c$) and far ($x_F>2c$) zone. There is a long and narrow focusing zone near the point F_2. The zone gradually widens since the point F_2 inward (along the –x) and outward (along the +x), yet the corresponding light intensity gradually weakens. But the area under the curve of the light intensity distribution perpendicular to x-axis is a constant C

$$C = \int_0^{r_0} |p_s(x_0, r_0)| dr_0 = |p_s(x_0)| W_{1/2} \tag{12}$$

Where $W_{1/2}$ is width of the beam . If consider the medium absorption, there

$$\int_0^{r_0} |p_s(x_0, r_0)| dr_0 = C \cdot \exp(-\alpha_a x_0) \tag{13}$$

Where α_a is the absorption coefficient of the light wave in the medium. The integral value depends on the radiation power.

Further analysis shows that, the distribution of the light beam in areas adjacent to the focus is nearly Gaussian-shaped distribution. The light intensity distribution curve in the focal plane should be represented as the following Gaussian normal distribution function.

$$|p_s|/|p_{so}| = e^{-(kr_0)^2} \text{ or } |p_s|/|p_o| = kFe^{-(kr_0)^2}, k = nk_0 \tag{14}$$

The other planes paralleled to the focal plane have the similar intensity distribution, especially in the areas adjacent to the focus.

Experiment on radiant energy distribution of fiber output

To manufacture RP products, the UV light should be firstly reflected from concave ellipsoidal surface. Then the convergent light which is coupled by the optical fiber is exported by a cylindrical fiber (glass rod). Finally the focused light irradiate to the surface of liquid resin[7].

In order to obtain the distribution of the UV radiant energy being couple to the liquid resin surface, the experiment analysis of the radiation distribution and the experiment plan is designed as shown in Fig. 3.

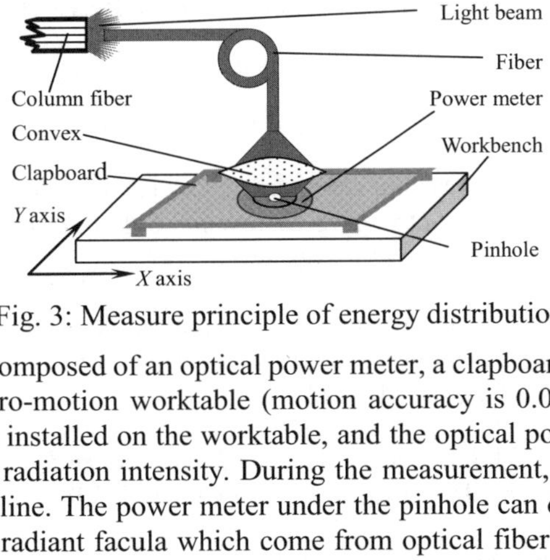

Fig. 3: Measure principle of energy distribution

Measurement devices is composed of an optical power meter, a clapboard with pinhole (D/0.5mm) and a two-dimensional micro-motion worktable (motion accuracy is 0.01mm). The optical power meter and the clapboard are installed on the worktable, and the optical power meter is placed under the pinhole to measure UV radiation intensity. During the measurement, the light is fixed, and the worktable is moved line by line. The power meter under the pinhole can detect the average radiated power of points on the UV radiant facula which come from optical fiber and is focused by lens. A curve surface as shown in Fig. 4 can be made up of the measured values. The curve surface represents the energy distribution of certain cross section perpendicular to the light axis.

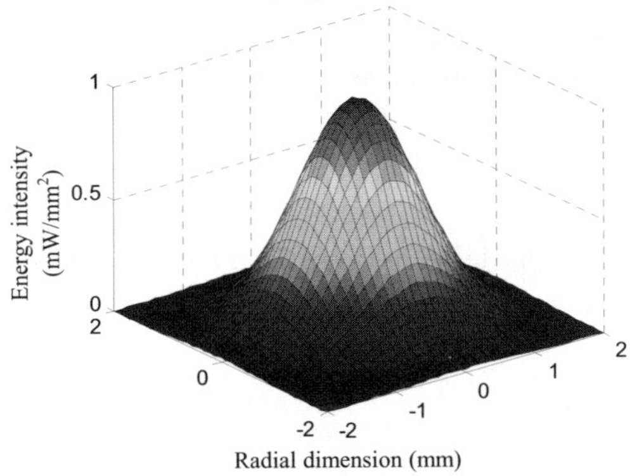

Fig. 4: Energy intensity at different position

The distribution curve surface (Fig. 4) shows that the light coupled to the surface of liquid resin is not an ideal light point, but is a dispersed facula in a certain region of space. The energy of facula is uneven, and the radiation energy distribution of the focused UV light is approximately Gaussian distribution. The energy density in the center is much higher than others of the facula, and the energy will decline sharply far away from the center. The fact is consistent with the theoretic analyse. Validity of the theoretical analysis is testified. It also showed that laser curing is analogized to UV-curing in both processes and results.

Conclusions

According to the analysis and the experiment above, theoretically, the research reveals the essential principle of a radiant point source reflected and focused from concave ellipsoidal surface. Experimentally, the research analyzes the distribution of the light energy, which has coupled and transmitted to the surface of liquid resin. It is essential theoretical basis and experiment basis for improving energy density, prototyping accuracy and effectively using light energy in the UV rapid prototyping technology.

Acknowledgement

The authors gratefully acknowledge the support of the Science Research Fund of Department of Education of Shaanxi Province's (09JK673) and the Science Research Fund of Xi'an University of Technology (102-210808)

References

[1] Wang Yi-qing, Zhao Wang-hua, Shi Le-ping, Le Guang, Lu Bing-heng: *Development of High Resolution Liquid Level Laser Detecting System for Stereolithography Apparatus*. Journal of Xi'an Jiaotong University, 42, 11(2008), pp. 1394-1397.

[2] Y. Abouliatim, T. Chartier, P. Abelard, C. Chaput, C. Delage: *Optical characterization of stereolithography alumina suspensions using the Kubelka–Munk model*. Journal of the European Ceramic Society, 29, 5(2009), pp. 919-924.

[3] Jae-WonChoi, Ho-Chan Kim, Ryan Wicker: *Multi-material stereolithography*. Journal of Materials Processing Technology Vol. 211 (2011), pp.318–328.

[4] Xie Jun, You Li-de, Hou Wen-jie: *Experimental Research of Ultraviolet LED Source System in Stereolithography Apparatus*. Opto-Electronic Engineering, 36, 2(2009), pp. 100-104.

[5] Wei Zhan-xian, Li Ji-cheng: *Fundamentals of Advanced Mathematics--Linear Algebra and Analytic Geometry*. Higher Education Press, Beijing, (2010).

[6] Zhao Da-zun, Fang Jun-yong: *Course in Advanced Optics*. Beijing University of Technology Press, Beijing, (2009).

[7] Zhao Xing-hai, Gao Yang, Xu Mei-jian, Duan Wen-tan, Yu Hai-wu: *Study on Power Delivery Characteristics of Fibers*. Laser Journal, 29, 1(2008), pp. 60-62.

Advanced Materials Research Vol. 279 (2011) pp 313-317
© (2011) Trans Tech Publications, Switzerland
doi:10.4028/www.scientific.net/AMR.279.313

Extraction Feature of Spindle Unbalance of Machine Tool Based on the Wavelet Transform

D. J. Chen[1, a], J. W. Fan[1, b] and F.H.Zhang[2, c]

[1]College of Mechanical Engineering & Applied Electronics Technology, Beijing University of Technology, Beijing, 100124, China

[2]Center of Precision Engineering, Department of Mechanical Engineering, Harbin Institute of Technology, Harbin,150001, China

[a]dongjuchen@gmail.com, [b]jwfan@bjut.edu.cn, [c]zhangfh@hit.edu.cn

Keywords: Spindle unbalance, flatness error, feature extraction, wavelet transform, power spectral density

Abstract. A new method for extracting spectrum feature of spindle unbalance of machine tool is proposed. The flatness error of workpiece surface includes much errors information, and the information contains high frequency signal and low frequency signal. For these errors information, a new identification method of turning errors of workpiece based on the wavelet transform and power spectral density analysis is proposed. According to the focal variation character of wavelet and the energy value of power spectral density analysis, the feature of spindle unbalance from the measured flatness error of workpiece is extracted and identified.

Introduction

Machining accuracy is the most important factor which determines the quality of workpiece, while own errors of machine tool are the direct cause that impacts the machining accuracy. Errors in all parts of machine tools reflect to the results of the workpiece in nonlinear way, and measurement of a single static error can not reflect the dynamic effects on the machine process. Machine tool has the dynamic errors that change with time and the static errors which do not change or change slowly over. Dynamic error is changed over time and is difficult to identify from the time domain, and thus this part of the error should be identified from the frequency domain. The flatness error of the workpiece is a comprehensive reflection of all the errors of machine tool. Since the errors of machine tool are in the test results, wavelet transform method is used to decompose the test results into various levels first, and then the corresponding errors of various components of machine tools are identified with the character of test result of workpiece. The premise of identification is processing the measured result firstly and shows up the important feature in the signal.

In this paper, wavelet transform is used to process the test results of the workpiece, and power spectral density is employed to analyze the flatness error of workpiece. Simulation results show the feasibility and validity of the proposed feature extracting method based on wavelet transform and power spectral density analysis.

Process the Flatness Error of Workpiece

Measurement of the Workpiece. A flat workpiece of radius 10mm has been machined by the large two-axis lathe (as shown in Fig. 1) in the laboratory. The material of the workpiece is LY16, the rotation speed is 110rev/min, the cutting depth is 15μm, and the feed rate is 2mm/min. Fig. 2 shows the measurement of the flatness of the workpiece by the profilometer PGI1240.

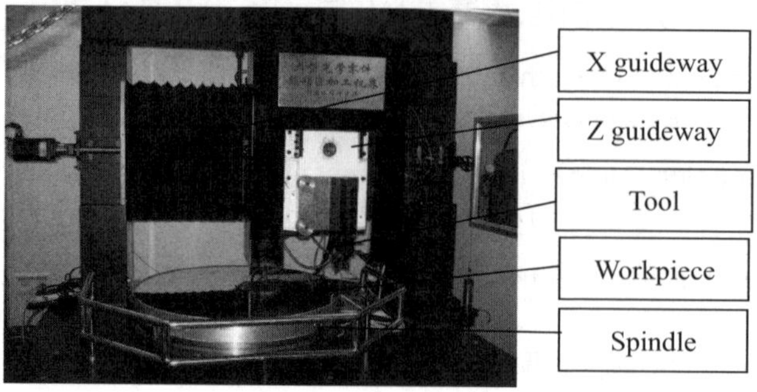

Fig. 1: Structure of the machine tool

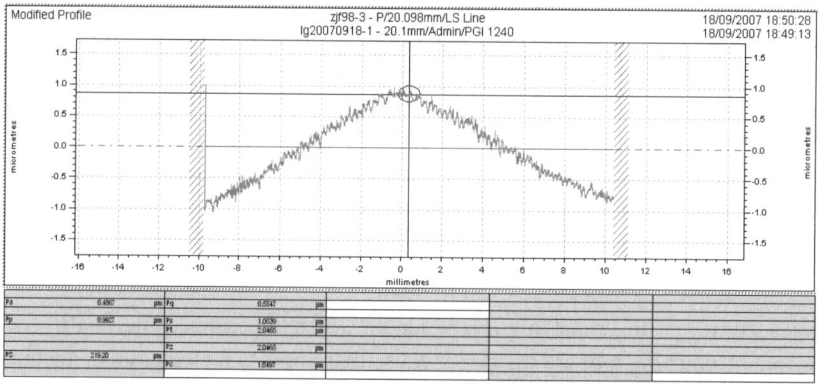

Fig. 2: Flatness error of workpiece

For the measured result of workpiece, it is a combined signal that contains a lot of error sources of machine tool. Each error source has the corresponding characteristics and specific frequency. An appropriate signal processing method is needed to separate the single frequency component, and the error source is identified.

The Discrete Wavelet Transform. The Discrete Wavelet Transform (DWT) is a special case of the WT that provides a compact representation of a signal in time and frequency that can be computed efficiently. The DWT is defined by the following equation:

$$W(j,k)=\sum_{j}\sum_{k}x(k)2^{-j/2}\psi(2^{-j}n-k) \tag{1}$$

where $\psi(t)$ is a time function with finite energy and fast decay called the mother wavelet. The DWT analysis can be performed using a fast, pyramidal algorithm related to multirate filterbanks [1].

In the pyramidal algorithm the signal is analyzed at different frequency bands with different resolution by decomposing the signal into a coarse approximation and detail information. The coarse approximation is then further decomposed using the same wavelet decomposition step. This is achieved by successive highpass and lowpass filtering of the time domain signal and is defined by the following equations:

$$y_{high}[k] = \sum_{n} x[n]g[2k-n] \tag{2}$$

$$y_{low}[k] = \sum_{n} x(n)h[2k-n] \tag{3}$$

where $y_{\text{high}}[k]$, $y_{\text{low}}[k]$ are the outputs of the highpass (g) and lowpass (h) filters, respectively after subsampling by 2. In our implementation, the 1 order wavelet family (DAUB1) proposed by Daubechies [2] is used.

Processing the Experiment Results. Daubechies 1 wavelet is used to decompose the signal into different frequency bands, and then the signal in ten levels can be analyzed. The result is shown in Fig. 3. The wavelet coefficients a1-a10 of low frequency signal and d1-d10 of high frequency signal in different scales are obtained, and they represent the error shape in different error band.

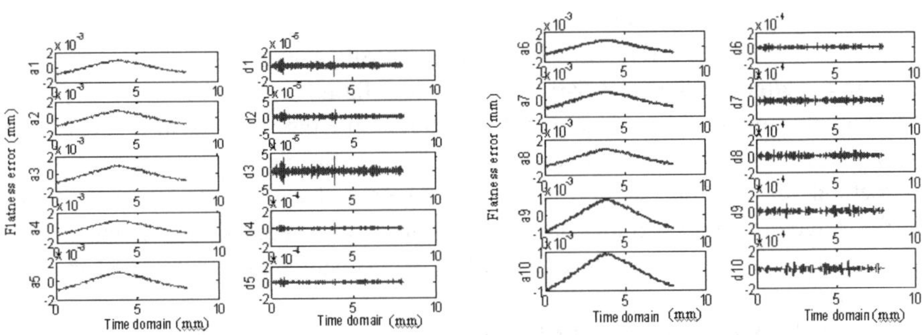

Fig. 3: Measured signal decomposed by wavelet

The power spectral density analysis is used to extract characteristics of the decomposed signal in each level, and then the main error source can be identified.

Analysis of Power Spectral Density

Power spectral density is a statistical method, and is defined as the "power" (mean-square value) within the unit frequency band [3]. It is statistical results of structure response excited by random dynamic loads and is a relationship curve with the values of power spectral density and frequency. $x(f)$ and $S_x(f)$ are the function which can describe the random signal $x(t)$. $x(f)$ is the amplitude of frequency spectrum, and $S_x(f)$ is its power spectral density.

Since the Fourier transform of random signal is not convergent and can not meet the Dirichlet conditions, it does not exist and its spectrum can not obtain directly. However, correlation function of the random signal whose mean value is zero can be convergent at $\tau \to \infty$, i.e. $R_x(\tau \to \infty)=0$. Therefore, it meets the conditions of Fourier transform, and the autocorrelation function $R_x(\tau)$ is integral absolutely according to Fourier transform, that is

$$\int_{-\infty}^{\infty} S_x(f)df = \int_{-\infty}^{\infty} \lim_{T\to\infty} \frac{|x(f)|^2}{T} df = \lim_{T\to\infty} \int_0^T \frac{x^2(t)}{T} dt \tag{4}$$

For time-varying dynamic error, it is very difficult to identify when it integrated with other errors, here identifying it from frequency domain. The signal decomposed by wavelet transform is analyzed with power spectral density, as shown in Fig. 4, where the abscissa is frequency of signal and ordinates is power spectral density of the signal. Fig. 4 (a) shows power spectral density of the signal in the 6-10 scales, which are expressed by d6-d10 on the left of the figure, and the maximum amplitude of each scale is appeared in the figure. Similarly, Fig. 4 (b) shows power spectral density of the signal in 1-5 scales. The frequencies of high amplitude in these 10 level signal are f1=1.85Hz, f2=3.9Hz, f3=7.8Hz, f4=11.7Hz, f5=15.6Hz, f6=25Hz, f7=35Hz, f8=99Hz, and f9=149Hz.

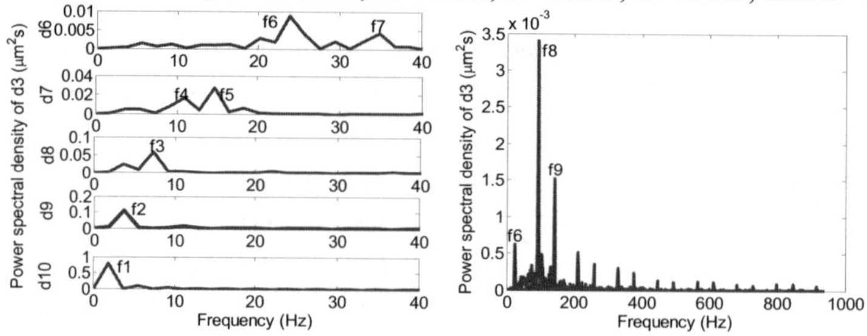

(a) d6-d10 (b) d1-d5

Fig. 4: Power spectral density analysis

Extracting the Frequency Characteristic of Vibration Signal

Vibration signals contain a lot of status information of the parts of machine tool. The vibration patterns will change when the state of machine tool changes. Spectrum analysis is the most common analytical method for vibration signal. One of the main tasks of extraction feature of signal is extracting the waveform with certain characteristics form the mixed signal. Usually, certain basis function is used to expand the waveform according with mathematical rules, and then extract the features of waveform.

Spectrum Sign of Unbalance of Spindle System. Unbalance of spindle is the universal problem of machine tool, and it is also one most common of the vibration fault. Forced vibration is generated due to the excitation by the periodical inertia during the rotation of the spindle.

The main spectrum feature of the spindle imbalance is the spectrum including the rotation frequency of spindle.

The rotation speed of spindle is 110rev/min. The corresponding base frequency (primary frequency) is

$$f_s = \frac{110}{60} = 1.833\,(\text{Hz}) \tag{5}$$

and the corresponding frequencies of the spindle system is shown in Table 1. In the table, the first order modal frequency i.e. the natural frequency of spindle is 33Hz. The rotation frequency is small, as 1.83Hz, and then the two and high frequency of spindle can be calculated based on the basic frequency respectively, as shown in Table 1.

Table 1: Frequencies of the spindle system

Frequency [Hz]						
Rotation frequency				Modal frequency		
Natural	Twice time	Three time	Four time	First order	Second order	Third order
1.83	3.66	5.5	7.33	33	65	99

To determine whether the spectral characteristics of the above analysis include spindle imbalance, and the spectrum which is related to natural frequency of spindle, we must first check the workpiece surfaces to know whether include these frequency components. According to power spectral density characteristics of the decomposed signal and the calculated spectral characteristics of the machine tool, the main errors can be further identified. The frequency with maximum amplitude in d10 scale (the top scale in Fig. 4 (a)) equals to the value of f1, and this scale signal shows the character of spindle imbalance. Fig. 5 shows the power spectral density in frequency domain, and there is a strong low-frequency signal component.

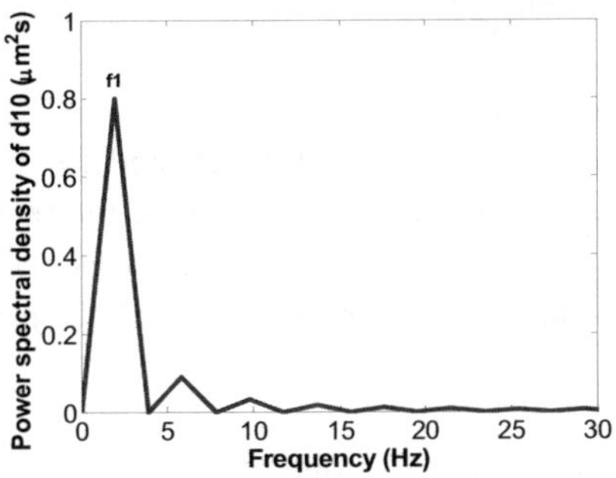

Fig. 5: Power spectral density of d10 signal

Conclusions

The measured results of the workpiece surface are the basis of this paper, and wavelet transform is the means to process the measured signal of workpeice. The difficult problem of extracting the low-frequency is resolved, and the error information of machine tool included in the measured result has been analyzed. All of these results provide a basis for the error identification of machine tool. The power spectral density is used to analyze the topology of workpiece, and the frequency characteristics of spindle unbalance of machine tool that related to vibration has been extracted out from measured signal, which provides an effective mean of identification and analysis the errors of machine tool.

References

[1] S.G. Mallat: *A theory for multiresolution signal decomposition: the wavelet representation.* IEEE Transactions on Pattern Analysis and Machine Intelligence Vol. 11 (1989), pp.674-693.

[2] I. Daubechies: *Orthonormal bases of compactly supported wavelets.* Communications on Pure and Applied Math Vol. 41 (1988), pp.909-996.

[3] X. S. Wen, X. Chen, B.Y. Tang: *Dynamic analysis and applications of mechanical systems.* National Defense University Press, Changsha, (1998), in press.

[4] K. D. Hurst, T. G. Habetler: *Sensorless speed measurement using Current harmonics spectral estimation in induction machines drives.* IEEE Transactions on Power Electronics Vol.11 (1996), pp.66-73.

[5] W. B. Lee, C. F. Cheung, S. To: *A microplasticity analysis of micro-cutting force variation in ultra-precision diamond turning.* Journal of Manufacturing Science and Engineering Vol.124 (2002), pp.170-177.

Advanced Materials Research Vol. 279 (2011) pp 318-323
© (2011) Trans Tech Publications, Switzerland
doi:10.4028/www.scientific.net/AMR.279.318

Dynamic Analysis of Elbow Joint in the Stroke of Tennis

Liang Han[a], Mengqian Tian[b] and Jingliang Gui[c]

School of Mechanical Engineering, Southeast University, Nanjing, China, 211189

[a]melhan@seu.edu.cn, [b]tianmq@seu.edu.cn, [c]guijingliang@126.com

Keywords: a multi-rigid model, dynamic analysis, impact force, tennis elbow

Abstract. The paper presents a multi-rigid model of arm for an impact between a tennis ball and racket, and makes dynamic analysis of it. The input to the model is the impact force caused by a tennis ball and racket. Its value comes from the deflection of the racket stringbed detected by a high speed camera and the stiffness of the racket stringbed is measured by a force gage. The effect of the impact force on the elbow is analyzed in order to research the pathogenesis of tennis elbow.

Introduction

Tennis is a graceful and keen sport, which is one of the most popular sports in the world [1]. Recently with the development of the national economy and the nationwide fitness campaign in China, tennis sport has been prospering in the masses, especially the appearance of "tennis hot" in many colleges and universities. However, tennis sport, like other sports, is not only good for people' health and spirit, but also brings about injuries. These injuries incorporate joints injuries, muscles injuries and ligament strain. Tennis elbow is the most typical tennis sport injury. More than 30% of professional players were diagnosed with tennis elbow [2]. It is generally thought that there are many factors causing tennis elbow, and the main one is the overloaded impulse force, which is caused by the impact between a tennis ball and racket. But the specific causes are not apparent [3-6]. Qianqi Jia produced a mechanics model of human's arm by pro/Engineer and ANSYS, researched the common elbow injury in high level occupation, and explained the damage mechanism of inner tennis elbow from the point of mechanics[7]. Haake and his research group produced the model of the impact between a tennis ball and racket, and developed the methods used to verify it experimentally [8]. The paper presents a multi-rigid model of arm for an impact between a tennis ball and racket, and makes preliminary dynamic analysis of elbow in striking tennis by a high-speed camera and a force gage, providing the basis of the revelation of tennis elbow's pathogenesis.

Experiment for Measuring Impact Force

As shown in Fig. 1, an experiment was developed to measure the deflection of the stringbed in the impact of the tennis and racket in order to research the impact force on the elbow of the arm. The tennis was ejected at an impact velocity of approximately 10m/s with an ejector and impacted against the racket stringbed directly, with whole impact process recorded by a high speed camera, and the pictures transmitted to computer for analysis. According to the pictures shown in Fig. 3, a deformation-time plot was gotten, as shown in Fig. 2. The data in the plot was interpolated and fitted and the function expression was gotten:

$$y = 0.021t^4 - 0.302t^3 + 0.783t^2 + 2.19t - 0.242 \quad (1)$$

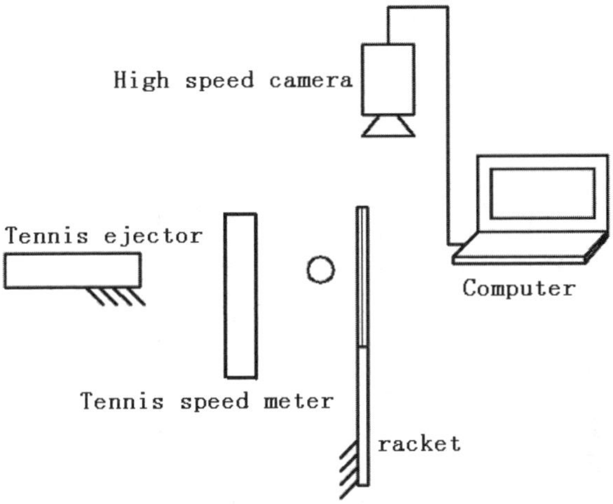

Fig. 1: Stringbed deflection measurement

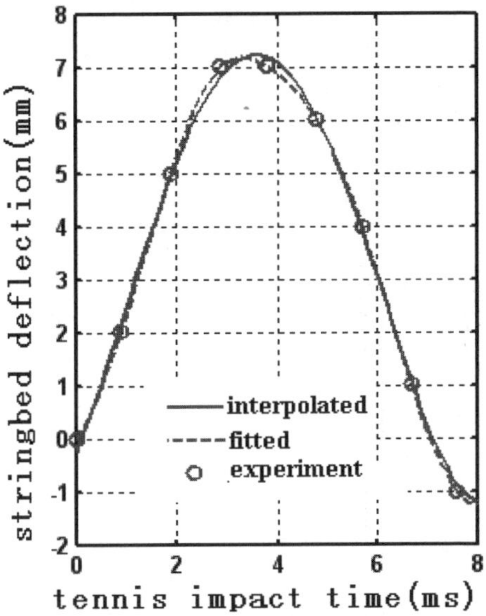

Fig. 2: Deflection-time plot

According to the Fig. 2, it was apparent that the curve by Eq. 1 is accordant with the experiment data and the data by cubic spline interpolation. Therefore Eq. 1 can be served as the function of the string bed deflection y about the impact time t.

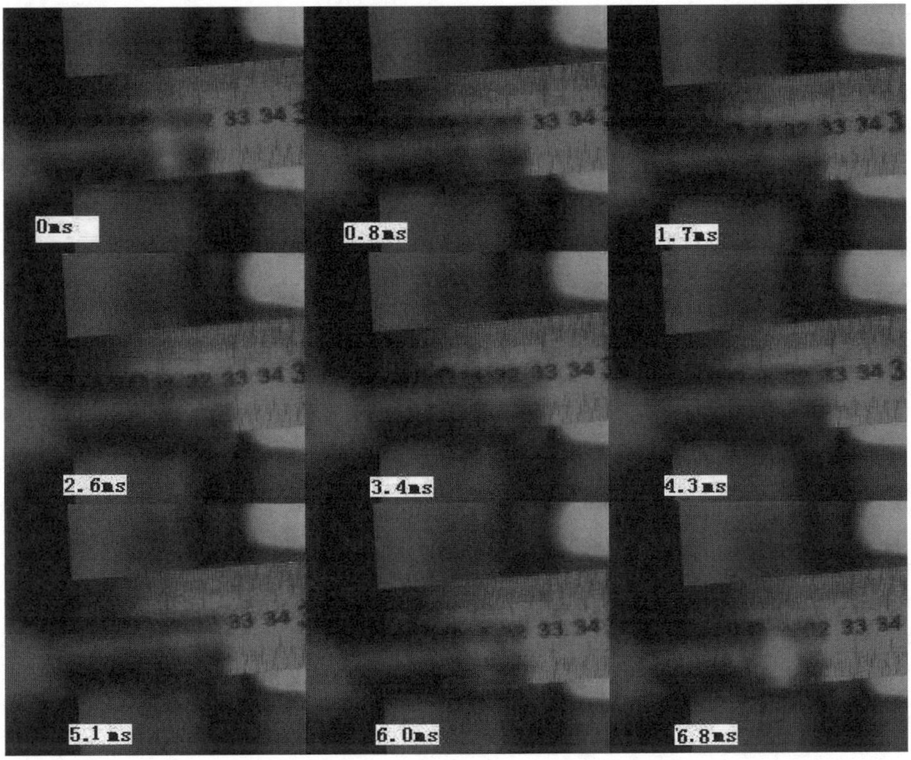

Fig. 3: Pictures of tennis impacting racket

The relation of the stringbed deflection and the force caused by the deflection was gotten with the experiment shown in Fig. 4, and the force is approximatively in direct proportion to the deflection, as shown in Fig. 5. The elastic coefficient ks=14.55KN/m, so the impact force F(t)= ksy, that is:

$$F(t) = 0.306t^4 - 4.393t^3 + 11.385t^2 + 31.86t - 3.523 \tag{2}$$

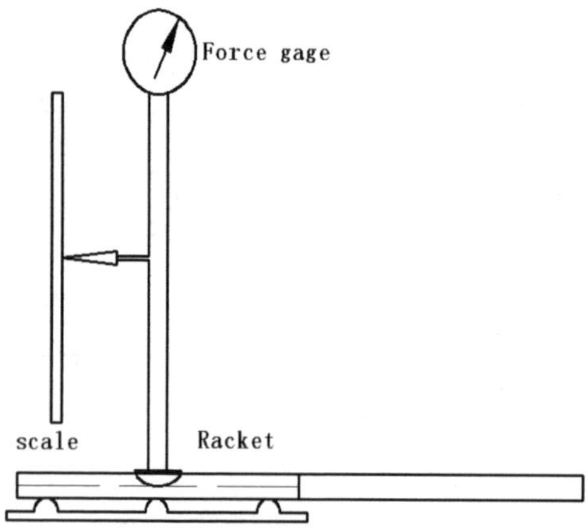

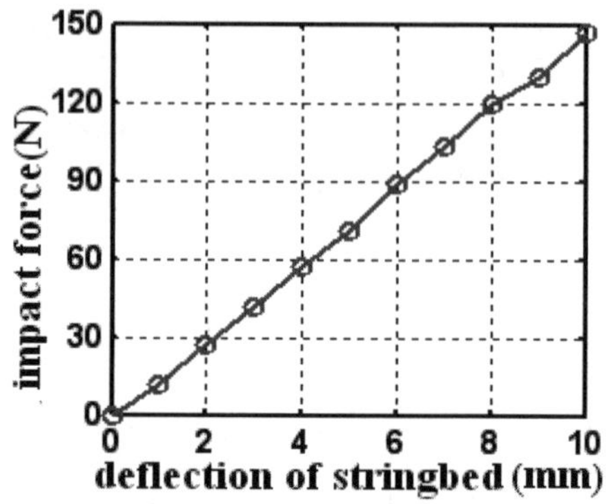

Fig. 4: String bed stiffness gauge experiment

Fig. 5: String bed deflection-force plot

Dynamic Model of Tennis Impacting Racket

Most of papers that study the impact of the tennis and racket simplified arm as rigid bodies with infinite mass, invariable elastic coefficient during the impact process. This paper simplified the forearm and the racket as a rigid body with finite mass, and simplified the upper arm as a rigid body which is fixed, the elbow regarded as an elastic damping joint to link the body B1 and the body B2, as shown in Fig. 6.

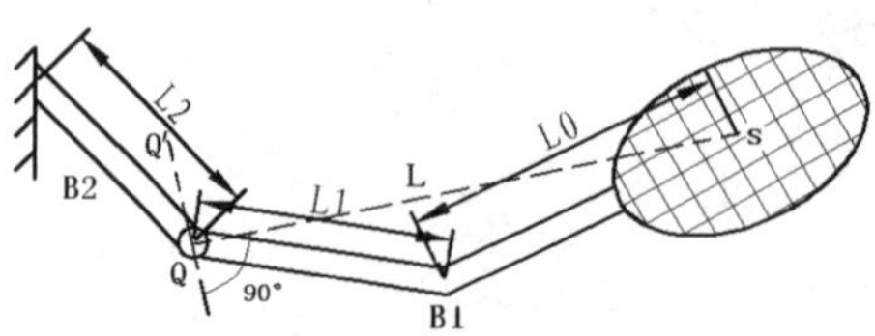

Fig. 6: Arm holding racket model

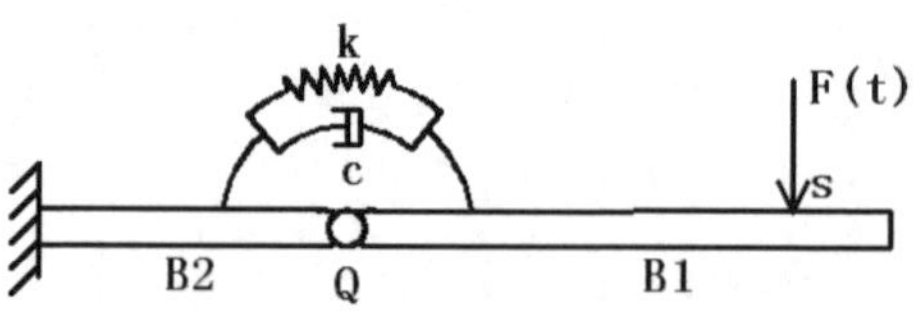

Fig. 7: Dynamic model of arm holding racket

The impact of the tennis and the racket string bed can be taken as an elastic impact process. It was assumed that the impact force was perpendicular to the string bed and also the upper arm, the forearm and the racket string bed in an erect plane. Then a dynamic model of the arm striking tennis under the impact of the force F (t) was obtained and it is shown in Fig. 7.

As the angular displacement φ of the joint Q is tiny, the projection L of the distance between the impact spot s and the center of the joint Q can be taken as a constant, and its value can be computed with the expression as follows:

$$L = \sqrt{L_0^{\,2} + L_1^{\,2} - 2L_0L_1\cos\theta} \tag{3}$$

With θ--the angle between the forearm and the handle of the racket.

The dynamic equation of the arm striking the tennis was worked out with the model shown in Fig. 7, which is:

$$F(t)L = k\varphi + c\frac{d\varphi}{dt} + J\frac{d^2\varphi}{dt^2} \qquad (4)$$

With J--the rotary inertia of the rigid body B1, which rotates about the axis QQ', and its value can be computed with the expression as follows:

$$J = (ML_{c1})^2 m_1 + (ML_1 + NL_{c0})^2 m_0 \qquad (5)$$

With $M = \frac{L_1^{\,2} + L^2 - L_0^{\,2}}{2L_1L}$, $N = \frac{L_0^{\,2} + L^2 - L_1^{\,2}}{2L_0L}$.

We could work out that the tennis begin to contact the string bed at t=0.1067ms with the Eq. 2, when $\varphi(t) = 0$, $\varphi'(t) = 0$, and $L_0 = 0.5m$, $L_1 = 0.25m$, $m_0 = 0.4kg$, $m_1 = 0.8kg$, $L_{c0} = 0.3m$, $L_{c1} = 0.15m$ with experiments. We assumed that $k = 3000N/m$, $c = 100Ns/m$ by experience. Eq. 2, Eq. 3 and Eq. 5 were substituted in Eq. 4. $\varphi = \varphi(t,\theta)$ was worked out, and then the flexural moment T of the joint Q was worked out: $T = \mathrm{k}\varphi(t,\theta) + c\frac{d\varphi(t,\theta)}{dt}$. Fig. 8 shows the relation of the flexural moment T and time t, the angular displacement $\varphi(t)$ and time t.

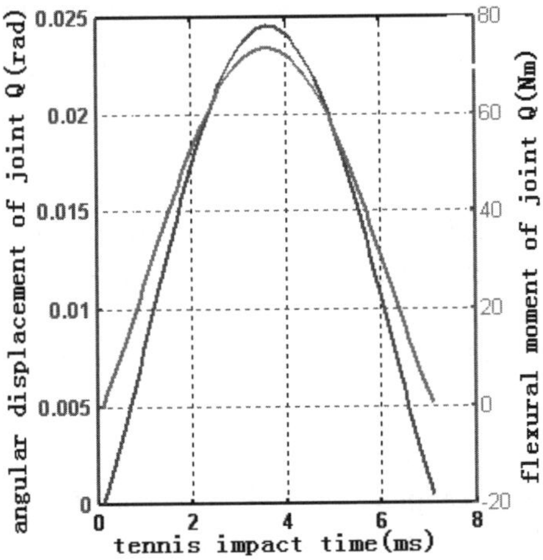

Fig. 8: Plot of T (t) and $\varphi(t)$ when $\theta = 3\pi/4$

Discussion

By analysis of the dynamic model, we can get that the maximum angular displacement φ_{max} of the joint Q is inversely proportional to the elastic coefficient k and independent of the damping coefficient c; the maximum moment T_{max} of the joint Q is independent of k and c.

The effect of θ on the angular displacement φ and the flexural moment T is shown in Fig. 9 and Fig. 10. Due to complexity of the structure of upper limb, the impact of the impact force F on arm is complicated. For that reason the model for analysis of the impact of F is limited.

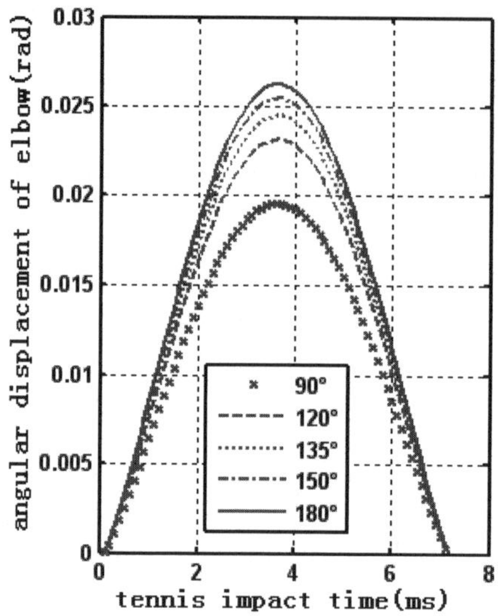

Fig. 9: The impact of θ on φ

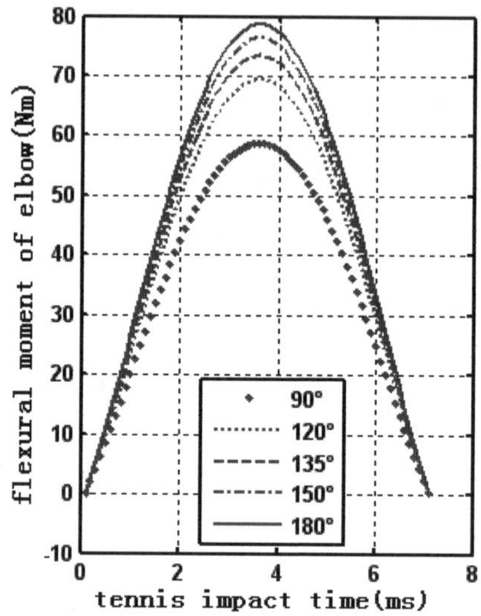

Fig. 10: The impact of θ on T

Conclusion

This paper is conducted to provide underlying basis for further revelation of elbow injuries in playing tennis, based on the impact of the impact force F caused by tennis striking racket. From the analysis above, it can be figured out that the impact of F on the elbow is closely related to the posture of holding the racket, rather than the stiffness of the elbow tissue.

References

[1] ZHU Lijun, SUN Ying, WU Yanping, ZHANF Yunchun: *Simple analysis of the positive impact of self-function of tennis sport on modern society*. Education Science & Culture Magazine Vol. 5 (2008), pp. 207.

[2] A. F. H. Renstrom: *Handbook of sports medicine and science Tennis*. First Edition, Wiley-Blackwell, (2000).

[3] FANG Xiaoqiang: *A brief talk about causes of and prevention measures against tennis elbow in tennis sport*. Journal of Changchun University of Science and Technology, 4, 2(2009), pp. 174-175.

[4] PENG Xue-han: *Causes of and prevention measures against elbow joint injury in tennis sport.* Journal of Fuzhou University, 29, 2(2001), pp. 132-134.

[5] Verhaar JA: *Tennis elbow. Anatomical, epidemiological and therapeutic aspects.* International Orthopaedics, 18, 5(1994), pp. 263-7.

[6] LUAN Li-xia, XU Xiang-feng: *Investigation and study on college tennis players' sports injuries.* Journal of Wuhan Institute of Physical Education, 39, 6(2005), pp. 92-95.

[7] Tianqi Jia: *Analysis on the damage mechanism of inner tennis elbow by using of finite element methods.* ICBBE 2009, Beijing, June 2009, pp. 1-3.

[8] S.R. Goodwill, S J Haake: *Spring damper model of an impact between a tennis ball and racket.* Proceedings of the Institution of Mechanical Engineers, Part C: Journal of Mechanical Engineering Science, 215, 11(2001), pp. 1331-1341.

III. Related Enlightening Topics

Advanced Materials Research Vol. 279 (2011) pp 327-332
© (2011) Trans Tech Publications, Switzerland
doi:10.4028/www.scientific.net/AMR.279.327

Studies on Methacryloyl guar gum: Surface morphology, Crystallinity, Biodegradability and Viscosity behavior of Semi-dilute solutions

Wenfa Xiao[1, a], Litao Dong[1, b]

[1]College of Chemistry and Chemical Engineering, Shanghai University of Engineering Science, Shanghai 201620, China

[a]xiaowenfa@yahoo.com.cn, [b]donglt369@163.com

Keywords: Methacryloyl guar gum, viscosity behavior, surface morphology, biodegradability, crystallinity

Abstract. Methacryloyl guar gum was investigated with respect to crystallinity, surface morphology, biodegradability and viscosity of semi-dilute solutions. It was studied that the shear-dependent viscosity behaviour could be well described by the Cross viscosity model, under various methacryloyl guar gum concentrations, salts addition and temperature. The zero-shear-rate viscosity was observed to change with different methacryloyl guar gum concentration and decease with the increase of temperature.

Introduction

For many years, various polysaccharides and their derivatives have been widely used in food, pharmaceutical, textile, cosmetics, agricultural and other industries, where they function as thickeners, emulsion stabilisers, gelling agents, film formers and texture modifiers [1-4]. This is mainly due to their viscosity or rheological properties. Therefore, a good understanding of rheological characteristics of these biopolymers in aqueous solutions is required.

Guar gum is a natural non-ionic hydrophilic polysaccharide derived from the seed of the guar plant. The chemical structure of guar gum consists of anhydro-D-mannose monomer units linked to each other by β-(1-4) linkages to form the backbone chain. Anhydro-D-galactose branches are joined by α-(1-6) bonds to the backbone. In general, the galactose branches occur on alternate mannose units. Guar gum applications as emulsifiers, thickeners, and stabilizers are approved for use in a wide range of foods, cosmetics, and pharmaceuticals. Although guar gum is abundantly available at low cost, its uncontrolled rate of hydration decreases its viscosity upon storage, and further microbial contamination limits its long term applications [5, 6]. So guar gum has been modified by derivatization, grafting and network formation to improve its property profile for a wide spectrum of end-uses [7-10]. The synthesis of methacryloyl guar gum had reported in previous article [11]. However, studies of the rheological behavior of MAG are scarce and thereby a wide rheological knowledge is not fully available. On the other hand, the knowledge of the rheological behaviour of guar gum derivatives in aqueous solutions is essential for meeting end-user requirements.

The purpose of the present study is to characterize the rheological properties of semi-dilute solutions of methacryloyl guar gum. The principal topics to be examined here are the shear-dependent viscosity behaviour under various conditions. In addition the surface morphology, crystallinity and biodegradability of methacryloyl guar gum (MAG) were also studied.

Experimental

Materials. Guar gum (GG, food-grade, Changxing Petroleum Chemical industry. Co.Ltd, Shandong, China) was commercial product and procured. Methacrylic acid (MAC), sodium hydroxide, thionyl chloride, were procured from Sinopharm Chemical Reagent. Co. Ltd (Shanghai, China), were of laboratory reagent grade and were used without any further purification. The other chemicals were analytical grade and commercially available products, used without further purification.

Methacrylation of GG. GG was dissolved in 18% sodium hydroxide solution and then extracted with 2-propanol. Sodium salt of GG was washed thoroughly with 2-propanol and then dried in oven at 40 °C. To it was slowly added methacryloyl chloride that was prepared by the direct reaction of MAC with thionyl chloride under controlled conditions in an ice bath. A thick gelatinous precipitate of methacryloyl guar gum (MAG) was obtained. This gelatinous precipitate was repeatedly washed with methanol to wash-off any unreacted methacryloyl chloride trapped in the gel. Then, the product was dried under vacuum. The D.S (degree of substitution) of the derivative products was estimated by evaluating the ester value of the sample [11, 12].

Scanning Electron Microscopy (SEM) Studies. Surface morphology of GG and MAG was analyzed in scanning electron microscopy (SEM) in powdered form (Model: S-2700N, Japan). Samples of the microspheres were taken on a copper stub and sputtered with gold for 2 min. The SEM micrographs of guar gum and MAG have been shown in Fig. 1.

X-ray Diffraction (XRD) Studies. The crystallinity of GG and MAG was powered form and evaluated by XRD (D/MAX-2550PC, Japan) measurements recorded. Scanning was done at the ambient temperature (25 °C) by varying the angle, 2θ, from 5°up to 90°.

Viscosity Measurements. All rheological measurements for semi-dilute aqueous MAG solutions were conducted with the help of a rotational rheometer (BROOKFIELD DV-III ULTRA, American) using a cylinder measuring system under various conditions. The following variables have been studied: shear rate (values in the range 1.0–70s^{-1}), MAG concentration (values in the range 0.5–1.5%), and temperature (values in the range 20–75 °C). In addition, the effects of the added salts (0.2% KCl, 0.5% KCl, 1.5% KCl, 1.5% $CaCl_2$) on the flow characteristics of the sample solution were also considered.

Resistance to Biodegradability. Resistance to biodegradability of partially GG and partially MAG has been measured in terms of viscosity and hence viscosity is calculated with the help of the rotational rheometer (BROOKFIELD DV-III ULTRA, American) at constant temperature i.e, at 20 °C.

Results and Discussion

MAG was synthesized in a facile process using the sodium guar gum as the precursor. The D.S of MAG was estimated 0.39. The polymer samples were prepared by dispersing the polymers in deionized water at a lower concentration. The samples were left to hydrate at least for 30 min.

Scanning electron microscopy (SEM) analysis. It is evident from the SEM micrographs of GG (Fig. 1a) and its grafted product (Fig.1b) that profound morphological changes, in form of transition from fibrous particles to flake structure, have taken place because of grafting of methacryloyl chains onto guar gum (GG). Thus, it is evident that the granular morphology of GG is lost after grafting and transformed into flake morphology.

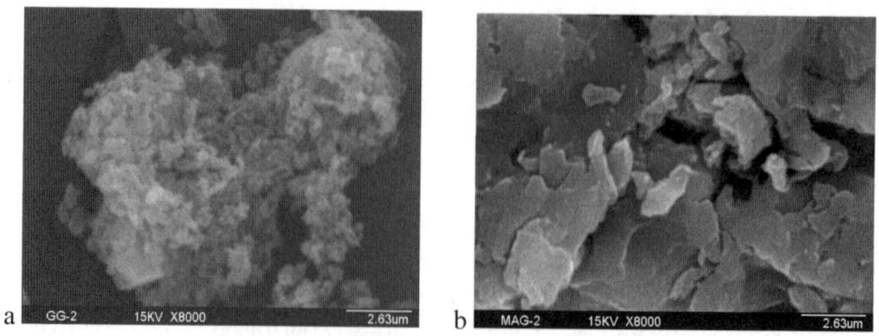

Fig. 1: SEM images of a (GG), b (MAG)

XRD Analyses. XRD patterns of GG and MAG were collected from5°-90°(2θ) and are shown in Fig. 2. It can be observed that GG shows a strong diffraction peak at 2θ=20°. After being grafted using MAC, the characteristic peak of MAG has no obvious change compared with GG (Fig. 2a), indicating that the crystal structure of GG has not been destroyed during methacrylation. However, a new peak at 2θ values of 40° was detected (Fig. 2b). This was a clear indication a new material had generated, further confirming MAC successfully grafted onto GG.

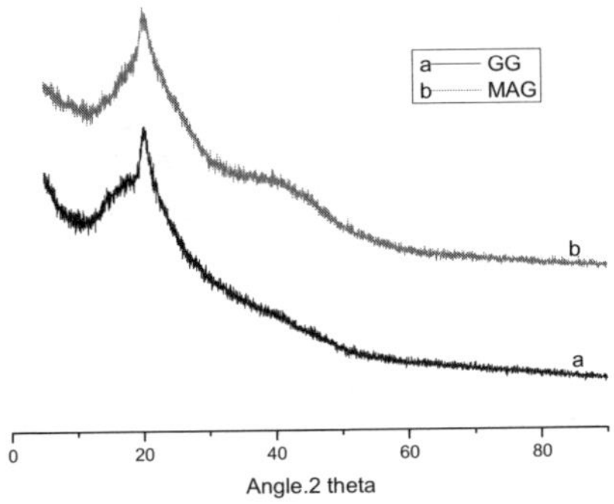

Fig. 2: X-ray diffraction studies

Viscosity Behavior Studies. Fig. 3(a-c) illustrates the effects of MAG concentration, temperature and added salts on the viscosities of aqueous MAG solutions under various shear rates, respectively. In all cases, the viscosity was observed to decrease with the increase of shear rate (γ), showing that these solutions behaved as typical non-Newtonian shear-thinning fluids. Similar behaviour has also been observed for aqueous solutions of GG [13,14] .Consequently, this shear dependent viscosity behaviour was examined using different models for non-Newtonian fluids, such as Ostwald-Dewaele model, Carrean model, Quemada model and the Cross model [15,16]. As a result, the following Cross viscosity model (Eq. 1) was found to describe all data more suitably than other models investigated:

$$\eta = \eta_1 + \frac{\eta_2 - \eta_1}{1 + \lambda\gamma^m} \quad (1)$$

in which η_2 is the zero-shear-rate viscosity, η_1 is the infinite-shear-rate viscosity, λ is a characteristic time and m is an exponent. Table 1 gives the values of Cross parameters obtained for aqueous MAG solutions under various conditions by non-linear curve fittings and corresponding regression coefficients (R^2). From Table 1, it was found that the η_2 value increased with the increase of MAG concentration when the other conditions were kept constant. The enhanced viscosity at higher MAG concentrations may be attributed to a progressive increase of physical entanglements of the macromolecular chains when the solution concentration increased. Oblonsek et al. [17] investigated the concentration dependence of the Cross parameter η_1 for concentrated solutions of guar gum found similar scaling law [8].

Added salts (KCl and $CaCl_2$) have some effects on the zero-shear-rate viscosity for aqueous MAG solutions with a given concentration of 1.0%. Compared with the MAG solution without added salts, the salt-containing MAG solutions have a slight decrease in the η_2 value. Moreover, an increase of KCl concentration from 0.2% to 1.5% resulted in a further decrease of the η_2 . The addition of 1.5% $CaCl_2$ caused a greater reduction in the η_2 than did the addition of 1.5% KCl.

The effect of temperature on the zero-shear-rate viscosity for aqueous MAG solution with a given concentration of 1.0% is shown in Fig. 4. With the increase of temperature, there is a decrease in the zero-shear-rate viscosity value. It seems that MAG has relatively good thermal stability when compared with GG [18].

Table 1: Regressed parameter values of the Cross model for aqueous MAG solutions under different conditions

C_{MAG}[gdL^{-1}]	T[°C]	C_{salt}[%]	η_2[Pa s]	η_1[Pa s]	λ	m	R^2
0.5	20	-	0.6650	0.0024	0.7865	0.5523	0.9975
0.8	20	-	1.7891	0.0051	0.4112	0.4578	0.9924
1.2	20	-	7.6432	0.0072	0.6623	0.5123	0.9921
1.5	20	-	15.5921	0.0087	0.6134	0.5002	0.9897
1.0	20	-	4.4361	0.0064	0.7013	0.6234	0.9946
1.0	45	-	2.1486	0.0048	0.5534	0.6123	0.9994
1.0	60	-	1.9128	0.0037	0.3398	0.5957	0.9995
1.0	75	-	0.9927	0.0019	0.1978	0.5672	0.9997
1.0	20	0.2[KCl]	3.1589	0.0094	0.2958	0.6754	0.9996
1.0	20	0.5[KCl]	2.9986	0.0056	0.3179	0.5997	0.9999
1.0	20	1.5[KCl]	1.7683	0.0068	0.4391	0.5885	0.9997
1.0	20	1.5[$CaCl_2$]	1.6238	0.0012	0.2578	0.5736	0.9996

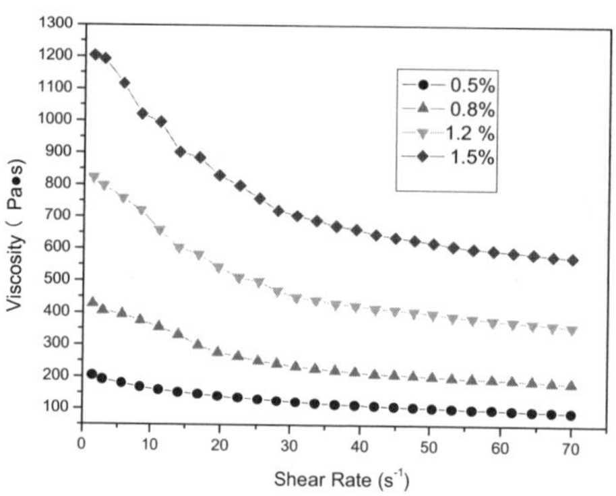

Fig. 3 (a)

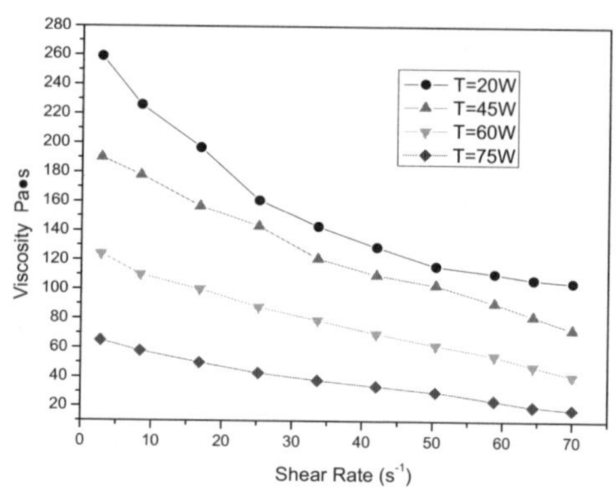

Fig. 3 (b)

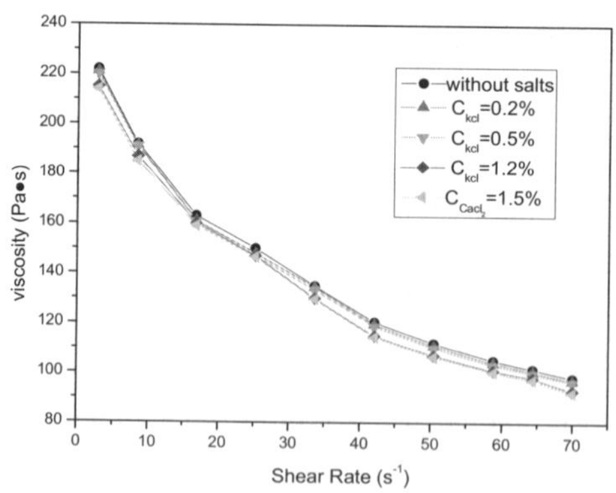

Fig. 3 (c)

Fig.3: The relationship between viscosity and shear rate for aqueous MAG solutions under various (a) MAG concentrations(20 °C), (b) temperatures(1.0% MAG) and (c) added salts(1.0% MAG)

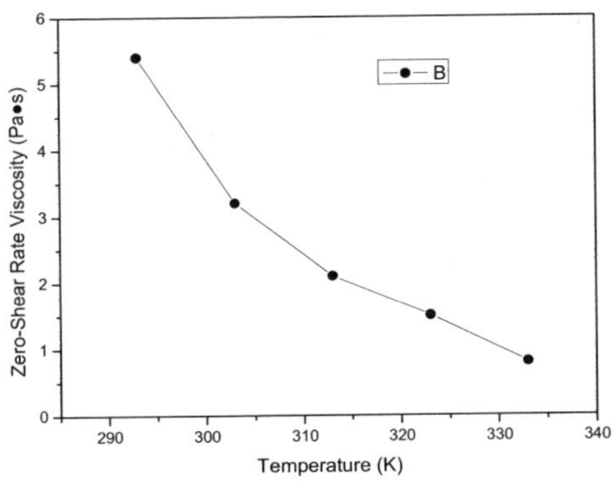

Fig. 4: The relationship between zero-shear -rateviscosity and temperature for 1.0% aqueous MAG solution

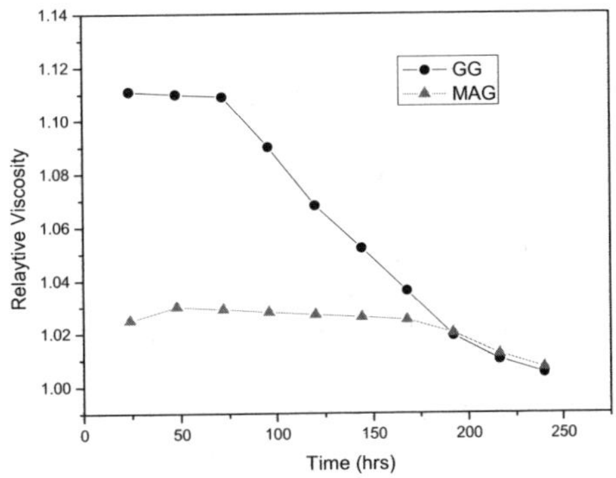

Fig. 5: Relative viscosity vs time period to study biodegradation

Resistance to Biodegradability. It has been observed that relative viscosity of partially MAG is lower than partially GG (Fig. 5). This might be due to presence of grafted groups which made the molecule more flexible and reduce the viscosity drastically. Partially GG, like other polysaccharide solutions, was highly prone to biodegradation [19], and it was found that its solution after 72 h of its preparation start degrading and during 10 days the solution showed considerable loss of viscosity. The graft polymer solution was subjected for same type of study for biodegradation, and it has been observed that MAG solution showed a few of loss of viscosity up to 10 days. These results show that the graft polymer is less susceptible to biodegradation. This is in an agreement with the fact that MAG can be made less susceptible to bacterial attack. Thus, it can be concluded that, by incorporation of methacryloyl groups graft onto partially GG, biodegradation can be minimized.

Conclusion

MAG was successfully synthesized in a facile process using the sodium guar gum as the precursor. The crystallinity of MAG samples had very small variation. SEM analysis indicated that the morphology of GG had changed when methacryloyl chains grafted onto GG. The biodegradation of MAG was obviously minimized. Semi-dilute aqueous MAG solution showed strong shear-thinning effects. Its shear-dependent viscosity behaviour under various conditions could be examined using the Cross model. With the increase of MAG concentration from 0.5 to 1.5%, the zero-shear-rate viscosity increased. With the increase of temperature from 20 to 75 °C, the zero-shear-rate viscosity deceased. In addition, added salt lowered the zero-shear-rate viscosity. The above advantageous of MAG that might be applications in the food, oilfield and other industries.

References

[1] Sandford P, Baird J: *The Polysaccharides*. Academic Press, New York, (1983).

[2] Feddersen RL, Thorp SN: *In Industrial Gums*. Whistler RL and BeMiller JN Academic Press, San Diego, (1993).

[3] RaoMa: *Rheology of Fluids and Semisolid Foods: Principles and Applications*. Aspen, Gaithersburg, (1999).

[4] Sopade PA , Kiaka K: *Rheology and microstructure of sago starch from Papua New Guinea*. J Food Eng, 50, 5(2001), pp. 47-57.

[5] Cavalieri, F, Chiessi, E.and M. F, Paradossi G: *Novel PVA-Based Hydrogel Microparticles for Doxorubicin Delivery*. Biomacromolecules, 12, 9(2008), pp.1967-1969.

[6] Ahmad Bahamdan, William H. Daly: *Hydrophobic guar gum derivatives prepared by controlled grafting processes–Part II: Rheological and degradation properties toward fracturing fluids applications*. Polym. Adv. Technol, 13, 18(2007), pp.660-672.

[7] Jian-Hua Zhu, Xiao-Quan Yang and Lei-Yan Wu: *Effect of guar gum on the rheological, thermal and textural properties of soybean β-conglycinin gel*. International Journal of Food Science & Technology, 24, 44(2009), pp.1314-1322.

[8] Daniela Risica, Cesare Cametti, Mariella Dentini: *Rheological properties of guar and its methyl, hydroxypropyl and hydroxypropyl-methy derivatives in semidilute and concentrated aqueous solutions*. Polymer, 23, 51(2010), pp.1972-1982.

[9] Li-Ming Zhang, Jian-Fang Zhou: *A comparative study on viscosity behavior of water-soluble chemically modified guar gum derivatives with different functional lateral groups*. Journal of the Science of Food and Agriculture, 5, 85(2005), pp.2638-2644.

[10] Li-Ming Zhang, Tao Kong, Peter S Hui: *Semi-dilute solutions of hydroxypropyl guar gum: Viscosity behaviour and thixotropic properties*. Journal of the Science of Food and Agriculture ,6, 87(2007), pp.684-688.

[11] Shiwali Thakur, J.-H. Ahn: *Synthesis of acryloyl guar gum and its hydrogel materials for use in the slow release of L-DOPA and L-tyrosine*. Carbohydrate Polymers, 8, 76(2009), pp.513-520.

[12] Mohan A. Shenoy, Dawid John D'Melo: *Synthesis and characterization of acryloyloxy guar gum.* Journal of Applied Polymer Science, 17, 117(2010), pp.148-154.

[13] Casas JA, Mohedano AF, Garcia-Ochoa F: *Viscosity of guar gum and xanthan/guar gummixture solutions*. J Sci Food Agric, 22, 80(2000), pp.1722-1727.

[14] Cross MM: *Rheology of non-Newtonian fluids a new flow equation for pseudoplastic systems*.

J Colloid Interf Sci, 9, 20(1965), pp.417-420.

[15] Wu QH ,Wu JA: *Polymer Rheology*. Higher Education Press, Beijing (2002).

[16] Oblonsek M, Sostar-Turk S, Lapasin R: *Rheological studies of concentrated guar gum*. Rheol Acta,11, 42(2003), pp.491-499.

[17] Y. Heo, R.G. Larson: *Universal Scaling of Linear and Nonlinear Rheological Properties of Semi-Dilute and Concentrated Polymer Solutions*. Macromolecules, 4, 41(2008), pp.8903-8915.

[18] Whit comb PJ, Gutowski J, Howland WW: *Rheology of guar solutions*. J Appl Polym Sci, 11, 25(1980), pp.2815-2827.

[19] Wenbo Wang, Aiqin Wang: *Preparation and swelling characteristics of a superabsorbent nanocomposite based on natural guar gum and cation-modified vermiculite*. J Appl Polym Sci, 6, 119(2011), pp.3675-3686.

Advanced Materials Research Vol. 279 (2011) pp 333-338
© (2011) Trans Tech Publications, Switzerland
doi:10.4028/www.scientific.net/AMR.279.333

Measurement for the damping coefficient in standard-beam balance oscillating system

TAN Hong[1, 3, a], GAO Jian[1, b, *], LI Shisong[2, c] and LI Zhengkun[3, d]

[1]College of Mechanical and Electrical Engineering, China Jiliang University, Hangzhou, China

[2]Department of Electrical Engineering, Tsinghua University, Beijing, China

[3]National Institute of Metrology, Beijing, China

[a]tan_hong2005@126.com, [b, *]gaoj@cjlu.edu.cn, [c]leeshisong@sina.com, [d]lzk@min.ac.cn

Keywords: Balance, damped oscillation, damping coefficient, external driving force, frequency

Abstract. Damping coefficient is an important parameter in the system of swinging balance. There are various factors, such as knife hysteresis and the air resistance around the swinging balance, which influence the damping coefficient during the swing process. The energy of the balance and the amplitude of the oscillation will decay with time. It will stop swinging in numerable period. This damping phenomenon prevents us from measuring the oscillation period. So we need to analyse and study the damping coefficient. In this paper, we firstly analyse the working principle of the swinging balance, and then we measure the damping coefficient in different driving frequencies using experimental method and spectrum analysis. In the end, we get the evolution of damping coefficient under different driving frequencies.

Introduction

Damping coefficient is an important parameter in the oscillation process. It is used to describe the energy dissipation[1]. It is one of the important factors affecting the dynamic characteristics of the system. So analysing the damping coefficient is an important issue in the field of oscillation study.

Damping contains all factors causing the energy dissipation in the oscillation system. Damping is the ratio of the maximum energy and energy dissipation of the system in oscillation process[1].

It is known that damping effect is a complex process. Not all systems can be solved correctly using the existing damping models. At present, damping models[2-4] such as structural damping model, hysteretic damping model and dry friction damping model have been mostly studied.

In terms of the swinging balance studied in this paper, there are many factors which affect the damping, such as the structure of the system, the load of the balance, the air resistance, the materials and so on. There is no precise damping model for this oscillation system.

In same conditions, the value of damping is different in repeated tests. So damping is a random variable. If we use different methods, the value is different. Damping is a fuzzy variable in some case. Although the damping is random, uncertain and variational, it is a necessary parameter in the control of the system.

In the following article, we mainly discuss the relationship between damping coefficient and the driving frequency. This study is helpful for the control of the balance oscillating system.

The working principle of the swinging balance

The swinging balance's force analysis in z-axis direction during the movement process at any position is shown in Fig. 1 below.

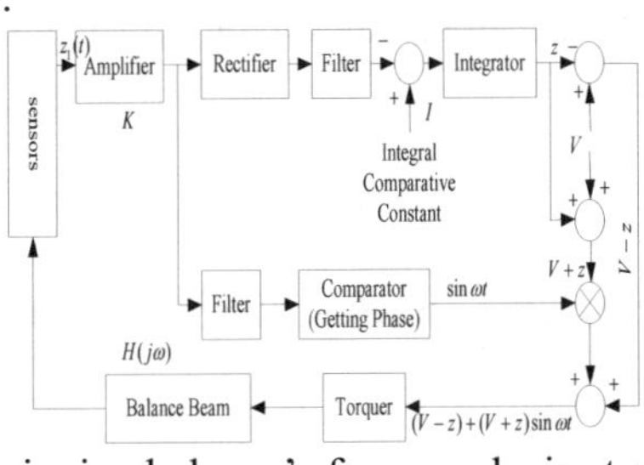

Fig. 1: The swinging balance's force analysis at any position

The l_1 and l_2 are the length of the left and right arms. The the quality of the tray are m_1 and m_2. The quality of the balance beam is M. Point G is the balance's center of gravity. Symbol δ is the restoring moment coefficient which is caused by the difference between the center of gravity of the balance and the fulcrum of the knife-edge. z_1, z_2 is the deviation of displacement, referenced the equilibrium position at z-axis direction. Symbol α is the swinging angle of deviation from the equilibrium position. The acceleration of gravity is g.

When the balance is at the equilibrium position in horizontal direction, according to the principle of moment equilibrium, we can get Eq. 1.

$$m_1 g l_1 = m_2 g l_2 \tag{1}$$

When external force f_e is acting on the balance, the oscillating angle is α. At this moment, damping coefficient is f. Moment of inertia is J. Tensile force on the arm are F_1, F_2. At this time, we can get the following relationship:

$$z_1 = -l_1 \sin\alpha \approx -l_1\alpha \tag{2}$$

$$z_2 = l_2 \sin\alpha \approx l_2\alpha \tag{3}$$

$$F_1 - m_1 g - f_e = m_1 z_1'' \tag{4}$$

$$F_2 - m_2 g - f_e = m_2 z_2'' \tag{5}$$

$$F_1 l_1 - F_2 l_2 - \delta M g\alpha - f\alpha' = J\alpha'' \tag{6}$$

Unite and simplify these equations and we can receive the differential equation when the swinging balance is at any state of motion. As shown in Eq. 7.

$$(m_1 + m_2 l_2^2/l_1^2 + J/l_1^2) z_1'' + (f/l_1^2) z_1' + (\delta M g/l_1^2) z_1 = -f_e \tag{7}$$

Assuming: $A = m_1 + m_2 l_2^2/l_1^2 + J/l_1^2$, $B = f/l_1^2$, $C = \delta M g/l_1^2$
So the differential equation can be simplified into Eq. 8:

$$z_1'' + (B/A) z_1' + (C/A) z_1 = -f_e/A \tag{8}$$

In the above equation, A is about the variable of the quality. B is about the variable of the damping. C is about the variable of the restoring moment coefficient on the balance's knife-edge.

The system of swinging balance can be equivalent to a second order spring-mass-damping system. Symbol ξ is the damping coefficient of the standard-beam balance oscillating system and symbol ω_n is the natural frequency of vibration. There are equations just as shown in the following:

$$B/A = 2\xi\omega_n \tag{9}$$

$$C/A = \omega_n^2 \tag{10}$$

Therefore, the damping coefficient and natural frequency are obtained as follows:

$$\xi = B/(2(AC)^{0.5}) \tag{11}$$

$$\omega_n = (C/A)^{0.5} \tag{12}$$

Assume $\lambda = \omega/\omega_n$, then the frequency characteristic of the system is as follows:

$$H(j\omega) = (1/C)/(1 - \lambda^2 + j2\xi\lambda) = |H(j\omega)| \angle\varphi \tag{13}$$

$$|H(j\omega)| = (1/C)/((1-\lambda^2)^2 + (2\xi\lambda)^2)^{0.5} \tag{14}$$

$$\varphi = -\arctan(2\xi\lambda/(1-\lambda^2)) \tag{15}$$

When the external driving force of the swinging balance is $f_e = f_0\sin\omega t$, the steady-state solution is obtained as follows:

$$z_1(t) = -[(1/C)/((1-\lambda^2)^2 + (2\xi\lambda)^2)^{0.5}]f_0\sin(\omega t - \varphi) \tag{16}$$

The following conclusions can be obtained by analysis. When $\omega=\omega_n$, $\lambda=1$, $\tan\varphi=-\infty$, the stead-state solution of the system is obtained as follows:

$$z_1(t) = (f_0/2\xi C)\cos\omega_n t \tag{17}$$

Clearly, the damping coefficient determines the amplitude of the system's steady-state solution. Therefore, as long as we analyse the amplitude in different frequencies, we can achieve the information of the damping coefficient.

In the following article, we will analyse the damping coefficient in different frequencies by experimental method.

The swinging balance of damped oscillation

The curves of damped oscillation obtained by experiment. In order to analyse the damping coefficient, the first step, we should get other parameters in Eq. 16, except ξ.

All parameters can be obtained by the experiments. It should be noted that the balance used in the experiments is form Shenyang Measurement Institute, and experimental data is processed based on MATLAB. The schematic diagram of experimental device is show in Fig. 2.

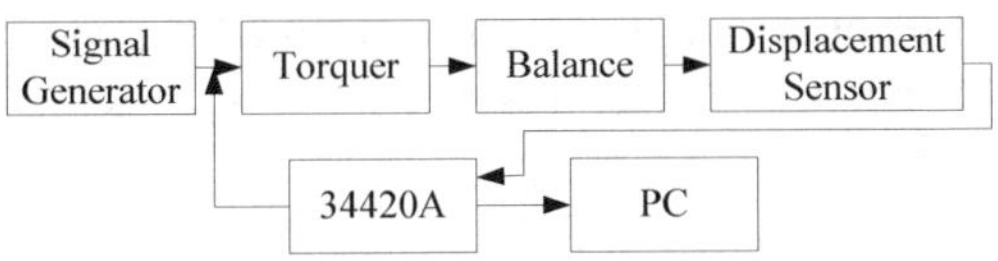

Fig. 2: The schematic diagram

Fristly, make the system balance at the horizontal position. When the period of the driving current is changed, the frequency of the vibration will change. If we stop giving the driving current to the system during the working process, the balance is to vibrate with a constant damping coefficient. And it will stop at the reference position. According to the damped oscillation, the period of the damped oscillation is the natural period of the swinging balance. And the constant damping coefficient is just the damping coefficient of free oscillation.The period of the driving current is set to 8second, 11second, 13second, and the amplitude is 0.371mA, and then stop driving. We can get the curves of the damped oscillation and the spectrum analyses charts, shown in Fig. 3.

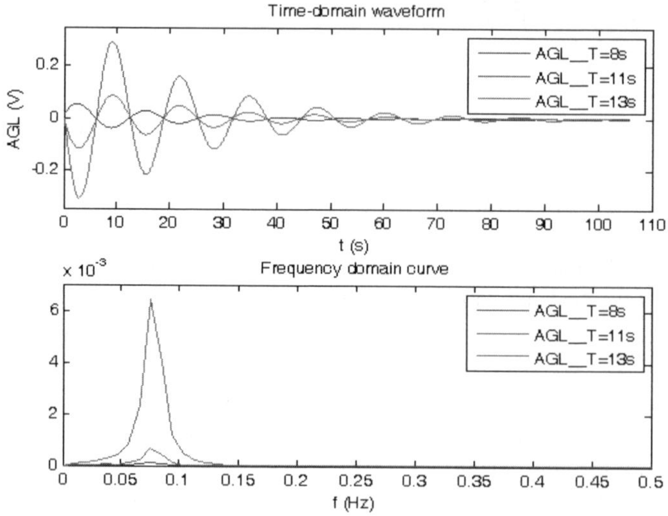

Fig. 3: The curves of the damped oscillation and spectrum analysis

According to the experimental data, the period of the free vibration is 13.243s, so the natural period of the balance is 13.243s. In particular, the natural period is different at different refference state.

The damping coefficient of free oscillation. In order to get the parameter C, we should solve the damping coefficient of the free oscillation first. Because we have got the curves of the damping oscillation, we can analyse the damping evolution of amplitude to achieve the parameter ξ, and then according to Eq. 17, we can get the value of the parameter C.

According to the trend of damping amplitude, we can fit an exponential function. The damping evolution of the fitting exponential function represents the damping coefficient of the system. The fitting curve is shown in Fig. 4.

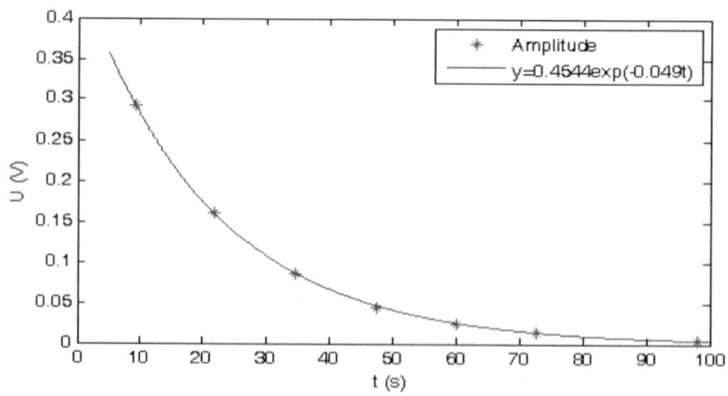

Fig. 4: The fitting curve of damping amplitude

According to the fitting curve, when the natural period of the oscillation is 13.243s, we can obtain the damping coefficient of free oscillation, ξ=0.1, because of, $\xi=0.049/\omega_n=0.049T_n/2\pi$.

In this case, we can achieve the parameter C. Because this parameter is depended on the structure of the balance, it is a constant in experimental conditions. At this point, we can find that C=0.0058219.

The oscillation under external driving force

The damping coefficient contains the influence of the knife hysteresis and the air resistance around the swinging balance. So, in different frequencies, the damping coefficient is different. Next, we will analyse the oscillation of the swinging balance and the damping coefficient in different frequencies.

The curves of oscillation in different frequencies. In the experiment, the driving current of the torquer is $f_e = 0.371\sin(2\pi t/T)$ mA. When we select different driving frequencies, we can find that the amplitude of the displacement is different. T is the driving period. Then we can get the curves of the displacement of the oscillation. They are shown in Fig. 5. And its spectrum analysis charts are shown in Fig. 6.

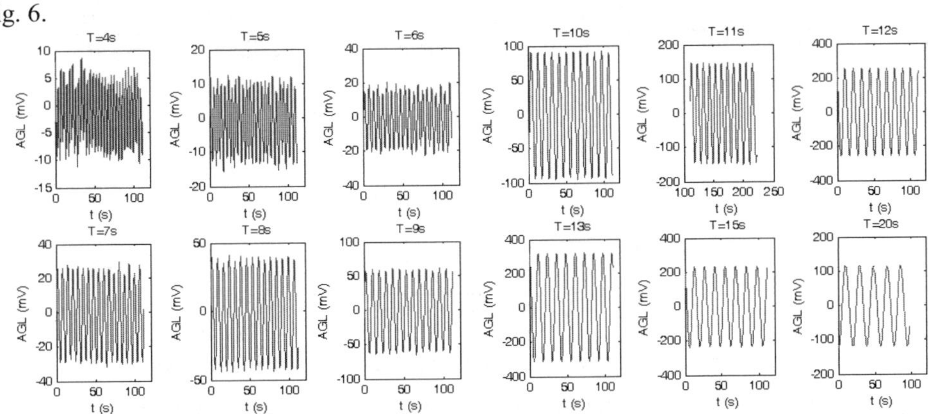

Fig. 5: The curves of the displacement of the oscillation in different driving periods

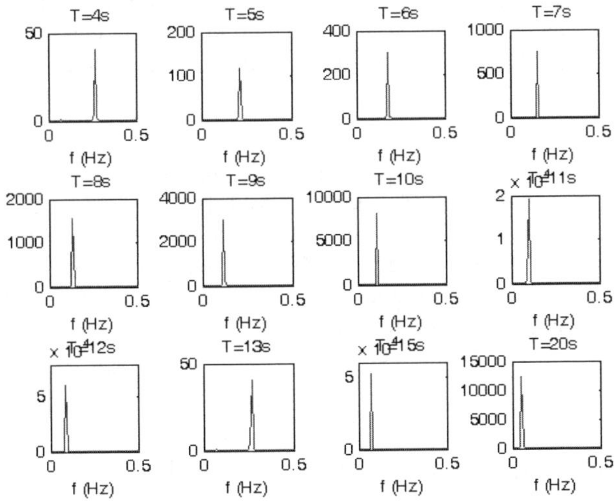

Fig. 6: The spectrum analysis chart of displacement of oscillation in different driving frequencies

So we can get the mean value of amplitude in different oscillation periods and the measurement periods by spectrum analysis chart. T is the driving period, and T_m is the measurement period of the oscillation. The results are shown in Table 1 as follows.

Table 1: The mean value of amplitude under different periods

T [s]	4	5	6	7	8	9
T_m [s]	3.702	4.627	5.553	6.533	7.404	8.543
Mean [mV]	5.5535	10.5655	16.8693	25.868	38.8745	58.9457
T [s]	10	11	12	13	15	20
T_m [s]	9.255	10.096	11.106	12.34	13.882	18.878
Mean [mV]	90.6448	145.8751	255.713	318.6238	231.826	115.9492

The damping coefficient in different frequencies. According to Eq. 16 and $\lambda = \omega/\omega_n$, $\omega=2\pi/T$, we can get the value of the damping coefficient in different oscillation periods. The results are shown in Table 2 as follows.

Table 2: The value of the damping coefficient in different periods

T_m [s]	3.702	4.627	5.553	6.533	7.404	8.543
ξ	0.3826	0.6843	0.5819	0.4679	0.4098	0.2884
T_m [s]	9.255	10.096	11.106	12.34	13.882	18.878
ξ	0.2713	0.2184	0.1427	0.0607	0.1361	0.1496

So we can obtain the curve about damping coefficient and the oscillation period. It is shown in Fig. 7.

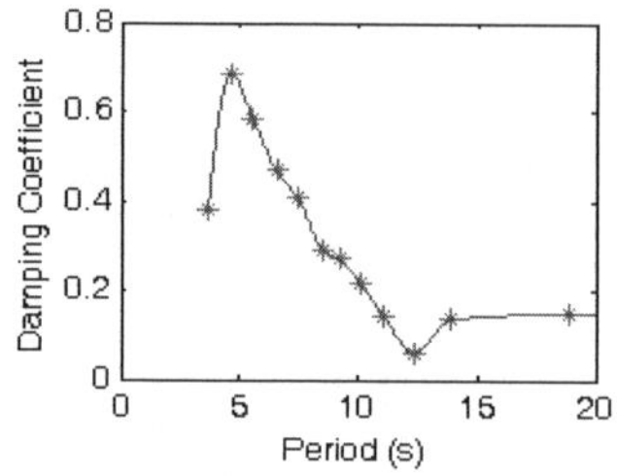

Fig. 7: The curve about damping coefficient and the oscillation period

According to the curve, we can achieve the following conclusion. When the period is between 4second and 13second, the relationship between damping coefficient and oscillation period is nearly linear. And when the period is greater than 13second, the damping coefficient is nearly a constant. That is to say, when the balance vibrates slowly, the damping coefficient is nearly a constant. In the study, we can select appropriate working period based on this conclusion.

Summary

This paper analyzed the working principle of the swinging balance, the period of free oscillation and the damping oscillation in different frequencies. And then we analyse the relationship between damping coefficient and oscillation by the experimental data. At last, we have obtained the following conclusions. Frist, When the period is between 4second and 13second, the relationship between damping coefficient and oscillation period is nearly linear. Second, And when the period is greater than 13second, the damping coefficient is nearly a constant. That is to say, when the balance vibrates slowly, the damping coefficient is nearly a constant.

References

[1] Myklestad N O: *The Concept of Complex Damping*. Journal of Applied Mechanics, 19, 3(1952), pp. 20-30.

[2] Canghey T K: *Classical Normal Modes in Damped Linear Dynamic System*. Journal of Applied Mechanics Vol. 27(1960), pp. 261-269.

[3] Adhikari S: *Damping modelling using generalized proportional damping*. Journal of Sound and Vibration Vol. 293(2006), pp. 156-170.

[4] Woodhouse J: *Linear damping models for structural vibration*. Journal of Sound and Vibration Vol. 215(1998), pp. 547-569.

[5] Hu Shousong: *Automatic Control Theory*. Science Press, Beijing, (2001).

[6] Tepes Onea Florin, Gelmambet Sunnai: *Evaluation of damping in dynamic analysis of structures*. International Journal of Mathematical Models and Methods in Applied Sciences, 4, 2(2010), pp. 124-131.

[7] F Weber, C Boston: *Energy based optimization of viscous-friction dampers on cables*. Smart Materials and Structures Vol. 19(2010), pp. 1-11.

Advanced Materials Research Vol. 279 (2011) pp 339-344
© (2011) Trans Tech Publications, Switzerland
doi:10.4028/www.scientific.net/AMR.279.339

Experimental Study of Automotive Aerodynamics in Headlamp Effect

Jiang Lanfang[1, a], Liu Hong[2, b] and Li Aiqi[2, c]

[1]Zhijiang College of Zhejiang University of Technology, Hangzhou, China

[2]College of Mechanical Engineering, Zhejiang University of Technology, Hangzhou, China

[a]jlf@zjc.zjut.edu.cn, [b]hliu@zjut.edu.cn, [c]zionmd@163.com

Keywords: Headlamp, Automotive aerodynamics, Wind tunnel, Drag coefficient

Abstract. The effect of headlamp modeling on automotive aerodynamics was studied by wind tunnel tests. Firstly, the effect of Reynolds number on drag coefficient of automotive scaled down models was studied under different velocity of flow to verify the rationality of selecting scale for scaled down model and setting inlet boundary condition. Secondly, drag coefficient of automotive scaled down models with different headlamp modeling design were measured. Thirdly, the distribution of surface pressure on central symmetry plane and headlamp was measured and analyzed. It also validated the validity of preceding numerical simulation. It is of importance to guide the headlamp modeling design and automotive modeling design.

Introduction

The fuel consumption is related to the automotive drag coefficient when it was driving. Due to environmental arguments and energy crisis, the improvement of automotive dynamic property and fuel economy by optimizing aerodynamic shape and reducing drag coefficient becomes the significant way [1].

Numerical simulation of external flow field around automobile foreside has been performed in ref. [2] to study the effect of different headstock shape on aerodynamic drag. The effect of different backlight geometry angel on aerodynamic drag has also been studied by numerical simulations in ref. [1, 3,] using Ahmed model [4]. Numerical simulation investigation of a new concept car was presented and the mechanism of low aerodynamic drag was discovered [5]. The automotive underbody flow characteristics were simulated to research the influence of automotive rough underbody shapes on automobile aerodynamic characteristics [6].

Headlamp is an important independent exterior decoration. It locates on the front face, side corner of headstock. When flows around headlamp, the airflow will change its motion path. And it will be different corresponding to different headlamp modeling. That is to say, the headlamp modeling has some influence on automotive aerodynamic drag [7]. Therefore, the effect of headlamp modeling on automotive aerodynamic drag should be studied. Numerical simulation has been performed in ref. [8]. This paper carried out the experimental study. It included:

1. Under different velocity of flow, 30m/s, 35m/s, 40m/s and 45m/s, wind tunnel tests with automotive scaled down model will be carried out to analyze the effect of Reynolds number on drag coefficient, which is used to verify the rationality of selecting scale for scaled down model and setting inlet boundary condition.
2. Measurement and analysis of drag coefficient of automotive scaled down model with different headlamp modeling design.
3. Measuring the distribution of surface pressure on central symmetry plane and headlamp under the velocity of flow 40 m/s, and analyzing the distribution rule.

Wind Tunnel Tests

Experimental Model. Two automotive scaled down models with different headlamp modeling design were selected called No. 1 and No. 2, which has been numerically calculated in ref. [8]. The headlamp modeling parameters are listed in Table 1. Considering the principle of similarity, the models were processed by rapid prototyping technology, as shown in Fig. 1. In order to measure the surface pressure and fix the pressure-measuring pipes, models were designed as hollow, with wall thickness about 8mm.

Table 1: Modeling parameters of scaled down models

Parameters	$r\,[R,\,r_1]$	β	γ	θ	x_1	y_1
No 1	0.300(0.376-5.000)	48°	10°	80°	0.060	0.120
No 2	0.300(0.376-5.000)	48°	46°	45°	0.060	0.000

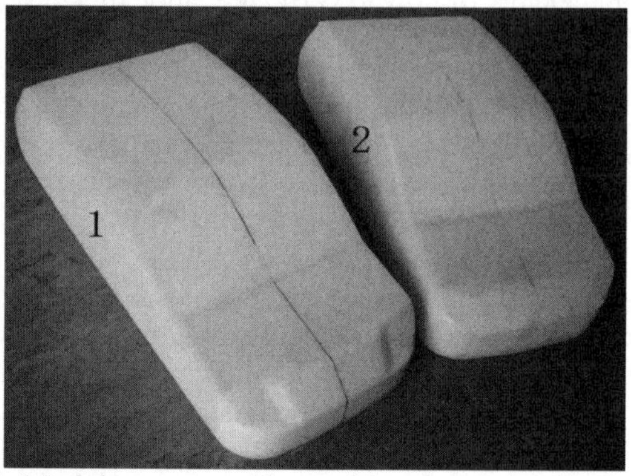

Fig. 1: Models processed by rapid prototype technology

The measurement of surface pressure was arranged after the measurement of drag coefficient. Before measuring the pressure, pressure-measuring pipes should be fixed on the surface of model. The distribution of pressure-measuring pipes should be good for evaluating the modeling characteristics and researching the distribution of surface pressure. The number of pressure-measuring pipes will be denser where it has big change in shape. Besides, the pressure-measuring pipes should be perpendicular to the surface of model. Model No. 1 was selected to measure the surface pressure. The distribution of pressure-measuring pipes is shown in Fig. 2. There were 14 points on central symmetry plane and 4 points on headlamp.

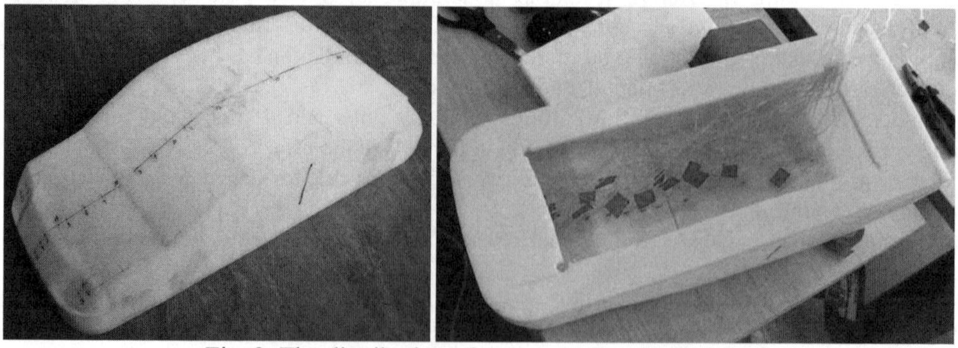

Fig. 2: The distribution of pressure-measuring pipes

Experimental Environment. The wind tunnel tests were carried out in scaled wind tunnel in Shanghai wind tunnel center for ground vehicles in Tongji University, as shown in Fig. 3. The wind tunnel is looped and open, and its basic parameters are listed in Table 2.

Table 2: The basic parameters of scaled wind tunnel

The basic size of experimental section	1515*1185*820
The area of inlet	0.12(433.3mm*283.3mm)
The maximum velocity	170(47.2m/s)
The length of airflow	1160
The power of engine	7.5

Fig. 3: The scaled wind tunnel

The bottom of scaled wind tunnel was considered as floor. Automotive scaled down model was set in the middle of round plant which has the best quality of flow flied. The yaw angle was 0 degree. Four cylinder brackets were added under the model. The height of cylinder bracket was equal to the space between the bottom of car to floor in numerical simulation. There were holes on the bottom of scaled wind tunnel, so the brackets and balance can be connected and fixed. In order to avoid the interference for flow field around the mode, the balance should not be exposed in experimental section, as shown in Fig. 4.

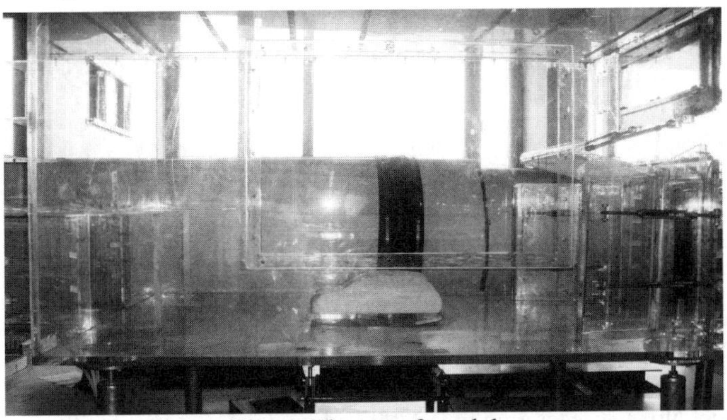

Fig. 4: The set of model

In drag-measuring experiment, the aerodynamic drag will be measured by high precision box strain sit-component of force balance. It could connect with PC by RS232 and collect and record data, as shown in Fig. 5. Mechanical scan valve pressure-measuring system was used for pressure-measuring experiment, as shown in Fig. 6.

Fig. 5: Equipment for drag-measuring

Fig. 6: Equipment for pressure –measuring

Results and Analysis

The Effect of Reynolds Number on Drag Coefficient. The experimental results of drag coefficient of two models under different velocity of flow are listed in Table 3. The relationship between drag coefficient and Reynolds number is shown in Fig. 7.

Table 3: Experimental results of drag coefficient

Velocity of flow [m/s]	30	35	40	45
Value of Re ($*10^6$)	0.863	1.007	1.151	1.295
No 1	0.4019	0.3872	0.3773	0.3743
No 2	0.3767	0.3557	0.3512	0.3500

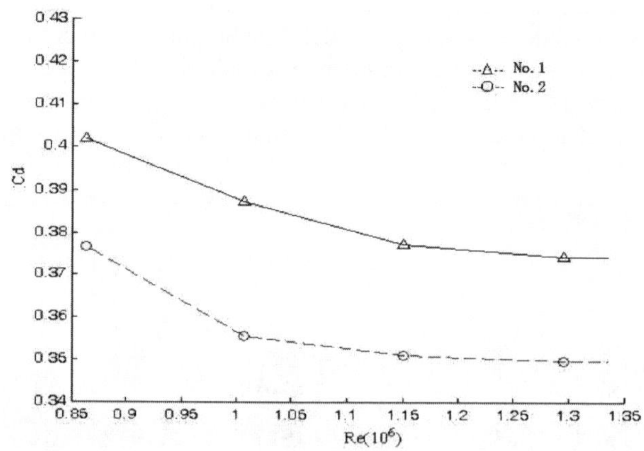

Fig. 7: Relationship between drag coefficient and Reynolds number

In Fig.7, different velocity of external flow field corresponded different Reynolds number, which affected drag coefficient of models. But when the Reynolds number was more than 106, as well as the velocity was more that 35m/s, the drag coefficient of models changed a little. In fig. 7, the drag coefficient of model No. 1 was tending to 0.3700 and model No. 2 was tending to 0.3500. That is to say, the Reynolds number had little effect on drag coefficient when the velocity was more than 35m/s. Therefore, the scale value 1/10 for scaled down model was feasible and the velocity 40m/s for inlet boundary condition was rational.

The Results of Drag-Measuring. The experimental drag coefficient of two models under the velocity of flow 40m/s are listed in Table 4. The numerical results are also listed in Table 4.

Table 4: Experimental and numerical drag coefficient

	Experimental drag coefficient	Numerical drag coefficient	Error
No 1	0.3773	0.3829	1.43%
No 2	0.3537	0.3640	2.83%

The drag coefficient of two models was different with different headlamp modeling design. That is to say, headlamp modeling had effect on automotive drag coefficient. Rational headlamp modeling design can reduce drag coefficient. Comparing experimental results and numerical results, it can be seen that the experimental results were less than numerical results. The errors are 1.43% and 2.83%, within rational error range. It also validated the validity of numerical simulation in ref. [8].

The Results of Pressure-Measuring. The experimental results of surface pressure of model No. 1 on central symmetry plane and headlamp under the velocity of flow 40 m/s are separately shown in Fig. 8 and Fig. 9. The numerical results are also shown in Fig. 8 and Fig. 9. It can see from the two figures that the distribution and trend are consistent. It also validated the validity of numerical simulation.

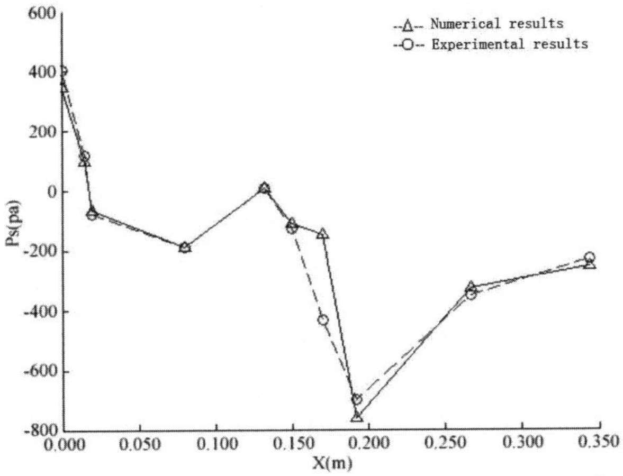

Fig. 8: Surface pressure of No. 1 on central symmetry plane

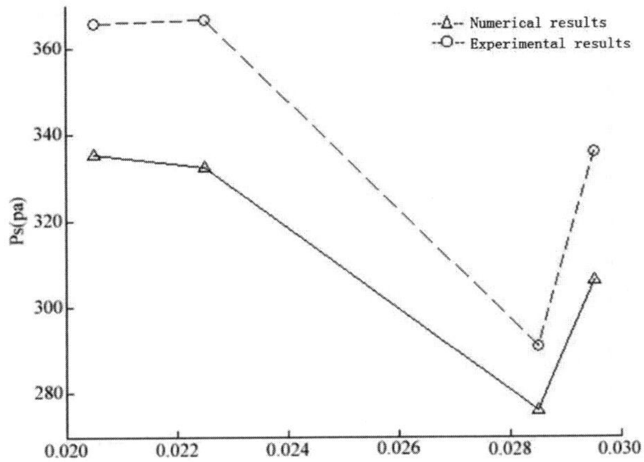

Fig. 9: Surface pressure of No. 1 on headlamp

Conclusion

Wind tunnel tests of scaled down models were performed with different headlamp modeling design. The effect of Reynolds number on drag coefficient was studied under different velocity of flow. Drag coefficient of automotive scaled down models with different headlamp modeling design were measured. The distribution of surface pressure on central symmetry plane and headlamp were also measured. It is of importance to guide the headlamp modeling design and automotive modeling design.

References

[1] G. Emmanuel: *Computational study of flow around a simplified car body*. Journal of Wind Engineering and Industrial Aerodynamic, Vol. 96(2008), pp. 1207-1217.

[2] Y.Q. Zhao, Q. Zhang: *Numerical simulation of external flow field around automobile foreside at high-speed*. Journal of China Mechanical Engineering, 18, 15(2007), pp. 1886-1889.

[3] F. Ehab: *Unsteady flow simulation of the Ahmed reference body using a lattice Boltzmann approach*. Journal of Computers & Fluids, Vol. 35(2006), pp. 940-950.

[4] S.R. Ahmed, G. Ramm, G. Faltin: *Some salient features of the time-averaged ground vehicle wake* [J]. SAE, 840300,(1984), pp. 1-31.

[5] Y.B. He, Z.Q. Gu, J.Wu, et al: *An experimental study on the aerodynamic characteristics of new concept car by wind tunnel test*. Automotive Engineering, 29, 4(2007), pp. 267-295.

[6] J.Y. Wang, C.G. Lai, X.J. Hu: *Influence of automotive underbody shape on aerodynamic performance*. China Journal of Highway and Transport, 21, 2(2008), pp. 111-115.

[7] L.M. Fu: *Aerodynamics of Road Vehicles*. China Machine Press, Beijing, (2005).

[8] L.F. Jiang, H. Liu. G.Z. Chai: *Parametric modeling method of automotive headlamps considering aerodynamics*. Journal of Computer-Aided Design & Computer Graphics, 22, 2(2010), pp. 293-298.

Advanced Materials Research Vol. 279 (2011) pp 345-349
© (2011) Trans Tech Publications, Switzerland
doi:10.4028/www.scientific.net/AMR.279.345

Parameters' Optimized Allocation of Spiral Feeder of RP953 Paver

YANG Yin-wei [1,a], LIU Xu[1,b]

[1]College of Mechanical and Electrical Engineering, China University of Mining and Technology, Xuzhou, China, 221116

[a]yywei-163@163.com, [b]liuxu123934@163.com

Keywords: paver, spiral, design, optimized allocation

Abstract. Paver is a kind of equipment used for high quality road alignment. As an important system of the paver, spiral feeder system consumes power accounted for about half the total power of the engine. Based on RP953 paver as an example, this paper analyzes the spiral feeder of the parameters of the points and optimizes configuration design basis, which can provide valuable information for the designing of the spiral feeder.

Introduction

Along with the improving requirements for quality of highway construction engineering, the work such as mixing the stabilized soil which was completed by land leveller or road-mix stabilizer before is now all finished by the paver. Spiral feeder (Fig. 1) portion is an important part of paver. Spiral feeder's working principle is that a drive shaft with helical vanes revolves within the feed spout surrounded by material, making the material go directly along the feed spout instead of rotating together with spiral because of gravity and the friction of leaves and the feed spout, which makes the material drop under the spiral, keeping feeding the screed at the same time. The key indicator deciding spiral feeder's performance is the productivity, meanwhile, the main parameters deciding productivity are the size of spiral, screw pitch, rotational speed, etc. Reasonably allocating these parameters has a decisive impact on the performance of paver.

In this paper, the impact caused by the friction of the material and the spiral feeder is ignored, and we start our research just from the similarity between the screw conveyor and paver. Using the research results of screw conveyor for analysing the working condition of the spiral feeder will be of great significance for the reasonable allocation of the spiral parameters.

Fig. 1: Spiral Feeder

Spiral Feeder's productivity

Most of the material transported to spiral feeder by scraper are continuously transported by width, and the rest is directly dropped onto the middle of width. Therefore, the productivity of spiral feeder is slightly less than the paver's productivity.

Spiral feeder's productivity is defined by(single spiral) :

$$Q_L = 60(\frac{\pi D^2}{4})\rho_1 s n_L K_l \quad (1)$$

Where D represents spiral diameter, m; S represents screw pitch, m ; n_L represents rotational speed, $r/\min$; ρ_1 represents mixture density, $\rho_1 = 1.75 t/m^3$; K_l represents filling coefficient. It can be seen from the equation, spiral feeder's productivity is connected with D , n_L , S , K_l , after allocation productivity Q_L ,we can allocation spiral diameter D , screw pitch S , rotational speed n_L ,filling coefficient K_l to meet its requirement.

Parameters allocation

As shown in Fig. 2, it is the stress analysis of material granule named A, which is r away from spiral axis, F_L is the resultant force, and F_z , F_r are components of F_L , respectively by the axis and circumference direction.

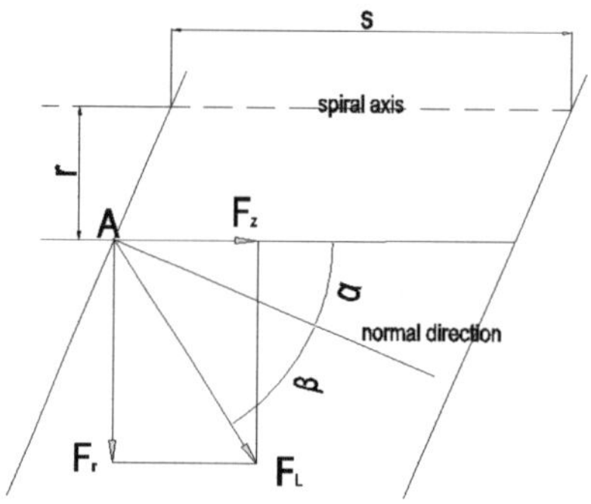

Fig. 2: Dynamics analysis

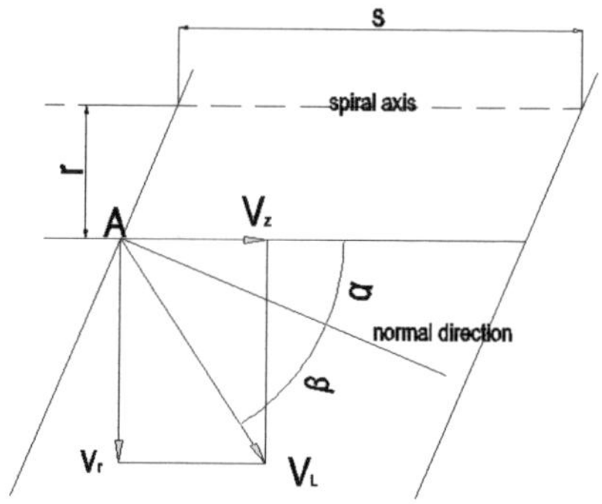

Fig. 3: Kinematics analysis

The following equations can be got from Fig. 2:

$$F_z = F_L \cos(\alpha + \beta), \quad (2)$$

$$F_r = F_L \sin(\alpha + \beta), \quad (3)$$

where α represents the pitch angle of spiral blade and is defined by:

$$\alpha = \arctan(S/_{2\pi r}) . \quad (4)$$

When the material is in transportation, its movement is not a simple linear motion, but a space movement. Material granule has the speed along the axial direction and the circumference direction, as shown in Fig. 3.

In theory:

$$V_z = \frac{s n_L}{60} \bullet \frac{1 - \mu s/_{2\pi r}}{1 + (s/_{2\pi r})^2}, \quad (5)$$

$$V_r = \frac{s n_L}{60} \bullet \frac{\mu + s/_{2\pi r}}{1 + (s/_{2\pi r})^2}. \quad (6)$$

In the equations, S represents screw pitch, m; n_L represents rotational speed, r/min; μ represents friction coefficient between material and helicoids, $\mu = \tan\beta$. Introducing function $K_v = {V_z}/{V_r} = \frac{1 - {\mu s}/{2\pi r}}{\mu + {s}/{2\pi r}}$ and making $K_v > 1$, we get the value of r when $V_z > V_r$, $r > \frac{s}{2\pi} \bullet \frac{1+\mu}{1-\mu}$, meanwhile, $\mu=\tan\beta=0.5$.

The data was imported into *Matlab* and the diagram of the relationship between V and r are made as follows.

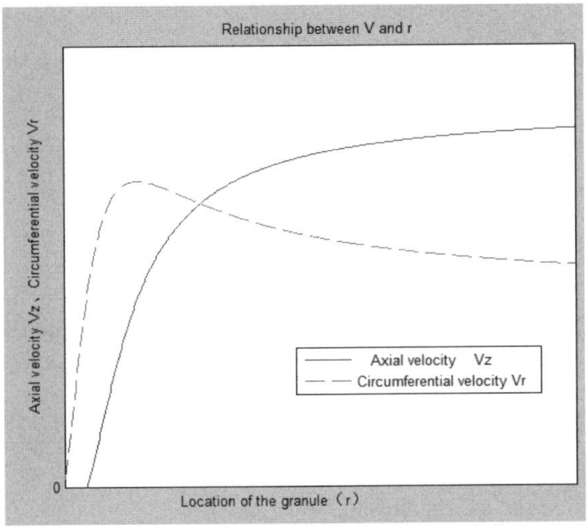

Fig. 4: Relationship between V and r

The value of the intersection should be:

$$r = \frac{s}{2\pi} \bullet \frac{1+\mu}{1-\mu} = 0.478s. \tag{7}$$

Thus, it can be seen, when $r > 0.478s$, $V_z > V_r$. It means that axial velocity is relatively large, and circumference speed is relatively small. It belongs to high-productivity-area, and should be made full use of, so we must take the conclusion into consideration in allocating filling coefficient.

Material's filling coefficient in the feed spout has very big effect on transportation and energy consumption. When the filling coefficient $K_t = {1}/{2} - {r}/{D}$ is relatively small, the corresponding r is relatively larger, and material piling height is lower. Most material lies near to spiral lateral, making the axial speed go up and circumference speed go down. The movement direction of material granule in the transport is much more remarkable than the movement along circumference direction, and the slip plane almost parallel to the transport direction, meanwhile, the additional material flow perpendicular to the transport direction, and the energy consumption reduced. Conversely, when filling coefficient is bigger, the corresponding r go smaller, material slip plane is very steep, and the movement along the circumference direction is much more remarkable than that along the transport direction, which will lead to transportation speed reduction and additional energy consumption. So, theoretically, it is more efficient to choose a smaller filling coefficient. But actually, if we chose a too small value without considering other impacts, spiral feeder's productivity will be very low, so we have to increase the speed to meet the requirements. However, it's not scientific, so we usually chose $K_t = {2}/{3} \sim {3}/{4}$, in this article, $K_t = 0.75$

Spiral speed

Rotational speed is an important measure of spiral feeder's productivity. When allocated the outer diameter of the spiral, the diameter of the shaft and the pitch screw, productivity is proportional to rotational speed, and it guarantees the feeder can get enough material when paving speed increased.

As with screw conveyor, the vanes of spiral rotate with the shaft together also, but materials are rotating and transported in the trough formed by material and keep dropping at the same time. As the circumferential speed V_r of the material closer to shaft is higher that of the outer, and at the same time, the axial velocity V_z is significantly reduced, so material in this layer is more easily to jump or rotate with the shaft and reach the surface early. This is the root cause of segregation of the material when the paver works.

Spiral speed is defined as follows. When the material granule which is at the outer diameter does not has the speed perpendicular to the transportation direction, the relationship of the maximum value inertia force of the granule with its own gravity should be as follows:

$$m\omega_{\max}^2 r = mg, \tag{8}$$

$$2\pi n_{\max} r / 60 = \sqrt{gr}. \tag{9}$$

Considering the influence of material for transportation, comprehensive coefficient K_S is introduced and then:

$$n_{\max} = \frac{30K_S}{\pi}\sqrt{\frac{2g}{D}}. \tag{10}$$

Also, the critical spiral speed can be defined by:

$$n_{\max} = \frac{A}{\sqrt{D}}. \tag{11}$$

In this equation, A is material characteristic coefficient and $A=50$.

In actual design process, the maximum rotational speed can be appropriately increased to ensure for the feeding for the lack of material caused by man, thus, it reduces shut-down time of paver caused by material.

Spiral diameter

In theory, the productivity of RP953 paver is 700t/h(two spirals), a single spiral's productivity is 350t/h, and the diameter of the shaft can be allocated from this. Spiral vanes' diameter is an important parameter of spiral feeder, directly relates to the size of feeder and the productivity.

Make A (in Eq. 11) 35% larger and apply it into Eq. 1, introducing the proportion coefficient of pitch screw and diameter k_1, $s = k_1 D$, then we get another equation:

$$Q_L = 64k_1 A\rho_1 K_l D^{\frac{5}{2}}. \tag{12}$$

Applying the productivity in the equation, therefore, D = 400mm.

Diameter of screw shaft

The size of screw diameter is related with screw pitch, since they determine the pitch angle of spiral blade together and the slipping direction of material and velocity distribution. Therefore, the friction relationship between spiral face and material, as well as the appropriate distribution of velocity component are considered to determine the most reasonable relationship between shaft diameter and screw pitch.

Diameter of screw shaft is defined by:

$$d = (0.2 \sim 0.35)D. \quad (14)$$

We choose $d = 0.35D = 140mm$. Applying the data in all the equations, we can obtain the screw pitch $s = 0.8D = 320mm$, rotational speed $n_L = 107r/\min$, making $n_L = 110r/\min$.

Applying the data in Eq. 1, $Q_L = 348.34t/h < 350t/h$, that means the parameters are reasonable.

Conclusion

(1) The speed of spiral feeder should match the rotational speed of scraper conveyor, the paving speed, paving width and thickness. Besides, the fluctuation of the speed is not allowed, or it will cause the isolation of the mixture or the mixture can't be promptly sent to both ends.

(2) The paving road conditions should be fully considered when allocating spiral parameters. We should do experiments as many as possible to obtain more design reference data, and then adjust the parameters according to the actual situation.

(3) The continuous working time of spiral feeder should be 80% of the working time of paver, and working too long may damage the components seriously.

References

[1] Y. Huaisan: *Design of Paver.* Construction Machinery, 7(2004), pp. 82-102.

[2] T. Xianwei, X. Weiguo: *Brief Talking about Design of Helical Spreader for Slipform-type Cement Concrete Paver.* Construction Machinery Technology&Management, 20, 2(2007), pp. 86-88.

[3] Z.Shiying, C. Yuanji: *Road Construction Machinery Engineering.* China Machine Press, Beijing, (1998).

[4] YAN. Shirong, WEN. Bangchun, GONG. Zhaomin: *Theorefical study of a nonlinear dynamics of vibratory compacting mechanism of paver.*China Journal of Highway and Transport, 7(2000), pp.123-126.

[5] LU. Zhaofeng, GUO. Xiaohong, QIN. Min, XIE. Rui: *Research on Transmission Mechanism of Spiral Feeder of Asphalt Paver and Movement of Aggregates.*Construction Machinery Technology & Management, 22, 1(2009), pp. 108-112.

[6] GUO. Xiaohong, GONG. Weimin, MI. Xue-yu: *Bituminous Mixture Movement Mechanism Analysis in Distributing Room for Paver.*Road Machinery & Construction Mechanization, 24, 11(2007), pp. 23-27.

[7] GUO. Xiaohong, LI. Rongqiang: *Research of Concrete Grain Speed Characteristics and Screw Distributor Structure and Rotate Speed.* Construction Machine & Construction Technology, 25, 4 (2008), pp. 38-41.

[8] LU. Zhaofeng, QIN. Min, GUO. Xiaohong: *Study on Moving Feature of Bituminous Mixture Flow in Distributing Room.* Journal of Chongqing Jiaotong University(Natural Science), 28, 1(2009), pp. 63-67.

Advanced Materials Research Vol. 279 (2011) pp 350-355
© (2011) Trans Tech Publications, Switzerland
doi:10.4028/www.scientific.net/AMR.279.350

Hydrodynamics of Coarse Coal Slime and Quartz Particles in a Liquid-Solid Fluidized Bed Separator

Sha Jie[1,a], Xie Guang-yuan[1,b], Peng Yao-li[1,c] and Shi Ben-xuan[1,d]

[1]School of Chemical Engineering and Technology, China University of Mining and Technology, Xuzhou, Jiangsu 221008, China

[a]43002893@qq.com, [b]xgywl@163.com, [c]397602141@qq.com, [d]125720230@qq.com

Keywords: hydrodynamics, coarse coal slime, quartz, minimum fluidization velocity, bed expansion ratio, mixed particles

Abstract. The hydrodynamics of particles of coarse coal slime and quartz of different sizes in a liquid-solid fluidized bed separator were investigated experimentally, including minimum fluidization velocities, bed expansion ratios, and segregation of mixed particles with two methods. Experimental parameters studied included density and size of particle, superficial water velocity and initial static bed height. The results provide a reference for choice of size scope on coarse coal slime separation by liquid-solid fluidized bed separators.

Introduction

Liquid-solid fluidized bed separators, also called teetered bed separators, are hindered settling based gravity separation equipments which are widely used for coarse coal slime separation[1]. In a liquid-solid fluidized bed separator, an upward current of water fluidizes a bed of particles, thus producing an autogenously medium suitable for separating a feed sample on the basis of density and size. The lower density and smaller particles report to overflow and the denser and larger particles report to underflow, and the other particles are distributed throughout the bed depending on their density and size. So separation on their density will be achieved if particles with small range of size are feed into the separator.

K.P.Galvin studied the dense medium separation of a coal feed slurry, -2.00+0.375mm in size. And the slurry consisted of the clean coal, the clean coal combined with the mineral matter, the clean coal and mineral matter combined with dense media, respectively[2]. And then he assembled sets of parallel inclined plates in the separator which was called reflux classifier. Separation of -2mm coal feed slurry by reflux classifier indicated the throughput is nearly 3 tines higher than for no-inclined plates TBS[3].

It is important to study the hydrodynamics of particles (especially coal and mineral matter) in the liquid-solid fluidized bed separator for the influence of operating parameters during separation process. In this paper, hydrodynamic characteristics of particles with different densities and sizes in a liquid-solid fluidized bed separator are studied, respectively, and then preliminary observation on segregation characteristics of mixed particles is carried out. The experiment is conducted in a liquid-solid fluidized bed separator of 0.3m internal diameter and 1.2m height. Coal particles of two sizes, quartz sand particles of four sizes and water have been used as the lower density solid, higher density solid and liquid phases respectively. Density and size of solid phases, superficial water velocity and static bed height are treated as the operating parameters.

Feed and experimental setup

The main purpose of coal separation is to separate gangue of higher density from clean coal of lower density. So clean coal of 1.34g/cm^{-3} density and quartz sand of 2.47g/cm^{-3} have been chosen to provide more reference for coal separation in liquid-solid fluidized bed separators.

Fig. 1 shows the abbreviated drawing of the experimental liquid-solid fluidized bed separator system, which consists of a fluidized bed assembly, a water supply system, and a density measured equipment. The fluidizer is a vertical cylindrical plexiglas column of 0.3m internal diameter and 1.2m height. There is a feed tube and an overflow tank at the top, and a tailing valve and a fluid distributor at the bottom of the column. Fluidized water is pumped out from storage tank, through the rotameter and fluid distributor, to the fluidized bed uniformly. The distributor is a column of 0.15m height filled with 2-3mm mineral particles, with a piece of 0.125 mm screen mesh covering it.

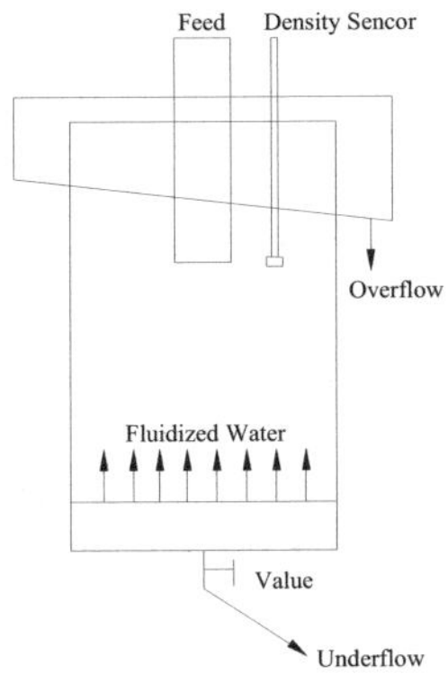

Fig. 1: The abbreviated drawing of experimental setup

Results and discussion

Experiments have been conducted with the superficial liquid velocity varying from 0 to 0.02m/s. To ensure steady state in operation at least 3 min have been allowed before each reading. The readings for the minimum fludization velocity, expanded bed height and layering of mixed bed are then noted. The experimental results have been presented graphically in this section.

Minimum fludization velocity. Minimum fludization velocity is the fluid velocity at which the bed of solid particles begins to be fludized. When fluid velocity is lower than minimum fludization velocity, the solid particles would keep steady. Along with the higher fluid velocity, the bed of solid particles is fludized until the the minimum fludization velocity.

In the present works, effect of size and density of particles, and initial static bed height to the minimum fludization velocity is studied experimentally. The minimum fludization velocity of different densities and sizes of coal and quartz at the static bed height of 50 mm is shown in Fig. 2. It has reported an increase in the minimum fluidization velocity with the increase in size at the same density, and an obvious increase in minimum fluidization velocity with the increase in density at the same size. We also see minimum fluidization of larger coal is higher than that of smaller quartz. It is because that an increase in density and size requires an increase in lifting force that corresponds to an increase in superficial water velocity to fluidize the particles.

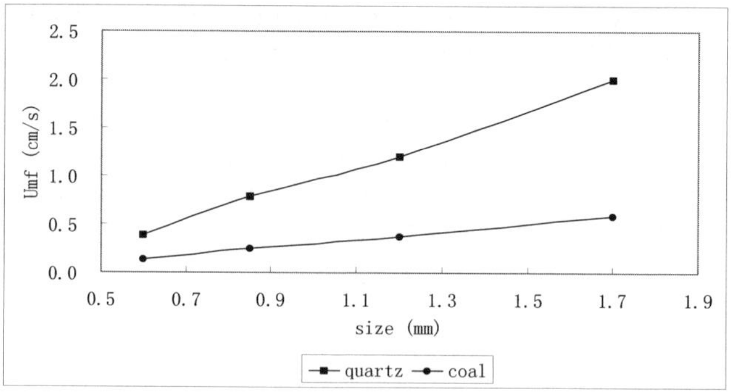

Fig. 2: Minimum fluidization velocity of quartz and coal with static bed height of 50mm

Fig. 3 shows the minimum fluidization velocity of 2-1mm coal particles at different static bed height. It has indicated a little decrease in the minimum fluidization velocity with the increase in initial static bed height. This may be true as the reasoning given by them is that an increase in static bed height corresponds to an increase in water flow, which in turn promotes the bed expansion and accelerates the fluidization process. But the phenomenon is not obvious in case of a liquid-solid system, thus the minimum fluidization velocity being practically un-affected by the static bed height.

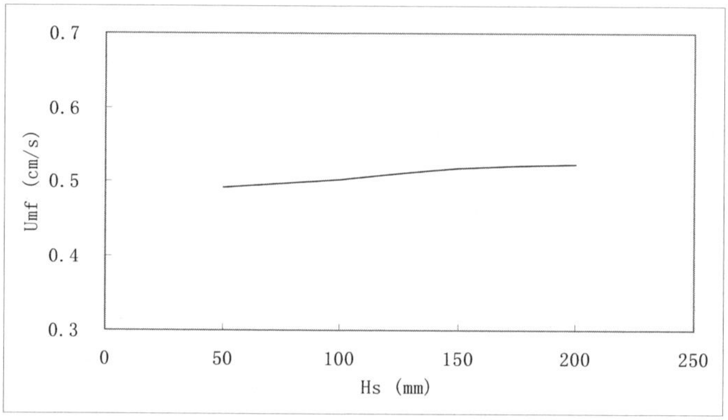

Fig. 3: Minimum fluidization velocity of particles of coal with the size of 2-1mm

Bed expansion ratio.

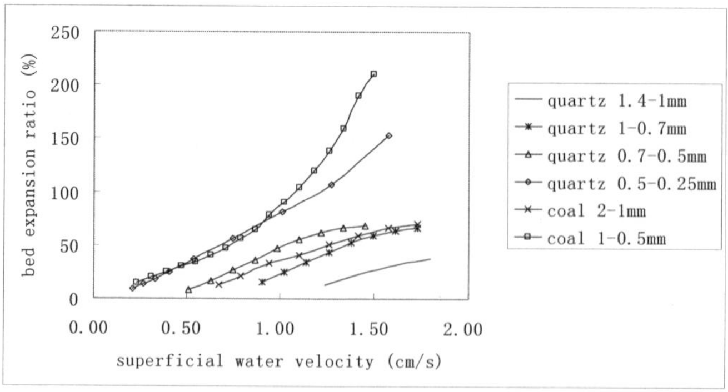

Fig. 4: Bed expansion ratio of particles of coal and quartz with the static bed height of 50mm

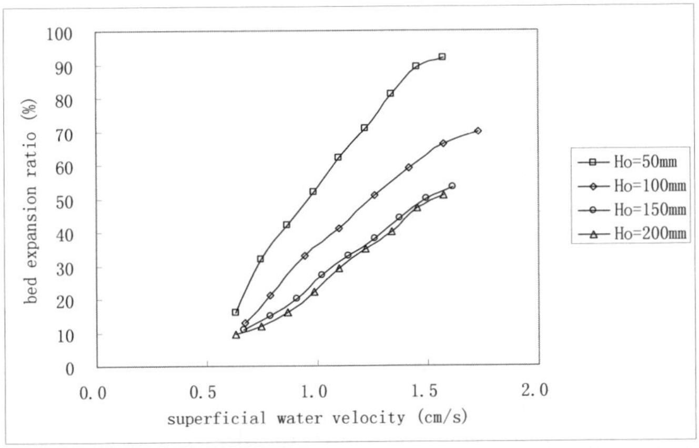

Fig. 5: Bed expansion ratio of particles of coal with the size of 2-1mm

When fluid water velocity reaches the minimum fludization velocity, bed of solid particles begins to expand, and it will expand to different heights with different fluid water velocity, while the balance of the solid remain in the fluidized state. Bed expansion ratio, which is regarded as the value of the bed height of anytime devides the static bed height[4], of coal and quartz of different sizes at the static bed height of 50 mm with the increase in superficial water velocity is shown in Fig. 4. It is indicated that bed expansion ratio of every size of coal and quartz increases with the increase in superficial water velocity, and the increase becomes slow when superficial water velocity is up to a certain degree, and the lager or the denser, the slower of the increase. We also see a decrease in bed expansion ratio with the increase in size at the same superficial water velocity. To different density, bed expansion ratio of coal of 2-1mm in size is between that of quartz of 0.7-0.5mm and of quartz of 1-0.7mm, and expansion tatio of coal of 1-0.5mm in size is close to that of quartz of 0.5-0.25mm at low superficial water velocity, but higher than it at high superficial water velocity. Fig. 5 shows bed expansion ratio of coal of 2-1mm in size at different initial static bed height with the increase of superficial water velocity. It is reported a decrease in bed expansion ratio with the increase of static bed height, but it is similar in bed expansion ratio when static bed height is higher than 150mm. It is because that more water velocity is required to lift more particles when static bed is higher. But the forces between particles have been evident when the static bed height up to a certain degree. Thus there is not quite different of bed expansion ratio at higher static bed height.

Segregation of mixed solid particles. Segregation experimets have been conducted with two method of mixed particles: method (1) is coal of 2-1mm in size and quartz of 1.4-0.25mm in size, and method (2) is coal of 1-0.5mm in size and quartz of 1-0.25mm in size. There are different phenomenons of segregationof mixed particles with the increase of superficial water velocity.

The phenomenons of Method (1) with different superficial water have been described in Table 1.

Table 1: The segregation phenomenon of mixed particles with method (1)

Superficial water velocity (cm/s)	Phenomenon
0.39	A little replacement has been occurred: coarse quartz particles get into the lower layer and coal into upper layer, and the fine quartz particles are over the whole bed, most of that intermix the coal particles in upper layer. The whole bed layer almost keeps still while only fine quartz particles at top of bed and in the void of bed begin to have peristalsis.
0.47	The peristalsis of fine quartz become to a little violent, the lower part of coarse particles of quartz stands motionless, the upper part of the coal particles have peristalsis. The whole bed has been devided into obvious two layer, upper fine quartz and coal mixed, and the lower mainly coarse quartz mixed with fine sand and a small amount of the coal, and there are mainly fine sand at the interface of two layer.
0.59	The whole bed has been fluidized, and it is more intense of lower layer than the upper layer. threr are almost coal and fine quartz in the upper layer, and mainly coarse quartz mixed with a few fine particles of quartz and coal.
0.79 and even higher	Fluidization becomes to more intense, but the segregation does not change.threr are almost coal and fine quartz in the upper layer, and mainly coarse quartz mixed with a few fine particles of quartz and coal, and mainly fine sand at the interface of two layer.

The reason of mixed of fine quartz and coal particles in the upper layer might be the hindered settling velocities, minimum fluidization velocity of those two particles are similar, and the bed expansion of fine quartz is higher. So it is necessary to decrease the upper limit of coal size or to increase the lower limit of quartz size to improve segregation. The scope of quartz size will decrease if the lower limit of its size is increased, and the observation of experiment will be more difficult. So we decide to adjust the limit of coal size to 1-0.5mm, and limit of quartz size to 1-0.25mm. The phenomenon has been shown in Table 2.

Table 2: The segregation phenomenon of mixed particles with method (2)

Superficial water velocity (cm/s)	Phenomenon
0-0.24	The whole bed has been devided into two layer, upper coal and a few fine quartz mixed, and the lower mainly coarse quartz mixed with fine sand and a small amount of the coal, and there are mainly fine sand at the interface of two layer. Not fluidization but peristalsis occurs in part.
0.31	There is peristalsis in the upper layer, and transitivity movement in the lower layer occasionalliy. A few of fine particles of quartz mix with coal in the upper layer.
0.39	The content of fine quartz has been increased. The transitivity movement in the lower layer would take some quartz into upper and bring some coal to the lower.
0.59	The transitivity movement, which becomes to intense and intermittent, could break the steady of upper layer. When upper layer keep steady, there are high content of fine quartz at the top of upper layer and low content at the bottom.
0.94 and even higher	Movement of bed becomes more intense, and the interface of two layer have been in intense fluctuation. The content of fine quartz is quite little on the top of upper layer. Replacement emerges constantly on the bottom of upper layer which could not steady.

It has been indicated that the fine particles of quartz taken into upper layer drop to the interface slowly, and there is few fine quartz on the top of the upper layer after the replacement, which is because of the intermittent transitivity movement of lower layer. If the feed content of particles is much more, more parts of the area with few quartz could exist on the top of upper layer, and then the overflow could be more pure of coal, and the replacement could only occur in the mid of fluidized bed.

With the increase of superficial water velocity, a lot of water which will get into the fluidized bed passing compacted quartz layer would get much block. So when one part of block is break, turbulence of water would occurs and the quartz of the part would be lifted. After that, the force of the quartz lifted from upper fine quartz and coal increases, and then particles of quartz lifted begin to drop. At the moment, the turbulence would transmit to the area surrounding, and the phenomenon will be repeat. Therefore we could observe the transitivity movement of lower layer.

Conclusions

The present study of the liquid-solid fluidized bed hydrodynamics has resulted in the following conclusions. The minimum fluidization velocity increases with the increase in particle size and density, but un-affected with the initial static bed height. The bed expansion ratio increases with the increase in superficial water velocity but decreases in the static bed height and particle size and density. When mixed particles, it would be better in segregation of coal and fine particles of quartz if the upper limit of coal size is decreased properly. Therefore in the coarse coal slime separation, it is better to use more narrow scope of particle size to get better effect of separation.

Acknowledgements

This work was subsidized by National Natural Science Foundation of China (Projects 50974120) and the Fundamental Research Funds for the Central Universities (2010QNB08).

References

[1]Biswajit Sarkar, Avimanyu Das, S P Mehrotra: *Study of separation features in floatex density separator for cleaning fine coal*. Int. J. Miner. Process Vol. 86(2008), pp. 40-49.

[2]K P Galvin, S J Pratten, S K Nicol: *Dense medium separation using a teetered bed separator*. Minerals Engineering, 12, 9(1999), pp. 1059-1081.

[3]K P Galvin, E Doroodchi, A M Callen: *Pilot plant trial of the reflux classifier*. Minerals Engineering, 15(2002), pp. 19-25.

[4]H.M.Jena, G.K.Roy, B.C.Meikap: *Hydrodynamics of regular particles in a liquid-solid semi-fluidized bed*. Powder Technology Vol. 196(2009), pp. 246-256.

Advanced Materials Research Vol. 279 (2011) pp 356-360

doi:10.4028/www.scientific.net/AMR.279.356

The temperature calculation of axisymmetric multilayer cylinder

Y. LI[1,a]

[1]The Chinese People's Amed Police Forces Academy, P.R.China, 065000

[a]liyu10_01@163.com

Keywords: multilayer cylinder, axisymmetric heat conduction, temperature, thermal stress

Abstract. A general temperature calculation method of axisymmetric multilayer cylinder was deduced according to the axisymmetric heat conduction of cylinder. Four types of combined boundary conditions were summarized, and temperature computational formulas of axisymmetrical monolayer cylinder were formed under each type. The intermediate interface temperatures and radial stresses of axisymmetrical multilayer cylinder satisfy tridiagonal linear system of eouations, which can be solved with pursuit method. The calculation method is applied universally. Meanwhile, the corresponding calculating program was developed.

Introduction

The axisymmetric temperature distributions of single and double layer cylinder belong to classic problem and have been reported [1-4]. But the temperature of multilayer cylinder was seldom covered yet. In practice, there are many kinds of long cylindrical furnaces. For example, blast furnaces and mixed iron cars etc. The temperature of these furnaces involves temperature and thermal stress analysis of multilayer cylinder. Based on the principle of axisymmetric thermal conduction of monolayer cylinder, a general temperature calculation method of multilayer cylinder was deduced. And the corresponding calculating program was developed.

Steady temperature distribution of axisymmetric monolayer cylinder

The temperature field in lateral section (Fig. 1) of axisymmetric cylinder only relates with radius r of point, $T=T(r)$. Under the polar coordinate, the steady heat conduction equation is

$$\left(\frac{d^2}{dr^2}+\frac{1}{r}\frac{d}{dr}\right)T=0 \tag{1}$$

$$T(r)=A\ln r+B \tag{2}$$

Where, A and B are coefficients determined by boundary conditions.

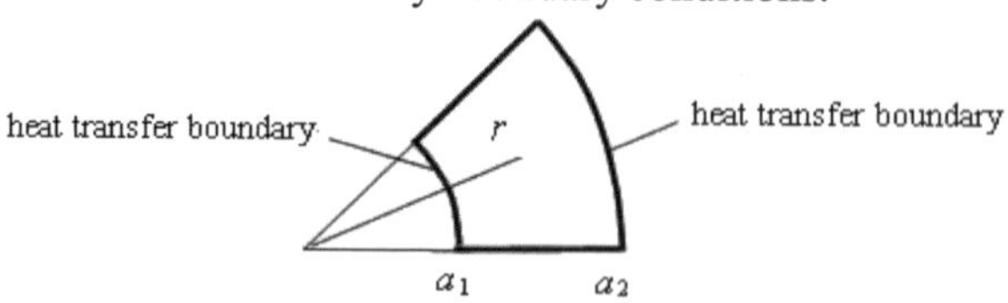

Fig. 1: Heat conduction of monolayer cylinder

Temperature gradient can be expressed as

$$G=-\text{grad}T=-\frac{dT}{dr}=-\frac{A}{r} \tag{3}$$

Heat flux density

$$q_r = -k\frac{\mathrm{d}T}{\mathrm{d}r} = -\frac{kA}{r} \tag{4}$$

Heat flux

$$Q = -2\pi r q_r \tag{5}$$

Supposed that the inside and outside free surfaces keep constant temperature or convective heat transfer, there are four combined boundary conditions: (1) temperatures of both boundary sides are constant; (2) inside boundary is fixed temperature and outside is convective heat transfer; (3) inside boundary is convective heat transfer and outside is fixed temperature; (4) Boundaries of both sides are convective heat transfer.

(1) Temperatures of both boundary sides are constant

$$\begin{cases} A = \dfrac{T_2 - T_1}{\ln\dfrac{a_2}{a_1}} \\ B = \dfrac{T_1 \ln a_2 - T_2 \ln a_1}{\ln\dfrac{a_2}{a_1}} \end{cases} \tag{6}$$

Where, a_1, a_2, T_1 and T_2 are inner surface radius, outer surface radius, inner surface temperature and outer surface temperature, respectively.

(2) Inside boundary is fixed temperature and outside is convective heat transfer

$$\begin{cases} A = \dfrac{T_w - T_1}{\ln\dfrac{a_2}{a_1} + \dfrac{k}{a_2 h_w}} \\ B = \dfrac{\dfrac{k}{a_2 h_w} T_1 - T_w \ln a_1 + T_1 \ln a_2}{\ln\dfrac{a_2}{a_1} + \dfrac{k}{a_2 h_w}} \end{cases} \tag{7}$$

Where, T_1, T_w and h_w are inside boundary temperature, outside fluid temperature and outside heat transfer coefficient, respectively.

(3) Inside boundary is convective heat transfer and outside is fixed temperature

$$\begin{cases} A = \dfrac{T_2 - T_n}{\ln\dfrac{a_2}{a_1} + \dfrac{k}{a_1 h_n}} \\ B = \dfrac{\dfrac{k}{a_1 h_n} T_2 - T_2 \ln a_1 + T_n \ln a_2}{\ln\dfrac{a_2}{a_1} + \dfrac{k}{a_1 h_n}} \end{cases} \tag{8}$$

Where, T_2, T_n and h_n are outside boundary temperature, inside fluid temperature and inside heat transfer coefficient, respectively.

(4) Boundaries of both sides are convective heat transfer

$$\begin{cases} A = \dfrac{T_w - T_n}{\dfrac{k}{a_1 h_n} + \ln\dfrac{a_2}{a_1} + \dfrac{k}{a_2 h_w}} \\ B = \dfrac{\dfrac{k}{a_1 h_n} T_w - T_w \ln a_1 + \dfrac{k}{a_2 h_w} T_n + T_n \ln a_2}{\dfrac{k}{a_1 h_n} + \ln\dfrac{a_2}{a_1} + \dfrac{k}{a_2 h_w}} \end{cases} \tag{9}$$

Steady temperature distribution of axisymmetric multilayer cylinder

As shown in Fig. 2, temperature distribution of *i*th layer of N-layer cylinder, referred to Eq. 2, is expressed as

$$T(r) = A_i \ln r + B_i \quad (a_i \le r \le a_{i+1}, i = 1,2,..., N) \tag{10}$$

Where, A_i and B_i are coefficients determined by boundary conditions. Suppose that a_i and T_i are radius and temperature of each midterm interface respectively. k_i is the thermal conductivity of each layer.

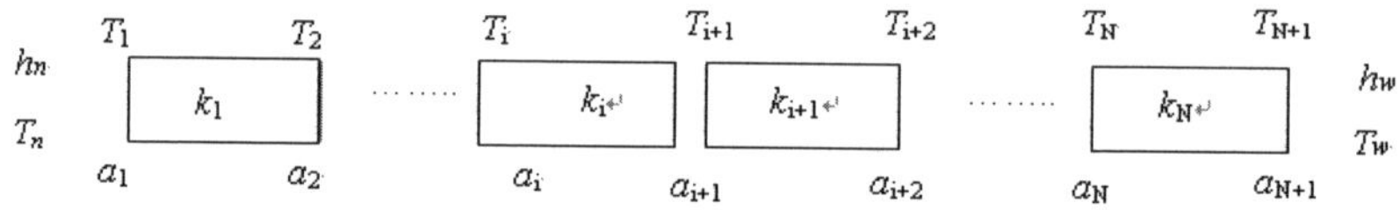

Fig. 2: Multilayer structure

For temperature of *i*th layer

Inside $T_i = A_i \ln a_i + B_i$ (a)

Outside $T_{i+1} = A_i \ln a_{i+1} + B_i$ (b)

For *i*+1th layer

Inside $T_{i+1} = A_{i+1} \ln a_{i+1} + B_{i+1}$ (c)

Outside $T_{i+2} = A_{i+1} \ln a_{i+2} + B_{i+1}$ (d)

From Eq. (a) ~ (d)

$$T_i - T_{i+1} = A_i(\ln a_i - \ln a_{i+1}) \tag{e}$$

$$T_{i+1} - T_{i+2} = A_{i+1}(\ln a_{i+1} - \ln a_{i+2}) \tag{f}$$

As the heat flux densitys gotten across the inside and outside of the midterm interface are equivalent

$$k_i \frac{dT(r)}{dr}\Big|_{r=a_{i+1}} = k_{i+1} \frac{dT(r)}{dr}\Big|_{r=a_{i+1}}$$

We can got

$$k_i A_i = k_{i+1} A_{i+1} \tag{g}$$

Combine Eq. (e)~(g)

$$\frac{T_i - T_{i+1}}{T_{i+1} - T_{i+2}} = \beta_i \quad (i=1,2,\ldots\ldots,N\text{-}1) \tag{11}$$

The Matrix type can be expressed as

$$\begin{bmatrix} 1+\beta_1 & -\beta_1 & & & \\ -1 & 1+\beta_2 & -\beta_2 & & \\ & \ddots & \ddots & \ddots & \\ & & & & \\ & & & -1 & 1+\beta_{N-1} \end{bmatrix} \begin{bmatrix} T_2 \\ T_3 \\ \vdots \\ \\ T_N \end{bmatrix} = \begin{bmatrix} T_1 \\ 0 \\ \vdots \\ \\ \beta_{N-1} T_{N+1} \end{bmatrix} \tag{12}$$

Where, coefficient β_i is

$$\beta_i = \frac{k_{i+1}}{k_i} \frac{\ln \frac{a_{i+1}}{a_i} + \omega_1 \frac{k_i}{a_i h_n}}{\ln \frac{a_{i+2}}{a_{i+1}} + \omega_2 \frac{k_{i+1}}{a_{i+2} h_w}} \quad (i = 1,2,3,\ldots, N-1) \tag{13}$$

When the inside boundary is convective heat transfer, $\omega_1=1$, $T_1=T_n$ and fixed temperature, $\omega_1=0$. When the outside is convective heat transfer, $\omega_2=1$, $T_{n+1}=T_w$ and fixed temperature, $\omega_2=0$.

The coefficients, A_i and B_i, can be expressed as

$$\begin{cases} A_i = \dfrac{T_{i+1} - T_i}{\omega_1 \dfrac{k_i}{a_i h_n} + \ln \dfrac{a_{i+1}}{a_i} + \omega_2 \dfrac{k_i}{a_{i+1} h_w}} \\ B_i = \dfrac{\omega_1 \dfrac{k_i}{a_i h_n} T_{i+1} - T_{i+1} \ln a_1 + \omega_2 \dfrac{k_i}{a_2 h_w} T_i + T_i \ln a_2}{\omega_1 \dfrac{k_i}{a_i h_n} + \ln \dfrac{a_{i+1}}{a_i} + \omega_2 \dfrac{k_i}{a_{i+1} h_w}} \end{cases} \quad (i = 1,2,3\ldots, N) \tag{14}$$

When inside or outside surface has heat radiation with environment, it can be approximately transformed to convective heat transfer in order to avoid computation nonlinear heat radiation directly.

Example

A 3-layers cylinder lining is composed by carbon brick 1, carbon brick 2 and shell. The interface dimensions are a_1= 2m, a_2 = 2.24m, a_3 = 2.44m, a_4 = 2.472m. The physical properties of layers are listed in Table 1. The boundary conditions are T_1= 1150°C, T4= 150°C. A general computer program was exploited with the method and the results are listed in Table 2 and 3.

Table 1: Physical properties of layers

Layer	k[$Wm^{-1}K^{-1}$]	E[Nmm^{-2}]	μ	α [$°C^{-1}$]$\times 10^{-6}$
1	16	37500	0.15	4.5
2	14	32000	0.15	5.5
3	50	192000	0.27	12

Table 2: Result (1)

Interface	T[°C]	p [MPa]	u_r [mm]	σ_r [MPa]
1(inside)	1150	0.6	6.8392	-0.6
2	649.8	4.67	7.9395	-4.67
3	218.4	1.81	8.3972	-1.81
4(outside)	150	0.1	8.4794	-0.1

Table 3: Result (2)

Interface	A_i	B_i	C_{i1} $\times 10^3$	C_{i2} $\times 10^3$	σ_θ [Mpa] inside	σ_θ [Mpa] outside
1	-4413.7	34698.3	1.40	8.08	-92.27	11.11
2	-5044.3	39562.3	1.39	10.8	-14.16	72.30
3	-1412.4	11234.7	1.08	14.1	101.64	158.01

By the data in the table, the temperature and thermal stress distribution of each layer can be written.

Conclusion

(1) Four types of combined boundary conditions were summarized, and temperature computational formulas of axisymmetrical monolayer cylinder were formed under each type.

(2) According to the axisymmetric planar problem, the displacement and thermal stress computational formulas of monolayer cylinder were obtained with bearing axisymmetric temperature and pressure.

(3) The intermediate interface temperatures and radial stresses of axisymmetrical multilayer cylinder satisfy tridiagonal linear system of eouations, which can be solved with pursuit method. The calculation method is applied universally.

Reference

[1] Zhang Hongji: *Heat Conduction*. Higher Education Press, Beijing, (1992).

[2] Yan Zongda: *Thermal Stress*. Higher Education Press, Beijing, (1993).

[3] Xu Zhilun: *Theory of Elastic Mechanics*. Higher Education Press, Beijing, (1979).

[4] Holman J P: *Heat Transfer*. Tsinghua University Press, Beijing, (2005).

[5] Zhang Tie. *Numerical Analysis*. Metallurgical Industry Press, Beijing, (2001).

[6] Haji-Sheikh A, Beck J V: *Steady-state heat conduction in multilayer bodies*. International Journal of Heat and Mass Transfer, 46, 4(2003), pp. 2363-2379.

[7] CHEN Liang-yu, LI Yu: *General temperature computational method of linear heat conduction multilayer cylinder*. Journal of Iron and Steel Research International, 17, 1(2010), pp. 33-37.

Advanced Materials Research Vol. 279 (2011) pp 361-366

doi:10.4028/www.scientific.net/AMR.279.361

Virtual Instrument for Weak Signal Detecting on Monostable Stochastic Resonance

YUAN Quan[1, a], SHEN Yan[2, b] and CHEN Liang[1, c]

[1]School of Mechatronics Engineering, University of Electronic Science and Technology of China, Chengdu, China, 610054

[2]School of Control Engineering, Chengdu University of Information Techology, Chendu, China, 610225

[a]springyuan129@163.com, [b]shenyan02@163.com, [c]chenliang72@uestc.edu.cn

Keywords: Virtual Instrument, LabVIEW, Mono-stable System, Stochastic Resonance,Weak Signal Detection

Abstract. Stochastic resonance (SR) is a nonlinear phenomenon which can be used to detect weak signal. The theory of SR in a biased mono-stable system driven by multiplicative and additive white noise as well as a weak periodic signal is investigated. The virtual instrument (VI) for weak signal detecting based on this theory is designed with LabVIEW. This instrument can be used to detect weak periodic signals which meets the conditions given and can greatly improved the power spectrum of the weak signal. The results that related to different sets of parameters are given and the features of these results are in accordance with the theory of mono-stable SR. Thus, the application of this theory in the detecting of weak signal is proven to be valid.

Introduction

Since Benzi and co-workers [1] discovered a phenomenon that they termed Stochastic Resonance (SR), this non-linear phenomenon has attracted the attention of scientists and engineers all over the world. It has been extensively investigated from both the theoretical and experimental points of view. In the diverse field as electronic and magnetic systems, climatology, chemistry, laser physiology neuroscience, biophysics, solid-state physics and even sociology [2-10], the phenomenon of stochastic resonance can be found.

Most of the studies on SR have been focus on the bistable potential with different types of noises. However, there are a lot of mono-stable systems in real occasions [11-13]. Dykman et al [11]. and Evstigneev et al [12]. studied the SR based on linear-response theory in a mono-stable monostable overdamped system. N.G. Stocks[14] and co-workers discovered a phenomenon of zero-dispersion stochastic resonance (ZDSR) in a underdamped mono-stable system. Guo and co-workers [15] investigated SR using the adiabatic approximation theory.

The theory of bistable SR can be applied in the field of signal processing, detection and amplification, as well as the mono-stable system. Guo and co-workers [15] have proved in the dissertation that the theory of mono-stable SR can also be utilized in the field of weak signal detection.

In this paper, a brief introduction of the theory of mono-stable SR was given. The virtual instrument of weak signal detecting based on this theory was designed with LabVIEW and was demonstrated. Comparisons of the output and input power spectrum densities are made concerning different sets of parameters. Examples of successful and failed detecting of weak signal were given in the last.

Theory of Mono-stable Stochastic Resonance

Consider a mono-stable over damped system driven by multiplicative noise, additive noise, as well as a periodic weak signal. The model is generally described by the following Langevin equation.

$$\frac{dx}{dt} = f(x) + x\xi(t) + \eta(t) + F(t) \tag{1}$$

$$f(x) = -ax^3 + r \tag{2}$$

$$F(t) = A\cos ft \tag{3}$$

Where $a > 0$, the constant b denotes the bias of the mono-stable system. Eq. 3 is the input weak signal which to be detected, and the amplitude A and frequency f should be far below 1. That is $A < 1, f \ll 1$. The noise terms are the Gaussian white noises which are described by their mean and variance.

$$\langle \xi(t) \rangle = \langle \eta(t) \rangle = 0 \tag{4}$$

$$\langle \xi(t)\xi(s) \rangle = 2D\delta(t-s) \tag{5}$$

$$\langle \eta(t)\eta(s) \rangle = 2Q\delta(t-s) \tag{6}$$

Where $D < 1$ and $Q < 1$ are respectively the intensities of the multiplicative and additive noise. According to Eq. 1-Eq. 6, the corresponding Fokker-Plank Equation is:

$$\frac{\partial \rho(x,t)}{\partial t} = [-\frac{\partial}{\partial x}A(x) + \frac{\partial^2}{\partial x^2}B(x)]\rho(x,t) \tag{7}$$

$$A(x) = f(x) + \frac{1}{2}B'(x) = Dx - ax^3 + r + F(t) \tag{8}$$

$$B(x) = Dx^2 + Q \tag{9}$$

The stationary distribution function can be derived from Eq. 7-Eq. 9, written as:

$$p_{st}(x) = \frac{N_{st}}{\sqrt{B(x)}} \exp\left[-\frac{V(x)}{D}\right] \tag{10}$$

Where N is the normalization constant, $V(x)$ is the rectified potential function and written as

$$V(x) = \int_{-\infty}^{x} D\frac{A(x)}{B(x)}dx = D\int_{-\infty}^{x} \frac{-U'(x) + r + F(t)}{Dx^2 + Q}dx \tag{11}$$

With

$$U'(x) = ax^3 - Dx \tag{12}$$

From Eq. 11 and Eq. 12, it can be discovered that the mono-stable system (Eq. 1) can be regarded as an equivalent bistable system in condition of $D \neq 0$. In this equivalent bistable system, $x_0 = 0$ and $x_\pm = \pm\sqrt{D/a}$ are the stable and unstable points. Thus, the transition rate out of $x_\pm$ based on the adiabatic condition can be obtained:

$$N_{\pm}(t)=\frac{\sqrt{|U''(x_0)U''(x_{\pm})|}}{2\pi}\exp\left[\frac{V(x_{\pm})-V(x_0)}{D}\right]=N_{\pm0}\exp\left(\mp qF(t)\right) \tag{13}$$

Where N_0 denotes the characteristic switching frequency of the equivalent bistable system when it is only driven by additive and multiplicative noise, given as:

$$N_0=\frac{D}{\sqrt{2}\pi}\exp\left[-\left(\frac{\Delta\Phi}{2}\pm rq\right)\right] \tag{14}$$

$$q=\frac{1}{\sqrt{DQ}}\arctan\left(\frac{D}{\sqrt{aQ}}\right) \tag{15}$$

$$\Delta\Phi=\left(1+\frac{aQ}{D^2}\right)\ln\left(\frac{D^2+aQ}{aQ}\right)-1 \tag{16}$$

This equivalent bistable system can be regarded as a two-state system, with $n_{\pm}$ denoting the occupation probabilities of the stable states $x_{\pm}$. $n_{\pm}$, the occupation probabilities, satisfied the following master equation [5]:

$$\begin{bmatrix} dn_+/dt \\ dn_-/dt \end{bmatrix}=\begin{bmatrix} -N_+(t) & N_-(t) \\ N_+(t) & -N_-(t) \end{bmatrix}\begin{bmatrix} n_+ \\ n_- \end{bmatrix} \tag{17}$$

The average autocorrelation function is given by

$$\langle\langle x(t)x(t+\tau_0)\rangle\rangle=\frac{\Omega}{2\pi}\int_0^{\frac{2\pi}{\Omega}}\lim_{t_0\to-\infty}\begin{bmatrix} x_+^2n_+(t+\tau_0|x_+,t)n_+(t|x_0,t_0)+x_+x_-n_+(t+\tau_0|x_-,t)n_-(t|x_0,t_0)+ \\ x_+x_-n_-(t+\tau_0|x_+,t)n_+(t|x_0,t_0)+x_-^2n_-(t+\tau_0|x_-,t)n_-(t|x_0,t_0) \end{bmatrix}dt \tag{18}$$

By applying the Fourier transform of the autocorrelation function, we can get the expression of the power spectrum $S(\omega)$

$$S(\omega)=S_1(\Omega)\delta(\omega-\Omega)+S_2(\omega-\Omega) \tag{19}$$

Where

$$S_1=\frac{4\pi Dv^2}{a(\beta^2+\Omega^2)}\quad,\quad S_2=(\omega,\Omega)=\frac{4D\beta}{a(\beta^2+\omega^2)}\left(\frac{1}{\cosh^2(qr)}-\frac{2v^2}{\beta^2+\Omega^2}\right) \tag{20}$$

$$v=\frac{DAq}{\sqrt{2}\pi\cosh(qr)}\exp(-\frac{\Delta\Phi}{2D})\quad,\quad \beta=\frac{\sqrt{2}D\cosh(qr)}{\pi}\exp(-\frac{\Delta\Phi}{2D}) \tag{21}$$

Here $S_1(\Omega)$ is the power density associated with the output signal. $S_2(\omega,\Omega)$ is the power spectrum coherent with the noise background in the adiabatic limit case.

$$\text{SNR}=\frac{S_1(\Omega)}{S_2(\omega=\Omega,\Omega)} \tag{22}$$

Virtual Instrument for Weak Signal Detecting

In this section, the virtual instrument for weak signal detecting on the theory of mono-stable stochastic resonance is demonstrated. It is designed by LabVIEW and is comprised of the front panel (Fig.1) and the block diagram (Fig. 2).

On the front panel, there are 2 main sections, the charts of waveforms and the list of different parameters. The 4 charts respectively reveal the time domain and power spectrum density of the input and output signal. The list consists of the feature parameters of the mono-stable system, the amplitude and frequency of the weak periodic system, the integration step, and the root-mean-square (RMS) and sampling number of noises. The 'STOP' button is used to terminate this process.

In the block diagram, an Express VI is used to generate a sine wave. The amplitude and frequency can be adjusted by the numeric control in the front panel. After that, two Gaussian white noises VIs are used to generate additive and multiplicative noises. The two noises are specified by RMS which can also be adjusted. The mixed input signal and the output signal were shown in the diagram chart 'INPUT SIGNAL' and the power spectrum density(PSD) was derived from the output of the FFT power spectrum VI and illustrated in the charts 'INPUT PSD'. Both of the charts are on the top of the panel.

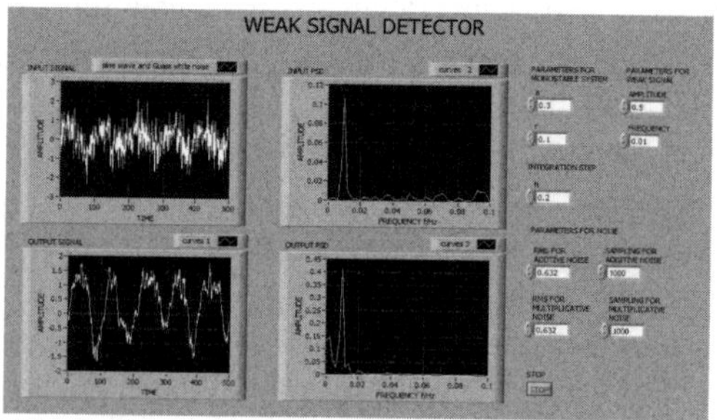

Fig. 1: Front panel of the weak signal detection VI

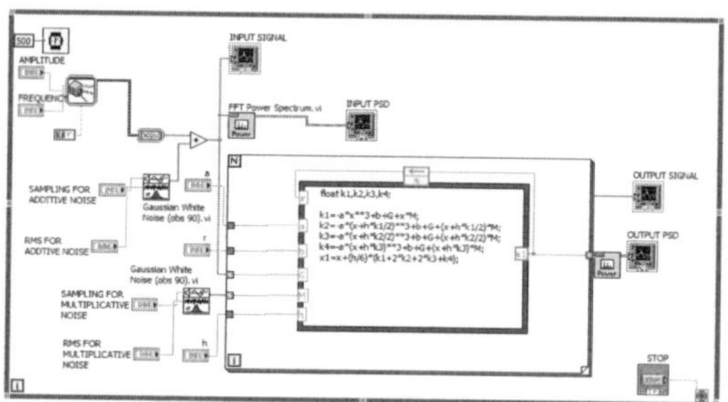

Fig. 2: Block diagram of the weak signal detection VI

According to the Langevin Equation (Eq. 1) given in the Theory section, the mono-stable system can be mathematically described in the way of a first order differential equation. Therefore, the solution of this differential equation is the result we try to get.

Here, we use the fourth order Runge-Kutta method to solve it. In the block diagram (Fig. 2), this can be reflected by the four loops and the formula node inside it. The time domain is shown in the chart 'OUTPUT SIGNAL' and power spectrum density is shown in the 'OUTPUT PSD' in the front panel, which were on the bottom of the panel.

Results and Discussion

In this section, the results of the detecting instrument are given and discussed. Then the mono-stable system is described by Eq. 2 and the parameters are a=0.3 and r=0.1. The input signal is described by Eq. 1 while the additive and multiplicative noises are described by Eq. 5 and Eq.6.

Table 1: results with various values of frequency

A	f [Hz]	D	Q	Iuput PSD	Output PSD
	0.2			0.43	0.23
	0.1			0.38	0.7
0.9	0.05	0.2	0.2	0.45	1.02
	0.01			0.38	0.7
	0.001			/	/

Table 1 gives information of the results under different frequency values. When the frequency is between 0.01 and 0.1, the detector can indicate the right frequency of the weak signal as well as a higher value of output PSD than the input. However, when the frequency came down to as small as 0.001 Hz, the instrument failed to detect. In another occasion which the frequency is as high as 0.2Hz, the PSD of input signal is 0.43 while that of output is 0.23. This indicates that the amplifying effect of stochastic resonance is failed. In conclusion, the instrument is valid when the frequency is in the range of 0.01 Hz to 0.1 Hz. This matches the theory of monostable stochastic resonance that it only applies to weak signal where the frequency is far below 1 Hz.

Table 2: results with various values of amplitude

A	f [Hz]	D	Q	Input PSD	Output PSD
0.1	0.05	0.2	0.2	/	/
0.2				0.031	0.18
0.5				0.125	0.42
0.9				0.45	1.02

At the same way, Table 2 demonstrates the situation that various values of amplitude of the weak signal are applied. In the case of A=0.1, the detector failed to give results. While the amplitude is between 0.2 and 0.9, the instrument could give appropriate outcome that the PSD of the output of mono-stable system were higher than that of FFT. To sum up, the instrument is available in cases that the amplitude is below 1, but not too low as 0.1. This is also in accordance with the theory of mono-stable stochastic resonance that it applies to weak signal where the amplitude should below 1.

Table 3: results with various values of additive noise intensity

A	f [Hz]	D	Q	Iuput PSD	Output PSD
0.5	0.05	0.1	0.2	0.16	0.62
		0.2		0.125	0.45
		0.5		0.12	0.3
		0.7		0.115	0.55
		1		0.12	0.49

Table 4: results with various values of mulitplicative noise intensity

A	f [Hz]	D	Q	Iuput PSD	Output PSD
0.5	0.05	0.2	0.1	0.14	0.58
			0.2	0.125	0.44
			0.5	0.115	0.6
			0.7	0.11	0.7
			1	0.12	0.5

In Table 3, the results are illustrated with various values of additive noise intensity. With the increase of noise intensity, the output amplitude through FFT kept decreasing. However, the output through the mono-stable system was not monotonously decreasing. It fell from 0.62 to 0.3 when D raised from 0.1 to 0.5, and then increase to 0.55 when D=0.7. After that, the output fell again when D rise to 1. This matches the theory of stochastic resonance that the output PSD does not fall monotonously but has a peak where the resonance happens.

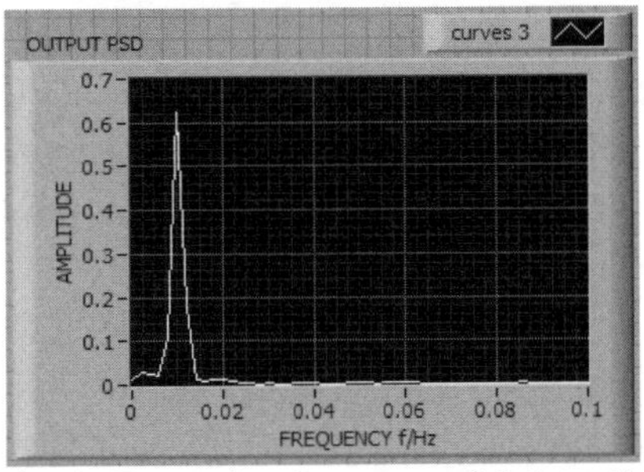

Fig. 3: Example of a successful detection

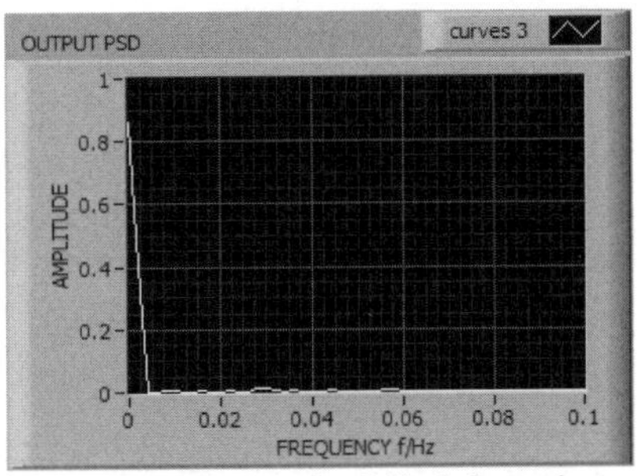

Fig. 4: Example of a failed detection

Similarly, Table 4 showed how the differences of multiplitive noise intensities affected the output amplitudes. The input amplitudes went up and down with a clear falling trend. The situation in the output signal was similar to that in the prior occasion, where the amplitude went down then up with the increase of noise intensity. This again showed the consensus of this result and the theory of mono-stable stochastic resonance.

A successful detecting example was illustrated in Fig. 3 where A=0.5, f=0.01, D=0.2 and Q=0.2.

When the frequency came down to f=0.001, or the amplitude went down to A<0.1 the instrument failed to detect the weak signal and the result is shown in Fig.4.

Summary

A virtual instrument for weak signal detecting which was demonstrated in this paper can be used to detect weak signal. The outcome of the instrument has reached a consensus with the theory of the mono-stable SR. Thus, the application of this theory in the detecting of weak signal is proved to be valid. In this process, the mono-stable system can be regarded as filter and amplifier at the same time. This is due to the phenomenon of SR, which would shift the power of noise to the periodic signal when it happens. Moreover, the feature of the unmonotonic relationship between the noise intensities and the output PSD is reflected.

References

[1] R.Benzi, A.Suterai:The mechanism of stochastic resonance Phys,A:Math.Gen14(1981), p453.

[2] Hu G, Haken H: A study of stochastic resonance without adiabatic approximation Phys. Lett. A Vol. 172 (1992), p.2030.

[3] Hu G, Haken H, Xie FG: Stochastic Resonance with Sensitive Frequency Dependence in Globally Coupled Continuous Systems Phys. Rev. Lett. Vol.77 (1996), p.1925.

[4] McNamara B, Wiesenfild K, Roy R: Observation of Stochastic Resonance in a Ring Laser Phys. Rec. Lett Vol.60 (1988), p.2626.

[5] McNamara B, Wiesenfild K: Theory of stochastic resonance Phys. Rev. A Vol. 39 (1989), p.4854.

[6] G. L, Hanggi P, Marchesoni F: Stochastic resonance Rev Mod Phys Vol.70 (1998), p223.

[7] Oya T, Asai T, Amemiya Y: Stochastic resonance in an ensemble of single-electron to competitive neural networks Chaos Solitons & Fractals Vol.32 (2007), p.855.

[8] Kang YM, Xie Y, Xu JX: Observing nonlinear stochastic resonance with piecewise constant driving forces by the method of momentsChaos Solitons & Fractals Vol.27 (2006), p.715.

[9] Jin YF, Xu W, Xu M: Stochastic resonance in an asymmetric bistable system driven by correlated multip and additive noiseChaos Solitons & Fractals Vol.26 (2005), p.1183.

[10] Babinec P: Stochastic resonance in an interacting-agent model of stock market Chaos Solitons & Fractals Vol.13 (2002), p.1767.

[11] Dykman MI, Luchinsky DG, McClintock: Stochastic resonance in perspective PVE Nuovo Cimento D Vol.17 (1995),p661.

[12] Evstigneev M,Reimann P, Pankov V: Dissipative Enhancement of the Response of a Bistable Noisy Oscillator to External Driving Europhys Lett Vol.65 (1997), p.343.

[13] Guo F, Zhou YR, Jiang SQ: Phys A: Stochastic resonance in a mono-stable system with multiplicative and additive noiseMath Gen Vol.39 (2006), p.13861.

[14] N.G. Stocks, N.D. Stein: Stochastic resonance in monostable system Phys. A:Math. Gen. Vol.26 (1993) p,385.

[15] Guo F, Gu TX Stochastic Resonance and Its Applications in Weak Signal Detection (2007).

Advanced Materials Research Vol. 279 (2011) pp 367-370

doi:10.4028/www.scientific.net/AMR.279.367

The Analysis and Control of Noise on Steam Header of Heat Exchange Station

Wang Changtian[1, a], Bao Huaiqian[1, b] and Zhang Wenhao[1, c]

[1]College of Mechanical and Electronic Engineering,
Shandong University of Science and Technology, Qingdao, Shandong, China, 266510

[a]kdw6255@163.com, [b]bhqian@163.com, [c]zhangwenhaoniu@126.com

Keywords: heat exchange, steam header, noise control, aerodynamic noise

Abstract. Based on the noise characteristics of heat exchange station, the reasons of the high noise are investigated by analyzing the mechanism of noise of steam header, simulating and calculating the internal flow field. Then the comprehensive noise abatement measures are proposed according to the characteristics of the equipment operating and laws of the noise spectrum. As a result, the working environment of staff was improved, and noise pollution to surrounding environment was reduced. In a word, there were a lot of good results by adopting the measures of noise control.

Introduction

Heating system is an important guarantee of the northern life in winter. It mainly consists of heat source, transporting heat pipe and heat dissipation devices. For ensuring long centralized heat supply and saving energy, the trunk pipeline generally uses steam transporting heat. Through heat exchange station raising water temperature, it can achieve the needs of resident heating. Therefore, heat exchange station is one of the important links in heating system. In internal heat exchange station, steam in main pipe needs to transport to two or more different heat exchangers, but during the running process it will produced great noise which exceeds the relevant standards prescribed by the state. According to the tests of the noise of some heat exchange station' steam header, the noise and the spectrum in different outlet pressure are shown in Table 1-3.

Table 1: Noise of steam header when outlet pressure is 0.8 Mpa

frequency spectrum[Hz]	A sound stage	31.5	63	125	250	500	1k	2k	4k	8k	16 k
Noise[dB]	108	97	95	101	104	98	105	97	92	86	85

Table 2: Noise of steam header when outlet pressure is 1.5 Mpa

frequency spectrum[Hz]	A sound stage	31.5	63	125	250	500	1k	2k	4k	8k	16 k
Noise[dB]	99	87	85	91	94	92	97	87	82	76	75

Table 3: Noise of steam header when outlet pressure is 2.7 Mpa

frequency spectrum[Hz]	A sound stage	31.5	63	125	250	500	1k	2k	4k	8k	16 k
Noise[dB]	87	77	75	81	84	78	85	81	72	76	75

The steam header noise mechanism analysis

Through the actual noise measurement of some heat exchange station' steam header, noise level is in inverse proportion to steam header pressure, that is, the higher the outlet pressure of steam header, the smaller the noise level; the lower the outlet pressure of steam header, the bigger the noise level. Bernoulli equation of fluid motion is as follows:

$$p_1 + \frac{\rho v_1^2}{2} = p_2 + \frac{\rho v_2^2}{2} + \lambda \frac{l}{d} \cdot \frac{\rho v_2^2}{2} + \xi \frac{\rho v_2^2}{2} \tag{1}$$

Where p_1 is the inlet pressure of the main steam pipeline, v_1 is flow velocity of the main steam pipeline, p_2 is the pressure of steam header, v_2 is flow velocity of steam header, λ is frictional resistance loss coefficient, d is diameter of the main steam pipeline, l is length of the main steam pipeline, ρ is steam density, ξ is local resistance loss coefficient.

Putting Bernoulli equation is as follows:

$$p_1 - p_2 = [\frac{\rho}{2} - \frac{\rho}{2}(\frac{A_1}{A_2})^2 + \lambda\frac{l}{d}\cdot\frac{\rho}{2} + \xi\frac{\rho}{2}]v_2^2 \quad (2)$$

Generally, the inlet pressure of the main steam pipeline is constant. When the pressure p_2 of steam header drops, velocity v^2 certainly rise and noise increase. This is airflow noise.

Airflow makes noises causing by two main reasons. One is structure borne noise caused by tubing inner wall or other components generating vibration and radiated noise by the action of airflow impact. It is mainly consisted of low frequency. The strength of the structure borne noise is increased by the fourth power of velocity. When the strength of pipe wall and other components are high and the velocity of the airflow is low, the noise can't afford main effect. At the same time, the noise is caused by the high-speed airflow of the pipe. Generally, it gives priority to mid-and-high frequency and is similar to dipole source radiation. The noise radiation intensity increases by the sixth power. When airflow velocity is above 20m/s, the position of noise spectrum peak can move from low frequency to mid-and-high frequency.

The steam header inner flow field simulation

In order to understand the state of steam header inner steam flow, pressure and velocity vector figure of steam header inner flow field can be obtained by the fluid simulation software in different outlet pressure condition, as shown form Fig. 1 to Fig. 6.

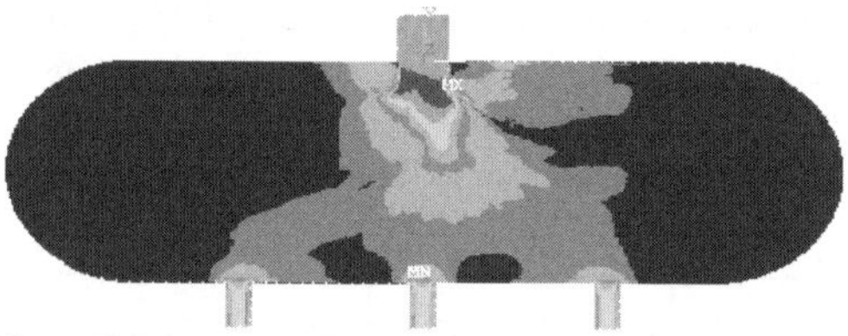

Fig. 1: The flow field pressure distribution when outlet pressure is 0.8MPa

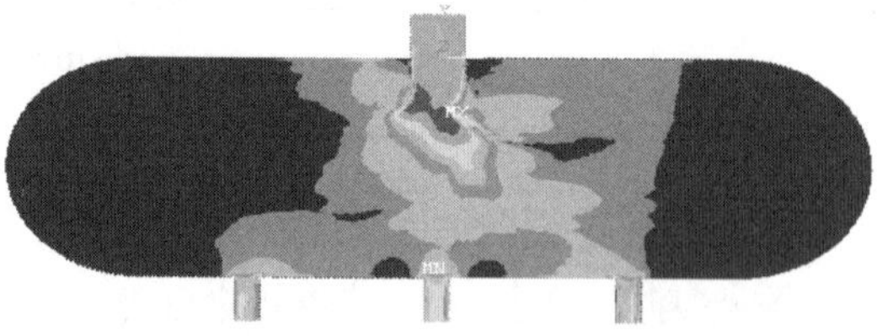

Fig. 2: The flow field pressure distribution when outlet pressure is 1.5Mpa

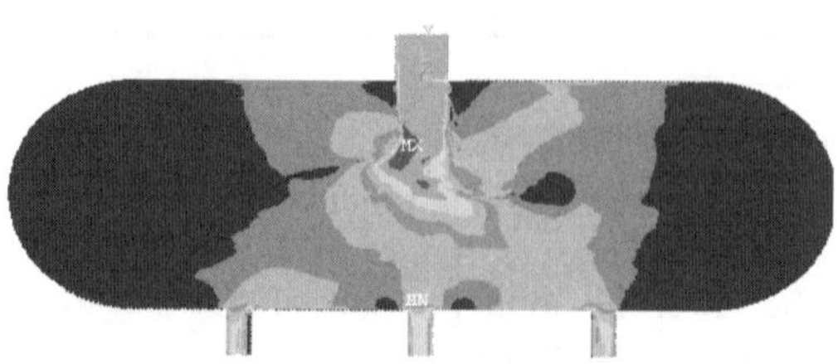

Fig. 3: The flow field pressure distribution when outlet pressure is 2.7MPa

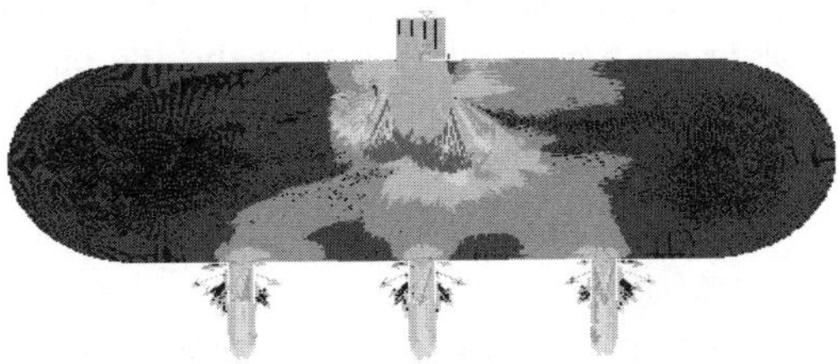

Fig. 4: The distribution of velocity vector when outlet pressure is 0.8MPa

Fig. 5: The distribution of velocity vector when outlet pressure is 1.5MPa

Fig. 6: The distribution of velocity vector when outlet pressure is 2.7MPa

It can be seen from Fig. 1 to Fig. 3 that with the outlet pressure increasing the stable area is larger in steam header inner flow field, and the whole steam header inner flow field is close to the dynamic balance state. Therefore, the higher the outlet pressure, the lower the noise value. Meanwhile, it can be seen from Fig. 4 to Fig. 6 that the lower the outlet pressure, the higher the steam header inner flow velocity, the greater the degree of turbulence and the higher the noise value. With the stream header pressure dropping, the direct feeling is noise obviously with screaming ingredients. It proves that noise spectrum peak position moves from low frequency to mid-and-high frequency with speed increasing.

Steam header noise control countermeasures

According to the above analysis, noise mainly consists of air flow noise. At present, there is not good theory to calculate this regenerative noise. Generally in engineering application, A-weighted sound level caused by turbulence noise can be estimated by the following empirical formula.

$$L_A = \beta + 60\lg \bar{u}(dB) \tag{3}$$

Where u is airflow velocity, with m/s plan; β is relevant with various complicated factors such as sound-absorbing material of surface roughness and main pipeline geometric parameters etc.

Therefore, the following measures can be taken to reduce steam header noise.

(1) Take technical reformation. In normal heating, it can ensure steam header pressure be greater than 2.7Mpa to control steam header noise being low level.

(2) Change steam header inlet and outlet pipe diameter to control steam flow velocity.

In addition, in order to reduce the steam header noise impacting to environment, part of passive noise control measures can be taken, such as bandaging the exposing outer steam pipe for sound insulation, changing heat transfer station room to sound insulation room, installing sound-absorbing structure in sound insulation room and eliminating reverberant sound, etc.

Conclusion

Based on the heat transfer station stream header noise measurement and analysis of stream header inner flow field characteristics, reasons of steam header high noise are found. Then, according to the characteristic of the stream header, countermeasures are put forward to reduce noise. It has a certain significance to analyze and control the similar noise source.

Acknowledgment

This work was financially supported by the China National Coal Association Science Foundation (MTKJ-2010-342), the support of a research grant from the scientific and technological development plan project of Qingdao Municipal (KZJ-50).

References

[1] Du Gonghuan, Zhu Zhemin, Gong Xiufen: *Fundamental of acoustics*. Nanjing University Press, Nanjing, (2001), pp. 239-243.

[2] Ma Dayou: *The engineering manual on noise and vibration control*. China Machine Press, Beijing, (2002), pp. 56-68.

[3] Gu Qiang, Wang Changtian: *The noise control engineering*. China Coal Industry Publishing House, Beijing, (2002), pp. 140-141

[4] Liu Liping, Zhu Zhenjie: *Discussion on effect of muffler on air flow regeneration noise*. Environmental Engineering, 22(6), (2004), pp. 54-56

[5] Fan Qianwang: *Acoustic performance simulation analysis and experiment study for exhaust muffler of four-cylinder diesel*. Kunming: Kunming University of Science and Technology, (2007), pp. 89-97.

[6] Li Zenggang: *SYSNOISE Rev5.6*. Beijing: Nation Defense Industry Press, (2005), pp. 17-78.

Advanced Materials Research Vol. 279 (2011) pp 371-376
© (2011) Trans Tech Publications, Switzerland
doi:10.4028/www.scientific.net/AMR.279.371

Study on Parameters Configuration for Ant Colony Optimization

Yi GAN[a], Lian-wang LAN[b] and Sheng LI[c]

College of Mechanical Engineering, University of Shanghai for Science and Technology, Shanghai, 200093, China

[a]ganyi@usst.edu.cn, [b]alone215@126.com, [c]lisheng0407@163.com

Keywords: Ant Colony Optimization, Parameters Configuration, TSP, JSSP

Abstract. To find the parameters' configuration relationship of the Ant Colony Algorithm, based on the ecological actions about ants, the distributing multiformity of ant colony pheromone, the pheromone updating strategy and the mutant of information difference were applied to Microhabitat Ant Colony Optimization (MACO). The parameters, α_0, β_0, k_α and k_β of MACO were configured by the orthogonal experiment to enhance the performance of the algorithm, in which the interactions of α_0 and β_0, k_α and k_β, α_0 and k_α, β_0 and k_β were also analyzed. Some benchmarks of TSP and JSSP were solved by MACO which showed significant optimize performance with configured parameters.

Introduction

Individual ant is very simple, but the whole community of ant demonstrates a highly structured group organization [1]. The idea of Ant Colony Optimization (ACO) comes from the ants' activity of looking for food, and uses pheromone as the basis for ants to choose the subsequent behaviors, and completes the optimization process by the ants' synergy. At present, the international study of ACO algorithm has expanded to a number of application areas from its original single Traveling Salesman Problem (TSP), and this algorithm has become a quite advanced intelligent optimization algorithm.

When the ant colony algorithm is used to solve a problem, the settings of values of the parameters are directly affecting the convergence of the algorithm. In this paper, the Orthogonal Experiment (OE) is used to obtain the optimal parameter configuration of Microhabitat Ant Colony Optimization (MACO) [2, 3] and to solve the Traveling Salesman Problem (TSP) [4, 5] and the Job Shop Scheduling Problems (JSSP) [6].

Microhabitat Ant Colony Optimization (MACO)

To improve the efficiency problem of the basic ant colony algorithm, MACO is improved from the diversity of ant pheromone, the pheromone update strategy and the mutation mechanism of Information Difference; and then it is used to solve the Dynamic Alliance Partner Selection Problem and JSSP.

In the initial stage of search, there is not enough available experience information in the microhabitat, so ants mainly use initial heuristic information to *explore*. With the evolution of search, the experience information of microhabitat are increased, while the ants' searches are mainly to *make use of* the experience information to speed up the convergence of solutions. And then in the later stage of search, it is necessary to *re-explore* based on the existing solution to avoid the algorithm premature problem and the stagnant phenomena, as well as to expand the search space so that it can bring out the better solutions.

There are two ways of updating pheromone for ants. They are partial update and global update. The partial update rule is that the ant leaves pheromones along its path each time when it moves to the next node; the global update rule is to update pheromones of all paths at the end of every search cycle (that is the end of search of all the colonies who seek the most optimal solution). Global update rule considers the search of optimize results of every colony, which implicates information feedback, and this makes the algorithm easier for optimization. In this paper, the global update rule is adopted.

In the MACO, the Information Difference is also generated to increase the diversity of information and to expand the search space. According to the characteristics of specific issue the Information

Difference cloud produced in different methods. For TSP, the Information Difference is generated by the genetic operations (crossover, genetic mutation) to search the solution of every generation, which can expand the search scope and ensure to obtain the better solution. For JSSP, the Information Difference is generated probability at the time of choosing the next node by the selective probability, which can make it jump out of local optimum, expand the search scope, and maintain the diversity of solutions.

The parameters of MACO that include α_0, β_0, k_α, k_β and so on are about the inspiration information and experience information. As the real ants affected by the environment, this parameters that reflect the ant optimization process is also affected by many factors. The values of these parameters are directly related to the optimization algorithm performance.

Orthogonal Experiment Design

In this paper, the MACO is used to solve the TSP and JSSP, and the orthogonal experiment is used to analysis the performance optimization impacting of the parameters: α_0, β_0, k_α, k_β.

Factors and Levels of the Orthogonal Experiment. In the MACO, an ant searching cycle is the process of finding a qualified process by all ants, which starts from the initial node. If the maximum number of cycles is too large, the convergence of algorithm would cost too much time. And if the maximum number of cycles is too small, the random of the result of the algorithm would increase and affect the optimization performance. Form counting many examples, the algorithm can be more stable when NC_m=500. Similarly, if the number of ants is too large, the global searching ability of the algorithm would expand, but the count time of the algorithm would exponentially increase. For the TSP, the ant number is 2/3 [7] of the number of city size n, and for JSSP, the ant number of each generation is taken as the number of part. Calculated by a large number of examples, in this article, for TSP, information evaporation rate $\rho = 0.1$, for JSSP, $\rho = 0.7$[8].

In the orthogonal experiment, select α_0 and β_0 increase at a step 1, and the value of $\lg k_\alpha$ and $\lg k_\beta$ is individually equaled to -4.699,-4,-3.699,-3, -2.699. Take the mean of 10 values as the experimental result, and the experimental program is shown in the Table 1. The information of the Information Difference is interrelated [2]. So the possible interactions between α_0 and β_0, k_α and k_β, α_0 and k_α , and β_0 and k_β were considered to find a good combination of factor level through Orthogonal Experiment. In this article, the simulation test software program is MatlabR2008b, the operating system is Windows XP, CPU 2.0GHz, RAM 1.99GB.

Table 1: Orthogonal Experiment Factors and Levels

Test Level	factor			
	α_0	β_0	$\lg k_\alpha$	$\lg k_\beta$
1	1	1	-4.699	-4.699
2	2	2	-4	-4
3	3	3	-3.699	-3.699
4	4	4	-3	-3
5	5	5	-2.699	-2.699

There are 4 factors and 4 levels in the orthogonal experiment, and taking 4 interactions into account, L50 (5^{11}) orthogonal experiment table is chosen. The header design is shown in Table 2.

Table 2: L50 (5^{11}) Header design

Col NO. / Factor NO.	1	2	3	4	5	6	7	8	9	10	11
5	α_0	k_β		k_α	$\alpha_0\times k_\alpha$	$k_\alpha\times k_\beta$		β_0	$\alpha_0\times\beta_0$	$\beta_0\times k_\beta$	

Experimental Results. Table 3 is the experimental result and analysis of ATT48 [9], which has 48 cities, and the test is based on the orthogonal table L50 (511). The Tour_ length is the relative error of travel distance, and the Iteration _Time is the relative error of convergence time.

The Range of each column is calculated form Table 3. The factor of maximum Range means that the experimental difference caused by different levels is large, thus identify the importance of the various factors on the MACO solving TSP, as shown in Table 4.

Set the objective function $f(x) = \min L$. Form Table 4, if L is Tour_ length, the order index of importance of various factors is $\alpha_0 > \beta_0 > k_\alpha > k_\beta > \alpha_0 \times \beta_0 > \beta_0 \times k_\beta > \alpha_0 \times k_\alpha > k_\alpha \times k_\beta$, and the difference between ρ and kα is small; if L is Iteration _Time, the order index of importance of various factors is $\alpha_0 > k_\alpha \times k_\beta > \beta_0 > k_\beta > k_\alpha > \alpha_0 \times \beta_0 > \alpha_0 \times k_\alpha > \beta_0 \times k_\beta$. And this also means that it cannot meet both objective functions optimal when solving the problem.

Table 3: The Result of Test

NO.	Column NO.								Test index	
	1	2	4	5	6	8	9	10	Tour_ length [%]	Iteration _Time [%]
	α_0	$\lg k_\beta$	$\lg k_\alpha$	$\alpha_0 \times k_\alpha$	$k_\alpha \times k_\beta$	β_0	$\alpha_0 \times \beta_0$	$\beta_0 \times k_\beta$		
1	1	1(-4.699)	1(-4.699)	1	1	1	1	1	2.14	1.47
2	1	2(-4)	2(-4)	2	2	2	2	2	0.11	2.28
3	1	3(-3.699)	3(-3.699)	3	3	3	3	3	1.63	2.19
4	1	4(-3)	4(-3)	4	4	4	4	4	1.07	2.40
5	1	5(-2.699)	5(-2.699)	5	5	5	5	5	3.55	1.58
6	2	1	3	4	5	2	3	4	4.33	2.03
7	2	2	4	5	1	3	4	5	6.32	3.90
8	2	3	5	1	2	4	5	1	2.39	3.20
9	2	4	1	2	3	5	1	2	4.48	2.64
10	2	5	2	3	4	1	2	3	2.95	2.49
11	3	1	5	2	4	1	3	5	3.85	5.91
12	3	2	1	3	5	2	4	1	5.02	3.92
13	3	3	2	4	1	3	5	2	3.67	8.76
14	3	4	3	5	2	4	1	3	0.87	3.89
15	3	5	4	1	3	5	2	4	3.73	6.07
16	4	1	2	5	5	3	1	4	12.76	6.35
17	4	2	3	1	1	4	2	5	3.69	5.38
18	4	3	4	2	2	5	3	1	2.52	5.95
19	4	4	5	3	3	1	4	2	3.16	6.19
20	4	5	1	4	4	2	5	3	4.56	7.74
21	5	1	4	3	2	3	2	1	10.46	8.32
22	5	2	5	4	3	4	3	2	5.79	4.44
23	5	3	1	5	4	5	4	3	4.76	3.84
24	5	4	2	1	5	1	5	4	3.61	2.26
25	5	5	3	2	1	2	1	5	3.37	3.14
26	1	1	4	5	4	2	5	2	1.43	3.19
27	1	2	5	1	5	3	1	3	2.11	3.42
28	1	3	1	2	1	4	2	4	2.62	3.71
29	1	4	2	3	2	5	3	5	2.09	2.56
30	1	5	3	4	3	1	4	1	4.82	1.46
31	2	1	1	3	3	4	5	5	8.34	2.79
32	2	2	2	4	4	5	1	1	9.31	5.51
33	2	3	3	5	5	1	2	2	2.49	3.61
34	2	4	4	1	1	2	3	3	1.43	5.37
35	2	5	5	2	2	3	4	4	1.95	3.14
36	3	1	3	1	2	5	4	2	2.85	5.65
37	3	2	4	2	3	1	5	3	3.72	2.66
38	3	3	5	3	4	2	1	4	2.33	4.09
39	3	4	1	4	5	3	2	5	0.62	3.85
40	3	5	2	5	1	4	3	1	2.78	8.33
41	4	1	5	4	1	5	2	3	11.38	9.31
42	4	2	1	5	2	1	3	4	3.19	2.26
43	4	3	2	1	3	2	4	5	2.43	3.48
44	4	4	3	2	4	3	5	1	3.76	2.95
45	4	5	4	3	5	4	1	2	4.58	3.85
46	5	1	2	2	5	4	4	3	14.86	4.72
47	5	2	3	3	4	5	5	4	9.76	2.63
48	5	3	4	4	3	1	1	5	8.21	1.86
49	5	4	5	5	2	2	2	1	5.21	3.74
50	5	5	1	1	1	3	3	2	4.33	4.75

Table 4: The result analysis of test index under all program

Test index	α_0	k_β	k_α	$\alpha_0 \times k_\alpha$	$k_\alpha \times k_\beta$	β_0	$\alpha_0 \times \beta_0$	$\beta_0 \times k_\beta$
Tour_ length Range	52.11	17.83	27.88	10.16	3.28	35.97	14.36	14.26
Iteration_Time Range	29.2	13.84	13.81	11.52	20.15	17.46	12.54	10.23

The relationship of the orthogonal level and factor is get from Table 4, as shown in Fig. 1. In Fig. 1, if L is Tour_ length, the best combination of various factors is $\alpha_0=1$, $\beta_0=3$, $k_\alpha=0.0002$, $k_\beta=0.001$; and if L is Iteration _Time, the best combination of various factors is $\alpha_0=1$, $\beta_0=1$, $k_\alpha=0.002$, $k_\beta=0.001$.

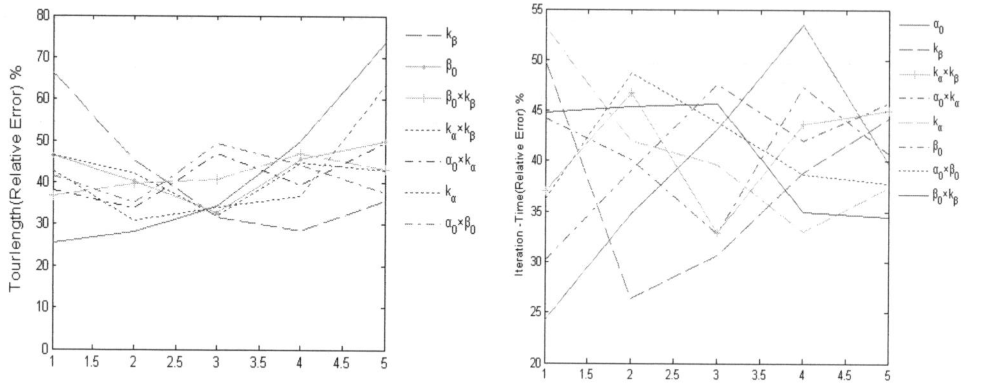

a)L is Tour_ length, the level of each factor b) L is Iteration _Time, the level of each factor

Fig. 1: The relationship of the orthogonal level and factor

Experiment and Result Analysis

Experiment. For TSP, considering the difference of city size, 4 TSP standard examples are used [9]: Bayg29, Att48, Kroc100, A180. Objective function $f(x) = \min L$, when L is Tour length, and the solution is obtained as shown in Table 5.

For JSSP, considering the difference of processing scale, 6 JSSP benchmark examples are used [10]: FT06, FT10, FT20, LA06, LA22, LA22. Objective function is a unit working hours, and the obtained solution is shown in Table 6.

Table 5: the optimal parameter configuration obtained by orthogonal experiment

Test Cases	**Tour_ length**			
	α_0	β_0	k_α	k_β
Bayg29	1	3	0.0002	0.002
Att48	1	3	0.0002	0.001
Kroc100	2	3	0.0002	0.001
A180	2	4	0.0001	0.002

Table 6: parameter configuration results

Test Cases	**Parameter**			
	α_0	β_0	k_α	k_β
FT06	3	2	0.0002	0.002
FT10	3	2	0.0001	0.002
FT20	2	3	0.0001	0.001
LA06	3	2	0.0002	0.002
LA22	3	2	0.0002	0.002
LA36	2	3	0.0001	0.001

Results Comparison. Table 7 gives when the objective function $f(x) = \min L$ (L is Tour_ length) the results of solving the above four examples, in which the orthogonal experiment is used.

Table 7: parameter configuration results of JSSP

Test Cases	N	Opt.	Best	Ave.	Orig.
Bayg29	29	9074.148048	9075.135023	9174.13457	11028.01608
Att48	48	33523.708507	33509.13456	33527.15879	35336.42677
Kroc100	100	20750.762504	20755.87235	20843.34625	22105.29421
A180	180	2586.769648	2588.45312	2607.43263	2820.54391

Note: N-city size, Optimum-the optimal solution known, Best result-the best solution obtained, Average-the average of optimal solution, Original-the result before parameter configuration(the average of 20 random).

As can be seen from Table 7, compared with the random solutions, the optimal solutions by the configuration parameters optimizing have been optimized, which are very close to the optimal solution as known. It can be certified that the parameters configuration through Orthogonal Experiment can obtain better optimization results in a short time. If the objective function $f(x) = \min L$ (L is Iteration-time), it also acquired similar conclusion.

For JSSP, the calculation solving of standard examples in different scales are shown in Table 8, in which the value of parameters through the orthogonal experiment configuration parameters are used. As can be seen from Table 8, when the scale of solving problem increases gradually, using the parameters configured MACO can obtain satisfactory solutions.

Table 8: the result of typical benchmark of JSSP

Benchmark	$n \times m$	Opt.	MACO			
			Best	% D	Ave.	SD
FT06	6×6	55	55	0	57	1.86
FT10	10×10	930	941	1.18	1008	31.65
FT20	20×5	1165	1165	0	1226	36.18
LA06	15×5	926	926	0	931	25.61
LA22	15×10	927	957	3.24	1064	43.18
LA36	15×15	1268	1315	3.71	1438	43.83

Note: n-number of parts, m-the number of equipment, Opt.- known optimal solutions, Best- the optimal solution of algorithm,% D-the deviation between BEST and Opt. Ave.- Average solution , SD-(Standard Deviation).

Fig. 2 shows the MACO evolution curve before and after configuration for solving the FT06 and FT10. As can be seen from Fig. 2, 100 iterations are used in the MACO to solve JSSP but generally 30 to 40 times in the convergence and with the increasing of the size it will gradually increase the number of convergence.

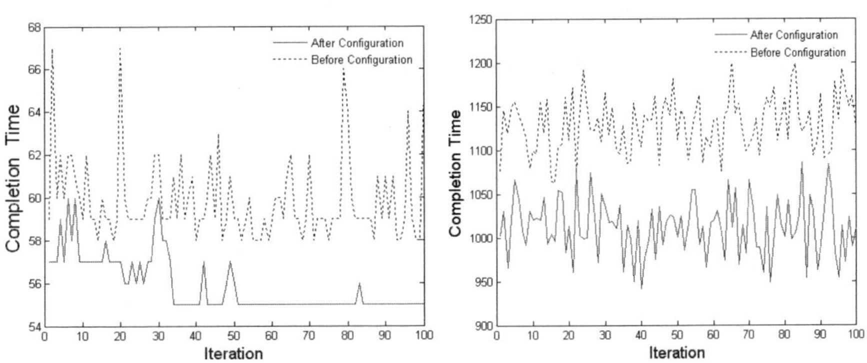

(a)The evolution curve for solving FT06 (b) The evolution curve for solving FT06

Fig. 2: the MACO evolution curve before and after configuration

Conclusion

Base on the existence of falling into stagnation of ant colony algorithm, the microhabitat ant colony algorithm is raised in this paper, and though the diversity pheromone of the ant colony, the update strategy of pheromone, and the mechanism of mutation of information differentiate the searching range expand and avoid convergence to local optimal solution. The orthogonal experiment is used to configure the parameters of MACO. The results obtained from the standard examples of TSP and JSSP shows that the parameter configuration of MACO algorithm is effective.

Acknowledgement

This work is supported by the National Natural Science Foundation of China (Grant No. 50505030), and "Dawn" Program of Shanghai Education Commission (Grant No.07SG51).

References

[1] Marco Dorigo, Thomas Stutzle: *Ant Colony Optimization*. Tsing-Hua University Press, Beijing, (2007).

[2] Macro D., Thomas stutzle: *ant colony optimization*. MIT Press, Cambridge, (2003).

[3] GAN Yi: *Studies and applications of the regional die and mould networked manufacturing enterprises dynamic*. School of Mechanical Engineering, Tongji University, (2004).

[4] Yi GAN: *Studies on selecting partners of networked manufacturing based on ACO*. Journal of Communication and Computer Vol. 5 (2007), pp. 15-21, 27.

[5] Dorigo M., Gambardella L. M.: *Ant Colony System: A Cooperative Learning Approach to the Traveling Salesman Problem*. IEEE Transactions on Evolutionary Computation Vol. 1 (1997), pp. 53-66.

[6] Garey M. R., Johnson D. S., Sethy R.: *The complexity of flow shop and job shop scheduling*. Mathematics of Operations Research Vol. 2 (1976), pp. 117-129.

[7] GAN Yi, DU Jitao: *Studies on JSSP based on Parallel ACO*. Manufacturing Automation Vol. 11 (2009), pp. 30-33.

[8] JIANG Ling-yan, ZHANG Jun, ZHONG Shu-hong: *Analysis of parameters in ant colony system*. Computer Engineering and Application Vol. 20 (2007), pp. 31-36.

[9] http://comopt.ifi.uni-heidelberg.de/software/TSPLIB95/

[10] http://people.brunel.ac.uk/~mastjjb/jeb/orlib/jobshopinfo.html

Advanced Materials Research Vol. 279 (2011) pp 377-381

doi:10.4028/www.scientific.net/AMR.279.377

Mechanical-electronic-hydraulic Integration: Light Load Linear Reciprocating Motion Device Having Variable Frequency, Displacement and Velocity

Mingzhe Zhang[1, a], Wei Wang[1, b], Dongning Su[1, c],

Kangmin Zhong[2, d], Zhiming Sui[1, e]

[1]School of mechanical engineering, University of Jinan, Jinan 250022, China
[2]School of Mechanical and Electronic Engineering, Soochow University, Suzhou 215021, China

[a]me_zhangmz@ujn.edu.cn, [b]me_wangw@ujn.edu.cn, [c]me_sudn@ujn.edu.cn
[d]zkm@suda.edu.cn, [e]szm9588@126.com

Keywords: light load linear reciprocating motion device, Stepping motor, rolling screw-nut mechanism, area effect stroke amplifying device

Abstract. The light load linear reciprocating motion device driven by a small-size pneumatic cylinder is most economical, but its velocity and location precision is not good. When substituting the drive for small-size hydraulic cylinder, the significant shortcoming is environmental pollution. If the motion is achieved through rolling screw-nut mechanism driven by stepping motor, the space layout is very inconvenient.

A light load linear reciprocating motion device integrating mechanical, electronic and hydraulic technology is designed. In this device, the motion is achieved through short-stroke rolling screw-nut mechanism driven by stepping motor, and then amplified by a splendid internal reflux area effect stroke amplifier. This device has the following noticeable advantages: (1) The location of output hydraulic cylinder is as convenient as common pneumatic or hydraulic cylinder. (2) The frequency, displacement and velocity of reciprocating motion can be altered convenient by programming. (3) Higher position precision. (4) The hydraulic fluid in the device is enclosed, so there is no environmental pollution causing by oil leakage and volatilization.

Introduction

The light load linear reciprocating motion device is widely used in industrial fields. In case of low displacement precision, pneumatic power transmission is selected generally. But if the displacement precision is higher, the rolling screw-nut mechanism driven by stepping motor is adopted, whereas the space layout is very inconvenient. It is not good as pneumatic cylinder or hydraulic cylinder.

The light load linear reciprocating motion device integrating mechanical, electronic and hydraulic technology presented in this paper has many advantages, such as the frequency, displacement and velocity of reciprocating can be altered convenient by programming. The location of output hydraulic cylinder is as convenient as common hydraulic cylinder.

Area Effect Stroke Amplifying Device

In the light load linear reciprocating motion device integrating mechanical, electronic and hydraulic technology, we use a kind of internal-reflux area-differential hydraulic stroke amplifying device. Three hydraulic stroke amplifying devices are discussed briefly as follows.

Traditional Area Effect Hydraulic Stroke Amplifying Device. The device is shown in Fig. 1. According to Pascal's law, any pressure of the liquid within sealed device will be distributed equally in all directions. If a small diameter piston pushes a big diameter piston, it will be a force amplifying device and a stroke reducing device simultaneously. Whereas, it is a stroke amplifying device and a force reducing device simultaneously as a big diameter piston pushes a small diameter piston, as shown in Fig. 1.

The stroke amplifying coefficient i_s is given as:

$$i_s = \frac{S_o}{S_i} = \frac{D_i^2}{D_o^2} \quad (1)$$

Where S_i and S_o is respectively the displacement of diving and output piston. D_i and D_o is respectively the diameter of driving and output piston.

The outstanding advantage of area effect stroke amplifying device is its relative large stroke amplifying coefficient i_s in condition of small structural dimension. For example, if $\frac{D_i}{D_o} = 8$, then $i_s = 64$. It is hardly conceivable to get this large stroke amplifying coefficient i_s for mechanical stroke amplifying device based on length effect principle.

The conspicuous defects of the device shown in Fig. 1 are its small diameter and insufficient rigidity of the output piston in condition of certain input piston diameter and stroke amplifying coefficient i_s when mounting the return spring. In this case, we have to enlarge the input piston diameter that may lead to a large device size and increase manufacture costs.

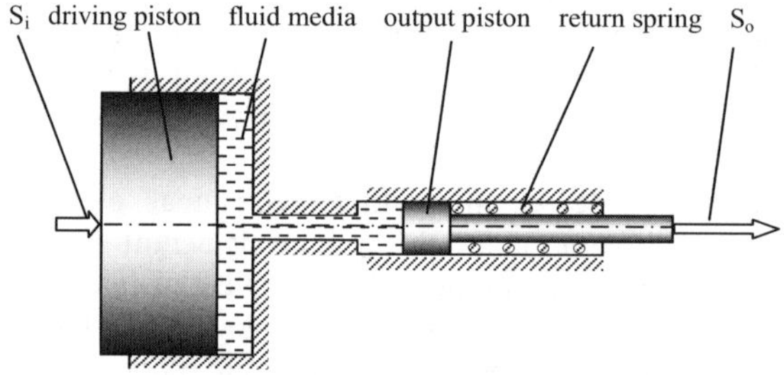

Fig. 1: Traditional Area Effect Hydraulic Stroke Amplifying Device

External-reflux Area-differential Hydraulic Stroke Amplifying Device. To solve the problem mentioned above, designers often use bypassing differential circuit as shown in Fig. 2. That means the fluid media in output hydraulic cylinder flow through bypassing oil circuit from rod chamber to non-rod chamber achieving area differential. The stroke amplifying coefficient i_s of area-differential device depends on input piston diameter D_i and output piston diameter d. The calculation formula is:

$$i_s = \frac{S_o}{S_i} = \frac{D_i^2}{d^2} \quad (2)$$

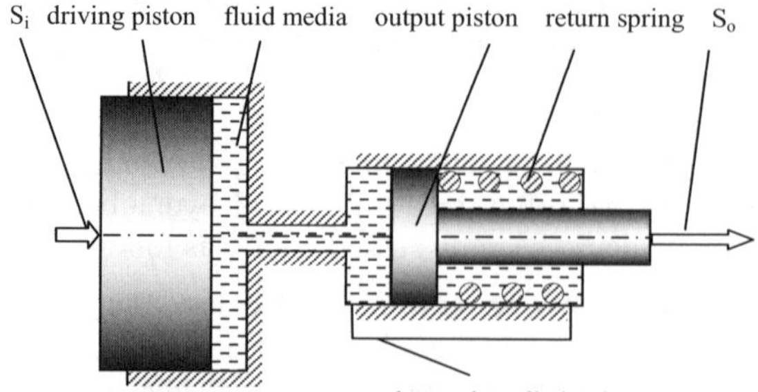

Fig. 2: External-reflux Area-differential Hydraulic Stroke Amplifying Device

It is obvious that the diameter of output piston d in Fig. 1 is equal to the output piston diameter D_o in Fig. 2. Thus the rigidity can be improved enormously. But the bypassing differential circuit has some disadvantages, such as the long reflux distance, high pressure loss, low transmission efficiency and complex structure. The whole device appears extremely not compact in structure.

Internal-reflux Area-differential Hydraulic Stroke Amplifying Device. Y.X. Dou and K.M. Zhong [1] have introduced a kind of internal-reflux area-differential hydraulic stroke amplifying device and its working principle is shown in Fig. 3. One or more reflux holes are drilled in the output piston, thus the fluid media in output hydraulic cylinder can flow through the reflux holes from rod chamber to non-rod chamber to achieve area differential. This function is absolutely the same as the bypassing differential circuit shown in Fig. 2. But the reflux distance is greatly shortened. The device given in Fig. 3 is simple and amazing no matter whether it concerns principle or structure.

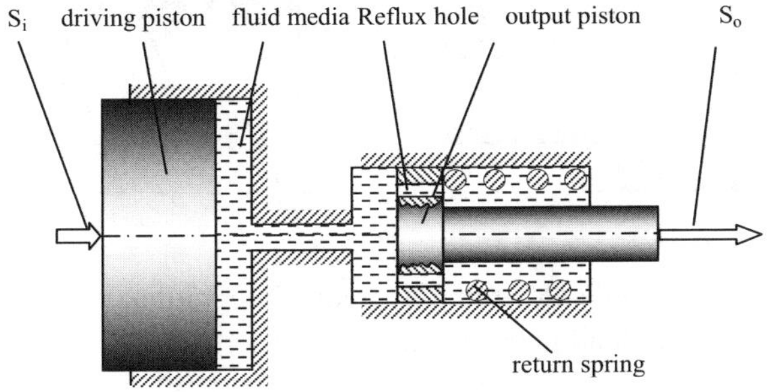

Fig. 3: Internal-reflux Area-differential Hydraulic Stroke Amplifying Device

The Light Load Linear Reciprocating Motion Device with Mechanical-electronic-hydraulic Integration

Working Principle. The working principle of this device is shown in Fig. 4. In this device, the stepping motor makes the ball screw rotate through shaft coupling, and the screw-nut change the rotation movement of the motor to linear movement of the driving piston. The cylinders of the driving and output piston are filled with fluid media and connected by pipeline.

While the driving piston moves to the left, the fluid media in the left chamber of the cylinder will be pressed to the output cylinder through pipeline. Because there are reflux holes in the output piston, the pressure of the left and right chamber in output cylinder is equal. Actually, the reflux holes have formed an internal differential circuit. The efficient area that the fluid media pressed on the output piston is exactly the piston area. By the action of the area-differential hydraulic forces, the output piston moves to the right and conquers the forces of return spring. Hence the output piston makes a linear movement. Meanwhile, the fluid media in the right chamber flow to the left chamber through reflux holes.

The stepping motor rotates counter-clockwise controlled by electro circuit and program, and the driving piston moves to the right operated by screw-nut. The output piston will move to the left with the help of return spring. In that situation, some of the fluid media in the left chamber will flow to the right chamber through reflux holes. Others will flow to the left chamber of the driving cylinder.

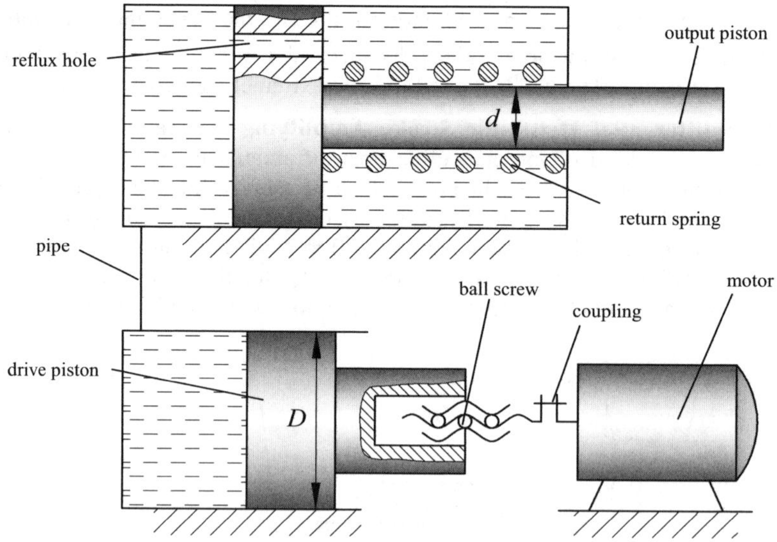

Fig.4: The mechanic-electronic-hydraulic integration light load linear motion device

Mechanics Calculation. The output displacement S of the piston can be calculated by the following formula:

$$S = \frac{\varphi p D^2}{2\pi d^2} \tag{3}$$

Where φ is total rotatory angle of stepping motor (radian), p is ball screw lead, D is diameter of driving piston, d is diameter of piston rod.

The formula for calculating piston rod velocity v can be described as follows:

$$v = \frac{\omega p D^2}{2\pi d^2} \tag{4}$$

Where ω is angular velocity of stepping motor (1/s).

The output force of the piston rod can be calculated by the follow formula:

$$F = \frac{2\pi T d^2 \eta}{P D^2} - F_S \tag{5}$$

Where T is output torque of stepping motor, F_s is spring force, η is total mechanical efficiency of the whole device.

Summary

The light load linear reciprocating motion device introduced in this paper has many advantages:

(1) Use pipeline to connect the driving and output device, so the mounting position is quite convenient for output device.

(2) The reflux holes in the output piston make amazing internal-reflux area-differential hydraulic stroke amplifying device. The displacement of driving piston is magnified to $\frac{D^2}{d^2}$ times. Compared with traditional area-differential stroke amplifying device, the rigidity of output piston rod has increased greatly.

(3) The frequency, displacement and velocity of linear reciprocating motion of piston rod can be altered convenient by programming. Its displacement accuracy is higher than those in pneumatic, hydraulic or mechanical device, but lower than those with ball-screw-nut driven by stepping motor.

(4) Compared with the similar device in reference [4], the stepping motor, ball screw and driving piston are mounted on a common axis in this device. Therefore the structure is more compact.

(5) The hydraulic oil is sealed in the device and no environmental pollution is caused.

The light load linear reciprocating motion device introduced in this paper adopt the merits of both mechanical-electronic-hydraulic and computer technology. Compared with traditional light load linear reciprocating motion device, it has many advantages, such as flexibility and high degree of automation. So it gets a very wide application foreground.

References

[1] Y.X. Dou and K.M. Zhong: *A Distance Amplifier Based on Area Differential Hydraulic Cylinder by Internal Reflux*. Chinese Hydraulics & Pneumatics Vol.12 (2008), pp. 74-75.

[2] X.Y. Song, in: *Pocket-size Hydraulic & Pneumatic Handbook*. Machinery Industrial Press, Beijing, (1995).

[3] M.D.Wang, K.M.Zhong, D.W.Zuo and M.Wang: *Green, Flexible, Intelligent: Clamping system based on thermal sensitive material with volume effect*. Material Science Forum Vol.532 -533 (2006), pp. 1028-1031.

[4] D.N. Su and Q. Song: *Light-load linear reciprocating motion device based on servo motor and small eccentricity crank mechanism*. Proceedings of ASME International Mechanical Engineering Congress and Exposition, Lake Buena Vista, Florida, USA, (2009).

Advanced Materials Research Vol. 279 (2011) pp 382-387
© (2011) Trans Tech Publications, Switzerland
doi:10.4028/www.scientific.net/AMR.279.382

Rotor Axis Orbit Monitoring Based on Virtual Instrument

Wei Yuan[a], Xu Wubin[b], Zeng Haijing[c],

Zhang Hongxian[d] and Zhang Zhouqiang[e]

Department of mechanical engineering, Guangxi University of Technology, Liuzhou, Guangxi, 545006 China

[a]weiyuan0315@sina.com, [b]xuwub@163.com, [c]617385058@qq.com; [d]20618616@qq.com,[e]1464644106@qq.com

Keywords: LabVIEW, Axis orbit, Oil film force, Measurement and analysis

Abstract. According to the typical rotating machinery faults and their corresponding rotor axis orbit characteristics, the using of eddy current sensor and the LabVIEW software platform constitute orbits measurement and analysis system, which is carried out for data acquisition, instrument control, measurement analysis and data display. The results show that orbits can be real-time display of motion, and two orthogonal vibration signals superimposed on the XY Graph.vi 'to generated the orbits and wave, calculating the quality of unbalance Block and the position of equilibrium rotor to make sure the system to reach equilibrium.

Introduction

Virtual instrument is a kind of instrument based on computerized measurement system, paired with the corresponding test function hardware as signal input and output interface which can carry out data acquisition, instrument control, measurement analysis and data display to complete a variety of measurement functions. The software technology is the core of virtual instrument, and LabVIEW, made by the United States NI company, is the most widely used computer language software presently. It is developed specifically for the test system, and its graphical programming model can greatly improve the speed of software development, so it has powerful data analysis, data control and data express skills when using it for principle research, design and test. Compared with traditional instruments, LabVIEW can improve work efficiency greatly, be maintained easily and cost less.

As the judgment of the rotating machinery fault symptom's important basis, axis orbit state of motion contain a large number of rotor system dynamic performance information which is an effective method to the diagnosis of rotor balancing problem. It has been an important research content in rotor dynamics. At present, many researchers use eddy current sensors in orbit test system and area vector of axle centre trail in quadrant and other methods to identify the orbits of the rotor rotating machinery fault [1-3], but these methods either measure rotor orbit in reality, or just in simulate measurement. In this paper, we adopt LabVIEW as a development platform which apply software instead of hardware for phase measurement. This platform can effectively solve the above problem, and avoid the measurement of temperature, vibration, noise and other external disturbances. Through the characteristic information of fault we can identify the running status of rotation mechanical and make fault diagnosis. Thereby improve the capability of monitoring and controlling.

Orbit Analysis

Axis orbit is trajectory of a point doing vortex motion on the rotor axis relative to the bearing seat. Orbit analysis is an important means to determine the system stability [4]. As shown in Fig. 1, assuming a journal from any starting point (xi, yj) in motion, under the action of oil-film force, according to Newton's law, it will accelerate the movement, and form the shape of a certain orbits.

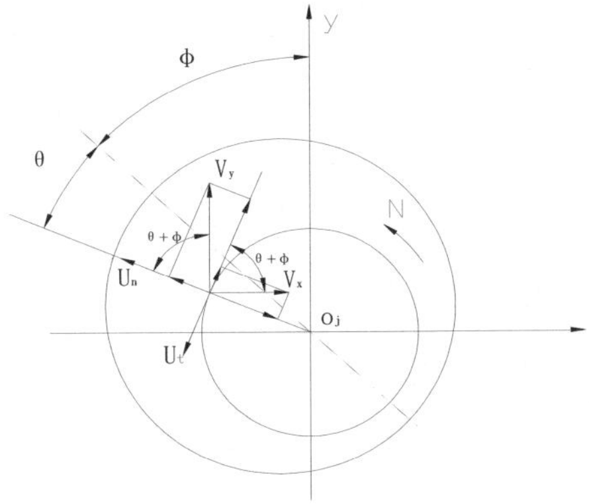

Fig. 1: The model graph for calculating the transient motion orbits of a journal center

The oil-film force Fbxi, Fbyi of journal in the x direction and y direction can be obtained by solving the Reynolds equation, and the instantaneous acceleration of the journal is calculated by the following formula:

$$\ddot{x}_i = F_{bxi} / m \tag{1}$$

$$\ddot{y}_i = (F_{byi} - F) / m \tag{2}$$

Where, F is loading of the bearing in steady state; m is the rotor mass.

The transient velocity and displacement of the axis can be calculated by the following expression:

$$\dot{x}_i = \ddot{x}_i \Delta t + \dot{x}_{i-1} \tag{3}$$

$$\dot{y}_i = \ddot{y}_i \Delta t + \dot{y}_{i-1} \tag{4}$$

$$x_i = \dot{x}_i \Delta t + x_{i-1} \tag{5}$$

$$y_i = \dot{y}_i \Delta t + y_{i-1} \tag{6}$$

Where, Δt is a small time interval, usually taken $\Delta t = \dfrac{120}{50N}$, N is the rotor speed (r / min). According to the theory of the journal given above we can calculate orbits at any moment.

System Hardware Components

Orbit monitoring system consists of eddy current sensor, NI cDAQ9172 chassis, NI 9203 analog current input module, NI 9211 thermocouple input module, PC, and the motor, rotor system, sensors, bearings, counterweight-plate and other components. Fig. 2 is the schematic diagram of experimental system, and adopt model of JX70-02-C-M10×1-80-05K's integrated eddy current displacement sensor for the measurement of axis orbit. As show in Fig. 2, No.6 is the eddy current sensor, and two sensors on the same cross section are installed perpendicularly. The displacement signal of sensor is obtained through the rotor. Then the sensor will detect the displacement current signal into the input LabVIEW, and ultimately get the rotor orbits.

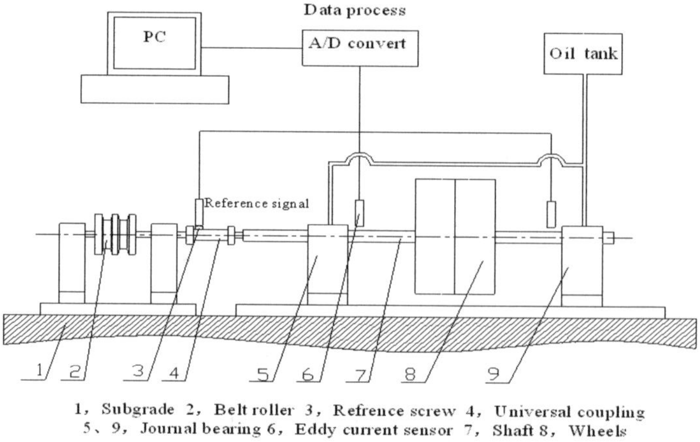

Fig. 2: Schematic diagram of rotor experiment

Software Design

System Block Diagram. Monitoring process need to measure the physical signals of rotor system by the eddy current displacement sensor. The total flow chart of acquisition system is as shown in Fig. 3.

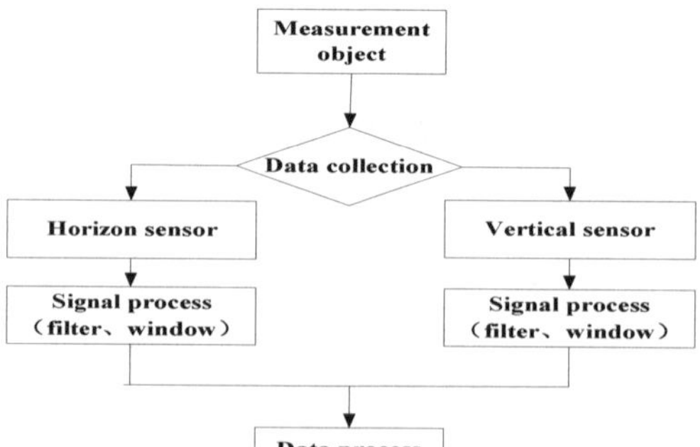

Fig. 3: LabVIEW flowchart of system

System Software. Utilizing the DAQ Assistant of LabVIEW for signal acquisition, the sampling rate and number of sample can be set according to the experiment. The data type of this module in output port can be converted into a required array type through data conversion module in experiment. One aspect of signal processing is windowing and filtering process, and the other is calculating amplitude and phase. On the basis of the standard features equation of the sensor, the transmission of current signal from the index array is translated into displacement signals at first. Then after 'Butterworth Filter.vi' filter processing, it will filter out interfering signals, and then use 'XY Graph.vi' to generate axis trajectory, as shown in Fig. 4.

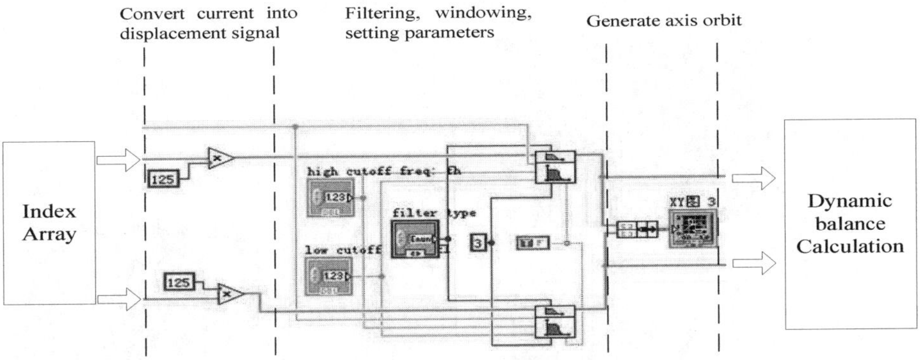

Fig. 4: Axis orbit test track program chart

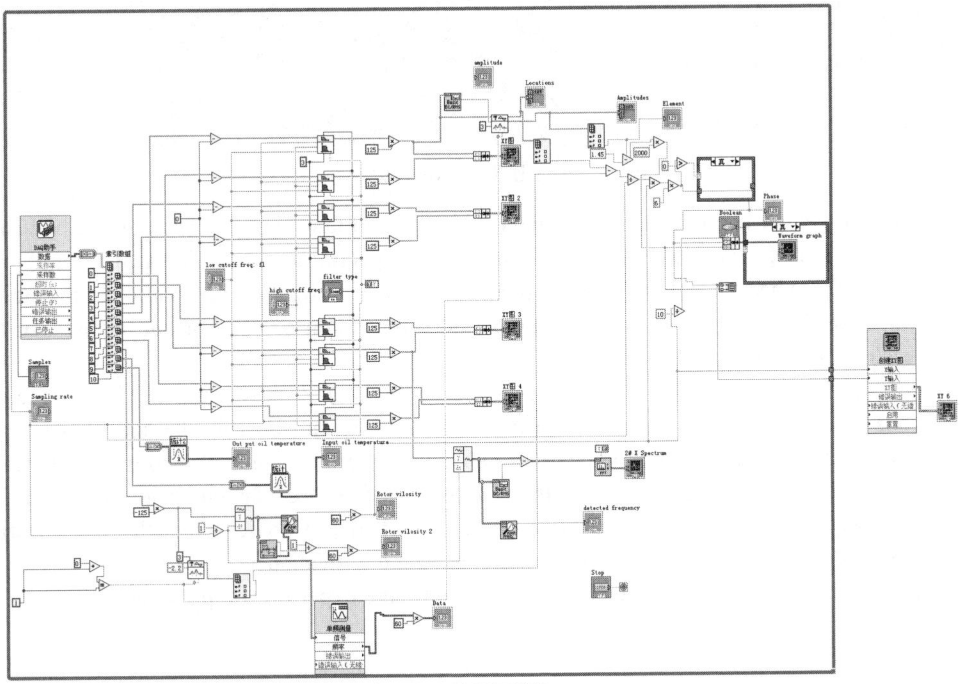

Fig. 5: LabVIEW program

Test Analyses

First we set up the parameters in the front panel, then collect signal, analysis process, display graphic, and save date.

Here we start the rotor balancing test motor, gradually adjust the speed, and run the program after setting up the parameters in the front panel. Then we can observe and analyze the waveform of displacement signal, and record the corresponding result, as shown in Fig. 6.

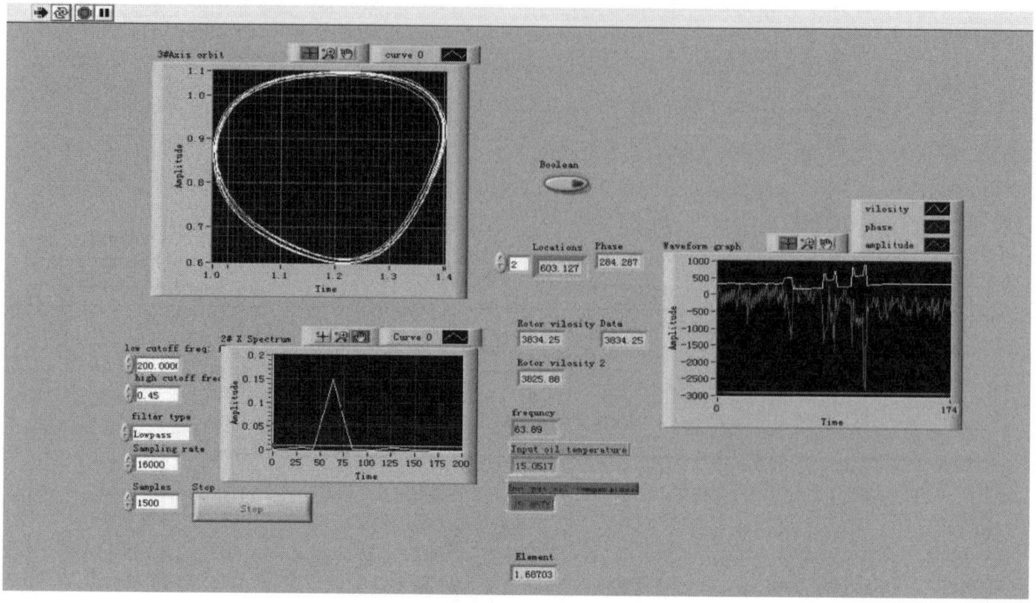

Fig. 6: LabVIEW measured trajectory

After several tests it is known that the balance of state and fault type in the rotor test bed can be judged according to the rotor axis orbit. And by further spectral analysis, the temperature, speed, amplitude and phase bearing rotor system of the bearing rotor system can be gotten. On this account we can calculate the quality of unbalance Block and the position of equilibrium to make sure the system reach equilibrium. [5].

The bearing parameters of the experiment test-bed is given as follow: fluid dynamic viscosity μ = 0.032 N • s/m2, rotor mass 13.43kg, bearing length L = 50mm, bearing diameter Db = 50mm, rotor diameter Dj = 49.25mm, bearing radius gap c = 0.375mm. Using the above orbit analysis method the orbit could be plotted when the system speed is 3830.5r/min, and at this time the temporality system is in stable and unstable critical point. It can be seen that the system critical speed is 3830.5r/min. The LabVIEW monitored data of the critical speed is 3834.25 r / min, which is consistent with the calculated critical speed under the same experimental parameters,.

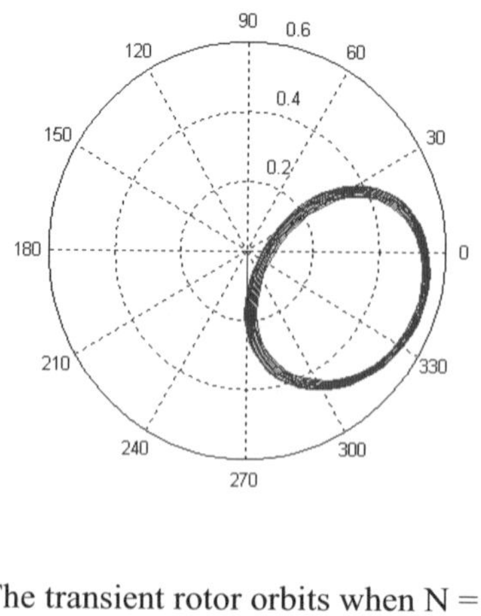

Fig. 7: The transient rotor orbits when N = 3380 rpm

Conclusions

The combination of the eddy current sensor and virtual instrument introduction are introduced to the orbits of the system, which can take full advantage of computer and software platform. It is easy for the system to maintain and implement the expansion or upgrade capabilities. Through theoretical analysis and comparison of the test results, the system, which provide an intelligent way to detect and remove the operation of the fault in the rotating machinery, can accurately indicate changes in the axis.

Acknowledgement

The research is supported by the National Natural Science Foundation of China (Project number: 50865001), the PhD scholarship of Staffordshire University and The Education Department of Guangxi Zhuang Autonomous Region (Project number: GJRC0826). The supports are gratefully acknowledged.

References

[1] Cheng Peng, Gao Yinhan, Han Yongqiang, et al.: *The Measurement of Journal Center Locus of the Main Bearing in Single-Cylinder Diesel Engine.* Chinese Internal Combustion Engine Engineering, 24, 1(2003), pp. 66-69.

[2] Wan Bo, Fu Guang Song, Liang Wei, et al.: *Feature Quantitative Extraction of Axle Centre Trail Character Parameters Based on Area Vector in Quadrant.* Science Technology and Engineering, 7, 23(2007), pp. 5977-5981, 5986.

[3] Wu, S. F. and W.H. Steyn: *Modelling and in-orbit calibration practice of a miniature 2-axis analogue sun sensor.* Aerospace Science and Technology, 6, 6(2002), pp. 423-433.

[4] Wu Gang, Li Taosheng, Yan Xinping: *Visualization of Sliding Radial Bearing Axes 0rbit and its Applications.* Science Technology and Engineering, 23, 1(2004), pp. 117-119.

[5] Lu Song Yuan: *Vibration of turbine-generator set.* China electric power press, Beijing, (2000).

[6] Métris, G., et al.: *LAGEOS spin axis and non-gravitational excitations of its orbit.* Advances in Space Research, 23, 4(1999), pp. 721-725.

[7] Li Rui, Zhou Bing, Hu Renxi et al.: *LabVIEW2009 Chinese version of the virtual instrument from the entry to the master.* China Machine Press, Beijing, (2010).

Advanced Materials Research Vol. 279 (2011) pp 388-393
© (2011) Trans Tech Publications, Switzerland
doi:10.4028/www.scientific.net/AMR.279.388

Research on the Describing for Products Information Evolution Based on TBS

Yi GAN[a], Ming ZHAO[b] and Zhi-wei ZHANG[c]

College of Mechanical Engineering, University of Shanghai for Science and Technology, Shanghai, 200093, China

[a]ganyi@usst.edu.cn, [b]zhaoming7622@126.com, [c]hapmi@163.com

Keywords: Top Basic Skeleton, Description for products information evolution, XML

Abstract. Based on the analysis of the product design and its process, by using Top Basic Skeleton (TBS) model as a link, and adopting eXtensible Markup Language (XML) to the product's top-down design and the description of the information evolution, this paper defines the transfer process and the structure of the document of the information transfer in TBS modeling process, and describes the transmission of the design information layer by layer. As a conclusion, the combination of the TBS and XML is beneficial to control the product design process, and to do variant design.

Introduction

Product assembly model is the product information with effective description and expression in the computer. The accurate and reasonable description of the product assembly information is required to meet the needs of the application of the assembly system [1]. The process of product design is a gradual process which is from abstract to concrete [2]. In the designing of a new product, top-down design is framed from the whole model, which can achieve good results.

Product of top basic skeleton (TBS) [3] is based on top-down design, which can express the main spatial location and spatial shape of the product model. It can response to topological relations between sub-products, and as well as their movement functions. TBS is found to be the interrelated middle bridge and tie of each subassembly from the relationship between the continuous process of development and the inheritance and correlation of TBS and its follow-up design. Researchers have innovated product design with TBS [3], but the lack of the evolution of product information exits during the current research on TBS. Besides, the changes of the information are difficult to describe in the design process. This paper will study on the description of product information evolution based on the process of assembly modeling of TBS.

As the product involves a wide range of information in the design process, it's difficult to describe a single and unified manner. The description of product information based on the process of TBS assembly modeling needs a means which is convenient to describe non-standard information.

XML(eXtensible Markup Language) is a simple data storage language, whose advantages are as follows [4, 5]. 1. Not only it can express the content of data, but also it can express the structure of data at the same time. 2. Developers can create a specific data type to meet specific application. 3. Heterogeneous structured data can be exchanged between different systems when using XML as an intermediary. 4. XML is beneficial to the integration of structured data and unstructured data. Comparing with structured tree and natural language, XML can achieve the standardization, generalization and simplification of data exchange better. Therefore this article describes the product information of assembly model based on XML language.

Modeling Based on TBS

TBS is to build a basic framework in the top floor of programming according to the basic functions and requirements of the product in the prime stage of product design. The location and space dimensions may be put in TBS as basic information, and then they are passed to each subsystem. Each subsystem has obtained the necessary information from the top assembly, and then they can design the details in the acquired skeleton [6-8].

In this paper, commonly used valve in industry is given as an example to illuminate the top-down design based on TBS method. Valve is a spool that closed parts(disc) move along the centerline of the valve seat, depending on the pressure of stem, keeping the spool sealing surface closed to valve seat sealing surface, and stopping the media from flowing. It is widely used on industry occasion when the media needs to be cut off or adjusted.

The sub-assembly is divided into two parts of valve seat components and spool components according to the basic function of valve. The basic framework of product top level is established. The assembly location, space dimensions and design basis of the valve is delimited, not including geometry shape and specific parts of any subassembly, as shown in Fig. 1. The location and thickness of valve cover, valve ring and valve stem are defined by datum plane 1, 2, 3.

Product level TBS is copied to each subassembly, and the required design information is obtained from top level assembly, then creating sub-assembly TBS, the design datum of product level TBS is inherited. The TBS model of valve seat components and valve spool components including the space size, shape, key dimensions and constraints are shown in Fig. 2 and Fig. 3.

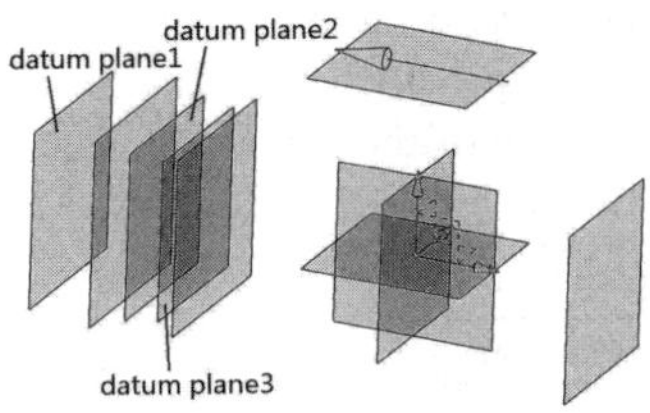

Fig. 1:TBS of valve

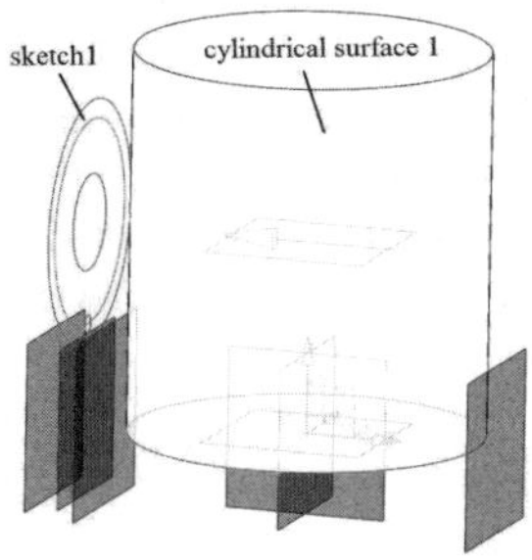

Fig. 2: TBS of valve seat components

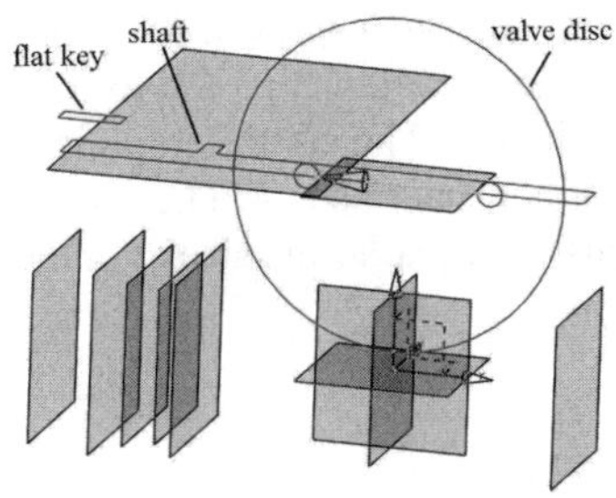

Fig. 3: TBS of valve disc components

"sketch 1" in Fig. 2 is the corresponding sketch of parts when the stretch feather is created. "cylindrical surface 1" is the basic surface of parts when the thicken feather is created. The sketches and curves in Fig. 3 include the shape information of shaft, valve disc and key groove. Each part inherits corresponding information, and then the corresponding solid model is obtained through stretching or rotation operation.

The valve seat components are composed of valve seat, valve cover and so on. The spool components are composed of valve disc, shaft, ring, flat key and so on. The information of subassembly level TBS in the components is required to be transmitted into the detailed design of each part. The detailed design of the parts can be carried out in the obtained subassembly framework.

Taking the seat components as an example, TBS model of valve seat components is copied completely when the valve seat is created. The raised part on the side face of the valve body is got by extruding the sketch1 in Fig. 2. The design of valve seat based on previous design can be completed through the operation such as revolving, stretch, arrays. It is shown in Fig. 4.

Because TBS in the link geometry of valve seat components is inherited by valve seat completely and the following detailed design is based on previous design, so they have characteristics with assembly relevance. The final valve model is shown in Fig. 5.

When the features in TBS are changed, the parts changed accordingly with inheriting associated relationship. The variant design is achieved, and it improves the efficiency and accuracy of the design. For example, changing the position of the datum plane 1, 2, 3 in Fig. 1 and the corresponding location and thickness of valve cover, ring, and stem changed automatically. Changing the skirt diameter of sketch 1 in Fig. 2, for the information of sketch 1 is inherited by valve cover and valve seat, the outside diameter of the boss on valve seat and valve cover changed at the same time, which is shown in Fig. 6.

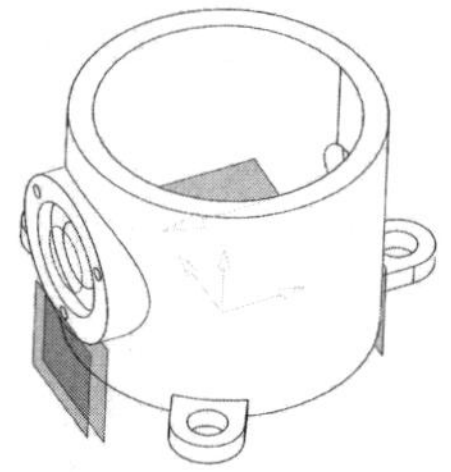

Fig. 4: valve seat

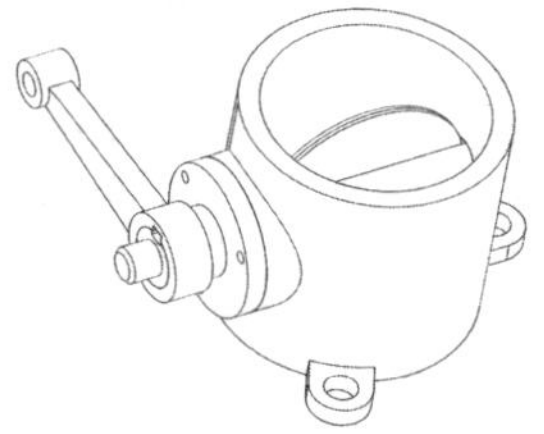

Fig. 5: valve

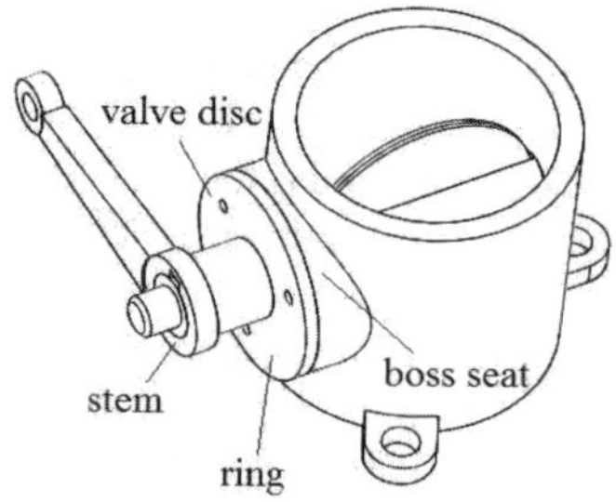

Fig 6: changing top basic skeleton, the model will make corresponding changes automatically

The Description for Product Design Process Based on TBS

The valve design process is shown in Fig. 7. Product TBS contains the following information: 1.The important datum of the product, such as coordinate origin, reference point, baseline, base-level and so on of product model. 2. The topology information of product. That is the location of geometry structure dimensions of the product parts and the relationship between the product parts, such as the location relationship of coaxial, parallel, collinear, and tangent between the main characters of each product. 3. The control parameters of the product. Such as the restriction requirements of the whole product space size. 4. The related information of product design. For example, the basic size of shaft must be equal to the sleeve in the assembly state. 5. The shared information of the product design.

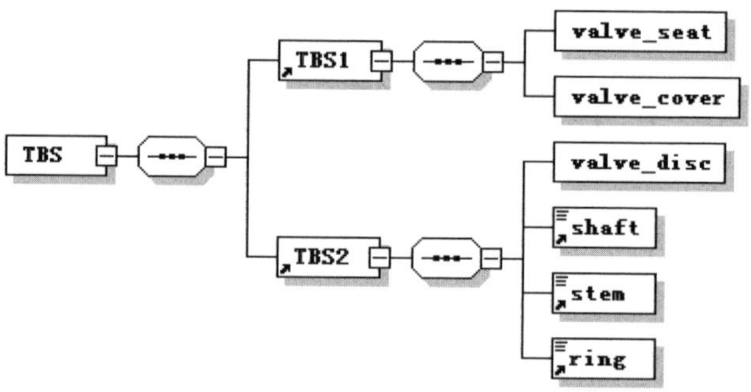

Fig. 7: the description information model of valve design process

TBS is file root element in Fig. 7. It contains two sub-elements: TBS1 and TBS2. TBS1 contains two sub-elements, valve seat and valve cover. TBS2 contains four sub-elements, ring, shaft, stem, and valve disc. The design information is passed from the top level TBS to sub-assembly level TBS1 and TBS2. Then the information is passed from sub-assembly level to the bottom valve seat, valve cover, ring and other parts.

TBS is product level TBS, as shown in Fig. 1 and Fig. 7. The information it contains can be described as follows by XML language:

<info>
<name>TBS<name>
<code>TBS<code >
<type>general assembly TBS<type>
<TBS information>
<DP>datum plane:DP1,DP2,DP3,DP4,DP5,DP6,DP7,DP8,DP9,DP10</DP>
<DA>reference axis:DA1</DA>
<E>parameter:D0,D1,D2,L1</E>
</TBS information>
</info>

The valve seat is a part of Fig. 4. It inherited design intention, design basis, topology and other useful information from TBS1, adding the detailed information of part layer. The information it contains can be described as follows:

<info>
<name>valve seat</name>
<code>valve</code>
<type>the parts</type>
<the inherited information>
<DP>datum plane:DP1,DP2,DP3,DP4,DP5,DP6,DP7</DP>
<DA>reference axis:DA1</DA>
<E>parameter:D1</E>
<S>sketch:S1,S2</S>
<SU>surface:SU1</SU>
</ the inherited information >
<the detailed information of the parts>
material: gray cast iron; roughcast: casting; heat treatment: heat aging, heat-temperature annealing; high temperature, corrosion, oxidation
</ the detailed information of the parts >
</info>

The other transfer process can be described with the same way.

Describe the Evolution of Product Design Information

In order to describe the information of valve design process based on XML language in Fig. 7 by the way of using a structured language, this paper uses XSL(eXtensible Stylesheet Language) to show XML data. The information evolution structure of valve based on TBS modeling process is shown in Table 1.

Table 1: The information evolution structure of valve based on TBS modeling process

name	type	The information of TBS	The inherited information	The detailed information of the parts	The changing design information
TBS	General assembly TBS	Datum plane: DP1,DP2,DP3,DP4, DP5,DP6,DP7,DP8, DP9,DP10 Reference axis:DA1 Parameter :D0,D1,D2,L1			DP4,DP9,DP10
TBS1	component level TBS	Datum plane: DP1,DP2,DP3,DP4, DP5,DP6,DP7,DP1 0 Reference axis:DA1 parameter:D0,D1,D 2,L1 sketch:S1,S2 surface:SU1	Datum plane: DP1,DP2,DP3,DP 4,DP5,DP6,DP7,D P8 Reference axis:DA1 Parameter :D0,D1,D2,L1		S1
Valve seat	parts		Datum plane: DP1,DP2,DP3,DP 4,DP5,DP6,DP7 Reference axis:DA1 Parameter :D1 sketch:S1,S2 surface:SU1	material: gray cast iron; roughcast: casting; heat treatment :heat aging,high-temperat ure annealing; high-temperature,cor rosion,oxidation	The outside diameter of the boss is changed

Valve cover	The parts		Datum plane:DP10 Reference axis:DA1 sketch:S1,S2	material: gray cast iron; roughcast: casting; heat treatment: heat aging,high temperature annealing; high-temperature,corrosion,oxidation	The thickness is changed; the outside diameter is changed
TBS2	Component level TBS	Datum plane: DP1,DP2,DP3,DP4, DP5,DP6,DP7,DP8, DP9,DP10,DP11 Reference axis:DA1 parameter:D0,D1,D2,L1 sketch:S3,S4,S5,S6	Datum plane: DP1,DP2,DP3,DP4,DP5,DP6,DP7,DP8,DP9,DP10 Reference axis:DA1 parameter:D0,D1, D2,L1		
Valve disc	The parts		sketch:S4,S5	roughcast: casting; heat treatment: heat-aging, high temperature annealing; high-temperature,corrosion,oxidation	
shaft	The parts		Datum plane:DP2,DP11 Reference axis:DA1 parameter:D2 sketch:S3,S4,S5,S6	material:45steel;roughcasting:forging;heat treatment: quenched and quenched; surface hardness:45~52HRC; high-temperature,corrosion,oxidation	
stem	The parts		Datum plane:DP8,DP9 sketch:S3,S6	material: malleable iron; roughcasting: forging; heat treatment :heat aging,high-temperature annealing	The thickness and the location are changed
The ring			Datum plane:DP8,DP10 sketch:S3	material: gray cast iron ;roughcast : casting ;heat treatment: heat aging,high-temperature annealing	The thickness and the location are changed

"TBS information" indicates the provided information from the current layer to the next layer, and "the inherited layer" indicates the inherited information passed from the top layer to the current layer in Table 1. First, the information is passed from the top level TBS to the sub-assembly level TBS1, and then the information is passed from sub-assembly level TBS1 to the bottom parts. The general assembly level TBS located on the top layer, and only has the information passed down. The parts level located on the bottom layer, and only can receive the information from the top level in the transfer process. The sub-assembly level located in the middle of the whole layers, and it can receive the top information as well as passing information to the bottom layer in the transfer process. The information expands gradually in the top-down transfer process. The sketch1 and sketch2 of TBS1 is the new added information of sub-assembly level TBS in Table 1. The parts layer becomes integral model by way of operating physical characteristics are based on the passed information from the top layer.

From the description of evolution information based on TBS model design process, it can be seen that for building on the same basis and passing the design intend in the whole assembly skeleton, which means many design teams or work teams can consult the basis information in the same layer, and the information is passed down "in time" when top layer is changed, one assembly can meet the combination of other parts with its development [10].

Conclusion

This paper describes the design information evolution based on the product top-down modeling process of TBS theory using the XML language. TBS includes the location and space size of important parts and other important information, and passes design intend in the whole assembly skeleton, then passes to each sub-assembly. Because of inheriting the mutual relation of TBS, when changing the characters of TBS, the variant design for parts can be achieved. This paper defines the transfer process and the document structure of information transmission in the modeling process of TBS when the XML language is used, and describes the transmission of design information layer by layer. It could be seen that for building on the same basis, the information is passed down "in time" when top layer is changed, and one assembly can meet the combination of other parts with its development.

Acknowledgement

This work is supported by the National Natural Science Foundation of China (Grant No. 50505030), and "Dawn" Program of Shanghai Education Commission (Grant No.07SG51).

References

[1] Zhang Fan, Zhang Youliang: *Reconfigurable assembly model based on XML*. Computer Integrated Manufacturing System Vol. 8 (2003), pp. 626-629, 644.

[2] Shui Junfeng, Gu Haitian: *Bridge crane girder design based on Top-down design patterns*. Mechanical Engineering and Automation Vol. 6 (2007), pp. 4-5.

[3] Qi Congqian, Cui Qiongyao: *Innovative Design Method for CAD based on Parametric Technology*. Chinese Mechanical Engineering Vol. 8 (2003), pp. 681-683.

[4] Huang Shuangxi, Fan Yushun, Zhao Dazhe, et al.: *enterprise application integration based on Web services*. Computer Integrated Manufacturing System Vol. 9 (2003), pp. 864-867.

[5] Wei Yinxing, Zhang Shensheng: *Research on Enterprise Application Integration Technology*. Computer Integrated Manufacturing System Vol. 8 (2002), pp. 593-596.

[6] Gan Yi, Qi Congqian: *Products Concurrent Design based on TBS assembly model*. Mechanical Design Vol. 5 (2006), pp. 5-6, 32.

[7] Qi Congqian, JIA Weixin: *Research on assembly modeling for product variant design*. Chinese Journal of Mechanical Engineering Vol. 1 (2004), pp. 38-42.

[8] Tang Linxin, Qi Congqian, Zhou Xiaoqing, et al.: *TBS evaluation method for product conceptual design*. Chinese Journal of Construction Machinery Vol. 3 (2007), pp. 357-360.

Advanced Materials Research Vol. 279 (2011) pp 394-399

doi:10.4028/www.scientific.net/AMR.279.394

Design modeling and solving implementation for product conceptual design in quality function deployment

Yanhua Yang[1,a], Zuping Zhu[1,b] and Ligang Yao[1,c]

[1]School of Mechanical Engineering and Automation, Fuzhou University, Fuzhou, Fujian, 350108, China

[a]yangyanhua1999@163.com, [b]zhuzuping@163.com, [c]ylgyao@fzu.edu.cn

Keywords: quality function deployment, conceptual design, entropy grey-correlation analysis, behaviour solving

Abstract. Considering the process of research in conceptual design, a framework model of Function-Behaviour-structure was put forward, and QFD was an effective tool for function conversion. Then a mixed solving technology combining case-based reasoning and behaviour solving reasoning was discussed to achieve design mapping. Finally, an automatic synthesis mapping system was developed and the conceptual design of clam collector was given as an example. Application results demonstrated that the proposed method was feasible.

Introduction

Conceptual design is perhaps most crucial and pivotal to accomplish task in product development cycle. The pertinent data demonstrated that conceptual design stage costs only 5% of the total investment, but the decision-making of this stage influences more than 75% of the total cost of the product [1]. In recent years, the conceptual design research has attracted numerous scholars' attentions on the theory and method of product conceptual design, the development of conceptual design system and modeling of creative design, etc. Literature [2] proposed eight fundamental processes involved in designing to extend the function-behaviour-structure framework, the eight fundamental processes are then reconstructed within this new framework to represent designing in a dynamic world. Literature [3] proposed an ontology-based approach for knowledge management that works along with the graphical modeling tool. Mapping solving studies emphasize the use of intelligence, coordination, integration and other new techniques and methods. Literature [4] studied the collaborative design and manufacture model based on constraint, and developed a computer-aided design/manufacture system (CAD/CAM). Literature [5] suggested a hierarchical co-evolutionary approach to support conceptual design. Higher-level functions are decomposed based on a set of grammar rules, and mappings between functions and their solutions, or means, are realized through a co-evolutionary computing process. Basically the research results of conceptual design solution could be divided into two categories, One is directly building mapping relationships between function field and structure field [4,5], and another is to build solution models of mapping relationships between three layers after behaviour field has been involved between function field and structure field [2]. Most conceptual design process models are expanded based on these two basic frameworks which are able to solve various kinds of product conceptual design function problems very effectively. However, current work and methods still aim at single product, and how to perform solving of conceptual design scheme by specific process and differences of innovation degree to complete conceptual design scheme intelligently and scientifically should be further discussed.

This paper makes a beneficial exploration, and it establishes a product conceptual design framework and process model based on Quality Function Development (QFD), and mixed reasoning and mapping technology was adopted to complete the product design scheme. Therefore, practical problems will be solved by designers under human-like reasoning processes to improve the competitiveness of product.

Conceptual Design Model Framework Based on Quality Function Deployment

Product modeling has been regarded as a key technique to develop conceptual design support systems [6]. Considering the process of research in conceptual design, a model framework of "Function-Behaviour-Structure" is proposed, as shown in Fig. 1. It contains two kinds of transform process: direct mapping for function-structure and indirect mapping for function-behaviour-structure, and this model has the following characters:

1) Consider the design functions retrieved by organizing requirements as input to map in different modes according to differences of product innovation. Two modeling frameworks as function-structure and function-behaviour-structure reflect different innovation degrees in new product design. Function represents rational design tasks and requirements; Behaviour describes the measures to implement function and principle structure describes physical structure of achievement behaviour as the specific materialization of behaviour, and it is the carrier of behaviour. During the process of product conceptual design, the design function of product is being decomposed till it is able to correspond to some behaviour, and then the principle structure could be constructed by finding corresponding behaviour carrier. Behaviour layer could be skipped to build function-structure direct mapping relationship when design experience accumulated to a certain stage.

2) QFD is an effective tool for the conversion of mapping between product designs attributes, and it brings customer requirements to the whole process of product design by a series of House of Quality with direct matrix framework expression to translate customer requirement to function technique character, structural character and manufacture character, etc. Conceptual design solving house of quality contains design function and its significance, structure solution, relation matrix, relative matrix and plan matrix to achieve the transformation of design from function space to structural space. QFD house of quality equals with the premise/conclusion matrix of expert system, and it is a framework with flexibility and criterion which every part of it is able to connect to other systems of the enterprise. Conceptual design composite solving system is based on QFD as its framework, and it mixed two kinds of transform process to achieve automatic matching of design function and principle structure and design scheme will be obtained automatically.

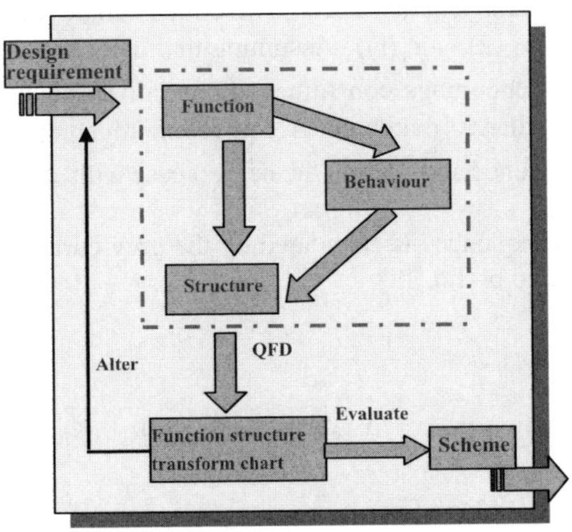

Fig. 1: Model framework of product conceptual design

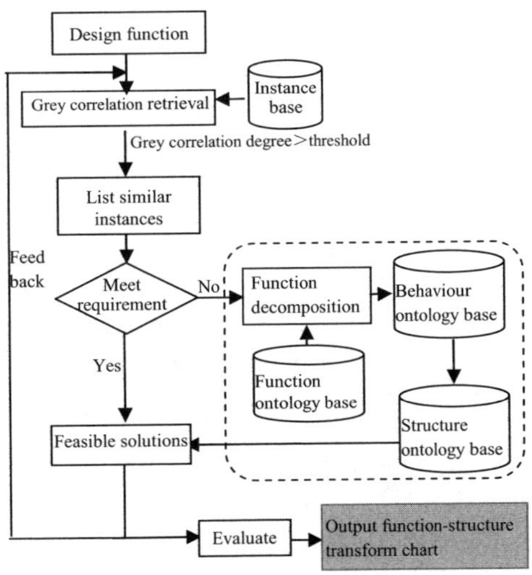

Fig. 2: Solving process for product conceptual design

Conceptual Design Solving Strategy Based on Quality Function Deployment

In automatic mapping from function to structure for product conceptual design, case search pattern and behaviour reasoning pattern are designed to establish the framework of intelligent conceptual design process. Fig. 2 shows the mapping search process from function to structure.

Product design knowledge base includes components case database and knowledge ontology base. Components case database can accomplish direct mapping from function to structure. Based on the process of product conceptual design, product design knowledge base includes function ontology layer, behaviour ontology layer and structure ontology layer. And knowledge base links is completed through the ontology mapping between layers. When users input function information into man-machine interface, the computer will search for mapping information in database, and the output result could be displayed in form of 'Function Structure Transform Chart'.

Case-based Reasoning. The search process of case-based reasoning is as follows:

(1) Case representation and encoding

In case reasoning, the case can be expressed as $CASE_i$={A,B,y}. Every instance is described by two sub-objects as function character and structure character; y is the reference frequency of this instance, which is the number of this instance listed as similar case greater than a threshold. During the process of reasoning, product function could be used as the input to run similarity discrimination reasoning, and the structural sub-object could be used as design scheme to return. The required input should be transformed into standard encoding form to create indexes of the case based on this encoding to apply on similarity measurement. Function encoding structure is shown in Table 1. For example, the function of collecting clams could be transformed to code "“E|01|M" for computing.

Table 1: Function Encoding Structure

Function Category Code	Function Semantic Code	Function of Object Code	Modifier Code

(2) Case retrieve.

Based on the input information of all design functions, the grey correlation grade of cases in the knowledge base with target functions is calculated and similar cases which grey correlation grade exceed a given threshold are retrieved.

Assuming there are l target functions, the serial encodings of these design functions will be formed into a reference sequence, remembered as $x_0 = (x_0(1), x_0(2), \cdots x_0(l))$; Assuming that the i th case in case database has m design functions, the functions encodings constitutes comparative sequence, remembered as $X_i = (x_i(1), x_i(2), \cdots x_i(m))$. The similarity matching is equivalent to comparing distance between index sequence and comparative sequence and finding the nearest case with l target functions points in the m dimension case space.

In grey correlation analysis, the distance between sequences is calculated by the grey correlation grade. The grey correlation coefficient can be calculated by Eq. 1.

$$\gamma(x_0(k), x_i(k)) = \frac{\min_i \min_k \Delta_{0i}(k) + \rho \max_i \max_k \Delta_{0i}(k)}{\Delta_{0i}(k) + \rho \max_i \max_k \Delta_{0i}(k)} \tag{1}$$

Where: $\Delta_{0i}(k)$ is the k th index differentia information in x_i to x_0, $\Delta_{0i}(k) = |x_0(k) - x_i(k)|$; ρ is resolution ratio, $\xi \in [0,1]$.

Traditional grey correlation analysis uses the average value of point grey correlation coefficient to calculate, and the grey correlation grade could be defined as $\gamma(x_0, x_i) = \frac{1}{n}\sum_{k=1}^{n} \gamma(x_0(k), x_i(k))$. However, every function index of the product is not as important as each other in conceptual design thus influences between the indexes should be considered and it is possible to reflect the degree of correlation in reality only by adding different index weight information to it.

Shannon information entropy is considered as the entropy theory of design function weight analysis. Shannon information entropy is the measure of potential information in system and it represents the uncertainty degree of the system. If there are m items of functions and the number of evaluation experts is n, the main steps to fix function weight using entropy are as follows:

1) Format every sequence value of evaluation matrix $R=(x_{ij})_{m*n}$ and calculate the weight of every index $p_{ij}=x_{ij}/\sum_{j=1}^{n}x_{ij}$, $\sum_{j=1}^{n}p_{ij}=1$, $i=1,2,\cdots,m$

2) Calculate Shannon information entropy of every function index $e_i=-\frac{1}{n}\sum_{j=1}^{n}p_{ij}\ln p_{ij}$, $i=1,2,\cdots,m$, $j=1,2,\cdots,n$.

3) Calculate the entropy weight of every function index:

$$w_i^{"}=\frac{1-e_i}{\sum_{i=1}^{m}(1-e_i)},\ i=1,2,\cdots,m \tag{2}$$

Then, the grey correlation grade between cases and target functions will be calculated by Eq. 3.

$$\gamma(x_0,x_i)=\sum_{k=1}^{n}\omega_k\gamma(x_0(k),x_i(k)) \tag{3}$$

Where: $\sum_{k=1}^{n}\omega_k=1$, $\gamma(x_0(k),x_i(k))$ is grey correlation coefficient, ω_k is the importance of design functions.

According to the value of grey correlation grade, similar instance could be retrieved. The larger the grey correlation degree, the closer this instance to current demand. The instance with higher reference frequency remains a priority. In actual use threshold should be set in advance and 3~6 instances bigger than threshold should be taken out according to the value of correlation grade.

The functions of similar instances are analyzed. The matched principle structure of satisfactory items can be abtained directly, and the unsatisfactory items which are unmatched or needed to be improved should be extracted to do another searching with different searching model.

Behaviour reasoning. Behaviour reasoning pattern takes target function as the input, divides functions into smaller and easier implemented sub functions, and retrieves related behaviours from product design knowledge base to map to corresponding principle structure. The retrieve process is as follows:

(1) The function which is unable to find matched structure through case search pattern must be decomposed by steps until it is made of final function base to form product function structure.

(2) Analyse each decomposed sub function to get corresponding behaviour of function; retrieve corresponding principle structure of this function base from behaviour ontology database and structure ontology database.

(3) Furthermore, the matched principle structure will be analyzed. The principle structures evaluated with satisfactions in knowledge base should be selected.

The behaviour search based on function decomposition build a design process framework covers function decomposition as well as behaviour solving. According to a solution route as function-behaviour-structure, the function-structure mapping could be implemented by using the decomposition and mapping of functions.

Comprehensive outputs. Combine all mapping relations from function to structure to get the ultimate function - structure transform chart.

Relationship matrix reflected the degree of influence from structure to function achievement and the relationship matrix in automatic transform system according to appearance frequency of every pair of mapping relationship could be calculated from Eq. 4.

$$r_{ij} = f_1 / N_1 \tag{4}$$

Where: r_{ij} is the relationship coefficient between the i th function and j th structure, $r_{ij} \in (0,1)$; f_1 is appearance number of certain structure; N_1 is number of instances with this function.

The relationship coefficient result will be put into relationship matrix of transform chart and the system will output a 'function structure transform chart' including whole mapping information from structure items to relationship matrix.

The relationship coefficient that is got automatically shows general and possible results of function solving, and it is valuable impersonal reference information in new product design. The magnitude of coefficient reflects successful probability of this function solving in the knowledge base, thus mapping relationship with high coefficient value can be selected firstly, and design time is saved when the scheme is choosen. But the design is a very complex process, and designer can not be replaced whatever computer technology is advanced. Designers still need to modify or evaluate manually the deficiencies in transform chart, and guide design process according to actual situation.

Case Study

This paper uses Visual Studio C# coding language and SQL Server database as developing tools to construct a conceptual design composite solving system to provide guidance and decision analysis for the designers. Here comes the example of clam collector to explain the system application.

A clam collector is designed to apply fully automation on collecting clam which including digging, cleaning and collecting of clams. Through initial filter of collecting design requirements, there are 6 main functions: provide power, transmission, pick up clam continuously, transport clam, clean up clam and collect clam. The weight of functions invested through entropy method evaluation is $\omega = (0.15, 0.15, 0.22, 0.16, 0.16, 0.17)$. By searching case, function item: "pick up clam" has not been converted. Through behaviour solving mapping, the suited structure could be acquired.

All the reasoning results above will be combined, and shown in the "function - structure transform chart", as shown in Fig. 3.

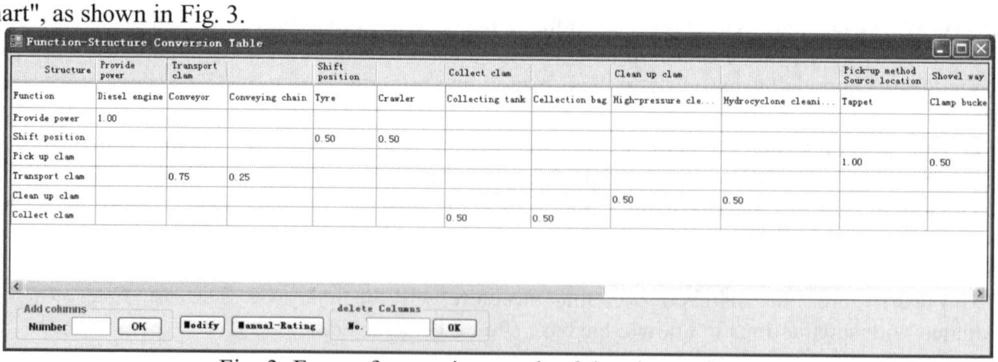

Structure	Provide power	Transport clam		Shift position		Collect clam		Clean up clam		Pick-up method Source location	Shovel way
Function	Diesel engine	Conveyor	Conveying chain	Tyre	Crawler	Collecting tank	Cellection bag	High-pressure cle...	Hydrocyclone cleani...	Tappet	Clamp bucke
Provide power	1.00										
Shift position				0.50	0.50						
Pick up clam										1.00	0.50
Transport clam		0.75	0.25								
Clean up clam								0.50	0.50		
Collect clam						0.50	0.50				

Fig. 3: Form of reasoning result of the clam collector

Through the analysis of feasible solutions, final design result could be achieved and the principle structure diagram of clam collector is shown in Fig. 4.

Clam collector part 1 connects body frame 7 by two pins on its top to keep clam collector part 1 rotating around these pins. It connect transmission system 4 on left side of its top, and the rotary hopper could rotate clockwise by the power provided by transmission system 4. Clam collector part 5 is installed inside body frame7 and tilt on some angle. Its front (bottom) besides the bottom of clam collector part 1, its height is higher than crawler driving part 8, and its back(top) tilts to the back of body frame 7 and higher than the back of body frame 7. It is in the form of higher back and lower front. Clam cleaning and collect part 6 is at the tale of body frame 7, and it is composed of cleaning equipment and collecting tank.

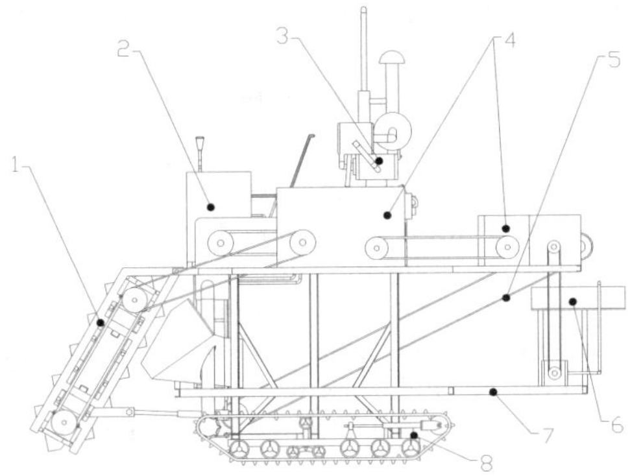

Fig. 4: Principle structure of clam collector

Conclusions

In this paper, design innovation is taken as the starting point of conceptual design study; a process framework of Function - Behaviour - structure based on Quality Function Development is proposed. The model solving is completed by combining case-based reasoning and behaviour reasoning method. Visual Studio C# coding language is used as its tool to develop conceptual design composite mapping system and it is proved to be efficient and exercisable by a case study. This method provides an feasible way to achieve automation for product conceptual design and it plays an importment role in the implementation of product development.

Acknowledgement

This work is supported by Natural Science Foundation of China (Grant No.50675035 and No.70372023).

References

[1] B. Loter: *Manufacturing assembly handbook*. Butterworths, Boston(1996).

[2] J.S. Gero, U. Kannengiesser: *The situated function–behaviour–structure framework*. Design Studies, 25, 4 (2004), pp.373-391.

[3] X. Chang. A. Sahin and J. Terpenny: *An ontology-based support for product conceptual design*. Robotics and Computer-Integrated Manufacturing, 24, 6 (2008), pp. 755-762.

[4] Y.L. Lai: *A constraint-based system for product design and manufacturing*. Robotics and Computer-Integrated Manufacturing, 25, 1(2009), pp. 246-258.

[5] Y. Jin, W.Li and S. C-Y. Lu: *A hierarchical co-evolutionary approach to conceptual design*. CIRP Annals-Manufacturing Technology, 54, 1(2005), pp.155-158.

[6] F.E.H. Tay, J.X Gu: *Product modeling for conceptual design support*. Computers in industry, 48, 2(2002), pp. 143-155.

Advanced Materials Research Vol. 279 (2011) pp 400-405
© (2011) Trans Tech Publications, Switzerland
doi:10.4028/www.scientific.net/AMR.279.400

Analysis of Lower Leg Injury of Occupant in Frontal Crash Tests Based on China NCAP Protocol

Zhixin Liu[1, a], Renjun Wan [1, b] and Yongwan Shi[1, c]

[1]China automotive technology and research center, Tianjin, China

[a] liuzhixin@catarc.ac.cn, [b] wanrenjun@catarc.ac.cn, [c] shiyongwan@catarc.ac.cn

Keywords: Vehicle Safety, Occupant Injury, Crashworthiness, C-NCAP

Abstract. With the popularization of passenger vehicle safety devices such as safety belt, airbag and so on, the chance that occupant's upper limbs were injured seriously was decreased significantly in frontal impact. However, the injury of occupant's lower limbs became more and more severe, especially on lower leg injury. 37 groups of test data of China NCAP crash tests including full-frontal rigid crash and 40% offset deformable barrier crash were investigated in this paper, and lower leg injury distributing characteristic of drivers and passengers in these two kinds of crash configurations were obtained. Finally the effect rules of characteristic parameters on lower leg injury were summarized.

Introduction

With the development of vehicle safety technology, especially with the popularization of passenger vehicle safety devices such as safety belt and airbag, the chance that occupant's upper limbs were seriously injured was decreased significantly in frontal impact. However, the injury problem of occupant's lower limbs especially the injury of lower leg became more and more severe. Statistical data of China's traffic accidents in recent years showed that about 72% leg injuries were resulted from frontal crash accidents and hereinto 33% lower leg injuries index exceeded AIS 2 level which stood for moderate severity. Although leg injury is rarely threatened to occupant's life, it will result in a long-term pain and need long time to cure and recuperate, and even lead to permanently physical disability, which will bring victims double suffering from physiological and mentally.

Considering lower leg injury was very common in frontal impact accidents, 37 groups data of C-NCAP crash tests including full-frontal crash and 40% offset frontal crash were investigated in this paper, then lower leg injury distributing characteristic of drivers and passengers in these two kinds of crash configurations were discussed, and finally the effect rules of characteristic parameters on lower leg injury were summarized.

Description of C-NCAP and its protocol

Introduction of C-NCAP. With the society developing and technology progressing, the safety performance of automobiles is attracting more attention than ever from the consumers, making it a focus of the manufacturers' R&D activities. Although all countries enforce mandatory impact safety standards such as ECE R94 and R95 of the EU, the US Federal Regulations and China's GB 11551 and GB 20071 etc, it just shows that the vehicles meet the minimum requirements of the vehicles' impact safety performance if they pass these safety regulations, and therefore consumers cannot distinguish which vehicle is safer. Thus, in order to provide consumers with accurate and reliable information in detail, and encourage manufacturers to improve the safety performance of their products, the United States and EU have adopted New Car Assessment Program (NCAP), by which an authoritative institution will regularly conduct impact tests on new vehicles for sale in the market by applying standards more stringent than the safety regulations stipulated by the government. Injuries caused to the occupant by more serious impacts are assessed, and the safety performance of the test vehicle is rated according to the severity of injuries to major body areas such as head, chest and leg. The test methods of the NCAP are basically the same as those for regulatory certification purposes, but cover more test items and apply more stringent standards. For instance, the NCAP tests not only include quantitative measurements of the injuries of occupants but also require measuring the deformation of the vehicle body, occupant compartment and the steering system. Then NCAP applies a star rating system that is more explicit and quantitative than the simple "pass" or "fail" approach which can give consumers much more safety information of vehicles. Now it is developed as a very popular and authoritative rating method to vehicle's passive safety in the world [1-3].

The experience of other countries indicates that NCAP can remarkably improve the safety performance. Therefore, on the basis of fully studying and borrowing from the NCAP development experience of other countries, China has launched China New Car Assessment Program in 2006, allowing for on the basis of China's national conditions, road traffic conditions and accident analyses. It consists of three test items, which are frontal impact test against rigid barrier with 100% overlapping, frontal impact test against deformable barrier with 40% overlapping and side impact test against mobile deformable barrier. A maximum score of 51 points in C-NCAP is available. The full score for each of the above tests will be 16 points, or a total of 48 points for all three of the tests. Safety belt reminder and side airbag (and side curtain airbag) are entitled to 2 points and 1 point respectively in extra. The sum of the points for the three tests plus the additional points (rounded to one decimal place) will be recorded as the total score and then be used to determine the star rating. Hereinafter we focused on introduction of the lower leg injury assessment criteria in the frontal tests of C-NCAP.

Lower leg injury mechanisms and assessment criteria. Lower leg injury is a frequent injury in traffic accidents; it is different in injury mechanism to that upper limb was subject to impact acceleration or continuous acceleration. It was very easy to result in mechanical injuries, such as bone fracture, muscle wrench, rupture and so on, even to bring about a cripple for life. Therefore, the load force and bending moment acting on the leg are adopted to assess occupant's leg injury in C-NCAP protocol. They are measured with the upper and lower tibia load cell installed in the lower leg of dummy and filtered with CFC 600. The instruments and the load force and bending moment acting on the leg are shown in Fig. 1. Because the load force perpendicular to tibia and the bending moment acting on both of upper and lower tibia are two important factors that result in bone fracture of lower leg and ankle and wrench of knee joint, tibia compression force Fz and Tibia index are selected as assessment parameters to assess injury of lower leg in C-NCAP protocol. Fz can be measured directly by tibia load cell and Tibia index can be obtained with Eq. 1, as follows:

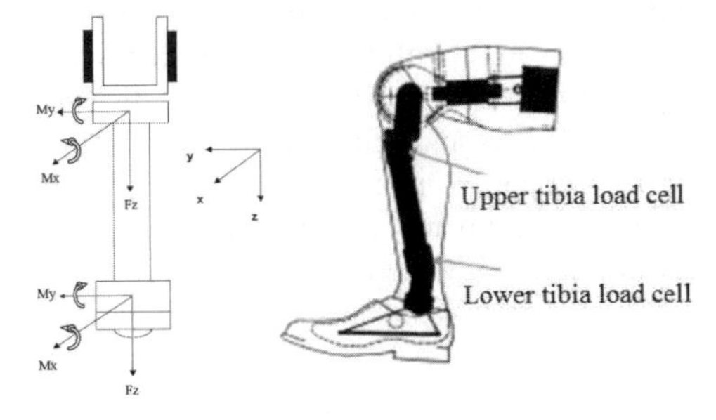

Fig. 1: Load force and torque acting on lower leg

$$TI = \left| \frac{\sqrt{M_x^2 + M_y^2}}{M_c} \right| + \left| \frac{F_z}{F_c} \right| \tag{1}$$

Where M_x is the bending moment about the x axis; M_y is the bending moment about the y axis; M_c stands for the critical value of the bending moment, which is set 225 Nm; F_z is the axial compressive force in vertical direction (z); F_c stands for the critical value of the axial force, which is set 35.9KN;

Lower leg can be assessed according to the procedure as follows: firstly, at the upper and lower of both the left and the right tibias, calculate the resultant bending moment from M_x and M_y after they have been filtered, then calculate the Tibia Index (TI) at the upper and lower tibia of each leg according to the Eq. 1, at last determine the highest value of the Tibia Index and highest value of the axial compressive force measured at either the upper or lower tibia. And the injury severity of lower leg can be determined according to the performance limits of two injury parameters [4-7].

Lower leg scoring rules of C-NCAP. For the full scale frontal crash, the maximum test score is 16 points. The dummy body areas including head, neck, chest, upper leg and lower leg can be awarded up to 5, 2, 5, 2 and 2 points respectively. The scoring is based on the injury criteria of the driver-side dummy. The scores of the passenger-side dummy may be validated only if they are less than the scores for corresponding areas of the driver-side dummy. The basic scoring principle for each dummy is: to set two limits for each parameter, a higher performance limit, threshold for maximum scores, and a lower performance limit, threshold for zero point, if a measurement value falls between the two limits, the score is calculated by linear interpolation and rounded to two decimal places. The lowest score will be validated for a body region where multiple assessment criteria exist.

As far as lower leg score is concerned, the maximum and minimum scores are 2 points and 0 point respectively. The scores for driver-side and passenger-side dummy's lower leg are generated by measuring relevant parameters of the dummy, and the final scores of lower leg depended on the minimum scores of two dummy. The assessment parameters of lower leg include the tibia axis compression force and tibia index. A maximum score of 2 points is available for each parameter. The higher and lower performance limits of tibia axis compression force are 2KN and 8KN respectively, and the higher and lower performance limits of tibia index are 0.4 and 1.3. The lower performance limit and the higher performance limit correspond to 0 point and 2 points respectively. Where a measurement value falls between the two limits, the score is calculated by linear interpolation.

For the lower leg scoring in frontal impact test with 40% overlapping, the assessment parameters and method are similar to those above-mentioned, but the maximum and minimum scores for lower leg are 4 points and 0 point respectively, and the scores for lower leg of the driver-side dummy will be subject to modification with 0~-1 for the excessive rearward and upward displacements of pedals. In the following are the details: if the rearward displacement of pedals is not in excess of 100 mm, there is no penalty. In the case of the displacement being up to 200 mm, there is a penalty of 1 point; if the upward displacement of pedals is not in excess of 72 mm, there is no penalty. In the case of the displacement being up to 88 mm, there is a penalty of 1 point; so the score for the driver-side dummy's lower leg is generated by measuring relevant parameters, then subtracting the penalties resulted from rearward displacement and upward displacement of the pedals.

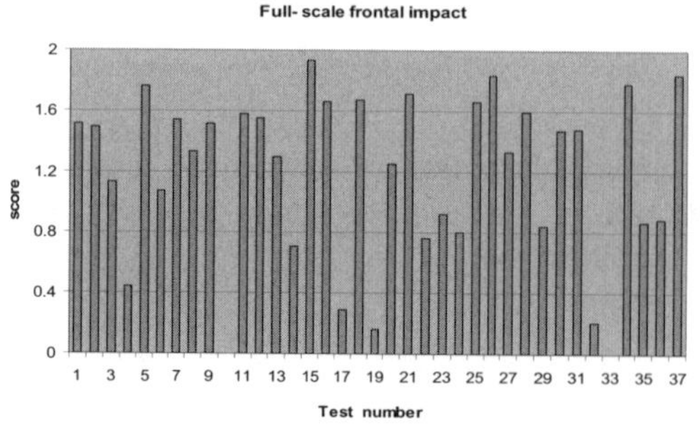

Fig. 2: Scores of 37 full frontal crash tests

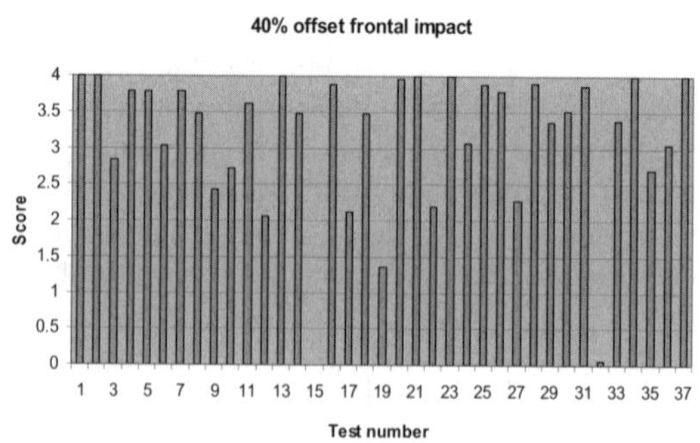

Fig. 3: Scores of 37 40% offset crash tests

Statistical characteristics of lower leg score

Analysis of overall scoring of two crash configurations. The test data of 37 groups of C-NCAP tests were investigated, including full scale frontal impact and 40% offset frontal impact. The scores of lower leg are shown in Fig. 2 and Fig. 3 respectively, and it was easily found from Fig. 2 that no vehicle model in 37 full-scale frontal impact tests got the full marks (2 points), and the average score of lower leg was only 1.18 points, which meant that the scoring rate of lower leg was only 59%. Comparatively speaking, from Fig. 3 it can get much more score and scoring rate in 40% offset frontal impact tests, and their averages which are 2.93 points and 73.25%. The main reason was the difference of energy absorption in these two crash configurations. In the full frontal crash test, the test vehicle frontally impact against a fixed rigid barrier; the impact energy was absorbed only by the test vehicle. But in 40% offset frontal impact test, the test vehicle impact against a deformable barrier, and the impact energy was absorbed by the deformable barrier and the test vehicle together. Therefore the injury to occupants in offset frontal impact test was alleviated.

Comparisons of scores in different brands and category.

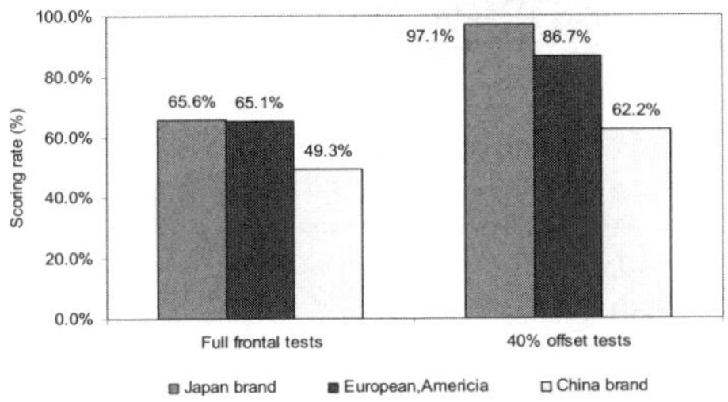

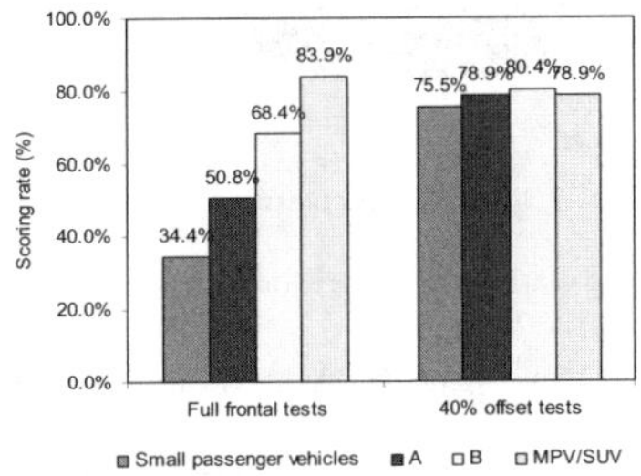

Fig. 4: Comparison of score in different

Fig. 5: Comparison of score in different

These test vehicles are classed as China brand, European and America brand, and Japan brand. European brand and America brand are thought of as one brand because of Chinese custom. The scoring rates of three brands were shown in Fig. 4. It showed that Japan brand ranked first in both of crash configurations, the scoring rates were 65.6% and 97.1%, and European and America brand ranked second, the scoring rates were 65.1% and 86.7%, whereas the scoring rates of China brand were the lowest, which were 49.3% and 62.2% respectively. It showed that there existed a sizable gap in the technology of body design and constrained system matching for these vehicles. Furthermore, one of reason that Japan brand got better scores than European and America brand is that J-NCAP has more similar test items to C-NCAP than US-NCAP and EURO-NCAP have.

Moreover, according to the vehicle classification of C-NCAP, these test vehicles are classed as small passenger car, category A passenger car, category B passenger car and MPV & SUV. Fig. 5 showed the score comparison of different categories vehicle. In the full frontal crash tests, the ranking of the scoring rate from high to low were SUV&MPV, category B, category A and small passenger car. The scoring rate was distributed from 49.3% to 65.6%, and the reason was that the bigger of the vehicle, the more space can be used to deformation, and therefore the impact energy could be absorbed commendably. In the 40% offset frontal impact crash tests, there were no evident differences in different categories; the scoring rate was distributed from 75.5% to 80.4%. But the scoring rate of small passenger car was still the lowest, because it was relatively difficult to realize the layout of crashworthiness structure due to its smaller space.

Analysis of different injury parts. Statistic analysis of the relation of different injury parts to final scores of lower leg were performed for two crash configurations and the results were shown in Fig. 6. From Fig. 6 it was significantly found in full frontal crash tests that the right leg of driver-side dummy and the left leg of passenger-side dummy were easier to be injured, and it accounted for 49% and 38%

respectively. The main reason was that driver's right foot was placed on acceleration pedal; the support reaction to the foot would be increased because of the deformation of vehicle during the crash process. Furthermore, the driver's right leg and passenger's left leg was more near to dashboard than other parts because of the deformation of center console.

In 40% offset crash tests, legs of driver-side dummy had more chance to be seriously injured than those of passenger-side dummy. And the injury ratio was 39% for left leg and 54% for right leg from Fig. 6. We thought there were two reason as follows: one was that the driver side was impact side, so the impact energy was mainly absorbed by this side so as to be seriously injured to driver-side dummy; the other reason was that offset crash test had higher impact velocity, thus it produced more impact energy to lead to more intrusion of driver side, which result in increase of the support reaction of pedal to right foot.

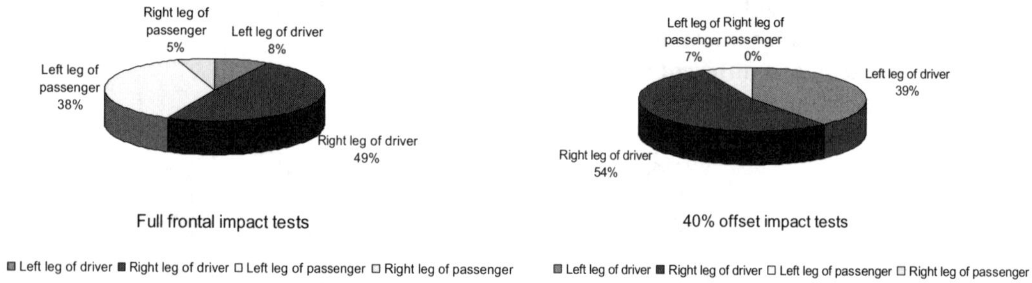

Fig. 6: Analysis of injury parts on two crash configurations

Influence of constrained system on scoring. WIC was firstly proposed in 1990 by Viano and Arepally [8], who were from General Motors Corporation. They adopted a weighting method to calculate a normalized value of injury assessment with HIC, chest resultant acceleration 3ms exceedence, chest compression deformation and femur load, and then used it to assess the protection performance of restraint system to occupants. And the smaller of the WIC value, the better of the performance of restraint system. It was defined as Eq. 2:

$$WIC = 0.6(\frac{HIC_{36}}{1000}) + 0.35(\frac{a_{3ms}}{60} + \frac{D_{fel}}{0.0762})/2.0 + 0.05(F_L^{femur} + F_R^{femur})/20 \tag{2}$$

Where HIC_{36} is the head injury criterion, a_{3ms} is the chest resultant acceleration 3ms exceedence, D_{fel} is the chest compression deformation and F_L^{femur}, F_R^{femur} stand for the axial forces of left and right femur.

Fig. 7 and Fig. 8 were the plots of WIC index and lower leg scores in full frontal and offset crash tests, respectively. From the trend lines of scores and WIC value, we found that the lower leg score would increase with the decline of the WIC value in two frontal crash configurations. The reason was as follows: the smaller of the WIC values, the better of protection performance of restraint system to occupants: therefore the restraining actions of safety belt and airbag to occupants were obvious; sequentially the probability of secondary impact between lower leg and vehicle was decreased.

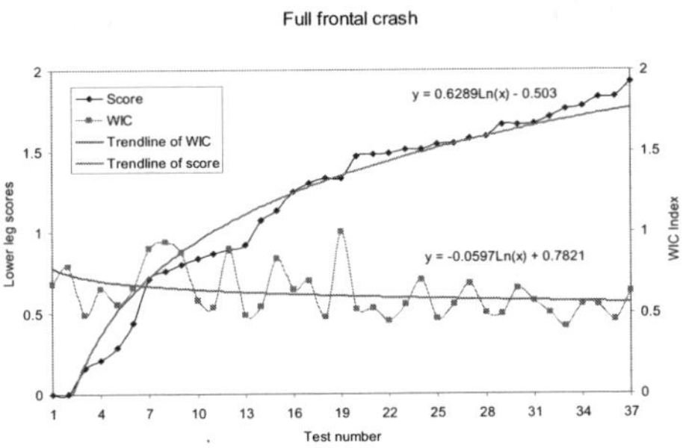

Fig. 7: Scores and WIC in full frontal crash tests

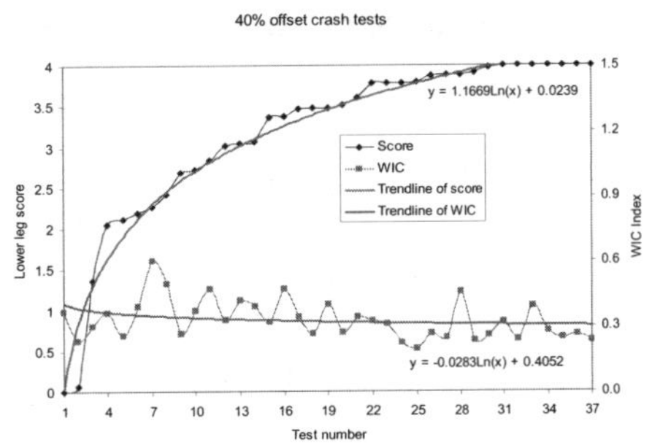

Fig. 8: Scores and WIC in 40% offset frontal crash tests

Summary

In this paper, the scores of lower legs in 37 groups C-NCAP frontal crash tests were statistically analyzed, and the effect rules of characteristic parameters on injury of lower leg were summarized as follows:

The scoring rate in full frontal crash tests was lower than that in offset crash tests.

Furthermore, there exists a sizable gap in the technology of vehicle safety between China brand, European and America brand, and Japan brand.

Thirdly, the dimension of vehicle would greatly affect the scoring rate of lower leg in full frontal crash tests. And there were no evident differences in 40% offset frontal crash tests.

Fourthly, there have different distributions of injury parts in different crash configurations. In full frontal crash tests the serious injuries concentrated in the middle of front seats that were right leg of driver and left leg of passenger. In offset crash tests the injury parts concentrated in driver side.

Finally a good occupant's restraint system would be helpful to get the good points.

References

[1] James Hurnall, Chris Coxon: *Influence of a passenger dummy on consumer rating in offset frontal NCAP crash testing*. SAE Paper No. 2001-06-0144.

[2] Kenichi Ando: *JNCAP: Developing overall rating protocol*. SAE Paper No. 2001-06-0156.

[3] Zhiqing Cheng: *Optimal Crash Pulse for Minimization of Peak Occupant Deceleration in Frontal Impact*. SAE Paper No. 2006-01-0670.

[4] Lawrence L. Hershman: *The U.S. new car assessment program (NCAP): Past, present and future.* SAE Paper No. 2001-06-0245.

[5] *C-NCAP Management Regulation*. China automotive technology and research center.

[6] Shashi Kuppa, Jing Wang: *Lower Extremity Injuries and Associated Injury Criteria*. SAE Paper No. 2001-06-0160.

[7] Wu Jianping, Guy Nusholtz and Sukhbir Bilkhu: *Optimization of Vehicle Crash Pulses in Relative Displacement Domain*. International Journal of Crashworthiness, Vol.7, No. 4, 2002.

[8] Yi L, Hailin L, Junyuan Z: *A new optimization approach for the driver restraint system for the frontal impact of a mini-bus*, International Journal of Vehicle Systems Modeling and Testing, Vol. 3, No.1-2 / 2008.

Advanced Materials Research Vol. 279 (2011) pp 406-411
© (2011) Trans Tech Publications, Switzerland
doi:10.4028/www.scientific.net/AMR.279.406

A Feature Recognition Approach with Fuzzy ART Neural Network

Cong Lu[1,2, a], Jun Zha[1,b]

[1]School of Mechatronics Engineering, University of Electronic Science and Technology of China, Chengdu, China

[2]State Key Laboratory for Manufacturing Systems Engineering, Xi'an Jiaotong University, Xi'an, China

[a]conglu@uestc.edu.cn, [b]zhajun315@126.com

Keywords: feature recognition, Fuzzy ART, neural network

Abstract. This paper proposes a feature recognition approach from a boundary representation solid model with Fuzzy ART neural network. To recognize the feature efficiently, some key technologies in Fuzzy ART neural network are used. The influence of the vigilance parameter on feature recognition is studied, and two learning modes, fast learning and slow learning are adopted and compared in feature recognition. Finally, a case study is given to verify the proposed approach.

Introduction

Feature recognition can extract the specific feature information from the components of solid models, so it can be used as the intelligent interface between CAD and CAPP. In recent years, neural network has become an important approach to feature recognition. Hwang and Henderson [1] firstly proposed a neural network approach to feature recognition. In this approach, a single-layer network was used, and it can recognize some common features, such as pockets and slots, but it failed to recognize more complex features, such as cross-slots. Prabhakar and Henderson [2] proposed an approach to translate the B-representation of a solid into face value vectors that can be input to neural network. But the network needs to be made more robust to feature definitions, so that the performance of feature recognition is not highly dependent on feature definitions. Onwubolu [3] used the BP neural networks for feature recognition, and the face complexity code is used to represent the manufacturing features. However, due to computational complexity of the recognition problem, much time and computational resource were required to train the network, which affects the efficiency of the recognition process. Chakraborty and Basu [4] used binary strings for defining engineering features, such as slots, steps and circular pockets. However, this approach was limited to other features. Kishore et al. [5] applied the adaptive resonance theory (ART) to solve the problem of feature recognition from boundary representation solid model. The ART could preserve past learning stably and it has good adaptability.

Although the above mentioned works have made some success in feature recognition, none of them is powerful enough, and they cannot study and recognize the feature effectively when it is in the test sample and is obviously different with those features in the training sample, which is called rare feature in this work. As shown in Fig. 1, the training sample contains four kinds of features and the test sample contains the cross-slot feature, but the cross-slot is obviously different with those features contained in the training sample, so the cross-slot is a rare feature in this case.

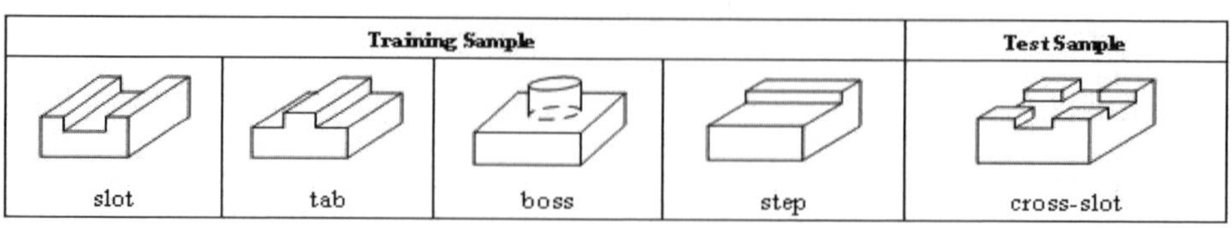

Fig. 1: A case of rare feature

To address the above problem, in this work, a fuzzy ART neural network is proposed to realize the efficient recognition of the rare feature.

Fuzzy ART Neural Network for Rare Feature Recognition

A fuzzy adaptive resonance theory (ART) neural network consists of two subsystems (show in Fig. 2 [6]): attentional subsystem and orienting subsystem. The attentional subsystem is a two-layer network architecture, where F_1 is the input layer to receive the input fuzzy patterns which is computed in F_0 layer (the components of F_0 layer are a_i, a_i is the normalized face value in this work) for transforming the input patterns, F_2 is the pattern expressing layer, and all fuzzy patterns in F_2 layer constitute a classification of the input patterns. The orienting subsystem includes a reset unit, which receives all information coming from F_1 layer, F_2 layer and F_0 layer. Let the connection weight between the node j in F_2 layer and the node i in F_1 layer be w_{ji}, and suppose the initial values of the connection weight w_{ji} as $w_{ji}(0)=1$ $(i=1,\cdots,2M; j=1,\cdots,N)$ and each node j in F_2 layer is uncommitted. Fuzzy ART has three parameters: (1) the choice parameter α, $\alpha>0$; (2) the learning parameter β, $\beta\in(0,1]$; (3) the vigilance parameter ρ, $\rho\in[0,1]$.

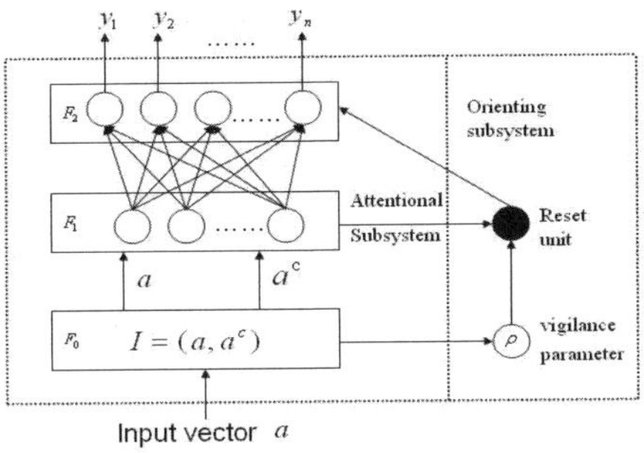

Fig. 2: Typical Fuzzy ART neural network

Complement Coding to Preprocess the Input Vector. In this work, when using fuzzy ART neural network for feature recognition, the input parameter is a matrix, where each row represents a normalized n-element face score vector and each column represents a feature. The face scores are calculated based on the geometric and topology information of a feature. In order to avoid the category proliferation problem that the input patterns cannot be accurately classified in feature recognition, the input vector (face score vector) should be preprocessed through complement coding, and the complement coding can be achieved by normalizing each input vector *a*, setting

$$I_0 = a/|a| \tag{1}$$

where the norm |.| is defined by

$$|a| = \sum_{i=1}^{M} |a_i| \tag{2}$$

When the normalized vector *a* (*M*-dimensional vector) is imported to the F_0 layer, the output of F_0, a $2M$-dimensional vector I can be derived as follows:

$$I=(a,a^c)=(a_1,\cdots,a_M,a_1^c,\cdots,a_M^c) \tag{3}$$

where, $a_i^c=1-a_i$, then

$$|I|=|(a,a^c)|=\sum_{i=1}^{M}a_i+(M-\sum_{i=1}^{M}a_i)=M \tag{4}$$

The above process is called complement coding, by which the fuzzy ART neural network is able to achieve good categorization of input patterns even if input vectors have not the same norm.

Category Choice and Resonance Reset. After the preprocessing, if the normalized face score vector I is inputted to F_1 layer, each node in F_2 layer has an output. For node *j*, the output value T_j is defined by the choice function as Eq. 5:

$$T_j = \frac{|I \wedge w_j|}{\alpha + |w_j|} \quad (5)$$

The T_j with the maximal value in F_2 layer will be chosen, and if more than one T_j is maximal, the node with smallest index is chosen. The $w_j = (w_{j,1}, \cdots, w_{j,2M})$ is adaptive weights of category j. Where, α is a choice parameter $(\alpha > 0)$. If $\alpha \geq |w_j|$ for node j ($j = 1, 2, \cdots, M$), the effect of $|w_j|$ in Eq. 5 can be neglected. So, the value of α should be the smaller the better, and α is usually suggested to be close to zero. In Eq. 5, the fuzzy intersection operation is defined by $(x \wedge y)_i = \min(x_i, y_i)$, If x represents one of the elements in I, then y is the corresponding element in w_j, the result of this fuzzy intersection operation is the smaller one of the element in I and the element in w_j.

The above process is called category choice. After category choice, the node in F_2 layer in fuzzy ART neural network receives the maximal value of T_j which represents the category of the features. If the match function meets the vigilance criterion as Eq. 6:

$$\frac{|I \wedge w_j|}{|I|} \geq \rho \quad (6)$$

where, ρ is the vigilance parameter, then the resonance occurs, and the learning occurs consequently.

In another condition, if Eq. 6 is not satisfied, then reset occurs and the value of the choice function is reset to 0 for the duration of input vector presentation to prevent its selection during the search. When a new node is chosen, the choice process continues until the node *j* in F_2 layer satisfies Eq. 6.

The vigilance parameter ρ allows user to control the degree of similarity of patterns placed in the same category. So, the influence of different vigilance parameters will be tested in this work.

Learning. To realize the recognition of features in the training sample, the network should have the ability of learning. If the node *j* satisfies the Eq. 6, the learning process starts and the weight vector will be updated as Eq. 7:

$$w_j^{(new)} = \beta(I \wedge w_j^{(old)}) + (1-\beta)w_j^{(old)} \quad (7)$$

where, β is a learning rate parameter and $\beta \in (0,1]$. In this work, the following two learning methods will be adopted and tested: (1) fast learning, where β is always equal to 1. Fast learning enables a system to adapt quickly to the inputs that occur only rarely and require immediate accurate performance. (2) slow learning, where β is smaller than 1, and with lower learning rates, neural network converge more slowly.

Overall Fuzzy ART Neural Network Algorithm for Rare Feature Recognition. The overall fuzzy ART neural network algorithm for rare feature recognition is proposed as follows:

Algorithm:

Step_1: Set the maximum iteration number *k*(max), choice parameter α and vigilance parameter ρ;

Step_2: Input the face score vector of features, derive the normalized vector using Eq. 1 and Eq. 2;

Step_3: Set all initial weights $w_{ji} = 1$;

Step_4: Complement coding the input vector, derive the 2M-dimensional vector using Eq. 3;

Step_5: Calculate T_j using Eq. 5, choose T_j that has the maximal value in F_2 layer, if more than one T_j is maximal, then choose the node with smallest index.

Step_6: If the node j in F_2 layer satisfies Eq. 6, go to Step 9; else go to Step 7;

Step_7: Reset occurs, and reset the choice function T_j to zero;

Step_8: Let $k = k+1$, if $k < k(\max)$, go to Step 4; else go to Step 10;

Step_9: Recognize the features in the training sample, and then update the weight vector of the network using Eq. 7;

Step_10: End the training process, and categorize the features in the training sample;

Step_11: Import the test sample to the network, if the network has an output value -1, go to Step 13; else go to Step12;

Step_12: Categorizes the test sample into one same category in the training sample;

Step_13: Creates a new category for the test sample.

Application of Fuzzy ART Neural Network for Feature Recognition

System Implementation. The proposed fuzzy ART neural network algorithm was coded in MATLAB R2008a. In this work, five features - slot, tab, boss, step and cross-slot are used, where the first four features are the training sample and the fifth feature is the test sample. The face score vector of the features and input vectors of the network are shown in Table 1 and Table 2, respectively.

Table 1: Face scores vector of five features

Feature	Face score vector
 slot	(1.5 1.5 1.0 1.25 0.5 1.25 1.0 1.5 1.5)
 tab	(1.5 1.0 1.0 1.25 1.5 1.25 1.0 1.0 1.5)
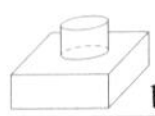 boss	(1.5 1.5 1.5 0.5 2.0 0.0 1.5 1.5 1.5)
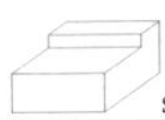 step	(1.5 1.5 1.33 1.5 1.0 1.5 1.33 1.5 1.5)
 cross-slot	(0.75 0.75 1.25 1.25 0.16 1.25 1.25 0.75 0.75)

Table 2: The input vector of five features

Feature	Input vector
Slot	(0.1364,0.1364,0.0909,0.1136,0.0455,0.1136,0.0909,0.1364,0.1364)
Tab	(0.1364,0.0909,0.0909,0.1136,0.1364,0.1136,0.0909,0.0909,0.1364)
Boss	(0.1304,0.1304,0.1304,0.0435,0.1739,0,0.1304,0.1304,0.1304)
Step	(0.1185, 0.1185, 0.1051, 0.1185, 0.0790, 0.1185, 0.1051, 0.1185, 0.1185)
Cross-slot	(0.0919, 0.0919, 0.1532, 0.1532, 0.0196, 0.1532, 0.1532, 0.0919, 0.0919)

Experimentation and Discussion. In the experiment, two learning methods, fast learning and slow learning, are adopted to verify the proposed approach to recognizing the rare features.

(1) Fast learning

In fast learning, set the choice parameter $\alpha = 0.1$.

Test 1: Set the learning rate $\beta = 1$, set the vigilance parameter ρ with different values, as shown in Table 3. The results show that when ρ=0.75, ρ=0.85, ρ=0.90, ρ=0.91, respectively, the Fuzzy ART neural network cannot categorize the training sample (slot, tab, boss, step) and the test sample (cross-slot), instead, the network clustered the five features together; when ρ=0.92, ρ=0.93, ρ=0.94, respectively, the network clustered the four features together, but it can recognize the cross-slot as a new feature; when ρ=0.95, ρ=0.96, ρ=0.97, respectively, the network can recognize the feature of boss, and the network recognized the cross-slot as a new feature; when ρ =0.98, the network created a new cluster for the tab; when ρ =0.985, the results show that four different clusters were created, and a new cluster was created as the new feature cross-slot. The above test results are shown in Table 3. The number 1 to 4 represents different features, respectively, and -1 indicates that the test sample is a rare feature.

Table 3: The Results of Fast Learning

Feature	$\beta=1$, $\alpha=0.1$											
	ρ =0.75	ρ =0.85	ρ =0.90	ρ =0.91	ρ =0.92	ρ =0.93	ρ =0.94	ρ =0.95	ρ =0.96	ρ =0.97	ρ =0.98	ρ =0.985
Slot	1	1	1	1	1	1	1	1	1	1	1	1
Tab	1	1	1	1	1	1	1	1	1	1	2	2
Boss	1	1	1	1	1	1	1	2	2	2	3	3
Step	1	1	1	1	1	1	1	1	1	1	1	4
Cross-slot	1	1	1	1	-1	-1	-1	-1	-1	-1	-1	-1

(2) Slow learning

In slow learning, set the choice parameter $\alpha = 0.1$.

Test 2: Set the learning rate β=0.5, set the vigilance parameter ρ with different values, the test results are shown in Table 4.

Test 3: Set the learning rate β=0.25, set the vigilance parameter ρ with different values, the test results are shown in Table 5.

Test 4: Set the learning rate β=0.1, set the vigilance parameter ρ with different values, the test results are shown in Table 6.

The results show that the system can also recognize the features (slot, tab, boss, step, cross-slot) when the value of β is different. However, the program running time increased when the learning rate decreased, as shown in Table 7, and it costs more time to recognize the rare feature than fast learning, which could be the reason that the network converge more slowly when the learning rate is lower.

Table 4: The results of slow learning ($\beta=0.5$)

Feature	$\beta=0.5$, $\alpha=0.1$											
	ρ =0.75	ρ =0.85	ρ =0.90	ρ =0.91	ρ =0.92	ρ =0.93	ρ =0.94	ρ =0.95	ρ =0.96	ρ =0.97	ρ =0.98	ρ =0.985
Slot	1	1	1	1	1	1	1	3	3	3	3	4
Tab	1	1	1	1	1	1	1	3	2	2	4	2
Boss	1	1	1	1	1	1	1	2	4	4	2	3
Step	1	1	1	1	1	1	1	1	1	1	1	1
Cross-slot	1	1	1	1	-1	-1	-1	-1	-1	-1	-1	-1

Table 5: The results of slow learning ($\beta=0.25$)

Feature	$\beta=0.25$, $\alpha=0.1$											
	ρ =0.75	ρ =0.85	ρ =0.90	ρ =0.91	ρ =0.92	ρ =0.93	ρ =0.94	ρ =0.95	ρ =0.96	ρ =0.97	ρ =0.98	ρ =0.985
Slot	1	1	1	1	1	1	1	1	3	4	4	3
Tab	1	1	1	1	1	1	1	1	3	2	2	2
Boss	1	1	1	1	1	1	1	2	2	3	3	4
Step	1	1	1	1	1	1	1	1	1	1	1	1
Cross-slot	1	1	1	1	-1	-1	-1	-1	-1	-1	-1	-1

Table 6: The results of slow learning($\beta=0.1$)

Feature	$\beta=0.1$, $\alpha=0.1$											
	ρ =0.75	ρ =0.85	ρ =0.90	ρ =0.91	ρ =0.92	ρ =0.93	ρ =0.94	ρ =0.95	ρ =0.96	ρ =0.97	ρ =0.98	ρ =0.985
Slot	1	1	1	1	1	1	1	1	1	3	4	4
Tab	1	1	1	1	1	1	1	1	1	3	2	2
Boss	1	1	1	1	1	1	1	2	2	2	3	3
Step	1	1	1	1	1	1	1	1	1	1	1	1
Cross-slot	1	1	1	1	-1	-1	-1	-1	-1	-1	-1	-1

Table 7: Program running time of the algorithms

Algorithm	Execution time (S)
Fast learning ($\beta=1$)	0.005353
Slow learning ($\beta=0.5$)	0.042134
Slow learning ($\beta=0.25$)	0.048680
Slow learning ($\beta=0.1$)	0.049802

The above tests show that fast learning is faster than slow learning in recognizing rare features, and with a higher vigilance parameter ρ, more accurate categories can be derived. So, the proposed fuzzy ART neural network approach with fast learning mode can be efficient to recognize the rare feature.

Conclusions

This paper proposes a fuzzy ART neural network approach to feature recognition. With some key technologies including the complement coding, category choice, resonance and reset, the influence of the vigilance parameter on rare feature recognition is investigated, and the fast learning is tested and verified to be effective learning method to recognize the rare feature with short learning time. So, the proposed fuzzy ART neural network approach can be used for efficient integration of CAD and CAM.

Acknowledgements

This work is sponsored by the Research Foundation from State Key Laboratory for Manufacturing Systems Engineering (Xi'an Jiaotong University); and the Fundamental Research Funds for the Central Universities.

References

[1] J. L. Hwang, M. R. Henderson: *Applying the perceptron to three-dimensional feature recognition.* Journal of Design and Manufacturing, 2, 4(1992), pp. 87-198.

[2] S. Prabhakar, M. R. Henderson: *Automatic form feature recognition using neural network based techniques on boundary representations of solid models*. Computer Aided Design, 24, 7(1992), pp. 381-393.

[3] G. C. Onwubolu: *Manufacturing features recognition using back propagation neural networks.* Journal of Intelligent Manufacturing, 10, 3-4(1999), pp. 289-299.

[4] S. Chakraborty, A. Basu: *Retrieval of machining information from feature patterns using artificial neural networks*. The International Journal of Advanced Manufacturing Technology, 27, 7-8 (2006), pp. 781-787.

[5] L. Kishore, C. Subrata, T. C. Chang: *Feature recognition using ART2: a self-organizing neural network.* Journal of Intelligent Manufacturing, 8, 3(1997), pp. 203-214.

[6] P. Y. Liu, H. X. Li: *Fuzzy neural network theory and application.* Series in machine perception artifical intelligence, World Scientifis, Vol. 59 (2004), pp. 53-60.

Advanced Materials Research Vol. 279 (2011) pp 412-417
© (2011) Trans Tech Publications, Switzerland
doi:10.4028/www.scientific.net/AMR.279.412

Sequencing Mixed Model Assembly Lines based on Genetic Algorithm Optimization

Yindi HUANG[1,a], Ronghua BIAN[1,b] and Zhen XU[1,c]

[1]Department of Automobile and Transportation Engineering, Nanjing Forestry University, Longpan Load 159, 210037, Nanjing, China

[a]huangyindi321@126.com, [b]xinqingface@126.com, [c]foreveropq@yahoo.com.cn

Keywords: Sequencing, Mixed Model Assembly, Genetic Algorithm, Optimization

Abstract. Firstly, genetic algorithm optimization method was put forward to solve optimization problems of sequencing in actual mixed model passenger car factory assembly line. Then genetic algorithm optimization methods and procedures of mixed model total assembly line production sequencing were studied. Finally, according to the characteristics of the mixed production line, the least total waiting time of vehicle was set to complete assembly as a sequencing optimization objective, the AutoMod simulation software was used to balance the allocation of resources and optimize the sequence by way of genetic algorithm, and the optimal sequencing solution was obtained. After optimization and balance, the total waiting time of mixed model vehicles in complete assembly line was reduced by 69.5%, which improves production efficiency greatly. This also proves the effectiveness of genetic algorithm optimization.

Introduction

With the increasing needs of the individual vehicles, automobile enterprises have to adopt the production mode of small batch products and mixed model of many varieties. In this way, the balancing and sequencing optimization of interflow assembly line are difficult problems. The characteristics of production in different enterprises determine different optimization decision objective. Optimization methods are continuously developed and improved, so the result of sequencing is also different [1].

There are many algorithms to solve scheduling problems of mixed model assembly line of production. For the production order problems, if objective function is different, by using different algorithms, the effect is not the same. The paper [2] was to balance the utilization of parts and assembly time, by using tracing algorithm, tabu search algorithm and artificial neural networks in these three different ways to solve this problem. Also there are methods using artificial intelligence [3]. Papers [4] used integer programming algorithm and heuristic algorithm to solve the production scheduling problem by setting the equilibrium rate of product output as the objective function.

To sum up these algorithms, they are respectively optimal algorithms, test algorithm and improved algorithm cycle[5].

Optimal algorithms include branch and bound method, dynamic rules method, and integer programming, and they can obtain the optimal solution of scheduling problems only in small-scale production.

In test algorithm, each input point is determined in accordance with the current state of the input current models. For the objective function as "a period to minimize the waiting time for all vehicles" the production scheduling problem, if the assembly line for more than one job field, time at each input point, its current status do not reflect the state of the entire assembly line, therefore it is difficult to ensure the test algorithm produced better approximate optimal solution.

Genetic algorithm is a kind of improved circulation algorithm. Compared with above algorithms, it has a high efficiency in the global optimization, and its established principle is based on the natural selection of biological evolution mechanisms and groups evolution.

Based on the above analysis, we use genetic algorithm to solve mixed models assembly line sequencing problems.

Taking small passenger car production enterprises as an example, the genetic algorithm optimization methods were studied to optimize the production scheduling of total passenger car assembly line in the actual factory based on the load balancing in the production line. Firstly, genetic algorithm optimization method was studied for total interflow assembly line production sequencing. Then the realistic goal for production sequencing optimization was set according to the characteristics of the mixed production line, and this is the least total waiting time of vehicle to complete assembly. Finally, production sequencing was optimized, so as to get the best programs to improve production efficiency.

Optimization problems of sequencing in actual mixed model passenger car factory assembly line

The focus of optimization sequencing in actual mixed model passenger car factory assembly line is to establish an optimization objective.

There are different targets for mixed model assembly line scheduling problem according to different business requirements and production management strategies [6].

This paper studies the general assembly line of passenger car, and the ultimate goal is to minimize the total waiting time online in a batch system. It indicate that the smaller the line waiting time, the smaller the total amount of time in the assembly line of the batch system, so enterprises can deliver faster and better products which ordered by users.

The objective function and constraint condition are expressed as follows.

$$g(x) = \min\left(\sum_{i=2}^{D'}\sum_{j=2}^{N} waitT_{ji} + \sum_{j=1}^{N} waitT_{j1} + \sum_{i=2}^{D'} waitT_{1i}\right) \quad (1)$$

$$\sum_{m=1}^{M} x_{mi} = 1 \quad (2)$$

Where i: production sequence number; j: Station code; D':The number of products in the production cycle time;N: The total number of stations in general assembly line ; $waitT_{ji}$: the waiting time before processing the i production and j station; x_{mi}: the i product is m-type product.

Genetic algorithm optimization methods of mixed models general assembly line sequencing

Scheduling optimization goal is to put into the production line of vehicles waiting time of ordering to a minimum. According to petri net model [7] built before, genetic algorithm optimized sequencing of the total assembly line.

Genetic algorithm uses group of global optimization search techniques. The search process mimics biological evolution, and it evaluates its initial individual fitness by using genetic operators that acting on the group.

Through use of selection, crossover and mutation and a series of genetic manipulation, it produces new generation of groups, and evaluates its individual. Then a new generation of community groups replaces the original one. Finally it repeats the previous generation until satisfactory results and iterative process[8]. Optimization process is as follows:

Parameter coding. There are two encoding ways of character encoding and real coding combined. *n* is the real-coded number which is the products number in the total assembly line in a cycle needed. All model types are codes as character encoding.

Different characters (such as *A, B, C*, etc.) represent different types of vehicles, real number *n* represent number of products. The number of vehicles of *A* class of model is a, and the number of *B* class vehicle's model is *b*, which means that the corresponding real number *1 ~ a* represent *A* class passenger cars and the real number *(a +1)* to *b* represent *B* class. So the above combination of the two encoding makes it easier to control crossover operation and mutation operation.

The fitness function. The fitness function is: $f(x)=c_{max}+100-g(x)$, where *f(x)* is the fitness function, *g(x)*is the objective function, 100 of the buffer value and it can guarantee that fitness function is non-negative. c_{max} can be an input value or the theoretical maximum, or can be the current all generation or the maximum of *g(x)* in most recent *K* generations. At this time c_{max} will change with the generation. It can be seen that the fitness function and objective function exists inverse relationship. The larger value of the fitness function and the higher degree of individual adaptation, the smaller value of the objective function. Accordingly, wait time of all vehicles on the assembly line within a planning cycle is minimum.

Selection mechanism. Selecting or copying operation is aimed at selecting good individuals from the current groups so that they have the opportunity to breed the next generation as a parent from children to grandchildren. The criteria for judging the good individual whether or not is their fitness value.

In this paper, the method is roulette (roulette wheel) way. First, the fitness value for each individual is calculated, and then proportion of the fitness value with the sum of the fitness value in the population is calculated. The proportion indicates the selected probability of the individual in the selection process. Selection process reflects the idea of the fittest survival in the process of biological evolution. It ensures that the individual who has good genes goes to the next generation.

Firstly, a group of initial solutions is generated randomly. With the implementation of genetic algorithms, M optimum individuals are kept as a sample group and to be selected. In the course of each generation computing, the probability of selected individuals is proportion to their relative fitness in the population. Then the individual's fitness or average fitness are compared and the larger the fitness, the larger the probability of being selected.

Genetic Operation.

1) Crossover

Step 1: Choose a mating area in string randomly, such as the two parent strings and the mate regional selected randomly as: Chromcode1 = 1 2 3 | 4 5 6...(n-4) | (n-3) (n-2) (n-1) n
Chromcode2 = n (n-1) | (n-2) (n-3) (n-4)...6 5 | 4 3 2 1

Step 2: Add Chromcode2 mating area in front of Chromcode1, then add Chromcode1 mating area in front of Chromcode2 , and obtains Chromcode1' and Chromcode2' as follows:

Chromcode1' = (n-2) (n-3) (n-4)...6 5 | 1 2 3 4 5 6... (n-3) (n-2) (n-1) n
Chromcode2' = 4 5 6... (n-4) | n (n-1) (n-2) (n-3) (n-4) (n-5)...6 5 4 3 2 1

Step3: In the Chromcode1 delete the same product code as in the mating area in turn to get the final two sub-strings:

Chromcode1''= (n-2) (n-3) (n-4) ...6 5 | 1 2 3 4 (n-1)n
Chromcode2''=4 5 6... (n-4) | n (n-1) (n-2) (n-3) 3 2 1

2) Mutation Technique

Since the selection mechanism is to keep samples, in order to maintain the diversity of individuals within populations, we adopt exchange mutation technique repeatedly. This method can make the feasible solution order change greatly. Exchanging variation is to change their values (or code) in the two choice strings, and at this time character code was used. For string Chrom1,

Chrom1 = A B A B | E G D C C ... A B D | C D (Total n)

If the change points are B, D, then after the exchange, Chrom1 is

Chrom1 = A B A D | E G D C C... A B B | C D

After each genetic operation, the offspring chromosome fitness are obtained and chromosome fitness of the offspring with the parent are compared. If the fitness of offspring chromosome is higher than the parent, the parent chromosome with the offspring chromosome is replaced; otherwise the parent chromosomes are reserved.

Algorithm flow. Algorithm flow chart is shown in Fig. 1.

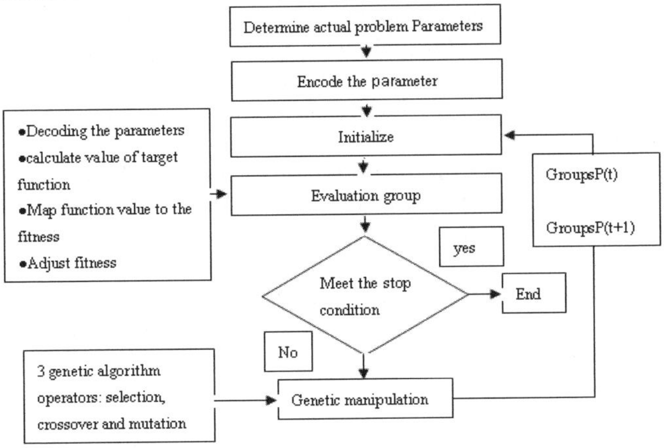

Fig. 1: Flow chart of genetic algorithm

Optimization results and analysis

We have got resource utilization of the process in the long run of mixed assembly line in the simulation software AutoMod environment [7]. We compare factory original plan with the new plan which are optimized through algorithm, and the impact on production schedule are found. The optimization objective is to minimum wait time of all vehicles within a period.

Sequencing production optimization under the existing resource allocation for mixed model assembly line.

Sequencing production under the original resources allocation. After programming and the optimization procedure, by entering the number of different models respectively, the program can find the common denominator automatically, that is, the number of cycles.

We take No. 20090801 production plan as an example to generate a production schedule. The numbers of vehicles are: A: 12, B: 28, C: 8, D: 2, E: 18, F: 7, G: 4. The actual production order show as follows, and the total waiting time are 37,843 minutes.

FFACCAABBBBBBCEBBBBBBBFBBBBBBBGGGGCADBAAEEABBFCEEFEEEBBCCBBBAFF AACDEEEEAAEEEEEE

Sequencing production genetic optimization under the original resources allocation. Sequencing production of final assembly of the vehicles is optimized by way of Genetic Algorithm under the current resource allocation. According to the number of vehicles we can see, just a cycle of optimization can be carried out. Sequencing of the optimized and the total waiting time are shown as below:

Statistics results: Maximum fitness is equal to: 63533; Minimum function value is equal to: 15964
The chromosome is encoded as:
ABBCBBCABABAABBABCBBBABAACGBDBAAABBCBBBBBBBBFDBFFGBAFCEGEEEBEGEE CEFEEGEEBFEEGEE

Under the current resource allocation, the total vehicle waiting time required to complete assembly of the actual production is 37,843 minutes, which is reduced to optimized 15,964 minutes, and the total waiting time is reduced by more than a half. Production optimization sequence is the chromosome code, and it is the company's production schedule.

Sequencing production optimization under the balanced resource allocation for mixed model assembly line. From the AutoMod simulation results under the original resources allocation [7], we know that there are some unreasonable in the allocation of original resources. After consultation and discussion with the craft workshop director, we adjusted allocation of resources of some stations and then resource utilization of each station is more balanced.

Sequencing production under the balanced allocation of resources. Under the condition of more balanced allocation of resources, the total vehicle waiting time to complete assembly is 22,209 minutes. The waiting time is shortened a lot than the original. However, it is still more than the time of the optimization of the original resource allocation.

Sequencing production genetic optimization under the more balanced allocation of resources. We have optimized the station more balance for the total assembly line by applying AutoMod simulation software. On this basis, we further optimized the total assembly line by way of Genetic Algorithm. Thus, we obtained the results of optimization for new allocation of resources, in which optimized sequencing of chromosome coding was shown as below.

Statistics results: Maximum fitness is equal to: 41634; Minimum function value is equal to: 10548
The chromosome is encoded as:
CABGBFBGBABBFABBBCBCBABBBBABAAABBABBABBFCCFDBAFBCFBEBGGGFECEABEB BEEBEEBDEEEEEEE

By contrast, total waiting time of passenger cars in this assembly line is 10,548 minutes, which is reduced by 69.5%, as shown in Table1. This significantly improves the production efficiency, and it helps to improve their market responsiveness, shorten lead times, and improve the overall competitiveness.

Summary

The paper optimized production sequence of passenger car balanced load of assembly line under the target of the least total vehicles' waiting time to complete the assembly. First, by comprehensive comparison methods of several main optimization algorithms, a genetic algorithm is proposed to optimize the production sequencing process of actual total mixed models passenger car assembly line. Then production sequencing optimization methods and procedures are studied for the general mixed models assembly line. Finally, according to the characteristics of the mixed models production assembly line, we compare with the results of production sequencing genetic optimization between original and more balanced resource allocation of stations by applying AutoMod simulation software, and obtain optimal solution of sequencing, as shown in Table1.

Table 1: Comparison of optimized results under the different resources allocation

Resources allocation (balanced) state	Assembly Line Sequencing status	Total waiting time of vehicle [minutes]	The total reduction time[minutes]	Reduce the percentage of time
the original resources allocation	Before optimization	37843		
	After genetic Algorithm optimization	15964	21879	57.8%
More balanced resources allocation	Before optimization	22209	15634	41.3%
	After genetic Algorithm optimization	10548	26293	69.5%

Compared with the original, balanced and optimized sequence, the total waiting time of all vehicles in all stations to complete assembly work is reduced by 69.5%. This plays a significant role on balancing production line, improving staff motivation, reducing business supply cycle, and improving the overall competitiveness. It also proves the optimization effectiveness by using genetic algorithm.

References

[1] Guo Jintian: *Car assembly line scheduling optimization and simulation*. Unpublished PhD thesis of Wuhan University of Technology, (2008).

[2] Leu Y Y, Matheson L A, Rees L P: *Sequencing mixed-model assembly lines with genetic algorithm.* Computers Industrial Engineering, 30, 4 (1996), pp.1027-1036.

[3] Takayoshi Tamura, Hong Long, Katsuhisa Ohno. *A sequencing problem to level part usage rates and work loads for a mixed-model assembly line with a bypass ssubline*. International Journal of Production Economics, 61(1999), pp.557-564.

[4] Caridi M, Sianesi A: *Multi-agent systems in production planning and control:an application to the scheduling of mixed-model assembly lines*. International journal of production economics, 68 (2000), pp.29-42.

[5] Min Qiang etc. *The basic theory and application of genetic algorithm*. Science press, (2002)

[6] Eastwood M: *Implementing mass customization*. Computers in industry, 30, 3(1996), pp.171-174.

[7] Huang Yindi: *Modeling and simulation analyzing of passenger car assembly line based on time colored petri net*. Proceedings of the 2010 International conference of logistics engineering and management, Vol.387 (2010), pp.3369-3376.

[8] Xu Zongben etc.: *Genetic algorithms recent developments in basic research*. J.Advances in Mathematics, Vol.4 (2004), pp.91-114.

Advanced Materials Research Vol. 279 (2011) pp 418-422
© (2011) Trans Tech Publications, Switzerland
doi:10.4028/www.scientific.net/AMR.279.418

A method about load distribution of rolling mills based on RBF neural network

Dong dong Liu
Engineering Colloge, LinYi University, LinYi City, Shandong Province, China
Bigbear1978@126.com

Keywords: RBF neural networks, prediction of thickness, prediction of rolling force, load distribution, hot strip mills

Abstract. Rolling mills process is too complicated to be described by formulas. RBF neural networks can establish finishing thickness and rolling force models. Traditional models are still useful to the neural network output. Compared with those finishing models which have or do not have traditional models as input, the importance of traditional models in application of neural networks is obvious. For improving the predictive precision, BP and RBF neural networks are established, and the result indicates that the model of load distribution based on RBF neural network is more accurate.

Introduction

The optimization problem of load distribution in the finishing-mill rolling process is how to get a serious of data about the slab thickness *hi*; and the data will make the whole system state most optimal. In general, the rolling schedule is a chart full of experiential data which people get from the load distribution of actual production[1]. The data relying on this method is reasonable, but not optimal. To get more accurate thickness value, the schedule must be optimized in the mill process. The conventional method of load distribution always establishes single target function under one of different conditions, and finds a serious of *hi*, through appropriate optimization methods, which minimize the target function in the range of definite slab thickness *hi*. Although there are many optimal methods for multi-target function, they do not fit the on-line calculation[2].

Based on RBF neural network model, a method about optimized load distribution is suggested. The result of simulation indicates that this method is more accurate than traditional conventional model method in the prediction to slab thickness and rolling force. The new method not only economizes the computer resource and predigests steps, but also improves efficiency, and thus opens a different way to new production's experiment and the optimal calculation of rolling schedule.

Basic algorithm about RBF neural network

As BP neural network is applied to function approximation, its weight adjusting is adopted negative gradient descending method. The main limitation of this method lies in slow convergence speed and local minimum. RBF (Radial Basis Function) neural network, compared with traditional BP neural network, obviously has more advantages in the ability to approximation, the ability to classification, learning speed and other fields. In recent years, RBF neural network was gained in-depth research and many training algorithms appeared. The combination between excellent capability of RBF and effective training algorithms makes it receive comprehensive application in solving practical problems.

In RBF network, it can be carried out the non-linear mapping $X \rightarrow \alpha_j(X)$ from nodes of input layer to nodes of hidden layer, while it is simple linear function from nodes of hidden layer to output layer. $\alpha_j(X)$ is adopted Gaussian kernel function and is shown as Eq. 1:

$$\alpha_j(X) = \Phi_j\left(\frac{\|X - C_j\|}{\sigma_j}\right) = e^{-\frac{\|X - C_j\|^2}{\sigma_j^2}} \tag{1}$$

where C_j is center point of the jth kernel function; σ_j determined the diversity of the input variables which surrounding the center of this kernel function, hat is to say standard deviation; $\|X-C_j\|$ is the vector which indicates the distance between X and C_j; Φ_j is a radially symmetric function, and there is a unique maximum at C_j. With the $\|X-C_j\|$ increasing, Φ_j will attenuate rapidly, until to zero. For definite input $X \in R^n$, there are only part of processors which near the center activated.

Linear mapping is realized from $\alpha_j(X)$ to output y in output layer and it can be given as Eq. 2

$$\begin{cases} y = W\alpha(X) \\ or \\ y = \sum_{j=1}^{N_k} \lambda_{ij} u_j - \lambda_0 = W_i^T U \end{cases}, i=1, 2, \ldots, m \tag{2}$$

where $u_j = e^{-\frac{\|X-C_j\|^2}{\sigma_j^2}}$, j=1, 2, …, N_k; $U=(1\ u_1\ u_2\ \ldots\ u_N)^T$; $W=(-\lambda_0\ \lambda_{i1}\ \lambda_{i2}\ \ldots\ \lambda_{iNk})^T$.

Thus using RBF algorithm can distinctly describe the input-output relationship in network, which can be seen in Eq. 3:

$$y = \sum_{i=1}^{N_k} \lambda_i \Phi(\|X - C_i\|) + \lambda_0 \tag{3}$$

where $X = [x_1, x_2, \cdots, x_n]^T \in R^n$ is n inputs of the network, N_k is the number of nodes in hidden layer, $\|\cdot\|$ is Euclide Norm, $\lambda_i (0 \le i \le N_k)$ is the weight in network, λ_0 is the bias of output, y is network output.

Convenient model of finishing mill load distribution

Model of thickness distribution. In fishing mill, to achieve the required reduction, final qualities and tolerances, several passes of rolling are executed by tandem rolling with six or seven successive stands. After passed all stands, the reheated steel slabs will be turned into strips from the thickness H_0 to h_n (n is the number of stand), so its total reduction is:

$$\Delta h = H_0 - h_n \tag{3}$$

In order to let computer know the parameters, such as unload gap and speed of strip mill stands, it is necessary to make the thicknesses of stands clear[3]. That is to say, for satisfying the limitation of equipments and techniques. It can be seen that $\Delta h = \sum_{i=1}^{n} \Delta h_i$ (n is the number of stand) depends on the slab thickness H_0, final thickness h_n, and other concerned parameters. According to the experiential data of energy consumption, convenient formula[4], as shown in Eq. 4, can be used to distribute thickness:

$$h_i = H_0 \exp(\frac{K2 - \sqrt{K_2^2 - 4K_1 E\Psi_i}}{2K_1}) \tag{4}$$

where K_1, K_2 are coefficients gotten from local Statistic data; i is the number of stand.

Model of rolling force distribution. Convenient rolling force is shown in Eq. 5[1,3]:

$$P_{mi} = B_c l_c' Q_P K \tag{5}$$

where B_c is the average strip wideness in rolling process, l_c' is the roll contact length considering roller's bending and arc, Q_p is stress status modulus, K is the planar yield stress, P_{mi} is the rolling force of *i*th stand.

For l_c', it is shown as Eq. 6

$$l_{ci}' = \sqrt{R_i' \Delta h_i}$$

$$R_i' = R\left(1 + 2.2\times10^{-4}\frac{P_i}{B\Delta h_i}\right) \tag{6}$$

where R is radius of roller, R' is radius of roller considering roller's bending and arc.

In fact, the indispensable task of most rolling force model is calculation to Q_p. There are many ways to calculate Q_p, while the most popular formula is SIMS formula. That is:

$$Q_P = b_1 + b_2\frac{l_c}{h_c} + b_3\varepsilon\frac{l_c}{h_c} + b_4\varepsilon \tag{7}$$

where ε is relative reduction, $\varepsilon = \dfrac{h_{i-1} - h_i}{h_{i-1}}$; l_c is roll contact length; h_c is average thickness, $h_c = \dfrac{h_{i-1} + h_i}{2}$, and here $b_1 = 0.8206$, $b_2 = 0.2376$, $b_3 = 0.1006$, $b_4 = -0.3768$ rely on actual stand.

The coefficient K is determined by deformational resistance σ, and can be written by $K = 1.15\sigma$. Therefore, deformational resistance σ is one of important parameters, and it expresses the effect not only from chemical constituents of metal material, but also from the physical condition in elastoplastic deformation process (deformation temperature, deformation rate, deformation degree). Its convenient formula is Eq. 8,

$$\sigma = \sigma_0 \exp(a_1 T + a_2)\left(\frac{u}{10}\right)^{(a_3 T + a_4)}\left[a_6\left(\frac{e}{0.4}\right)^{a_5} - (a_6 - 1)\frac{e}{0.4}\right] \tag{8}$$

where e is the deformation degree, $e = \ln\dfrac{h_{i-1}}{h_i}$; u is the deformation rate, $u = \dfrac{v_i}{l_c}e$; v_i is the roll rate based on second-flow formula, $v_i = \dfrac{v_n h_n}{h_i}$; $T = \dfrac{t + 273}{1000}$, t is the roll temperature; σ_0 is the deformational resistance when $t = 1000^oC$, $u = 10s^{-1}$, $e = 0.4$; $a_1 \sim a_6$ are regression coefficient, and determined by steel grade.

Simulation

We combined 470 rolling data to establish and 16 data to check up the predict thickness model. Basic environmental parameters are $B = 1500mm$, $Q235$, $R_7 = 365mm$, $H_0 = 38mm$, $h_7 = 7.55mm$. The process can be realized in two steps.

Step one: Establish two RBF neural networks for thickness model and rolling force model. The first has convenient model as an input variable, and the second not. Fig. 1(a, b, the same with others) shows that joined convenient model is propitious to reduce error. So in next step, this mode is accepted.

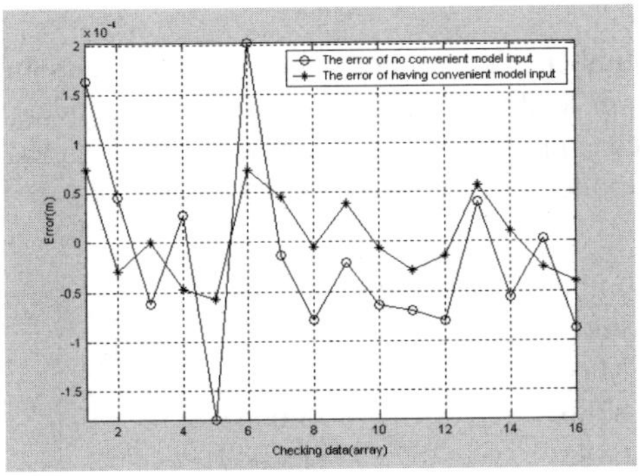

Fig. 1a: Thickness model comparison

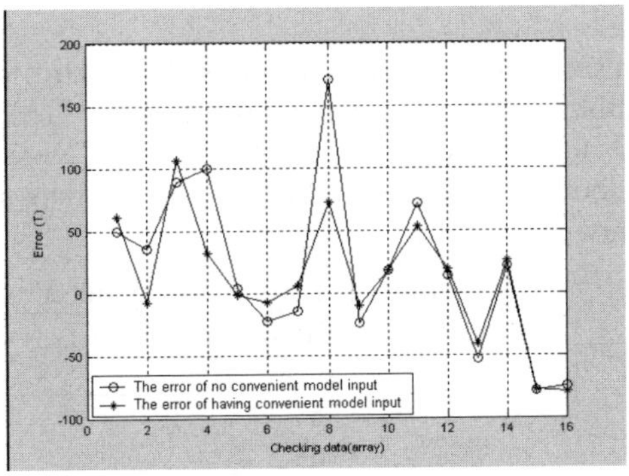

Fig. 1b: Rolling force model comparison

Step two: Establish two neural networks with BP algorithm and RBF algorithm for predict thickness and rolling force model. BP network adopts 4-3-1 structure, training echo is 5000, and training error is 0.005. In RBF network, the max number of neural nodes in hidden layer is appropriate, increasing one each training, and training error is also 0.005. The result is shown in Fig. 2 and Fig. 3. Through contrast between Fig. 2 and Fig. 3, it is clear that the effect of RBF network training excels BP network.

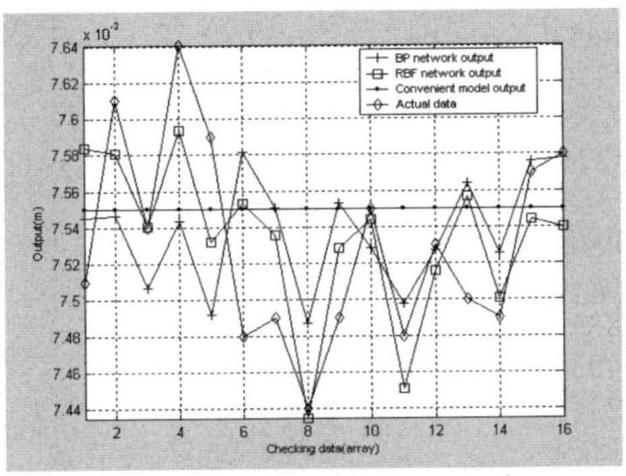

Fig. 2a: Thickness model comparison by BP, RBF

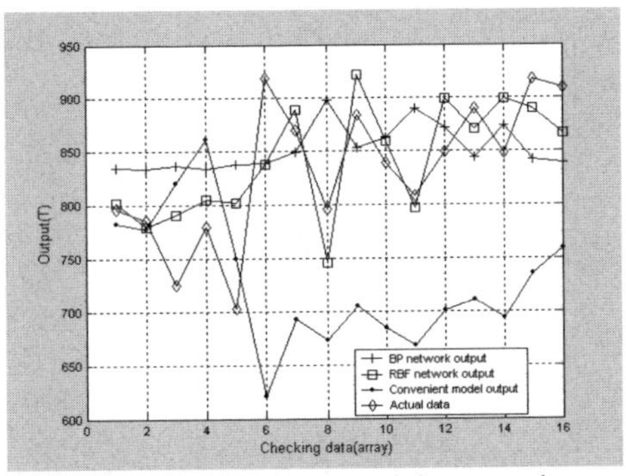

Fig. 2b: Rolling force model comparison by BP, RBF

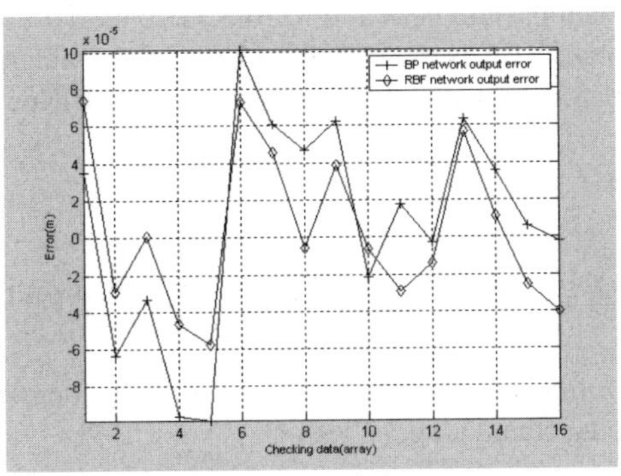

Fig. 3a: Thickness model output error by BP, RBF

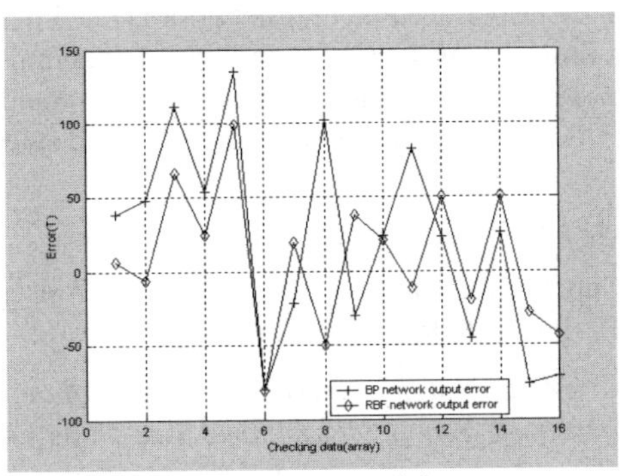

Fig. 3b: Rolling force model output error by BP, RBF

Note

In simulate process, the result is affected by max number of neural nodes in hidden layer, as shown in Table 1. It is not true that more nodes in hidden layer lead to a higher precision. If the model pays most attention to higher precision, it will give much emphasize to individuality while neglect its commonness. So once checking data outputs, the model usually educes unreasonable error. This phenomena is called over-fitness.

Table 1: Error made by maximum number of nodes in hidden layer

Max number of nodes in hidden layer	Max error-Max(e)	Mean squared error-Mse(e)
5	15.8786	38.6041
6	14.6885	41.2049
7	10.3676	29.9227
8	14.3119	51.8820
9	14.0644	52.3481
10	18.7143	60.8778
11	18.7107	60.8796
12	17.9297	60.6117
13	18.7699	63.8498
14	15.5737	61.3873
15	16.8654	76.5206
16	8.4375	27.3036
17	8.3985	26.7262
18	8.3278	28.7202
19	8.3345	19.5786
20	7.3948	17.1468
21	14.6280	47.0248
22	14.8359	58.9630
23	14.4036	61.3393
24	25.2211	110.5435
25	44.1184	299.8826

Summary

In this paper, RBF neural network is used in establishing thickness and rolling force model of hot strip finishing mills. Compared with BP neural network, RBF is proved that has quicker calculated speed and more predictive precision. For nonlinear system with many uncertainties, RBF neural network is considered as the topic of future study.

References

[1] Sun Y K: *Foundation of mathematical model on hot strip rolling.* Metallurgical Industry Press, Beijing, (1979).

[2] Cui Z L: *Design and Implementation about the Optimization of the Rolling Schedules Based on Improved Genetic Algorithms.* ShanDong University, ShanDong, (2008).

[3] Cai Z X: *Intelligent control:Foundation and application.* National Defence Industry Press, Beijing, (2004).

[4] Sun Y K: *Models and control of hot strip rolling.* Metallurgical Industry Press, Beijing, (2002).

[5] Wang X M, Lü C, Wang G D, et al: *Artificial neural network and mathematical model for the rolling force prediction.* Journal of Northeastern University (Nature Science), 20, 3(1999), pp.319-321.

Advanced Materials Research Vol. 279 (2011) pp 423-428
© (2011) Trans Tech Publications, Switzerland
doi:10.4028/www.scientific.net/AMR.279.423

Fractional $PI^{\lambda}D^{\mu}$ Control for Steer-by-wire System

Tian Jie[1, a], Wu Jin[1,b] and Chen Ning[1,c]

[1]College of Automobile & Traffic Engineering, Nanjing Forestry University, Nanjing, China.

[a]njtianjie@163.com, [b]nuliwujin@126.com, [c]chenning@njfu.com.cn

Key words: Fractional calculus, Fractional $PI^{\lambda}D^{\mu}$ controller, Steer-by-wire system, Oustaloup method

Abstract. According to the design demands of the steer-by-wire system, a $PI^{\lambda}D^{\mu}$ controller based on fractional calculus was proposed. Aligning controller and steering controller were respectively designed to achieve the aligning and steering function of the front wheel steering module, which can ensure the robust of the steer-by-wire system during the special ranges of frequency. The five design parameters of fractional $PI^{\lambda}D^{\mu}$ controller were achieved by optimization method. Oustaloup method was used to approximate the fractional $PI^{\lambda}D^{\mu}$ controller and simulation model was achieved, which can be used in Matlab/Simulink. Computational simulations of the control system were carried out and simulation results showed the effectiveness of the control method to improve the robust of the steering-by-wire system.

Introduction

Steer-by-wire system, which completely does away with the steering column and shaft, can design the steering ratio freely and overcome the shortcomings of previously fixed steering ratio. So it recently becomes the research focus of domestic and foreign experts and scholars [1,2]. Now most academic studies to steer-by-wire have been theoretical and validated only in control strategies of active steering, variable ratio and road feeling [3]. In fact, it is more important to achieve the normal steering and aligning for steer-by-wire. So the author will pay more attention on them.

PID controllers are widely used and enjoy significant popularity, because they are simple, robust and effective. Fractional PID controller which is the generalization of PIDs, namely the $PI^{\lambda}D^{\mu}$ controller, has been recently received a considerable attention. Many design methods and tuning rules for fractional controller have been developed in recent years. Due to the extension of derivation and integration order from integer to non integer numbers, the fractional $PI^{\lambda}D^{\mu}$ provides a more flexible tuning strategy and therefore an easier achieving of control requirements with respect to classical controllers [4]. The goal of this paper is to develop a fractional $PI^{\lambda}D^{\mu}$ for achieving the normal steering and aligning as well as improving the robust of steer-by-wire.

Fractional calculus

Fractional order calculus, whose order of calculus can be a negative number, a fraction, an irrational, or even an imaginary or complex number, is the generalization of integer order calculus. Its complete definition is [5]:

$$ {}_{c}D_{x}^{\nu} = \begin{cases} d^{\nu}/dx^{\nu} & R(\nu)>0 \\ 1 & R(\nu)=0 \\ \int_{c}^{x}(d\zeta)^{-\nu} & R(\nu)<0 \end{cases} . \tag{1} $$

Where ν is the operator order of fractional calculus; x and c are respectively the upper and the lower limits.

Fractional calculus has two usual definitions: Riemann-Liouville (RL) definition and Grunwald-Letnikov (GL) definition, which are respectively given by

$$ {}_cD_x^{\nu} f(x) = \frac{1}{\Gamma(n-\nu)} \left(\frac{d}{dx}\right)^n \int_c^x (x-\varsigma)^{n-\nu-1} f(\varsigma)d\varsigma. \tag{2}$$

$$ {}_cD_x^{\nu} f(x) = \lim_{h\to 0} \frac{1}{\Gamma(\nu)h^{\nu}} \sum_{k=0}^{(x-c)/h} \frac{\Gamma(\nu+k)}{\Gamma(k+1)} f(x-kh). \tag{3}$$

Where (n-1)<v<n, $\Gamma(\cdot)$ is the Euler Gamma function, h is the computer step.

Steer-by-wire system

Steer-by-wire system is composed of steering wheel module, controller and front-wheels steering module. In this paper, the way to achieve the steering and aligning function of the front wheel steering module was debated.

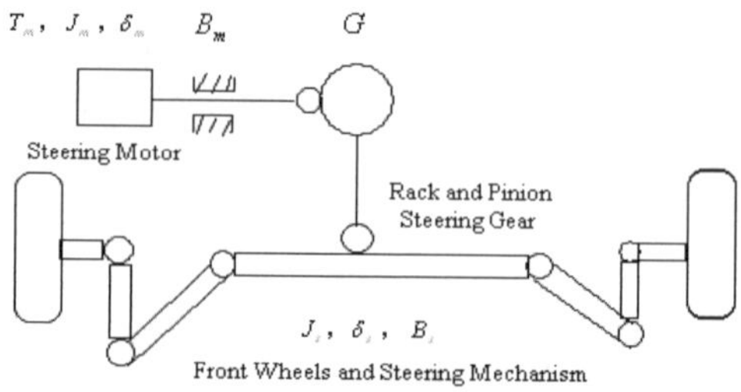

Fig. 1: Dynamics model of front wheel steering module

The front wheel steering module shown in Fig. 1 is described by the following equations:

$$\begin{cases} T_m - \dfrac{T_a}{G} = J_m\ddot{\delta}_m + B_m\dot{\delta}_m \\ U_a = Ri_a + L\dot{i}_a + k_e\dot{\delta}_m \\ T_a - T_r = J_s\ddot{\delta}_s + B_s\dot{\delta}_s \end{cases} \tag{4}$$

where J_m, B_m, δ_m and G are the moment of inertia, the damping coefficient, the steering angle and the gear reduction ratio of the steering motor; J_s and B_s are the moment of inertia and damping coefficient of the road wheels and steering mechanism; δ_s is the pinion turning angle; T_a is the torque of steering motor acting on the pinion; T_r is the equivalent load torque caused by the tire self-aligning moment; U_a, R, L and k_e are respectively the operating voltage, the terminal resistance, the inductance and the back-EMF constant of the steering motor; T_m is the actuator torque, which can be written in terms of motor constant, k_t, and motor current , i_a ,that is $T_m = k_t i_a$.

If the load torque T_r is neglected and $\delta_m = G\delta_s$, the system transfer function between the pinion turning angle $\delta_s(s)$ and the control voltage $U_a(s)$ can be described by

$$G(s) = \frac{\delta_s(s)}{U_a(s)} = \frac{Gk_t}{(Ls+R)\left[(G^2J_m + J_c)s^2 + (G^2B_m + B_c)s\right] + G^2k_ek_ts}. \tag{5}$$

Realization of aligning function

When a vehicle with traditional mechanical steering system undergoing a turn, tire forces acting on the steering system tend to resist steering motion away from the straight-ahead position. But the reduction mechanism of the steering motor has a lock function, so the front wheels of the steer-by-wire can't return to the straight-ahead position without the aligning controller.

Dynamics model of aligning moment. The portion of aligning moment due to the tire pneumatic trail may be directly described by [3]

$$M_{ZV} = \frac{mv^2 b}{\left|L^2 + m\dfrac{k_2 b - k_1 a}{k_1 k_2} v^2\right|}\left(\xi' + \xi''\right)\delta. \tag{6}$$

Where M_{zv} is the portion of aligning moment due to the tire pneumatic trail; ξ′ and ξ″ are the tire pneumatic trail and mechanical trail; m is the vehicle mass; v is the vehicle speed; L is the wheelbase; a and b are distances of the front and rear axles to the center of gravity; k_1 and k_2 are the cornering stiffness of the front and rear tires; δ is the steering angle of the front wheel.

The portion of aligning moment due to the kingpin inclination may be described by

$$M_A = \frac{QD}{2}\sin 2\beta' \sin\delta. \tag{7}$$

Where M_A is the portion of aligning moment due to the kingpin inclination; Q is the tire load; β′ is the kingpin inclination angle; D is the kingpin inclination distance.

From Eq. 6 and Eq. 7, the aligning moment can be described by

$$M_Z = M_{ZV} + M_A. \tag{8}$$

From Eq. 8 and taking into the moment of inertia and the damping of the steering system, the front wheel torque in the mechanical steering system is described by

$$M_h = \left|J_h\ddot{\delta}_h + B_h\dot{\delta}_h + \frac{M_{zv} + M_A}{i_l}\right| / v_L. \tag{9}$$

Where Mh is the front wheel torque in the mechanical steering system; Jh and Bh are the moment of inertia and the damping coefficient of the steering wheel; il is the steering ratio; δh is the front wheel steering angle and δh=ilδ, vL is the assisting coefficient.

The vehicle parameters employed for computer simulations are: m=1818.2kg, Iz=3885kg.m², β′=0.14rad,k1=62618N/rad,k2=110185N/rad, ξ′+ξ″=0.07m, a=1.9435m, b=1.1045m, D=0.2m.

Fig. 2 and Fig. 3 respectively show the simulation result and the test result of front wheel torque at a constant speed of 40km/h. It's easy to find the simulation curve is very close to the test curve. And we can draw a conclusion that the calculation of the front wheel torque is correct and the dynamics model of aligning moment is effective.

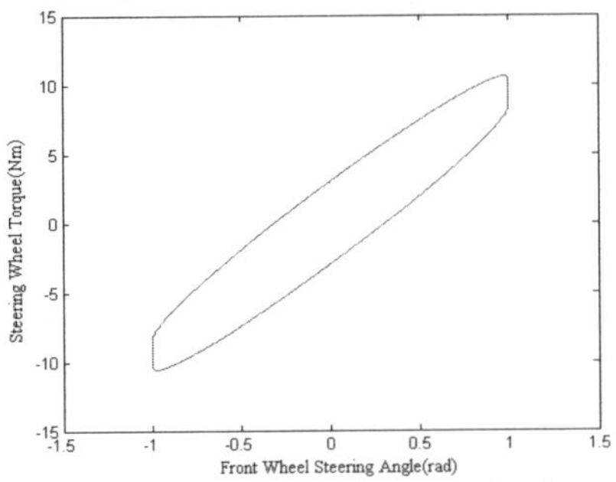

Fig. 2: Simulation result of front wheel torque

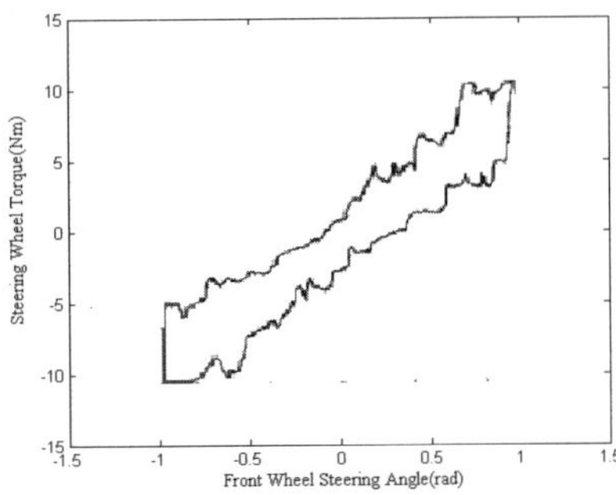

Fig. 3: Test result of front wheel torque [6]

Aligning controller based on fractional calculus. The aligning control block diagram of the front wheels steering module is shown in Fig. 4. Its purpose is to control the steering motor to achieve the aligning of the front wheels.

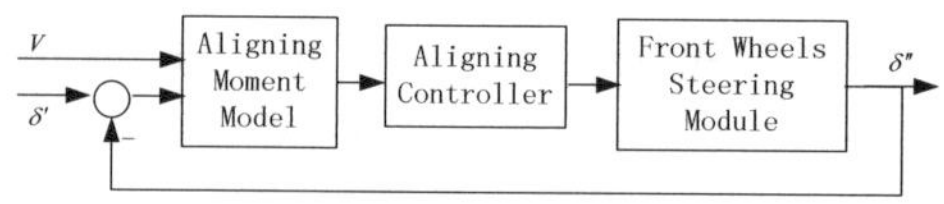

Fig. 4: Block diagram of aligning control

Fractional PI$^{\lambda}$D$^{\mu}$ controller. Fractional PI$^{\lambda}$D$^{\mu}$ controllers are the generalizations of PIDs: their output is a linear combination of the input, a fractional derivative of the input and a fractional integral of the input [5]. If both λ and μ are 1, the result is a usual PID.

$$G_{cf}(s)=\frac{U(s)}{E(s)}=K_P+\frac{K_I}{s^{\lambda}}+K_D s^{\mu}. \tag{10}$$

Fractional PI$^{\lambda}$D$^{\mu}$ tuning by minimization. Fractional PI$^{\lambda}$D$^{\mu}$ may be tuned by finding parameters that satisfy the following conditions [7] ($G_{cf}(s)$ being the controller and $G(s)$ the plant):

$$\left|G_{cf}(\omega_{cg})G(\omega_{cg})\right|=0dB. \tag{11}$$

$$-\pi+\phi_m=\arg\left[G_{cf}(\omega_{cg})G(\omega_{cg})\right]. \tag{12}$$

$$\left|\frac{G_{cf}(\omega_h)G(\omega_h)}{1+G_{cf}(\omega_h)G(\omega_h)}\right|<H. \tag{13}$$

$$\left|\frac{1}{1+G_{cf}(\omega_l)G(\omega_l)}\right|<N. \tag{14}$$

$$\frac{d}{d\omega}\arg\left[G_{cf}(\omega_{cg})G(\omega_{cg})\right]\Big|_{\omega=\omega_{cg}}=0. \tag{15}$$

Where ω_{cg} is the gain-crossover frequency; φ_m is the phase margin, ω_h and ω_l are specified frequencies; H and N are specified gains.

To satisfy them all, the use of numerical optimization algorithms is proposed: the condition in Eq. 11 is assumed as the objective function, conditions in Equations Eq. 12 to 15 are assumed as constraints. This is effective but allows local minima to be obtained.

Numerical simulation for fractional calculus. Transfer functions such as (10) are not easy to implement for computational purpose. Simulations are usually carried out with software prepared to deal with integer transfer functions only. So it is necessary to find integer order approximations of fractional transfer functions. One of the best-known approximations is due to Oustaloup and is given by [8]

$$s^{\nu}_{[\omega_l\ \ \omega_h]}\cong K\prod_{i=-N}^{N}\frac{1+s/\omega_i}{1+s/\omega_i'},\quad \omega_i'=\omega_a\left(\frac{\omega_b}{\omega_a}\right)^{\frac{i+N+0.5(1+\nu)}{2N+1}},\quad \omega_i=\omega_a\left(\frac{\omega_b}{\omega_a}\right)^{\frac{i+N+0.5(1-\nu)}{2N+1}},\quad K=\left(\frac{\omega_b}{\omega_a}\right)^{-\frac{\nu}{2}}\prod_{i=-N}^{N}\frac{\omega_i}{\omega_i'}. \tag{16}$$

To use this approximation, the frequency range of interest [ω_l; ω_h] and the number of zeros and poles N must be fixed in advance. When the integer order approximation of fractional transfer functions was achieved, the simulation model can be easily built by MATLAB.

Simulation results of aligning controller. The parameters employed for computer simulations are: ω_{cg}=25rad/s, ω_l=0.01rad/s, ω_h=100rad/s, φ_m=45.9°, H=-10dB, N=-20dB. The simulation result is obtained for a low speed operating condition, as illustrated in Fig. 5 at a constant speed of 20Km/h and the front-wheels steering angle input is set at 20°. Shortly after the driver looses the front wheel, the front wheels under the control of the aligning controller come back to the straight-ahead position.

In order to verify the robust of the fractional $PI^{\lambda}D^{\mu}$ controller, all of the parameters are changed with the amplitude of ±20% and the system response curves of the fractional $PI^{\lambda}D^{\mu}$ controller are almost the same. But the system responses of the PIDs are very different, as illustrated in Fig. 6. And it shows the parameters' changes greatly influence the performance indicators of the system responses under the control of normal PIDs.

The front wheel aligning moment can be treated as a disturbance on the steer-by-wire system. So, in order to verify the robust of disturbance, we added constant disturbance to the system,such as 0Nm, 5Nm, 10Nm and 15Nm, and the system responses of the fractional $PI^{\lambda}D^{\mu}$ controller are almost the same. But the system responses of the PIDs are also a little different, as illustrated in Fig. 7. And it shows that the disturbance's changes greatly influence the performance indicators of the system responses under the control of normal PIDs.

Therefore, we can draw a conclusion that the fractional $PI^{\lambda}D^{\mu}$ controller has good robust.

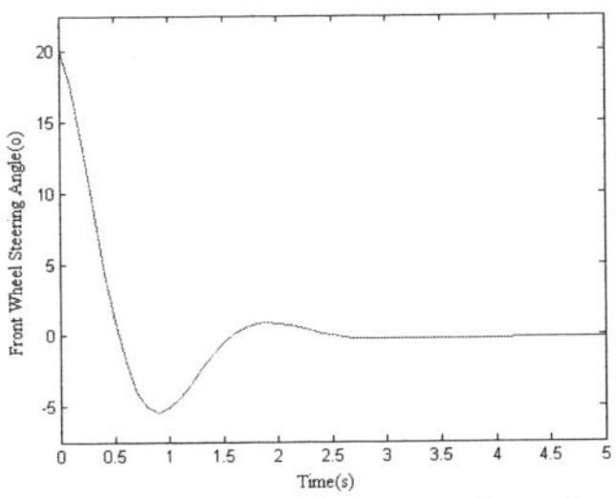

Fig. 5: Response of fractional $PI^{\lambda}D^{\mu}$ controller

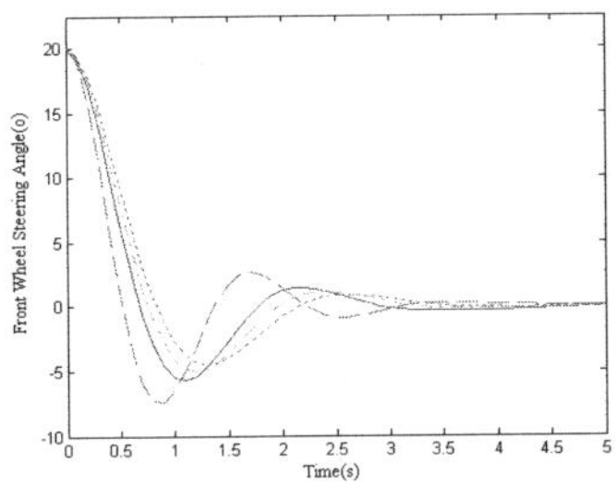

Fig. 6: Responses of PIDs with different parameters

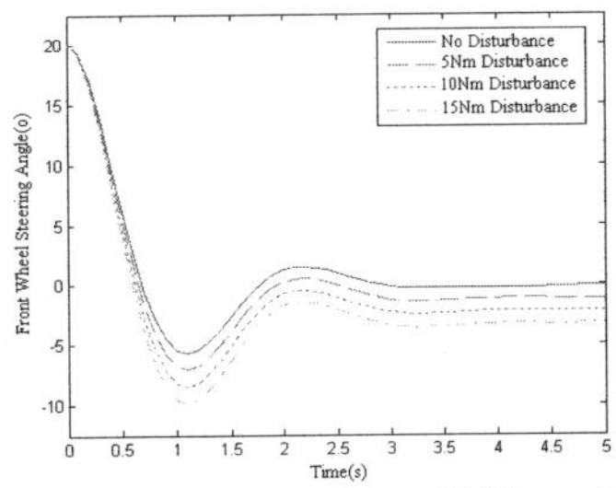

Fig. 7: Responses of PIDs with different disturbances

Realization of steering function

In order to achieve the normal steering function, it is necessary to know the target steering angle of the front wheel and according to this angle to control the actuation of the steering motor quickly. Here we assumed the target steering angle of the front wheel known.

Steering controller based on fractional calculus. The steering control block diagram of the front wheels steering module is shown in Fig. 8. Its purpose is to control the steering motor to achieve the steering of the front wheels.

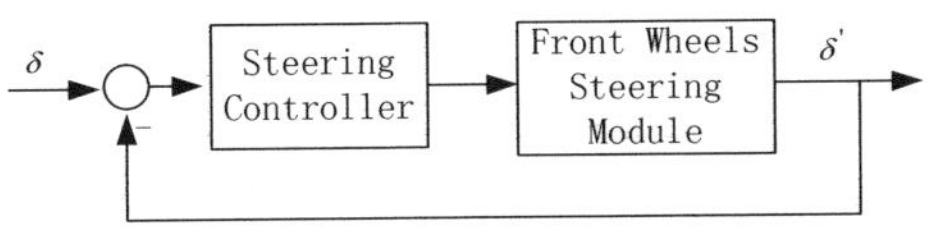

Fig. 8: Block diagram of steering control

Simulation results of steering controller. The parameters employed for computer simulations are: ω_{cg}=0.99rad/s, ω_l=0.001rad/s, ω_h=100rad/s, φ_m=45.9°, H=-10dB, N=-20dB. The simulation result is obtained for a low speed operating condition, as illustrated in Fig.9 at a constant speed of 20Km/h and the front-wheels steering angle input is set at 20°. Shortly after the driver operates the front wheel, the front wheels under the control of the aligning controller come back to the target position.

In order to verify the robust of the fractional $PI^{\lambda}D^{\mu}$ controller, all of the parameters were changed with the amplitude of ±20% and the system response curves of the fractional $PI^{\lambda}D^{\mu}$ controller are almost the same. But the system responses of the PIDs are very different, as illustrated in Fig. 10. And it shows that the parameters' changes greatly influence the performance indicators of the system responses under the control of normal PIDs.

In order to verify the robusteness of disturbance, constant disturbance to the system, for instance, 0Nm, 1Nm, 3Nm and 5Nm are added. And the system responses of the fractional $PI^{\lambda}D^{\mu}$ controller are almost the same. But the system responses of the PIDs are also a little different, as illustrated in Fig.

11. And it shows the disturbance's changes greatly influence the performance indicators of the system responses under the control of normal PIDs. Therefore, a conclusion can be drawn that the fractional $PI^{\lambda}D^{\mu}$ controller has good robust and can track the signal very well.

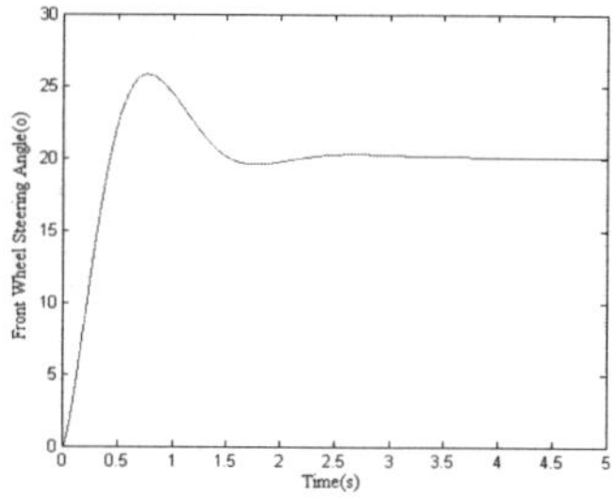

Fig.9 Response of fractional $PI^{\lambda}D^{\mu}$ controller

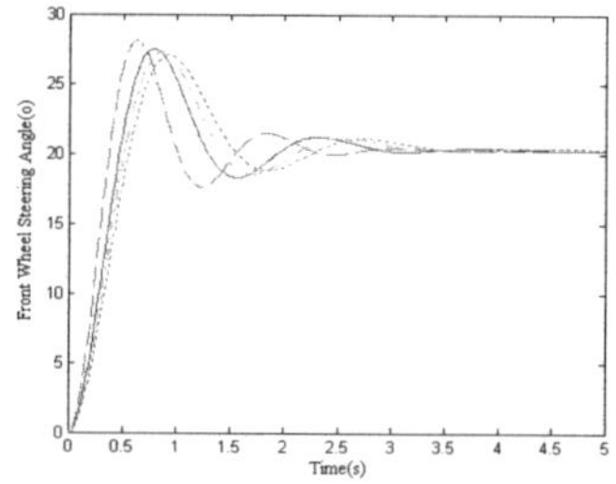

Fig.10 Responses of PIDs with different parameters

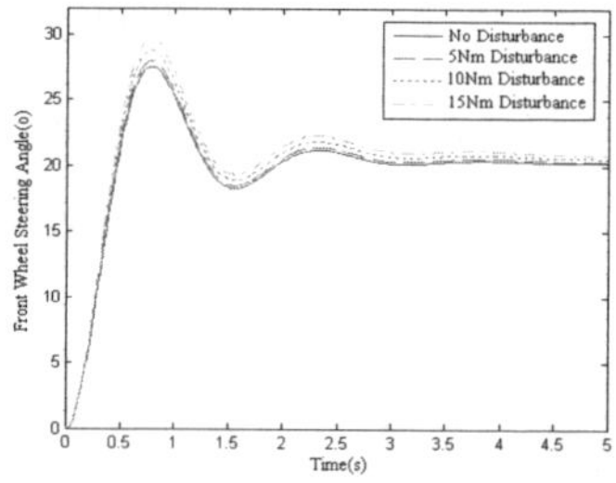

Fig.11 Responses of PIDs with different disturbances

Conclusion

A high performance and reliable steering motor fractional robust control system for the front wheels steering actuator has been successfully designed. This controller has excellent tracking characteristics and robust stability in regard to the target steering angle variation. Moreover, this control system is able to take into account environmental factors, aging deterioration and variations in mass production tolerance.

Acknowledgement

This work was supported by Natural Sciences Foundation of Jiangsu Province universities and colleges (No.163040037) and Scientific and Technological Innovation Foundation of Nanjing Forestry University (No.163106007)

References

[1] Wang Jian, Gao Yong: *Research on Highprecision Auto Electron Steering Wheels Based on DSP.* Mechanical Science and Technology, Vol.25 (2006), pp. 1085-1087.

[2] Yung-Hsiang,Christian Gerdes: *Stabilization of a Steer-By-Wire Vehicle at the Limits of Handling Using Feedback Linearization.* Proceedings of IMECE2005 ASME International Mechanical Engineering Congress and Exposition Vol.11 (2005), pp. 5-11.

[3] YU Lei-yan, LIN Yi, SHI Guo-biao: *Summarization of Automobile Steer-by-Wire Technology.* Agricultural Equipment & Vehicle Engineering,Vol.5 (2007), pp. 3-6.

[4] Petras I, Vinagre B M: *Practical application of digital fractional-order controller to temperature control.* Acta Montanistica Slovaca,Vol.7 (2002), pp. 131-137.

[5] Podlubny I: *Fractional order systems and $PI^{\lambda}D^{\mu}$ controllers.* IEEE Transaction Automatic Control,Vol.44 (1999), pp. 208-214.

[6] Luo Shi; Shang Gaogao: *A Research on Steering Wheel Force Feedback Control Model for Steering by Wire System.* Automotive Engineering, Vol.28 (2006), pp. 914-918.

[7] Monje,C.A., Vinagre,B.M., Chen,Y.Q., et al.: *Proposals for fractional $PI^{\lambda}D^{\mu}$ tuning.* First IFAC workshop on fractional differentiation and its Applications. Bordeaux: IFAC, 2004.

[8] Alain Oustaloup: *Frequency-Band Complex Noniteger Differentiator: Characterization and Synthesis.* IEEE Transactions on circuits and systems 1: Fundamental theory and applications, Vol.47 (2000), pp. 25-39.

Advanced Materials Research Vol. 279 (2011) pp 429-433
© (2011) Trans Tech Publications, Switzerland
doi:10.4028/www.scientific.net/AMR.279.429

Development of a Multi-fingered Massaging Robot Hand

Liang Han[1, a], Mengqian Tian[1, b], Jinzhong Chen[1, c]

[1]School of Mechanical Engineering, Southeast University, Nanjing, China, 211189

[a]melhan@seu.edu.cn, [b]tianmq@seu.edu.cn, [c]chenjinzhong1987@163.com

Keywords: Chinese massage, multi-fingered, robot hand, servo motor

Abstract. This paper proposes a multi-fingered robot hand for the application of Chinese massage, which has a long history and is an important part of Chinese medicine. It is very helpful for human body conditioning and therapy. This research is about a new and feasible application of multi-fingered robot hand. The robot hand which has six degree of freedom (6-DOF) realizing 3translation-3 rotation (3T-3R) can automatically do the full massages process including pushing, picking-up and kneading actions to get the best therapy effect. In order to imitate a real human's hand, we have designed four fingers for the robot hand. A single motor and connecting rod driving are chosen for fingers and closed-loop control is adopted for the robot hand by collecting and processing the signal of encoder and force sensor.

Introduction

Chinese massage which is a physical treatment has a long history [1]. There are many kinds of Chinese massage techniques, which mainly include the pinching method, fluttering method, pressing method, kneading method and rolling method[2]. The different methods have different effects in the local pain area of human body. From the medical books and literatures [3], we know the Chinese massage can contribute lots of good physiology effects to our bodies, such as following effects: (1) Increasing the flowing influence of blood and lymphoid, (2) Relief muscle pain, (3)Relief pressure, (4) Relief muscle tension and relaxing the muscles, (5) Clearance of the chronic edema, (6)Increasing expandable degree of connective tissue, (7) Increasing the joint movable degree, (8) Increasing flexibility of muscle, (9) Improving the function of the self-disciplined nerve, (10) Improving the function of visceral organ etc..

With the more and more violent competition in the industrial society, nervous living, large work pressure, long time sitting and lack of exercise cause the pains all over the human body and the self-discipline nervous system unbalance. For the above patients, the physiotherapy of massaging kind is really needed. Also the current society already tends to the aging society, and many elders do need the physiotherapy. Massage is already recognized to be effective physiotherapy by the medical science field.

In recent decades, massaging robot has been studied by many researchers, such as percussive massage robot by Chul-goo Kang et al [4] and massage robot for medical therapy by K.C. Jones et al. [5]. However, these massage robots have no fingers. Multi-fingered robot hands have been developed for almost three decades and a lot of research on multi-fingered robot hand has been studied. A multi-fingered robot hand can perform the grasping and manipulating motions to handle objects, and can imitate the movements of a human hand. With the rapid evolution of the humanoid robot research, the robots are expected not only to interact with human beings but also to have dexterous manipulation abilities. To this end, several sophisticated robot hands that have anthropomorphic configuration and sensory equipments are developed. An impressive multi-sensory dexterous robot hand, DLR/HIT Hand I [6] and II [7], is developed by DLR (German Aerospace Center) and HIT (Harbin Institute of Technology). The DLR/HIT hand is designed with high modularity and integrality, it has identical modular fingers and all the actuators and electronics are integrated in the hand body.

While the robot hand commonly operated as a grasping tool, it also can be used as a functional massage means instead of a massage specialist or a massage chair. Thus, in this paper, we present a multi-fingered massage robot hand system with closed-loop control. It is especially suitable for the control of a massage motion, since both the applied force and the precise control of the robot hand need to be governed simultaneously during the massage proceedings [8].

Overview of the Multi-fingered Massaging Robot Hand

We divide the multi-fingered Chinese massage robot hand into two main parts. The first part is the manipulator platform with 4-DOF as a base to integrate with the connecting human-like mechanical palm and then be able to reach the three-dimensional space range of massage onto the patients. The other main part is robot hand with four fingers and one palm. The finger tips can approach and reach any massage starting location, and the flexible pressure sensors on palm and fingers are used to real-time monitor the massage force during the therapy onto the patients.

4-DOF Manipulator Platform Design. In order to enable the massage manipulator platform to process the massage in three-dimensional space with no restriction, we design the platform to be a 4-DOF manipulator platform. It is mainly consist of mechanical frame and mechanical arm with mechanical arm fixed on the mechanical frame. The mechanical frame provides mechanical arm with 2-DOF realizing 2 translations—up and down, left and right. The mechanical arm provides robot hand with 2-DOF realizing 1 transtation-1 rotation. The whole mechanical structure designed with aluminum alloy is to reduce the weight of the mechanism but to support necessary mechanical stiffness and effective cost of materials. We use Solid Works software to design the 3D drawing as shown in Fig. 1.

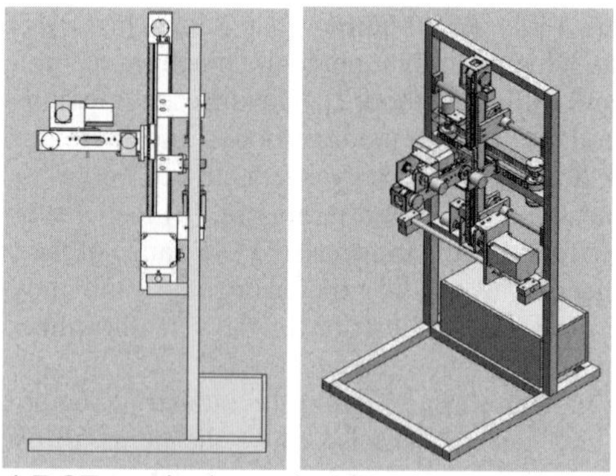

Fig. 1: 4-DOF manipulator platform side and vertical views

The implementation of the massage manipulator platform is shown in Fig. 2. Please refer to the following figure as the real implementation of the manipulator platform.

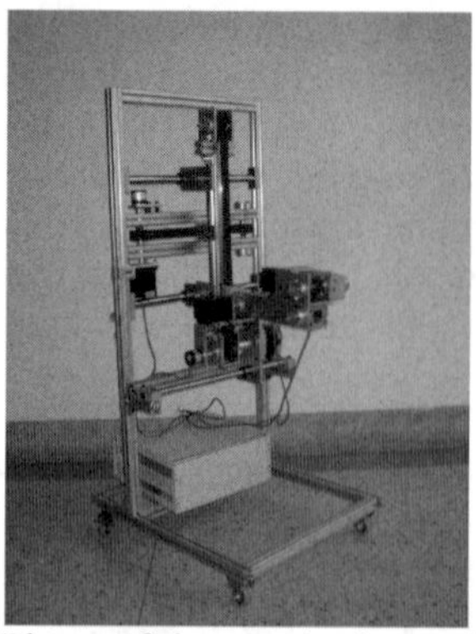

Fig. 2: Photo of the manipulator platform

Multi-fingered Robot Hand Design. The purpose of this research is to design a human-like robot hand. In order to imitate a real human's palm with finger action and to accomplish the massage process as "pushing, picking-up, and kneading", the most decisive technology is also lying on the design of the fingers [9]. Most of earlier developed robot hands are cable-driven where the actuators are located apart from the hand and control the fingers through a set of cable tendons. An alternative approach is to actuate the fingers directly by servo motors which are built in the palm and fingers [8]. Single motor and connecting rod driving is also common used approach [9]. Compared with the motor built-in mechanism, the approach of single motor can reduce the weight of robot hand effectively. Especially, it can reduce the load of mechanical arm when the robot hand fixed on the mechanical arm. Connecting rod is designed for coordinating with single motor driving several movable points at the same time. The advantage is that it will reduce the weight of robot hand to lightest weight. The disadvantage is that all of the finger points must do some action at the same time. Then, the degree of freedom of the robot hand is restricted seriously. So, the activities of a single finger which are needed by working mode are unable. The motor built-in mechanism possesses advantages of more degree of freedom. So the hand can do more complex action, such as action of single finger. But the disadvantage is obvious that the weight of the whole mechanism will increase and the expense of the mechanism will be higher because of more motors. Herein, the mechanism of single motor with connecting rod is adopted. Fig. 3 shows schematic draws of the appearance and unfolded poses of the designed robot hand. Fig. 4 shows schematic draws of the appearance and flexed poses of the designed robot hand.

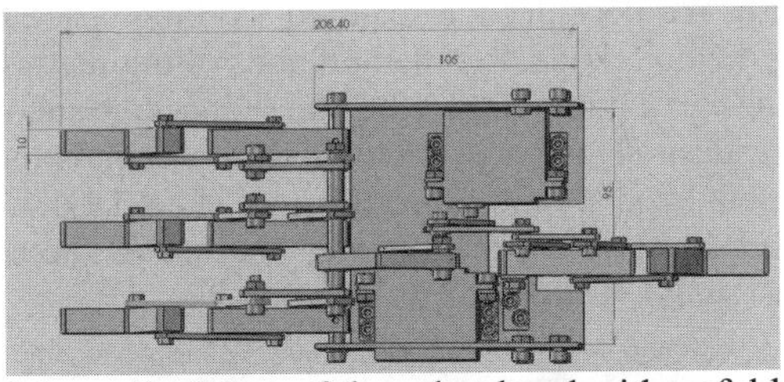

Fig. 3: Schematic draws of the robot hand with unfolded pose

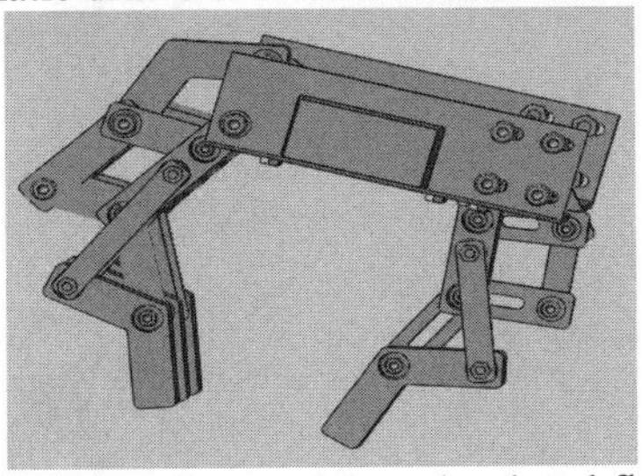

Fig. 4: Schematic draws of the robot hand flexed pose

Design Concept. The principal consideration of the robot hand design is the high output force for massage application and human-like design. The common massage force is about 20N at least, and this is much higher than the output forces that most existed multi-fingered robot hand can offer. To generate higher output forces, steering engines with high output force are chosen. Since the robot hand here is used to perform massage motions which are similar to power grasps, the adduction/abduction function of the finger is not needed so the finger possesses only the flexure/extension function. By simplifying the mechanism, the robot hand becomes more rigid and compact.

Actuation. For a high output torque, the steering engine with size in 54×20×43mm is used. The maximum output force of the steering engine is 13.6kg.cm and the maximum output force generated at the fingertip is about 26N with the phalange length of 115mm. This is relative high compared with other multi- fingered robot hand. As shown in Fig. 2, all fingers are actuated by two steering engines which are embedded in the palm. The single thumb is driven by a steering engine while the three other fingers are driven by the other steering engine.

Size. As is shown in Fig. 3, the total length of the robot hand from the wrist to the finger tip is about 209mm and the extended width is about 95mm. As is shown in Fig. 5, the total length of each finger from finger tip to finger root is 115mm and the width of each finger is 10mm. For an anthropomorphic robot hand the size especially the total length of the hand and the total length of each finger seems to be a little bit large, and it is an unavoidable trade-off due to the requirement of high output forces. However, for the use of a massage means, the robot hand size is acceptable. Fig. 5 is the skeletal draw of a finger module.

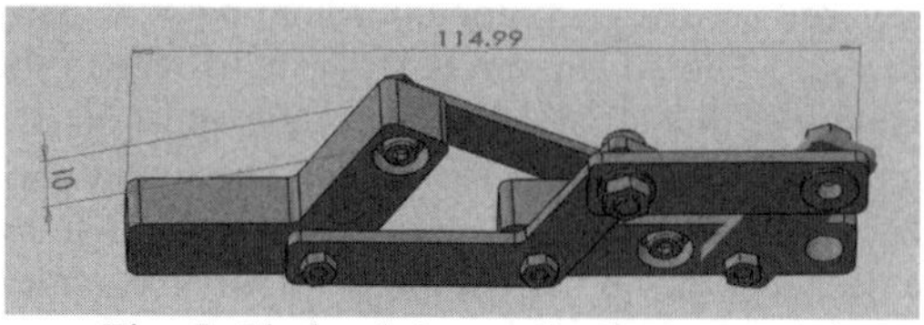

Fig. 5: Skeletal draw of a finger module

Safety Consideration. Safety is a critical consideration when the robot is used as massage applications. In addition to the force feedback, the designed robot hand is covered with soft material to provide gentle contacts on human body.

Overall Mechanism Design. We consider the required working space and the "pinching method, fluttering method, pressing method, kneading method and rolling method" actions for the Chinese massage process, and then design a multi-fingered Chinese massage robot hand with 6-DOF within the 3D space to realize the desired massage therapy with the planning trajectories. The 4-DOF manipulator platform and robot hand are introduced above. The 4-DOF manipulator platform is also implemented. Fixed the robot hand on the manipulator platform is the overall mechanism. We use Solid Works software to design the 3D drawing as shown in Fig. 6.

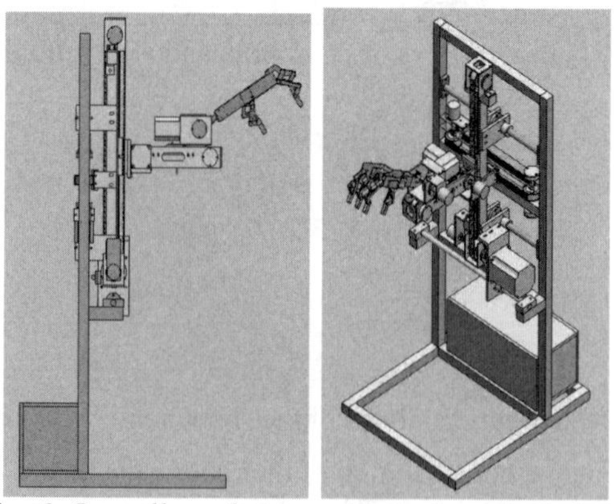

Fig. 6: Overall mechanism side and vertical views

Control System Design

Based on the design requirements of robot hand, in order to implement massaging action the motion of four servo motors and two steering engines are coordinated with each other. The signals of rotary encoder and force sensors which are processed by a series of processing circuits after collected are send to MCU (Micro Controller Unit). After MCU process all signals including encoder signals and force signals, the MCU outputs control signals to servo motors and steering engines . The signals are generated by the motion of servo motors and steering engines. By collecting and processing the signal of encoder and force sensor, closed-loop control of robot has been achieved. The robot hand implements different massaging action under the MCU controlling. The overall schematic draw of control system is shown in Fig. 7.

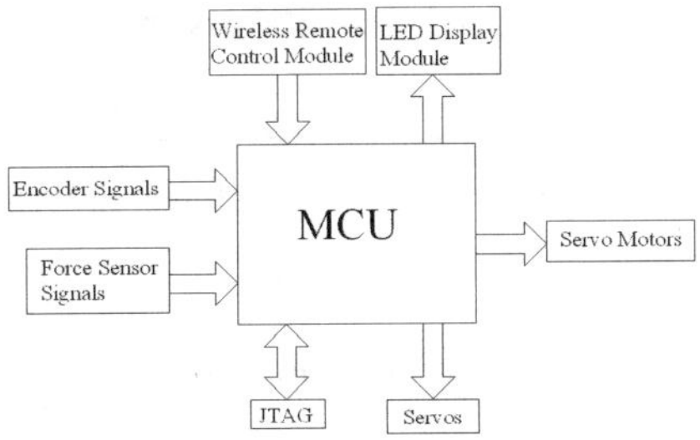

Fig 7: Overall schematic draw of control system

The control of force is very important, so ultra-thin pressure sensor is chosen. It is resistance effect type of pressure sensor. It is very thin and flexible. There is high degree of linear relationship between load and resistance. The view of ultra-thin pressure sensor is shown in Fig. 8.

Fig. 8: Ultra-thin pressure sensor view

Conclusion

In this paper, we present a multi-fingered Chinese massaging robot system where a multi-fingered robot hand is designed to be the massage manipulator. The massage motion control use closed-loop control method which takes care of the massaging force effectively. In the near future, we will make efforts on control strategy which makes the massaging action of the robot hand more like the action of masseur. There are still lots of improvement we can keep on doing to contribute to human life.

References

[1] Zhihong Wang: *Chinese Massage Manipulation*. People's Medical Publishing Press, (2001).

[2] Yin Xiaoqin, Xu Yonggen: *Design and Simulation of Chinese Massage Robot Based on Parallel Mechanism*. 2010 IEEE Conference on Mechanic Automation and Control Engineering, Wuhan, June 2010, pp. 2512 – 2515.

[3] Renfa Cao: *Chinese Massage*. People's Medical Publishing Press, (2006).

[4] Chul-goo Kang et al.: *Design of a Percussive Massage Robot Tapping Human Backs*. The 16th IEEE International Conference on Robot & Human Interactive Communication, Aug. 2007, pp. 962-967.

[5] K.C.Jones, Winncy Du: *Development of a Massage Robot for Medical Therapy.* Proceedings of the 2003 IEEE/ASME International Conference on Advanced Intelligent Mechatronics (AIM 2003) Vol. 2, (2003), pp. 20-24.

[6] H. Liu et al.: *The modular multi-sensory DLR-HIT-Hand.* Mechanism and Machine Theory, 42, 5(2007), pp. 612-625.

[7] H. Liu et al.: *Multi-sensory five-finger dexterous hand: The DLR/HIT Hand II*. IEEE/RSJ Int. Coif. on Intelligent Robots and Systems, Sept. 2008, pp. 3692-3697.

[8] Ren C. Luo et al.: *Impedance Control on a Multi-fingered Robot Hand Based on Analyzed Electromyographic Information for Massage Applications*. IEEE International Symposiumon Industrial Electronics. Seoul Olympic Parktel, Seoul, Korea, July 5-8, 2009, pp. 1228-1233.

[9] Chih-Cheng et al.: *Development of Intelligent Massage Manipulator and Reconstruction of Massage Process Path using Image Processing Technique*. 2010 IEEE Conference on Robotics Automation and Mechatronics, Singapore, June 2010, pp. 551-556.

Advanced Materials Research Vol. 279 (2011) pp 434-439
© (2011) Trans Tech Publications, Switzerland
doi:10.4028/www.scientific.net/AMR.279.434

A Study on the Vision-Based Flexible Vibratory Feeding System

L. Han[1, a], C.B. Li[1, b] and G.P. Hu[1, c]

[1]School of Mechanical Engineering, Southeast University, Nanjing, China, 211189

[a]melhan@seu.edu.cn, [b]LCB@seu.edu.cn, [c]hgping668@163.com

Keywords: machine vision, vibratory feeding, decoupled vibratory feeder, low-cost

Abstract. This paper describes a vision-based flexible vibratory feeding system (VBFVFS) for agile manufacturing, which mainly consists of a decupled vibratory feeder, a machine vision system, displacement sensors, power amplifiers and a computer system. Its major feeding parameters such as the vibration angle, frequency, acceleration amplitude, and phase difference all can be easily adjusted online by software means. Consequently, the system not only can feed a wide range of parts without any retooling of the feeder, but also can feed various parts made of different material in a way close to the optimal state. The VBFVFS also has intelligence. It can find out the natural frequencies of the decoupled vibratory feeder by means of automatic frequency response analysis and hence determine the best frequency for the driving signals. It can also find out any parts jamming and eliminate it by making the parts move in reverse directions. Prototypes of the decoupled vibratory feeder have been successfully developed. Experimental investigation is carried out based on the prototypes and the results agree well with the simulation results. Moreover low-cost vision technique is also discussed for the commercialization of the vision-based flexible vibratory feeding system.

Introduction

Studies have shown that custom-built vibratory feeders are responsible for about 50% failures in automation systems [1]. The major problem of the conventional vibratory feeders is lack of flexibility. Despite the fact that conventional parts feeders perform very effectively when dealing with a specific type of parts, they do have limitations where families of parts, similar in shape but variable in size, are handled in the same system [2]. It is quite common that the feeder must be taken off-line and retooled whenever a new batch of parts or a new orientation is introduced. The current design and retooling of these feeders, usually time consuming, are "black art", based merely on modifications to previous designs and empirical debugging rather than on theory and automated design [3].

Agile manufacturing is a term that has seen increased use in industry over the last decade, which can be defined as the ability to accomplish rapid changeover between the manufacture of different assemblies utilising essentially the same facilities in a workcell [4]. In order to suit agile manufacturing, machine vision has been introduced into vibratory feeding systems and some sensor-based vibratory bowl feeders have been developed [5-8]. Through the use of traditional camera vision system or optical sensors, the system views, records and stores the image of a "perfect" part — size, shape and orientation — in its computer memory, and compares every part presented subsequently to this stored image before deciding whether to accept or reject it. Being equipped with the machine vision system, a sensor-based vibratory bowl feeder may be able to feed a wide range of parts without any retooling of the feeder. However, there is a problem to be dealt with. According to the vibratory feeding theory [9], the conventional vibratory feeder has a fixed vibration angle and this angle can only be optimal for a certain part material, with a certain value of friction coefficient. When parts with different friction coefficients are fed in the feeder, the feeding performance may be quite different and even very poor. In addition, the frequency of the driving signal is also fixed at 50 Hz (or 60 Hz). The natural frequency of the vibratory feeder is usually specially designed to suit this fixed frequency with the mass of the parts to be fed taken into consideration. If the feeder is used to feed another kind of parts with a different mass, the natural frequency of the vibratory feeder will be different and the feeding performance may also be severely affected. Considering the fact that the only difference between the conventional vibratory feeders and the sensor-based vibratory feeders is the orienting technique, the existing sensor-based vibratory bowl feeder can not feed parts made of different materials always in the optimal state, though it can feed a wide range of parts without any retooling of the feeder. In order to improve the performance of the sensor-based vibratory feeders and hence better adapt them to the fast changing need of the market today, a vision-based flexible vibratory feeding system (VBFVFS) for agile manufacturing, which is more flexible and intelligent, is hence proposed and investigated in this research. It mainly consists of a decoupled vibratory feeder, a machine vision system, displacement sensors, power amplifiers and a computer system. Its major feeding parameters such as the vibration angle, frequency, acceleration amplitude, and phase difference all can be easily adjusted online by software means. Consequently, the system not only can feed a wide range of parts without any retooling of the feeder, but also can feed various parts made of different material in a way close to the optimal state. The VBFVFS also has some embedded intelligence. It can find out the natural frequencies of the decoupled vibratory feeder by means of automatic frequency response analysis and hence determine the best frequency for the driving signals. It can also find out any parts jamming and eliminate it by making the parts move in reverse directions. Prototypes of the decoupled vibratory feeder have been successfully developed. Experimental investigation is carried out based on the prototypes and the results agree well with the simulation results. Moreover, low-cost vision technique is also discussed for the commercialization of the vision-based flexible vibratory feeding system.

Existing Vision Application in Vibratory Feeding System

Vision-Based vibratory feeding system can recognize correctly oriented parts even when some are contiguous or overlapping. Hardware consists of a vibratory bowl feeder, vision system, and control system. There are two kinds of common used vision systems for vibratory feeding system. One type system is based on the CCD camera [6, 10, 11] and it describes the 3D view of a part by the images which capture by the CCD camera. For example, a machine vision system was designed for determining the peanut pod ripeness [10]. As shown in Fig. 1, peanut was fed from a vibratory bowl feeder through a singulation chute to a measurement chamber where they entered free-fall. An IR detector sensed the presence of the peanut pod leaving the singulation chute and triggered a B/W CCD camera to capture the image. Algorithms were developed to determine the pod ripeness color for three views of the pod. The other type system is based on the optical fiber [5, 8], and it captures images by optical fiber. Such as a system was designed by Gary P.Maul and Nebojsa I. Jaksic, which consists of a linear row of optical fiber in the side wall and track of the bowl, as shown in Fig. 2. This configuration creates a 3D silhouette of the part and an algorithm is developed to recognize parts in the orientation specified by the user [5].

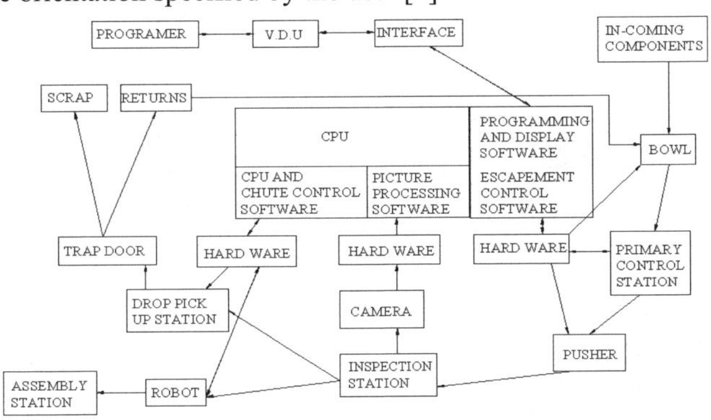

Fig. 1: System block diagram

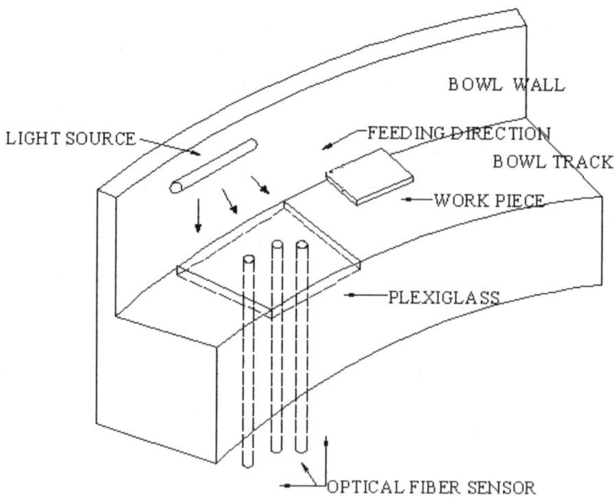

Fig. 2: Mounting of optical fiber window in track of vibratory bowl feeders

System Construction of the VBFVFS

Fig. 3 schematically shows the vision-based flexible vibratory feeding system (VBFVFS) for agile manufacturing proposed to solve the problem concerned. The full system would consist of a decoupled vibratory feeder, an air jet system, a computer system, a power amplifier, a machine vision system and displacement sensors. The machine vision system, which may be based on camera vision or optical grid, is used to identify or recognize the right orientation of the right part and to measure the conveying velocity as well. If it is a right part in the right orientation, it is allowed to move to the exit of the trough, otherwise, the air jet will be activated by the computer to blow the part back into the container below. The displacement sensors are used to measure the horizontal vibration and vertical vibration of the track. Based on the signals from the machine vision and the displacement sensors, the computer can generate control signals, which are amplified through the power amplifiers and then sent to the decoupled vibratory feeder so as to make the feeder work in the best state and feed the parts at the desired feed rate.

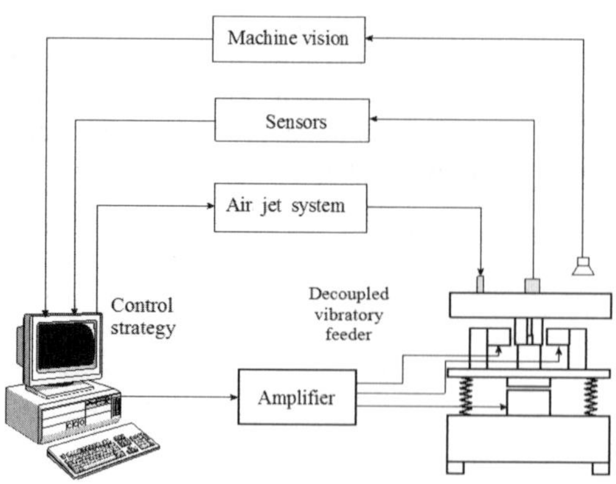

Fig. 3: Vision-based flexible vibratory feeding system

The decoupled vibratory feeder, sharing the same general principle as the out-of-phase vibratory feeder [12, 13], is the key mechanical component of the system. Mechanically, the decoupled linear vibratory feeder has a two-stage structure as schematically shown in Fig. 4. The trough along which parts travel is supported by suspension spring 2-1 and suspension spring 2-2, while the electromagnet 2-1 and the electromagnet 2-2 together with the two suspension springs are mounted on the link-board. The link-board is supported by suspension spring 1. The electromagnet 1 and suspension spring 1 are mounted on the base.

When the electromagnet 1 is activated, the link-board together with all the elements mounted on it will move vertically, so that the linear vertical vibration can be obtained. Similarly, when the electromagnet 2-1 or the electromagnet 2-2 is activated, the trough will move horizontally, and thus the linear horizontal vibration can be obtained. It can be seen that the horizontal vibration and the vertical vibration can be generated and controlled independently and hence the vibration angle ψ can be easily adjusted by means of changing the ratio of the vibration amplitudes of the vertical vibration and the horizontal vibration. Moreover, when the electromagnets 2-1 and 1 are activated together, the part on the trough may travel rightward (say forward), while it may travel leftward (or backward) if the electromagnets 2-2 and 1 are activated instead.

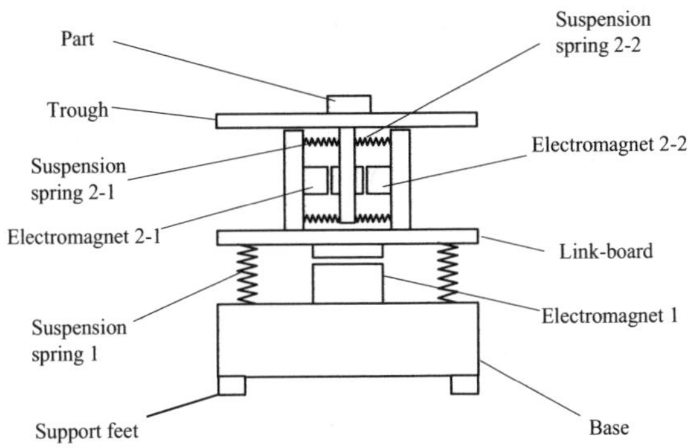

Fig. 4: Decoupled linear vibratory feeder

Two prototypes of the decoupled vibratory feeders have been successfully developed, as shown in Fig. 5, where (a) is the linear type and (b) the bowl type. Both basically consist of a decoupled vibratory feeder, a computer system, a power amplifier and two sets of vibration sensors. Experiments have been carried out on the two prototypes and experimental results are promising.

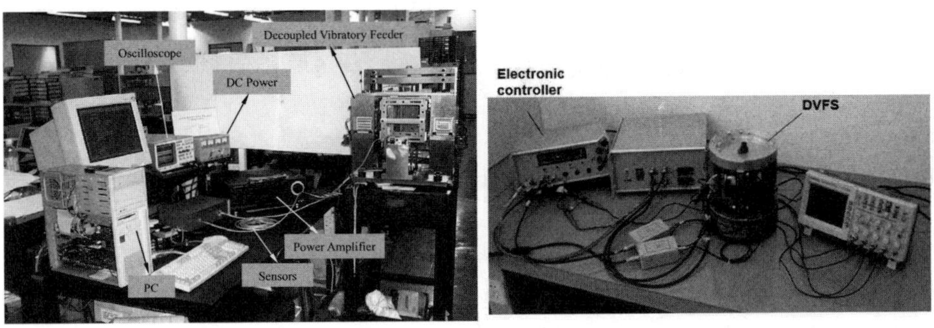

(a) (b)

Fig. 5: Two Prototypes of decoupled vibratory feeders

The major attractive features of the VBFVFS include:

(1) Highly flexible — The frequency for the driving signals, the vibration angle, the acceleration amplitude and the phase difference between the horizontal vibration and vertical vibration can all be adjusted by means of software. This guarantees that the VBFVFS not only can feed a wide range of

parts without any retooling, it can also feed parts made of different material in a way always close to the optimum state.

(2) Intelligent — VBFVFS can find out the current natural frequencies by way of obtaining the frequency response and hence determine the best frequency for the driving signals automatically. Moreover, it can detect part jamming with the aid of machine vision and can overcome it by controlling the parts to move in the reverse direction.

(3) Controllable — VBFVFS is a closed-loop digital control system where various control strategies can be easily applied according to performance requirements.

Commercialization Issue Discussion

As related above, VBFVFS will be the trend in order to suit agile manufacturing. Though both the decoupled vibratory feeding technology and machine vision technology are successful and available, VBFVFS is still far from real application due to the high cost. However, it is encouraging to see the fast development of vision technology as well as that of the computer technology, and more and more lost-cost vision products, such as the webcams and other CCD products are available on the market. The authors believe that the VBFVFS is the right way to go, and will take the place of the conventional vibratory feeder, which means that design and retooling of vibratory feeders will not be the time consuming "black art" anymore.

Conclusions

To adapt the vibratory feeding techniques to agile manufacturing, a vision-based flexible vibratory feeding system (VBFVFS) is proposed in this research, which not only can feed a wide range of parts without any retooling of the feeder, but can also feed parts made of different materials in a way always close to the optimal state (i.e., high conveying velocity and low power consumption). The VBFVFS is a closed-loop control system based on computer operation, so that various control strategies can be easily applied by software to guarantee the desired track vibration and hence almost optimal feeding performance. The VBFVFS also has built-in intelligence. It can find out the natural frequencies of the decoupled vibratory feeder by means of automatic frequency response analysis and hence determine the appropriate frequency for the driving signals. In the event of parts jamming, it can be eliminated by controlling the parts to move momentarily in the reverse direction. The authors believe that the VBFVFS is the right way to go, and will take the place of the conventional vibratory feeder, which means that design and retooling of vibratory feeders will not be the time consuming "black art" anymore.

Acknowledgments

This work is supported by the NSFC, Project No.:50975048.

References

[1] D. Boehlke: *Smart Design for Flexible Feeding*. Machine Design Vol. 12 (1994), pp. 132-134.

[2] Andrew Cokayne: *The Way of the World — A Review of Current Practice in Automatic Parts Recognition, Feeding and Orientation*. Assembly Automation Vol. 11 (1991), pp. 29-32.

[3] Berkowitz, Dina R. Canny, John: *Designing parts feeders using dynamic simulation*. Proceedings - IEEE International Conference on Robotics and Automation Vol. 2 (1996), IEEE, Piscataway, NJ, USA, 96CB35857, pp. 1127-1132.

[4] Quinn, R. D., Causey, G. C., Merat, F. L., et al.: *Design of an Agile Manufacturing Workcell for Light Mechanical Applications*. Proceedings - IEEE International Conference on Robotics and Automation Vol. 1 (1996), Minneapolis, Minnesota, pp. 1-6.

[5] Maul, Gary P. Jaksic, Nebojsa I: *Sensor-based solution to contiguous and overlapping parts in vibratory bowl feeders*. Journal of Manufacturing Systems, 13, 3(1994), pp. 190-195.

[6] Heginbotham, W. B., Barnes, D. F. Purdue D.R. and Law, D. J.: *Flexible Assembly Module with Vision Controlled Placement Device*. Robot Vision, 1983, IFS (Publications) Ltd. UK, pp. 246-251.

[7] Cronshaw, A. J., Heginbotham, W. B. and Pugh, A.: *A Practical Vision System for Use with Bowl Feeder*. Robot Sensors: *Volume l-Vision*, 1986, IFS (Publications) Ltd. UK, pp. 147-156.

[8] Maul, Gary P. Chao, Ou-Yang: *Predicting the Cycle Time for a Sequence of Parts in a Sensor-based Vibratory Bowl Feeder*. International Journal of Production Research, 25, 12(1987), pp. 1705-1714.

[9] Boothroyd G. Polio C, Murch L E, (1982) "Automatic Assembly" Marcel Dekker, New York.

[10] E. Jay Williams and Stephen D. Adams: *A 3D Vision System for Peanut Pod Maturity*. SPIE Conference on Optics in Agriculture, Boston, MA (USA), 7-8 Nov. 1990, pp. 236-245.

[11] Kok-Meng Lee, YIfei Qian: *A vision-guided fuzzy logic control system for dynamic pursuit of a moving target*. Microprocessors and Microsystems, 21, 9(1998), pp. 571-580.

[12] Redford A. H. and Boothroyd G.: *Vibratory Feeding*. Proceedings of the Institution of Mechanical Engineers Vol.182 (1967), pp. 135-152.

[13] Mozer, Donald Thomas: *Vibratory Feeders*. Syracuse University (0659) Degree: PhD Date: 1972, p. 104.

Advanced Materials Research Vol. 279 (2011) pp 440-444

doi:10.4028/www.scientific.net/AMR.279.440

Research on Key Technologies of Manufacturing Resources Sharing Platform Supporting SMEs Based on Web Services

Junying Wei[1, a], Peisi Zhong[1, b] and Chunfen Guo[1, c]

[1]College of Mechanical and Electronic Engineering, Shandong University of Science and Technology, Qingdao 266510, China

[a]jdwjy0726@163.com, [b]pszhong@163.com, [c]guochunfen@163.com

Keywords: Small and Medium-size Enterprises (SMEs), Function Analysis, Manufacturing Resources Sharing Platform, Key Technologies, Web Services

Abstract. Manufacturing resources sharing is a strong support for the development of SMEs. By analyzing the development and current situation of SMEs, the function, architecture and the function model of the manufacturing resources sharing platform supporting SMEs are analyzed and established. The key technologies of the platform are discussed, such as the definition, classification, the information model, integration and the security management of the sharing manufacturing resources. The platform provides a strong support for the survival and development of the SMEs.

Introduction

SMEs are the basic power in driving the development of national economy and promoting social stability. Especially since the reform and opening-up, SMEs have had rapid development in China, which play a more and more important role in assuring measurable growth of national economy, alleviating the pressure on employment, making development through education, and optimizing economic structure [1].

The development of information technology has had a tremendous impact on the SMEs. With the help of its development, SMEs can achieve rapid product design and manufacturing, online ordering and book facilities with resources that the network provides. However, due to a great quantity and a large variety of online resources, the uneven distribution of manufacturing resources and the low level of information, it is necessary to provide a resources sharing platform supporting SMEs, making full use of design resources, equipment resources, software resources, knowledge resources and other resources which are distributed in the regional research institutions and various manufacturing enterprises to meet the demands of SMEs for requesting resources and speed up the product development cycle.

Function Analysis of the Manufacturing Resources Sharing Platform

In order to integrate the regional manufacturing resources, organize and manage these resources to provide a strong support for SMEs, the platform should have the following function modules.

Resources Integration Module. This module systemizes the resources which are distributed in the regional research institutions, universities and various manufacturing enterprises, and then stores them in the data base of the platform to facilitate the users (SMEs) to view.

Resources Searching Module. This module provides the functions of both keyword search and advanced search, so that users of SMEs can inquire the resources that they need such as software, equipment, part and so on.

Resources Management Module. The resources provider uploads the resources information which can be shared and processes classified management to facilitate the resources demander to search. In this module, resources information can be published, edited, deleted and examined.

Register and Logging in Module. This module completes the functions of the registration of new users and the security login of registered users.

Other Modules. Other modules provide some other functions such as extent of permissions for different users and so on.

Manufacturing Resources Sharing Platform

Manufacturing Resource Sharing Platform Architecture Supporting SMEs Based on Web Services. Web services is a web-based, distributed and modular components that can be described, published, searched and called in the network. With many good characteristics such as perfect encapsulation, loosely coupling, using basic protocol criterions, high integration and crossing the firewall, web services have broad application prospects. Because of the above mentioned, the manufacturing resources sharing platform supporting SMEs adopts the architecture based on web services shown as Fig. 1, which includes five layers. (1) The layer of user interface. This layer provides a platform on which users can interact with the resources sharing system. It's a web client application. (2) The layer of applied logic. According to the demands of users, this layer can provide the function of calling the service logic to complete the requests of users and then returning the requests to users. This layer includes the module of resources integration, resources management, resources choosing, register and logging in and son on. (3) The layer of manufacturing resources data. This layer includes kinds of manufacturing resources data base to provide data service for the layer of applied logic. (4) The layer of data transport. This layer provides the infrastructure support for the platform. (5) The layer of basic data base. This layer provides the manufacturing resources which are distributed in the region.

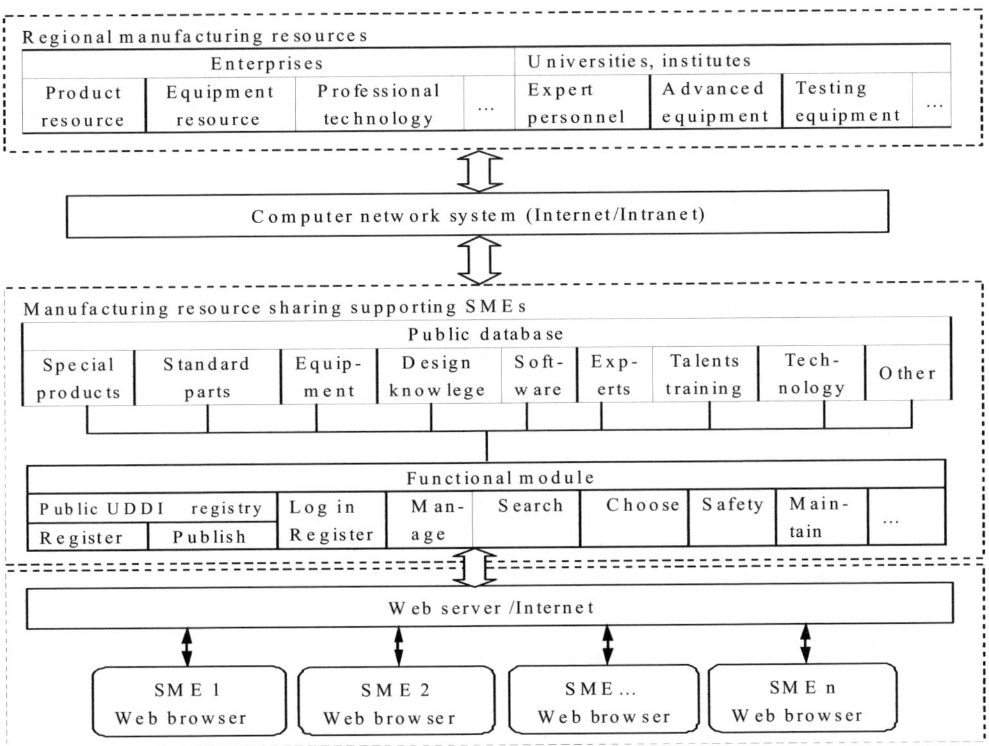

Fig. 1: Manufacturing Resources Sharing Platform Architecture Supporting SMEs Based on Web Services

Manufacturing Resources Sharing Platform Function Model Supporting SMEs Based on Web Services. Regional manufacturing enterprises as well as universities and research institutes register and publish manufacturing resources information on the sharing platform[2] which provides information management and sharing function (as Fig. 2 showing) in order to facilitate the SMEs to obtain the required manufacturing resources from the platform.

Key Technologies of the Platform

Definition and Classification of Manufacturing Resources. Manufacturing resources are the sum of the all production activities' physical elements for the enterprise to produce products in the whole life cycle [3]. Manufacturing resources can be divided into narrow and extensive manufacturing resources. Extensive manufacturing resources are the sum of physical, informational and technical elements for the enterprise to produce products in the whole life cycle including all the hardware and software elements involved in the process such as the design, manufacture, maintenance and other related activities. Narrow manufacturing resources mainly refer to the material elements needed to manufacture one component. Being the bottom manufacturing resources in a manufacturing system [4], narrow manufacturing resources include machine, tools, fixtures, gauges and materials and so on.

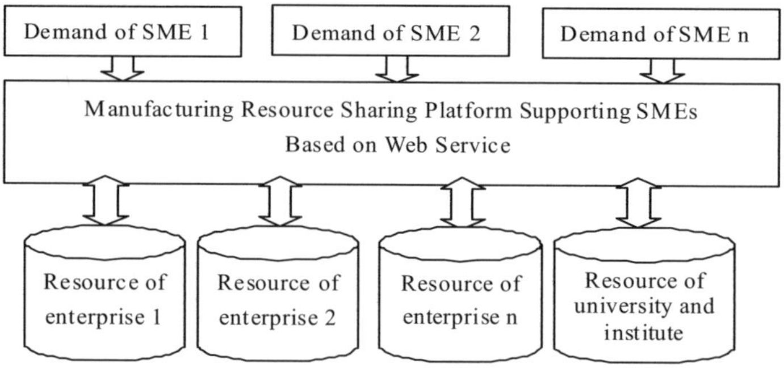

Fig. 2: Manufacturing Resources Sharing Platform Function Model Supporting SMEs Based on Web Services

The manufacturing resources are distributed in the different departments of the regional research institutions, universities and various manufacturing enterprises, and therefore these distributed resources data must be reasonably organized and managed in order to provide a strong support for resources sharing platform.

The classification of manufacturing resources refers to grouping some manufacturing resources which have common attributes or characteristics together. Resource classification is an important part of manufacturing resource modeling, so a uniform standard for the classification must be adopted to make the manufacturing resource model consistent in the description, expression and understanding [5]. In order to organize and manage manufacturing resource management system and to integrate and share resources, the regional resource situation must be considered. According to the phase properties of the manufacturing resources and by defining them in the extensive conception, the manufacturing resources are divided into nine categories as follows: (1) Special-specific products; (2) Standard parts resources; (3) Equipment and material resources; (4) Design knowledge resources; (5) Software system resources; (6) Experts human resources; (7) Training resources; (8) Science and technology information resources; (9) Other resources. These resources correspond to the layer of manufacturing resources data as shown in Fig. 1.

Information Model of Manufacturing Resources Sharing. XML (Extensive Makeup Language) is a set of specifications created by W3C, which is a recommended markup language used to generate other markup languages and describe the various types of data. Having advantages in data storage format, expandability, highly formatting and easier transfer, XML has become the new standards for data representation and data exchange on internet.

Using XML to describe the information of regional manufacturing resources can solve the resources information exchange between different departments, enterprises and heterogeneous systems. Fig. 3 shows the XML-based information model of manufacturing resources sharing. Manufacturing resources' providers (such as regional manufacturing enterprises and universities, research institutes) generate XML data of the sharing resources by the XML conversion interface. The resources' demanders (all the SMEs) convert the desired XML data into the available data for the application by the XML parser.

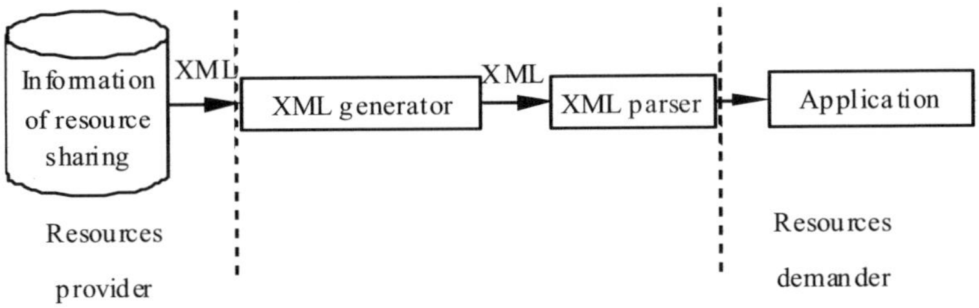

Fig.3 Information Model of Manufacturing Resources Sharing

Manufacturing Resources Integration Based on Web Services. The resources integration is realized by web services [6]. Regional enterprises, universities and research institutes register the sharing resources information to the public UDDI (Universal Description, Discovery and Integration) center on the platform and integrate the manufacturing resources. The SMEs complete calling service by accessing the platform according to their manufacturing resources' requirements. The main steps include: manufacturing resources' providers describe the sharing manufacturing resources information in WSDL documents; then the WSDL document is mapped to the public UDDI center.

Safety Management of the Platform. Five levels of user roles are defined on the platform including the guests, individual users, enterprise users, system administrators and power users, each role having its permissions to access the platform. The unregistered users and the users not logged in are guests, who can browse information about the platform only. Individual users can inquire, browse information of manufacturing resources and download some public data. Enterprise users manage their own resources information, such as the data operation of adding, deleting, and changing and so on. The system administrators are in charge of the daily management and the associated operations of the platform. The power users are mainly responsible for the management of user permissions.

Conclusion

The manufacturing resources sharing platform supporting SMEs based on web services can maximize the advantages of manufacturing resources in regional enterprises, universities and research institutes to meet the demands of SMEs for sharing manufacturing resources, thus providing a strong support for their survival and development.

Acknowledgement

This research is supported by the Science and Technology Development Program of Shandong Province (No. 2010GGX10408) and the Applied Basic Research Program of Qingdao, China (No. 09-1-3-51-jch).

References

[1] Project Team of development research for SMEs: *Development Report 2005 of China's Growing SMEs*, Shenzhen, China (2005) (In Chinese).

[2] Q.L. Chen, Z.M. Fang: *Research on Manufacturing Resource Sharing Platform for the Small and Middle Die Enterprises Based on ASP Model.* Electromachining & Mould, No.2 (2007), pp.50-53 (In Chinese).

[3] B.Y. Sheng: *A Model for the Integration of the Constantly Changing of Manufacturing Resources of Enterprises*. Journal of Wuhan Automotive Polytechnic University, 22, 2(2000), pp. 19-21 (In Chinese).

[4] W.S. Jiang: *Enterprise Information Modeling Technology Based on Object-oriented in Agile Manufacturing.* Manufacturing Automation, 21, 4(1999), pp.13-16 (In Chinese).

[5] Z.C. Wang: *Study on Several Key Problems for Networked Manufacturing Resources Integration Platform.* Zhejiang University, China (2009) (In Chinese).

[6] J.Z. Chen, L. Tian, B.S. Tong: *Research on Enterprise Information Integration Platform for Collaborative Product Commerce.* Computer Integrated Manufacturing Systems, 11, 8(2005), pp.1083-1087 (In Chinese).

Advanced Materials Research Vol. 279 (2011) pp 445-450
© (2011) Trans Tech Publications, Switzerland
doi:10.4028/www.scientific.net/AMR.279.445

Period-varying Sampling Scheme for networked control systems

Shen Yan[1, a], Guo Bing[2, b]

[1]School of Control Engineering, Chengdu University of Information Technology, Chengdu, China, 610225

[2]School of Computer, Sichuan University, Chengdu, China, 610054

[a]shenyan02@163.com, [b]guobing@scu.edu.cn

Keywords: Networked control systems, Period-varying sampling, BP neural network, Time delay, Network utilization.

Abstract. In the networked control systems, control performance and network performance are closely related to the sampling period influenced by the network-induced delay. So in order to improve the network performance by adjusting the current sampling period, a period-vary sampling scheme is proposed based on an adjuster for the sampling period designed. In this scheme, the adjuster consists of a monitor, which acquires network resources utilization and the executive time of data packet, and a predictor, which uses BP neural network to predict the next sampling period by utilizing network resources utilization and data packet executive time. The simulation results show that the proposed scheme can alleviate the influence of time delay and improve the performance of the networked control systems.

Introduction

Advances in digital computation and communication have boosted the development of networked control systems in which sensors, actuators and controllers are interconnected by communication networks. The networked control systems have many attractive advantages, such as low cost, simple installation, maintenance and high reliability, so that the networked control systems are now widely applied in remote control, automobiles, aircraft and robotics.

However, the insertion of the communication network makes the analysis and design of a networked control system complicated. One of the important issues is the network-induced delay which occurs during data exchanging among the devices across the network. It is well-known that the time delay degrades the performance of the networked control systems [1].

Various methodologies have been proposed in order to deal with the issue induced by the time delay in the networked control system. Zhang et al. [1] investigated the stabilization problem of NCS with random delay in discrete-time domain, and modeled the sensor-to-controller delay and the controller-to-sensor delay as two Markov chains. Li et al. [2] regarded the time-varying delay as the sum of the mean delay and uncertain delay, and presented a new control law based on an iterative linear matrix inequality approach. Beldiman et al. [3] separately designed an open-loop structure predictor and a closed-loop structure predictor for estimating the plant outputs between two successive transmission times, and improved NCS performance without affecting system stability. Wang et al. [4] designed a new undelayed plant state estimator and predicted current control signals in every sampling interval to compensate the long delay by using an event-driven actuator and a time-driven actuator simultaneously. Wu et al. [5] adopted a generalized predictive control approach and established an error predictive model based on BP neural network. The error predictive value was used to compensate output predictive value. Hur et al. [6] presented a predictive controller based upon stochastic methods to compensate the time-delay. The predictive controller estimated future outputs using a linear prediction function and a probability function in terms of previous outputs.

The control action at each sampling interval cannot be expected to function exactly because the approaches above basically adopted a fixed sampling rate. Motivated from the concept of variable-period sampling, Huang et al. [7] designed a network-induced time delay compensator based on variable-period sampling. Lin et al. [8] proposed the mathematical model of networked control systems whose network induced delay is longer than one sampling period. Furthermore, Markov characteristics of the transfer delay were discussed. Based on Markov chain theory, the infinite horizon controller was also designed. Yi [9] proposed that the time delay was taken as the variable sampling period. The time delay could be predicted based on the BP neural network. Meanwhile, the BP neural network was designed to further update according to its prediction error after each prediction in order to improve its prediction accuracy. However, such method could not be generalized and applied in the practical condition except for Echelon Lonworks and Internet-based robots.

In this paper, we provide a mathematical model for networked control systems with uncertain time delay. Based on this model, an adjuster is designed to acquire two parameters, i.e., the network utilization and the executive time of data packet, and then predict the sampling period by using a BP neural network so that we utilize this value to adjust the current sampling period. In the end, we provide detailed simulation results.

Modeling the Networked Control System

The block diagram of the networked control system with uncertain delay is shown in Fig. 1. Providing that the sensor is time-driven and the controller and actuator are all event-driven, when one sampling period begins, the data sampled by the sensor is sent into the controller via the computer network, and then after the controller computes the data-sampled, the control data is sent into the actuator via the computer network. The actuator acts on the plant in no time once it receives the control data. Therefore there are two sources of delays from the network: sensor-to-controller τ_k^{sc} and controller-to-actuator τ_k^{ca}. The total delay $\tau_k = \tau_k^{sc} + \tau_k^{ca}$ occurs during the sampling interval.

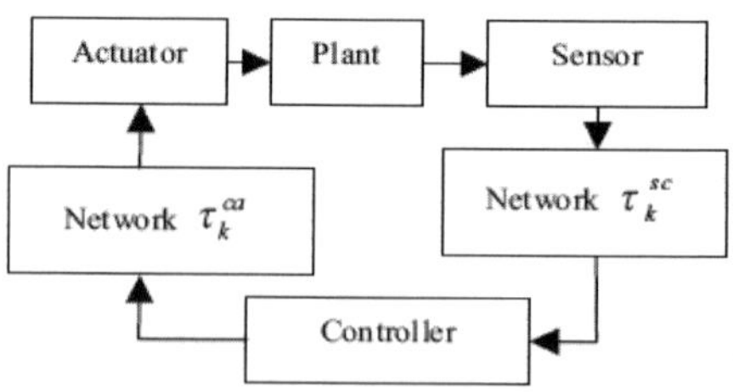

Fig. 1: The block diagram of the networked control system with uncertain delay

The model of the plant is depicted in the continuous form:

$$\begin{aligned} \dot{x}(t) &= Ax(t) + Bu(t) \\ y(t) &= Cx(t) \end{aligned} \quad (1)$$

And the controller is expressed in the following discrete form:

$$u(kT) = -Kx(kT), k = 0,1,2,\cdots \quad (2)$$

where $x(t) \in \Re^n, u(t) \in \Re^n, y(t) \in \Re^n$ are separately the system state variable, the system input variable, and the system output variable, and A, B, C, K are the system parameters and feedback control gain. They are of compatible dimensions.

Sampling the system mentioned above with sampling period of the system $T_k = \frac{C_k}{U_k}$, the following discrete equations can be gotten:

$$x(k+1) = \varphi_a(k)x(k) + \Gamma_a(k)u(k)$$

$$y(k) = Cx(k) \tag{3}$$

$$u(k) = -Kx(k)$$

Where $\varphi_a(k) = e^{AT_k}, \Gamma_a(k) = \int_0^{T_k} e^{As}Bds$, U_k denotes network utilization, C_k is the executive time of data packet, which consists of τ_k and the processing time of data packet.

The augmented state vector $z(k)$ is defined as

$$z(k) = \begin{bmatrix} x(k) \\ u(k-1) \end{bmatrix} \tag{4}$$

Therefore the augmented closed-loop system can be expressed in the following form

$$z(k+1) = \Phi(k)z(k) \tag{5}$$

Where

$$\Phi(k) = \begin{bmatrix} \varphi_a(k) - \Gamma_a(k)K & 0 \\ -K & 0 \end{bmatrix} \tag{6}$$

Sampling Period Prediction Based on Feedback control and BP Networks

In order to adjust the sampling period, an adjuster for the sampling period is designed. The adjuster is composed of a monitor and a predictor as illustrated in Fig. 2. The monitor acquires the network utilization and the executive time of data packet in the monitoring period and transmits these values into the predictor. The predictor predicts the sampling period based on the neural networks by using these values. According to the predictive sampling value, the current period is being replaced.

In the predictor, the neural networks are designed as a three-layer structure, which contains an input layer, hidden layer and output layer. The topological structure of the neural networks for the delay prediction is illustrated in Fig. 3. The neuron number of the input layer and the neuron number of the hidden layer are respectively selected to be 3. The input to the input layer is a vector of the known historical sampling period values. The output layer has only one neuron, and its output is just the predicted sampling period.

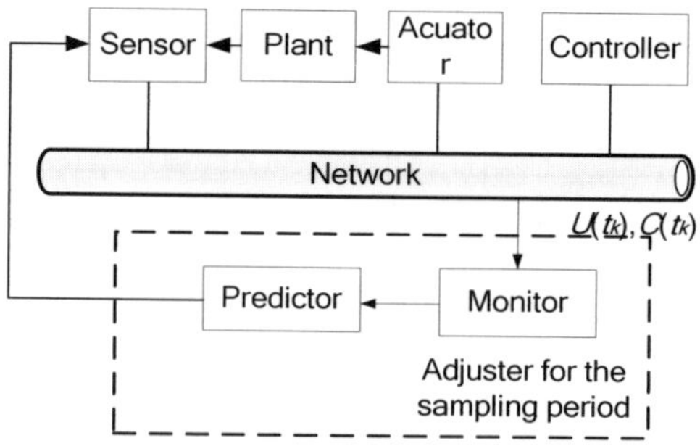

Fig. 2: The scheme for adjusting the sampling period

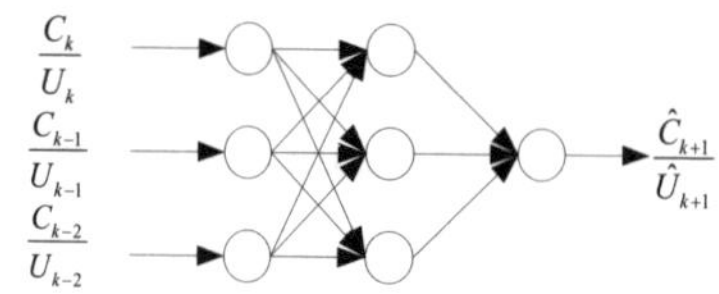

Fig. 3: BP neural network configuration for sampling period prediction

Providing the input I_i of the input layer neuron i, the weight w_{ji}^1 from the input layer neuron i to the hidden layer neuron j, and the threshold θ_j^1 of the hidden layer neuron j, the output H_j of the hidden layer neuron j is calculated by

$$H_j = f\left(\sum_i w_{ji}^1 I_i + \theta_j^1\right) \tag{7}$$

Where $f(\cdot)$ denotes the following sigmoid function:

$$f(x) = \frac{1}{1+e^{-x}} \tag{8}$$

The predicted delay value of the output layer

$$O = f\left(\sum_j w_j^2 H_j + \theta^2\right) \tag{9}$$

Where w_j^2 : the weight from the hidden layer j to the output layer, θ^2: the threshold of the output layer.

The updating values are as follows:

$$\begin{aligned}
w_j^2 &= w_j^2 + \alpha\left(O_p - O_r\right)O_p\left(1-O_p\right)H_j \\
\theta^2 &= \theta^2 + \beta\left(O_p - O_r\right)O_p\left(1-O_p\right) \\
w_{ji}^1 &= w_{ji}^1 + \alpha\left(O_p - O_r\right)w_j^2 H_j\left(1-H_j\right)I_i \\
\theta_j^1 &= \theta_j^1 + \beta\left(O_p - O_r\right)w_j^2 H_j\left(1-H_j\right)
\end{aligned} \tag{10}$$

Where α , β are the learning parameters.

Simulations and Discussion

Let the state space equation of the plant be described by

$$\begin{bmatrix} \dot{x}_1 \\ \dot{x}_2 \end{bmatrix} = \begin{bmatrix} 0 & 1 \\ 0 & -20 \end{bmatrix} \begin{bmatrix} x_1 \\ x_2 \end{bmatrix} + \begin{bmatrix} 0 \\ 640 \end{bmatrix} u \tag{11}$$

A continuous-state feedback controller can be obtained by $u = -Kx$, where K = [0.0617 -0.0174]. Suppose that the initial state of the plant is [0, 2], and the control goal is [0, 0]. Meanwhile, the network utilization is 60%. The maximum sampling period is $0.1s$.

We compare the performance of the system with the fixed-period sampling scheme with the performance of the system with the variable-period sampling scheme when the initial sampling period $T_k = 0.1s$. In order to simplify the discussed problem, we assume that the processing time of data packet is constant in the simulation, and the predictive executive time of data packet converts to predicting τ_k. The compared results are shown in Fig. 4. Fig. 4 shows that the variable-period sampling control achieves better result.

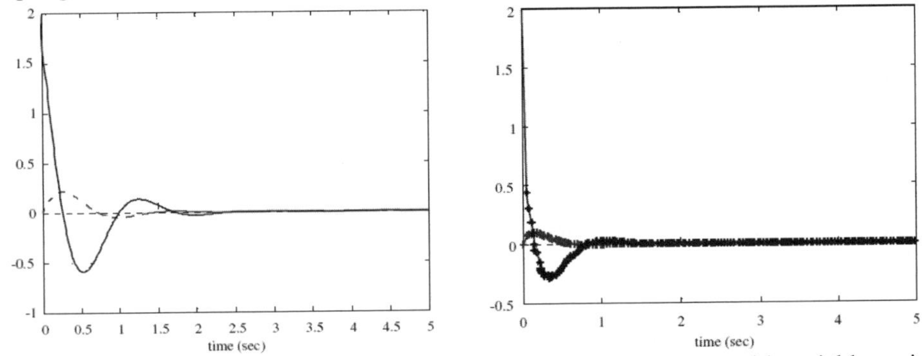

(a) state response with fixed-period (b) state response with variable-period

Fig. 4: Simulation results with the initial period $T_k = 0.1s$

Fig. 5 represents the state response with the fixed-period and the state response with the variable-period when the initial sampling period $T_k = 0.05s$. The variation curve of sampling period is shown in Fig. 6. From Fig. 5(a), under the fixed-period, the two state variables of the plant still keep oscillating. From Fig. 5(b), although the result is not as good as that in Fig. 4(b), it converges to the goal in about 2 s, much better than Fig. 5(a). Nevertheless, the variable-period sampling shows a satisfactory performance.

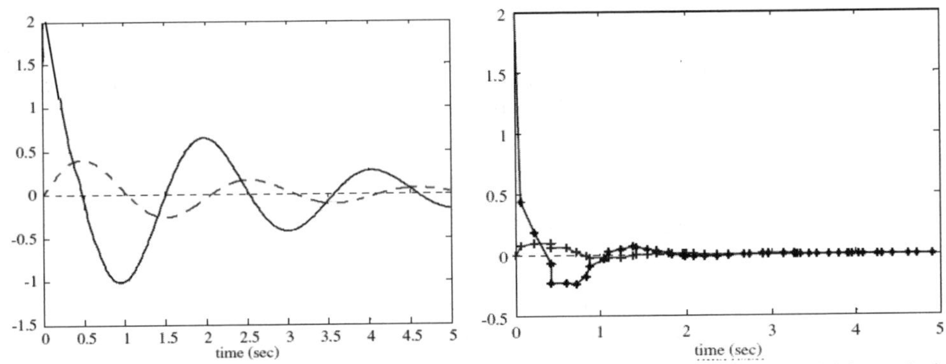

(a) state response with fixed-period (b) state response with variable-period

Fig. 5: Simulation results with the initial period $T_k = 0.05s$

Summary

In this paper, we first model for networked control systems. Motivated from the concept of variable-period sampling, a new sampling scheme is presented based on feed control and the neural network. The sampling period is predicted by using the neural network. It is possible that the variable-period sampling control for networked control systems become true. Networked control simulations indicated that this new approach decreases the influence of the time delay to the greatest extent and improves the performance of the control system.

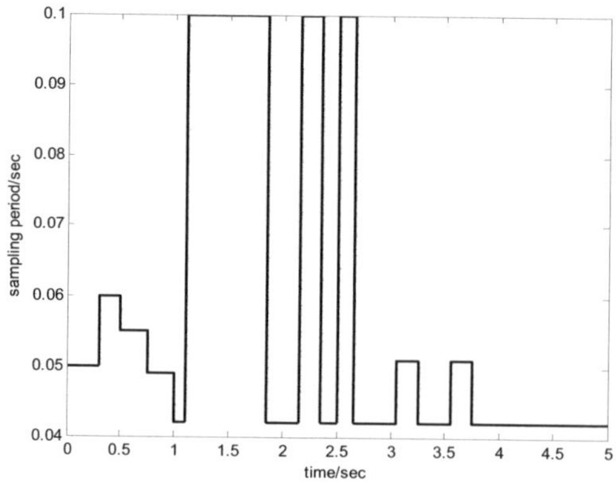

Fig. 6: Variation curve of sampling period

Acknowledgements

This research was supported by the Scientific Research Foundation of CUIT under Grant No. KYTZ201105

References

[1] L. Zhang, Y. Shi, T. Chen and B. Huang: *A New Method for Stabilization of Networked Control Systems with Random Delays*. Proceedings of American Control Conference Vol.1 (2005), p. 633.

[2] S. Li, Z. Wang and Y. Sun: *Observer-based Compensation Design for Networked Control Systems with Long Time Delay*. Proceedings of 30th Annual Conference of IEEE Industrial Electronics Society Vol.1 (2004), p. 678.

[3] O. Beldiman, G.C. Walsh and L. Bushnell: *Predictors for Networked Control Systems*. Proceedings of American Control Conference Vol.4 (2000), p. 2347.

[4] Z. Wang, J. Yang, D. Tan and X. Wang: *Compensation for the Networked Control Systems with the Long Time Delay*. Proceedings of IEEE International Conference on Systems, Man and Cybernetics Vol.4 (2003), p. 3170.

[5] A. Wu, D. Wang and J. Liang: *Application of Generalized predictive Control with Neural Network Error Correction in Networked Control System*. Proceedings of Fifth World Congress on Intelligent Control and Automation Vol.2 (2004), p. 1386.

[6] H. Hur, J. Lee, S. Lee and M. Lee: *Compensation of Time Delay Using a Predictive Controller*. Proceedings of IEEE International Symposium on Industrial Electronics Vol.3 (1999), p. 1087.

[7] J. Huang, P. Wu, Z. Xiu, S. Ma and Z. Chen: *Variable-sampling Process in Internet-based Motion Control System*. Control and Decision Vol.20 (2005), p. 755.

[8] C.Ma and H. Fang: *Research on mean square exponential stability of networked control systems with multistep delay*. Applied Mathematical Modeling Vol.30 (2006), p. 941.

[9] J. Yi: *BP Neural Network Prediction-based Variable-period Sampling Approach for Networked Control Systems*. Applied Mathematics and Computation Vol.185 (2007), p. 976.

Advanced Materials Research Vol. 279 (2011) pp 451-455
© (2011) Trans Tech Publications, Switzerland
doi:10.4028/www.scientific.net/AMR.279.451

Effect of sulfosalicylic acid on kinetics of Ce^{3+}/Ce^{4+} electrode reaction

Zhipeng Xie[1, a], Fengjiao Xiong[1,b] and Debi Zhou[1,c]

[1]College of Chemistry and Chemical Engineering, Central South University, Changsha 410083, China

[a]zhpxie_2011@126.com, [b]xfjwfj@126.com, [c]dbzhou_09@126.com

Keywords: Cyclic voltammetry, Rotating disc electrode, Redox flow battery, Cerium electrolyte

Abstract. The effect of sulfosalicylic acid (SSA) on the electrochemical behaviour of Ce^{3+}/Ce^{4+} redox couple was studied using voltammetry at stationary electrode and rotating disc electrode (RDE). When the SSA was added to the Ce^{3+}/Ce^{4+} electrolyte there was change to the peak current and peak potential, i.e. the increase of peak current and the decrease of peak potential difference. The diffusion coefficient of Ce^{3+} is 6.5×10^{-6} cm^2 s^{-1} and the standard rate constant of the Ce^{3+}/Ce^{4+} redox reaction is 3.2×10^{-4} cm s^{-1} in the presence of SSA, which are all larger than that in the absence of SSA (3.6×10^{-6} cm^2 s^{-1}, 2.0×10^{-4} cm s^{-1}). Therefore, the SSA as the additive in the supporting electrolyte for a cerium redox flow battery (RFB) application will result in a high voltage efficiency and will satisfy an important requirement for a RFB electrolyte.

Introduction

The extensive utility of the renewable energy sources helps prevent the environment from deteriorating even further. Energy storage technology plays a significant role in renewable energy sources usage. RFB stores energy in the electrolyte solutions in contrast to conventional batteries. Many RFB systems [1] have been fabricated and developed. Among them, the all-vanadium RFB is the most practical candidate for its excellent performance, where is no decrease in capacity caused by the cross mixing of positive electrolyte and negative electrolyte.

A Zn/Ce system using organic acid as positive electrolyte media was patented by Clarke *et al.*[2]. The cell voltage of the Zn/Ce system after full charging can reach 2.5 V, which helps result in a greater energy storage capacity. The electrochemical behaviours of the Ce^{3+}/Ce^{4+} in sulfuric acid [3], methanesulfuric acid [4] and nitric acid[5] were reported by different researchers.

In the present work, the effect of SSA on the electrochemical behaviour of Ce^{3+}/Ce^{4+} redox couple was investigated using cyclic voltammetry (CV), linear scan voltammetry (LSV) and rotating disc electrode (RDE). This study demonstrated that, as the SSA was added into the Ce^{3+}/Ce^{4+} electrolyte, an increase in peak current was observed for the oxidation and reduction, when a cyclic voltammogram was taken on a platinum electrode; at the same time, a decrease in peak potential difference was also observed. Results obtained from LSV and RDE are in a good agreement with that from CV.

Experimental

All reagents were of AR grade unless stated otherwise. Sulfosalicylic acid and cerium carbonate are obtained from Alfa Corp., TianJin, China. Cerium solutions in sulfuric acid were prepared by neutralizing cerium carbonate in water by adding concentrated sulfuric acid. Desired concentrations of sulfuric acid were obtained by adding the required amount of H_2SO_4 to this solution. Doubly distilled water was used throughout.

Voltammetry measurements were performed using a CHI electrochemistry station. A 0.5 cm^2 platinum sheet and a 0.12 cm^2 rotating platinum disc electrode were used as working electrodes, a 5 cm^2 titanium sheet as counter electrode and an SCE as reference electrode. Prior to test, the platinum electrode was cycled in 3 M H_2SO_4 solution between – 0.3 and 1.8 V *vs.* SCE for 15 min at a scan rate of 20 mV s^{-1}.

Results and discussion

Cyclic voltammogram. Previous investigations [3,6] of the Ce^{3+}/Ce^{4+} redox couple using cyclic voltammetry confirmed the slow kinetics of the Ce^{3+}/Ce^{4+} redox couple in sulfuric acid media. Quite large peak potential separations, more than 400 mV in sulfuric acid at 293 K, were observed by them. They proposed that the especially slow kinetics of the Ce^{3+}/Ce^{4+} redox reaction was the result of an especially large structural change associated with a conversion between Ce^{3+} and Ce^{4+} in sulfuric acid media. Wei et al.[7] observed a rapid kinetics of the Ce^{3+}/Ce^{4+} couple in high concentration nitric acid solutions.

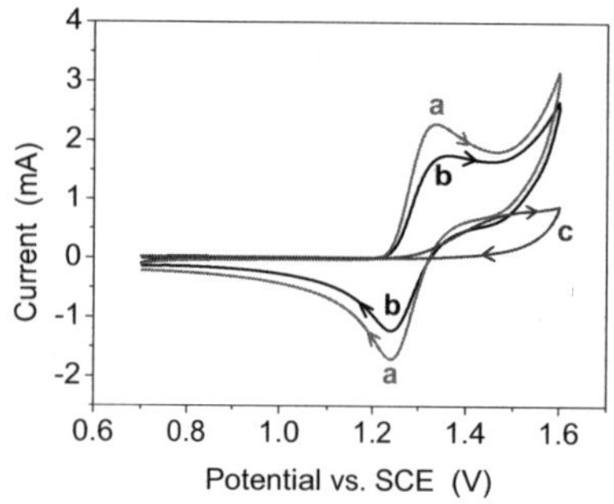

Fig. 1: Effect of SSA on cyclic voltammogram for Ce^{3+}/Ce^{4+} redox couple. Concentration: $[Ce^{3+}] = [Ce^{4+}] = 5$ mmol dm^{-3}, [SSA] = 0.5 mmol dm^{-3}. $[H_2SO_4] = 1$ mol dm^{-3}. **a**. in the presence of SSA; **b**. in the absence of SSA; **c.** for 0.5 mmol dm^{-3} SSA in 1 mol dm^{-3} H_2SO_4 solution without cerium. Scan rate: 10 mV s^{-1}.

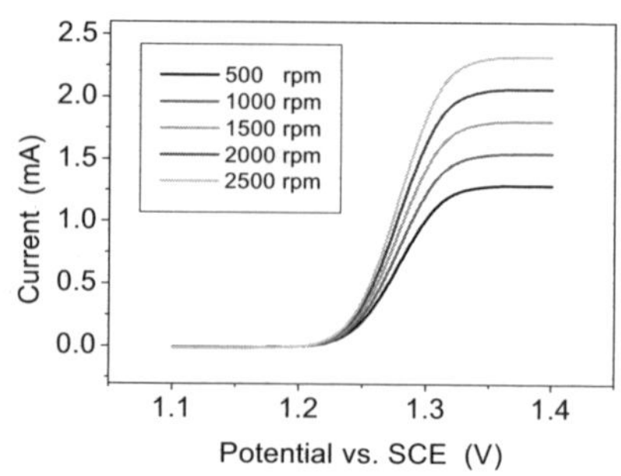

Fig. 2: Voltammograms at a rotating platinum disc electrode as a function of rotation rate. Concentration: $[Ce^{3+}] = 5$ mmol dm^{-3}. $[H_2SO_4] = 1$ mol dm^{-3}. [SSA] = 0.5 mmol dm^{-3}. Scan rate: 5 mV s^{-1}.

The effect of SSA on cyclic voltammogram for a solution containing 1 mol dm^{-3} H_2SO_4 as well as 5 mmol dm^{-3} each of Ce^{3+} and Ce^{4+} at platinum electrode is shown in Fig. 1. From Fig. 1, it is easy to observe that the cathodic and anodic peaks for the Ce^{3+}/Ce^{4+} redox couple are located between 1.2 and 1.4 V *vs*. SCE. However, the distinct anodic peak could not be observed for the oxidation of SSA at this potential range; furthermore, the voltammetry of SSA lacks the return peak for the reduction of oxidized SSA between 1.2 and 1.4 V *vs*. SCE. So it can be concluded that the increase in peak current for the oxidation and reduction of Ce^{3+}/Ce^{4+} redox couple in the presence of SSA is not caused by the oxidation and reduction of SSA. It is due to the improvement of kinetics of the Ce^{3+}/Ce^{4+} redox couple by the SSA.

Table 1: Cathodic and anodic peak current and peak potential obtained from Fig. 1

Additive	Peak current [mA]		I_{pc}/I_{pa}	Peak potential [mV]		
	Anodic I_{pa}	Cathodic I_{pc}		Anodic	Cathodic	Separation
No	1.742	1.403	0.81	1363	1240	123
SSA	2.300	2.037	0.89	1337	1240	97

Some data obtained from Fig. 1 are summarized in Table 1. As seen in Table 1, both the anodic peak current and the cathodic peak current increase remarkedly in the presence of 0.5 mmol dm^{-3} SSA; at the same time, the ratio of cathodic peak current to anodic peak current increases from 0.81 to 0.89, and the peak potential separation decreases from 123 to 97 mV, indicating an improvement in reversibility of Ce^{3+}/Ce^{4+} electrode process by SSA.

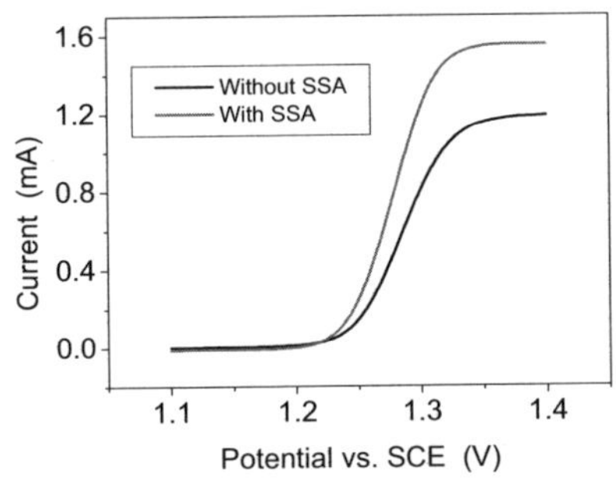

Fig. 3: Effect of SSA on the voltammogram of the Ce^{3+}/Ce^{4+} redox couple at RDE. Concentration: $[Ce^{3+}] = 5$ mmol dm^{-3}. $[H_2SO_4] = 1$ mol dm^{-3}. [SSA] = 0.5 mmol dm^{-3}. Scan rate: 5 mV s^{-1}. Rotating rate: 1000 rpm

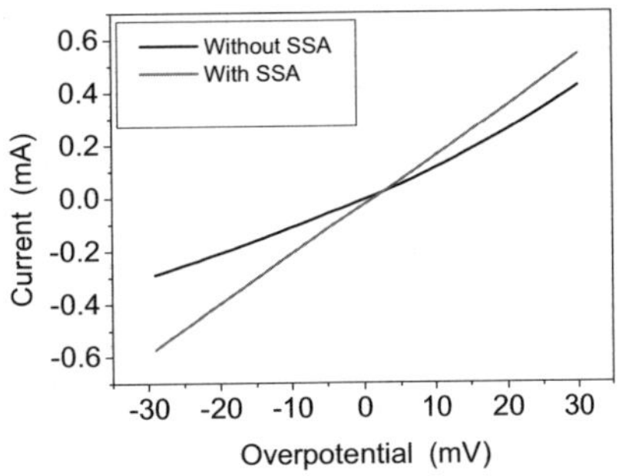

Fig. 4: The plot of current vs. overpotential. Concentration: $[H_2SO_4] = 1$ mol dm^{-3}. [SSA] = 0.5 mmol dm^{-3}. $[Ce^{3+}] = [Ce^{4+}] = 0.1$ mol dm^{-3}. Scan rate: 1 mV s^{-1}. Open circuit potential: 1.253 V *vs.* SCE. At platinum electrode.

Rotating Disc Voltammetry. The dependence of the voltammograms on the rotating rate of the platinum disc electrode in the solution containing 5 mmol dm^{-3} Ce^{3+}, 1 mol dm^{-3} H_2SO_4 and 0.5 mmol dm^{-3} SSA is shown in Fig. 2. The limiting current depends strongly on the rotation rate of the disc and a well-formed oxidation wave is observed. Indeed, a plot of the limiting current, i_l *vs.* $\omega^{1/2}$ is linear, confirming that the oxidation of Ce^{3+} becomes mass transport controlled and the Levich equation was used to obtain a diffusion coefficient for Ce^{3+} in this medium; the value calculated was 6.5×10^{-6} cm^2 s^{-1} which is larger than that in the absence of SSA (3.6×10^{-6} cm^2 s^{-1}, obtained using the same method).

The effect of SSA on the voltammogram of the Ce^{3+}/Ce^{4+} redox couple at a rotating platinum disc electrode is shown in Fig. 3. As shown in Fig. 3, the limiting current of the oxidation of Ce^{3+} in the presence of SSA is obviously larger than that in the absence of SSA. It is agreement with the result obtained using CV.

Current-overpotential Curve. At very small overpotential η, the current-overpotential relation is virtually linear (see Fig. 4). It can be expressed [8] as

$$i = -i_0 F\eta / RT \quad (1)$$

Since

$$R_{ct} = -\eta / i \quad (2)$$

$$j_0 = i_0 / A \quad (3)$$

we have

$$j_0 = RT / AR_{ct}F \quad (4)$$

$$k_0 = j_0 / FC^* \quad (5)$$

where R is the universal gas constant; T, the Kelvin temperature; A, the area of the electrode; F, the Faraday constant; i_0, exchange current; j_0, exchange current density; R_{ct}, the charge-transfer resistance, which is the negative reciprocal slope of the current *vs.* overpotential curve ($-\eta/i$).

Some kinetic parameters of the Ce^{3+}/Ce^{4+} redox reaction calculated using Eq. 4 and Eq. 5 are shown in Table 2. As seen in Table 2, the exchange current and standard rate constant of the Ce^{3+}/Ce^{4+} redox reaction at platinum in the presence of SSA are significantly larger than that in the absence of SSA. The comparison of our value in sulfuric acid media with literature data is good. For example, the standard rate constant of the Ce^{3+}/Ce^{4+} redox reaction in 1 mol dm^{-3} H_2SO_4 at platinum electrode was reported as 1.6×10^{-4} cm s^{-1} [9], and we found 2.0×10^{-4} cm s^{-1}. In addition, Wei *et al.* reported 2.6×10^{-2} cm s^{-1} for the Ce^{3+}/Ce^{4+} couple in 6 mol dm^{-3} HNO_3 media, which is obviously larger than in sulfuric acid media. In sulfuric acid media, the Ce^{3+}/Ce^{4+} couple [6] can form stable complexes with SO_4^{2-}, however, it can not form stable complexes with NO_3^- in nitric acid media, which may be the main reason for the difference of standard rate constant between in sulfuric acid media and in nitric acid media.

Table 2: Kinetic parameters of the Ce^{3+}/Ce^{4+} redox couple

Supporting medium	j_0 [mA cm^{-2}]		k_0 [cm s^{-1}]	
	This work	Literature	This work	Literature
Sufuric acid	1.95	--	2.0×10^{-4}	1.6×10^{-4} [9]
Sufuric acid +SSA	3.13	--	3.2×10^{-4}	--
Nitric acid	--	--	--	2.6×10^{-2} [7]

Conclusions

The electrochemical behaviour of the Ce^{3+}/Ce^{4+} redox reaction in the presence of SSA on platinum electrode was studied using voltammetry.

The results for cyclic voltammetry study of Ce^{3+}/Ce^{4+} couple in the presence of SSA illusrtate a more reversible reaction electrode process on platinum than that in the absence of SSA. The cathodic peak current, the anodic peak current and the ratio of the cathodic peak current to the anodic peak current increase remarkedly in the presence of 0.5 mmol dm^{-3} SSA; at the same time, the peak potential separation decreases from 123 to 97 mV.

RDE results also confirmed the positive effect of SSA on the kinetics of Ce^{3+}/Ce^{4+} redox electrode process. The diffusion coefficient of Ce^{3+} is 6.5×10^{-6} cm^2 s^{-1} in the presence of SSA, which is larger than that in the absence of SSA (3.6×10^{-6} cm^2 s^{-1}).

The standard rate constant of the Ce^{3+}/Ce^{4+} redox reaction at platinum electrode in the presence of SSA is calculated to be 3.2×10^{-4} cm s^{-1} from the plot of current *vs.* overpotential, which is larger than that in the absence of SSA (2.0×10^{-4} cm s^{-1}).

Hence the suggestion is that a RFB employing the SSA as the additive in the supporting electrolyte for the Ce^{3+}/Ce^{4+} couple will have a high voltage efficiency and will satisfy an important requirement for a RFB electrolyte.

References

[1] C. Ponce de León, A. Frías-Ferrer, J. González-García, D.A. Szánto and F.C. Walsh: *Redox flow cells for energy conversion*. Journal of Power Sources Vol. 160 (2006), pp. 716-732.

[2] R.L. Clarke, B. Dougherty, S. Harrison, J.P. Millington, S. Mohanta: *Lanthanide Batteries*. WO 03/017408 A1, (2003).

[3] A. Paulenova, S.E. Creager, J.D. Navratil, Y. Wei: *Redox potentials and kinetics of the Ce^{3+}/Ce^{4+} redox reaction and solubility of cerium sulfates in sulfuric acid solutions*. Journal of Power Sources Vol. 109 (2002), pp. 431-438.

[4] T. Vijayabarathi, D. Velayutham, M. Noel: *Influence of aromatic reactants and products involved in the two stage electrochemical oxidation on the voltammetric behaviour Ce(IV)/Ce(III) redox couple*. Journal of Applied Electrochemistry Vol. 31 (2001), pp. 979-986.

[5] L. Jelinek, Y.Z. Wei, K. Mikio: *Electro-Oxidation of Concentrated Ce(III) at Carbon Felt Anode in Nitic Acid Media*. Journal of Rare Earths Vol. 24 (2006), pp. 257-263.

[6] Y. Liu, X. Xia, H. Liu: *Studies on cerium($Ce^{4+/3+}$)-vanadium($V^{2+/3+}$) redox flow cell—cyclic voltammogram response of Ce^{4+}/Ce^{3+} redox couple in H_2SO_4 solution*. Journal of Power Sources Vol. 130 (2004), pp. 299-305.

[7] Y. Wei, B. Fang, T. Arai, M Kumagai: *Electrolytic oxidation of Ce(III) in nitric acid and sulfuric acid media using a flow type cell*. Journal of Applied Electrochemistry Vol. 35 (2005), pp. 561-566.

[8] A.J. Bard, L.R. Faulkner: *Electrochemical Methods: Fundamentals and Applications*. Wiley & Sons, New York, (2001).

[9] P. Modiba, A.M. Crouch: *Electrochemical study of cerium(IV) in the presence of ethylenediaminetetraacetic acid (EDTA) and diethylenetriaminepentaacetate (DTPA) ligands*. Journal of Applied Electrochemistry Vol. 38 (2008), pp. 1293-1299.

Advanced Materials Research Vol. 279 (2011) pp 456-460
© (2011) Trans Tech Publications, Switzerland
doi:10.4028/www.scientific.net/AMR.279.456

An image encryption method based on Tent chaotic map in DCT domain

Jianhua Song[a], Qun Ding[b]

School of Electronic Engineering, Heilongjiang University, Harbin 150080, China

[a]98dg_sjh@163.com, [b]dingqun@263.net

Keywords: Chaos sequence, Tent map, DCT, Image encryption

Abstract. In recent years, the chaos based cryptographic algorithms have suggested some new and efficient ways to develop secure image encryption techniques. In this paper, the sequence was generated using chaotic system, combined with high efficiency and better security in transform domain. An image encryption method based on Tent chaotic map in DCT domain was proposed. The simulation results show that this method has characteristics such as high efficiency and good safety, and has a certain practicality and objectivity.

Introduction

At present, Internet has become an indispensable tool in people's daily life. According to statistics about 70% of the network information is in the form of digital images. These images of the transmission are often artificially attacked and the information owners cause huge losses [1]. Therefore, the way to ensure information security of the images has been extensively concerned.

Cryptogram technology is one of the main means of protecting information security, and chaotic encryption technology is a newly developed encryption technology, which uses chaotic sequence generated by chaotic system to encrypt message. Chaotic sequence possesses many attractive features such as random-like, extreme sensitivity to initial conditions, etc, so it can be constructed using encryption system. In addition, the chaotic encryption system is simple, fast, and suitable for encryption of digital image with large amount of data.

Chaotic encryption methods can generally be divided into one-dimensional chaotic encryption [2] and multi-dimensional chaotic encryption [3]. In addition, the encryption methods based on chaotic sequences can also be divided into the encryption in spatial domain [4] and in transform domain [5, 6]. Because the encryption methods in transform domain have many advantages such as high-efficiency, good safety performance, etc, an image encryption method based on Tent chaos map in DCT domain was proposed in this paper and got good results.

Tent Chaotic Map

Chaotic phenomena is a deterministic and random-like process in nonlinear dynamical system, this process is non-periodic, non-convergence but bounded, and it is extremely sensitive to the initial state of chaos [7], that is, only minor differences in the initial state of the two chaotic systems with the same structure will produce two completely different and unrelated chaotic sequence value in a short period of time. Due to the properties of sensitivity to initial conditions and parameters, the biggest advantage of chaotic system is that it can provide a large number of keys.

Tent chaotic map belongs to one-dimensional discrete chaotic system, and the mapping equation is shown as Eq. 1.

$$x_{n+1} = \begin{cases} \dfrac{x_n}{a}, & 0 < x_n < a \\ \dfrac{1-x_n}{1-a}, & a \le x_n \le 1 \end{cases} \tag{1}$$

When the control parameter $a\in(0,1)$, Eq. 1 works in the chaotic state, compared with the Logistic map. Tent map has faster iteration speed and better convergence property[8], and variable density of Tent map is stable than Logistic map[9]. Take the initial value $x_0=0.32$, control parameter $a=0.43$, chaotic sequence is shown in Fig. 1.

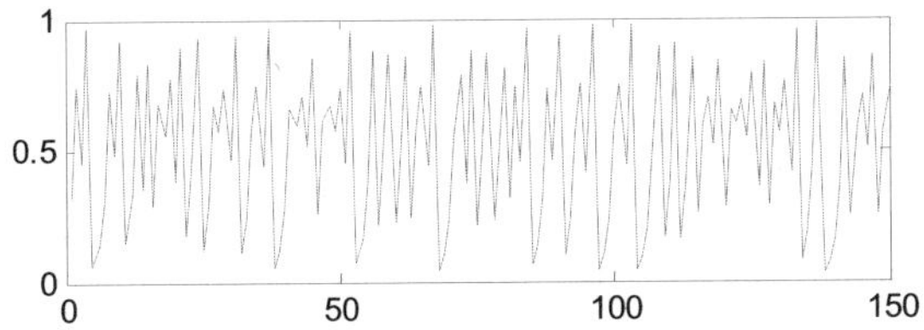

Fig.1: Tent chaotic sequence

From Fig.1, Tent chaotic map has a certain mathematical expression, but its statistical ergodicity is equivalent to the white noise. So Tent map has a simple form, is sensitive to initial conditions and has white noise statistics properties and so on many merits.

Discrete Cosine Transform (DCT)

Discrete cosine transform (DCT) is an orthogonal transformation, and DCT was defined by Ahmed and Rao in 1974, given the sequence $f(n)$, $n=0,1,\ldots,N-1$. DCT is defined as

$$F(0)=\sqrt{\frac{1}{N}}\sum_{n=0}^{N-1}f(n) \tag{2}$$

$$F(k)=\sqrt{\frac{2}{N}}\sum_{n=0}^{N-1}f(n)\cos\frac{(2n+1)k\pi}{2N} \qquad k=1,2,\cdots,N-1 \tag{3}$$

From Eq. 2 and Eq. 3, kernel function of DCT is

$$C_{k,n}=\sqrt{\frac{2}{N}}g_k\cos\frac{(2n+1)k\pi}{2N} \qquad k,n=0,1,\cdots,N-1 \tag{4}$$

$C_{k,n}$ is a real number, where the coefficient is

$$g_k=\begin{cases}\frac{1}{\sqrt{2}} & k=0\\ 1 & k\neq 0\end{cases} \tag{5}$$

Therefore, if $f(n)$ is real number, then its DCT will be the real number. In Fourier transform, if $f(n)$ is real number, generally its DFT is complex number. It can be seen that DCT avoids the complex operation.

The IDCT is

$$f(n)=\sqrt{\frac{1}{N}}F(0)+\sqrt{\frac{2}{N}}\sum_{k=0}^{N-1}F(k)\cos\frac{(2n+1)k\pi}{2N} \qquad n=0,1,\cdots,N-1 \tag{6}$$

The whole image is divided into $N\times N$ pixel blocks in image DCT, and then these pixel blocks were DCT processed one by one. Since most images have smaller high-frequency components, correspondingly the coefficients of high-frequency components are often zero. In addition, the human eyes are insensitive to the distortion of high-frequency components. The quantization of high-frequency components is coarser. Therefore, the transmission rate of transform coefficients is much less than the transmission rate of digital image pixels. The transform coefficients will be IDCT transformed after they have reached the receiver. Although there will be some distortions, they can be acceptable by the human eyes.

Image Encryption Algorithm

The algorithm principle. The basic method of chaotic encryption is to use aperiodicity and pseudo-randomness of chaotic sequences, and the chaotic sequences as key stream are performed bit by bit XOR or other operations with plaintext sequences which are ciphertext. Compared to the traditional block cipher algorithm DES and IDEA, the cryptographic system based on chaos has many excellent features. First of all, its data size of the raw data is not fixed. The greater the amount of raw data, the larger for the key space, to attack that it is more difficult, and more secure for the cryptographic system. The advantages make it suitable for mass data encryption such as image and video stream. Second, scrambling encryption key is entirely controlled by the user, so it shows better security than the fixed scrambling processing and diffusion treatment of the traditional cryptosystem. Third, the encryption process and decryption process is symmetric, and easy to implement, which makes it particularly suitable for image encryption.

Most of the energy is concentrated in a small number of low-frequency and DC components for the image DCT, so a small part of data encryption can achieve the security of the whole image. Encryption algorithm model in this paper is shown in Fig. 2.

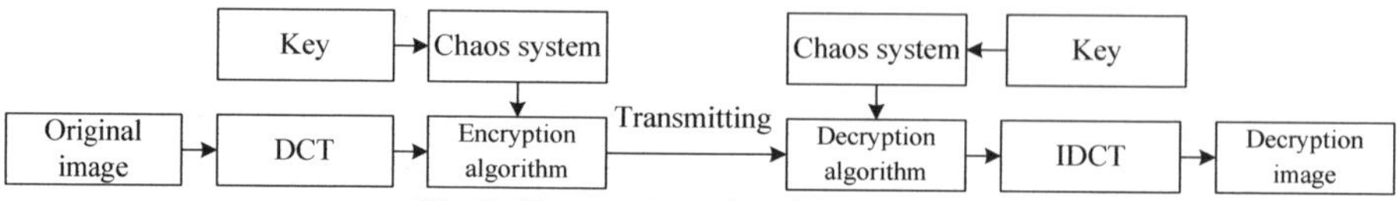

Fig.2: Encryption algorithm model

The algorithm implementation. The grayscale image $I(i, j)$ of size $M\times N$ is used as the experimental image, $0 \leq i \leq M-1$, $0 \leq j \leq N-1$, and the grayscale value of each pixel will be gat in terms of the order of rows and columns. $E(i,j)$ is the encrypted image, and the encryption process is shown in Eq. 7:

$$E(i,j)= \mathrm{mod}(((1-\lambda)Z(i,j)+\lambda G(i,j)),256) \tag{7}$$

Decryption and encryption is the inverse process, as shown in Eq. 8:

$$G(i,j)= \mathrm{mod}(((1-\lambda)Z(i,j)-E(i,j)/\lambda),256) \tag{8}$$

Where $G(i,j)$ is the DCT matrix processed by $I(i,j)$, $Z(i,j)$ is the chaotic sequence matrix generated by the Tent chaotic map, take x_0 and K as a key to encrypt the image. The primary process of the encryption algorithm is to construct a transformation process that the chaos character of Tent chaotic map is passed to each pixel of the image, so as to get the image encryption.

The specific steps of encryption algorithm are as follows:

Step1 get M and N value from the original image $I(i,j)$.

Step2 The gray-level image $I(i,j)$ is firstly DCT transformed and to obtain transform coefficients, then it is divided into 8×8 blocks to generate $G(i,j)$.

Step3 Take the initial value x_0 to Tent mapping model, after K times iterations is $x_{(1,1)}$.

Step4 Set the value of control parameter a, according to the new initial value $x_{(1,1)}$, using formula(1) to generate chaotic sequence matrix $Z(i,j)$.

Step5 Algebraic operations are performed to each pixel in $G(i,j)$ and the corresponding element of $Z(i,j)$, and get the final encrypted image $E(i,j)$;

When user enter the correct keys x_0 and a, and reverse operation from step2 to step5, the original image $I(i,j)$ will be recovered from $E(i,j)$.

Experimental results

In order to confirm the effectiveness of the encryption algorithm, the standard grayscale image 'Lena' with size of 256×256 is used as the test image, as shown in Fig.3 (a). The iterative equation of Tent map is shown in Eq. 1. Here the keys are x_0=0.33 and a=0.42. The encrypted image is shown in Fig.3 (b), and when the correct keys are inputted, the decrypted image is shown in Fig.3 (c).

(a) Standard grayscale image (b) The encrypted image (c) The decrypted image
Fig.3: Image encryption and decryption

The following tests were executed for the security of encryption algorithm. Because it is very sensitive to the initial value of chaotic sequence, when the keys are changed, it will generate a completely different chaotic transformation matrix and result in erroneous decryption results. When the initial value x_0 is equal to 0.33000001, the decrypted image is shown in Fig.4 (a), and when the control parameter a is 0.42000001, the decrypted image is shown in Fig.4 (b). From Fig. 4, although the key has only a very slight variation, the encrypted image still can not be decrypted correctly.

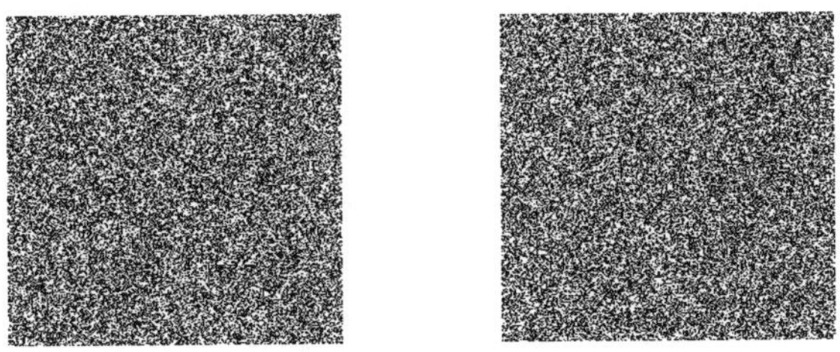

(a) Using the secret keys 'x_0=0. 33000001' (b) Using the secret keys 'a= 0.42000001'
Fig.4: The decrypted image for error key

To validate and compare the relationship between the encrypted image and the original image, here the histogram is analyzed before and after encryption, as shown in Fig.5.

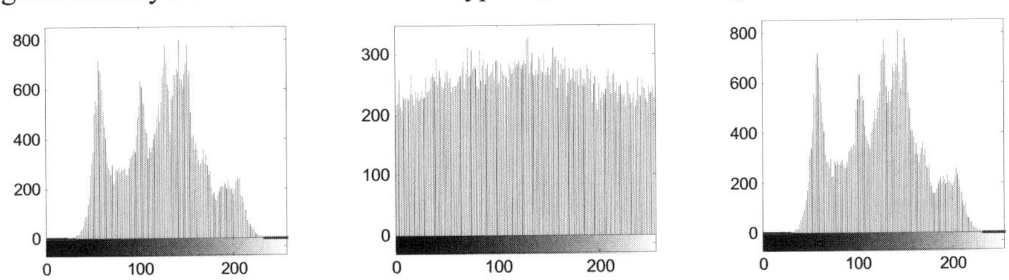

(a) Histogram of Lena (b) Histogram of encrypted image (c) Histogram of decrypted image
Fig.5: Image Histogram

From Fig.5 (b), it can be seen that the histogram of encrypted image and original image are very different, and the former shows a certain degree of uniformity. Because it is encrypted for the DCT coefficients in the transform domain, the corresponding calculation amount is reduced. Therefore, the encryption algorithm based on DCT domain is more efficient than the algorithm based on spatial domain.

Conclusion

Image encryption based on chaotic sequences can be carried out in transform domain or in spatial domain, and it has the advantages of high efficiency and good safety comparing to the algorithm in spatial domain. In this paper, an encryption algorithm was proposed based on Tent chaotic map in DCT domain. The experiment results show that the encryption method achieves higher efficiency and generality, and it is simple and easy to implement. However, from the transform domain algorithm it can also be seen that numerical value transform and its inverse transform must be resolved between spatial domain and transform domain, because it introduces a complex operation. In addition, the transform domain algorithm is likely to lead to information loss and distortion of image. Therefore, how to coordinate these contradictions is a problem which should be in-depth study.

Acknowledgment

This work is supported by the scientific and technological project of Key Laboratory of Electronics Engineering, College of Heilongjiang Province, China (No. D22D20100032), Science Fund for Youths of Heilongjiang University (No. QL200913).

References

[1] K. S. Ntalianis, S. D. Kollias: *Chaotic video objects encryption based on mixed feedback, multiresolution decomposition and time-variant S-boxes*. Proceedings of 2005 IEEE International Conference on Image Processing (ICIP 2005), pp. 1110-1113.

[2] W. Z. Yu, H. G. Ma, L. H. Wang: *Method for encrypting image with chaos series*. Journal of Computer Applications Vol. 25 (2009), pp. 141-143.

[3] G. R. Chen, Y. B. Mao, C. K. Chui: *A symmetric image encryption scheme based on 3D chaotic cat maps*. Chaos, Solitons & Fractals Vol. 21 (2004), pp. 749-761.

[4] K. Zhang, D. H. Wang: *Research on Compound Encryption of Digital Image Based on Logistic Chaotic Sequence*. Computer and Modernization, (2005), pp. 66-69.

[5] D. X. Qi, J. C. Zou: *A novel scrambling transformation and its application in Image information hiding*. Science in China (Series E) Vol. 30 (2000), pp. 440-447.

[6] K. X. Yi, X. Sun, J. Y. Shi: *An Image Encryption Algorithm Based on Chaotic Sequences*. Journal of Computer-Aided Design & Computer Graphics Vol. 12 (2000), pp. 672-676.

[7] Pecora L. M., Carroll T. L.: *Synchronization in Chaotic Systems*. Phys. Rev. Lett. Vol. 64 (1990), pp. 821-824.

[8] L. Shan, H. Qiang, J. Li: *Chaotic optimization algorithm based on Tent map*. Control and Decision Vol. 20 (2005), pp. 179-182.

[9] Xiang T, Wong K, Liao X: *An improved chaotic cryptosystem with external key*. Communications in Nonlinear Science and Numerical Simulation Vol. 13 (2008), pp. 1879-1887.

Keyword Index

Author Index